ALIENS AND RELIGION: WHERE TWO WORLDS COLLIDE

Assessing the Impact of Discovering Extraterrestrial Intelligence on Religion and Theology

JONATHAN M.S. PEARCE
DR. AARON ADAIR

FOREWORD BY
JAMES FODOR

AFTERWORD BY
ED BUCKNER

Aliens and Religion: Where Two Worlds Collide
Copyright © 2023 Jonathan M.S. Pearce & Aaron Adair

Published by Onus Books

All rights reserved. No part of this publication may be reproduced, stored in a retrieval system, or transmitted in any form by any means, electronic, mechanical, photocopy, recording, or otherwise, without the prior permission of the publisher, except as provided for by UK copyright law.

Cover design: Onus Books
Image credits: Space background—ESA/Hubble & NASA. Main image—created by AI at DALL E (https://openai.com/blog/dall-e/).

Trade paperback ISBN: 978-1-8382391-5-2

OB 21/37

Praise for this book:

As thousands of planets are discovered beyond our solar system and the Search for Extraterrestrial Intelligence intensifies on many fronts from UFOs to distant regions of our Galaxy, the question of the effect of an alien discovery on terrestrial religions and theology becomes ever more urgent. This well-written and provocative book is a substantial contribution to studies of the societal impact of astrobiology, and especially to the new field of astrotheology. Fair warning: you may experience cognitive dissonance and perhaps even change your worldview!

- Steven J. Dick, former NASA Chief Historian, author of *Astrobiology, Discovery, and Societal Impact*

I love definitive treatments of a subject. This is a definitive treatment of its subject. The impact and significance to religion of even the possibility of alien civilizations is much in need of a thorough look. Pearce and Adair cover every angle, and well.

- Richard Carrier, Ph.D., author of *Jesus Christ from Outer Space*

The stories of mysterious cults, quasi-religions, and other ancient rites and modern conspiracy theories are balanced by quantitative reconsiderations of the Drake Equation and other statistical tools. While they show that the scriptures of most religions can accommodate the existence of aliens, they imaginatively point out that if aliens with advanced biology could generate life eternal here on earth, practical impact on current religions might result—why wait for forever?

- David E. Pritchard, Physics Professor of Physics, MIT, and editor of *Alien Discussions*

When the aliens arrive from the heavens, what happens to religion —on earth? Two of today's most daring and incisive thinkers, philosopher/journalist Jonathan M.S. Pearce and physicist/biblical scholar Dr. Aaron Adair, take on all the questions raised by close encounters of the theological kind. An invaluable and highly entertaining resource for science fiction writers and readers alike, for both believers and skeptics, and anyone who enjoys wrestling with high-end thought experiments.

- David Fitzgerald, author of *The Complete Heretic's Guide to Western Religion Series*

I've never been much of a fan of science fiction or even of thinking much about aliens, but *Aliens and Religion* has changed that–delightfully.

I've long been a fan of Jonathan MS Pearce while ignorant of Dr. Adair– but *Aliens and Religion* seems to show that at least one of them is a form of extra-terrestrial extreme intelligence. An absorbing and original read of the first order.

Adair and Pearce are top guns at showing us the way to original new thinking about religion and aliens–brilliantly.

I have no desire to journey to outer space–but I'll go gladly if Pearce and Adair are the pilots. *Aliens and Religion* is gripping good stuff.

- Ed Buckner, former President, American Atheists

This is an impressive work! It's an ingenious attempt to reach believers who are otherwise impervious to reason, which has a good chance of succeeding. Bravo!

- John Loftus, author of *God and Horrendous Suffering* and *The Case Against Miracles*

Are we alone in our galaxy? In the universe? What are the chances that we could one day come in contact with intelligent forms of extraterrestrial life? What would such life forms be like? What are the chances that such life forms are already out there looking for us? And how would contact with civilizations beyond our own world potentially affect our thoughts about gods, sacred texts, hallowed traditions, and the problem of evil? In this masterful tour de force, Jonathan M.S. Pearce and Dr. Aaron Adair guide readers on a fascinating journey through these and other questions. This book is an accessible crash course on scientific and data-driven ways of thinking about extraterrestrial intelligence, but it is also a thoughtful, erudite, and deep dive into the types of theological and philosophical conundrums that the existence of such intelligence would raise. Highly recommended for anyone who has ever found themselves wondering about the possibilities hidden among the stars, and about what might happen if those possibilities came to Earth.

- Eric Vanden Eykel, Associate Professor of Religious Studies, Forrest S. Williams Teaching Chair in Humanities, Ferrum College

As a former diehard, anti-science, evangelical, this book presents the complex physics in a fun and laymen-friendly way while also taking a deep dive into the philosophical way the confirmation of extraterrestrial life would impact religious theology and praxis. Adair and Pearce meet the reader with an education and an excitement about the questions and unknowns. If you grew up having your curiosity shamed or punished, this book welcomes your "what ifs" with open arms.

- Jesseca R., a.k.a. "Heathen Queen", public scholar on social media and religious studies graduate student, University of Chicago Divinity School

About the authors:

Jonathan M.S. Pearce is a philosopher, journalist (for *OnlySky*, the first news media company for and by the nonreligious), author, and philosopher from south Hampshire, UK, who has dedicated many years to studying all manner of things philosophical and theological. A philosopher with a marked interest in religion, he became a founder member of the *Skeptic Ink Network* (SIN) before moving to write for *Patheos Nonreligious* (with his blog *A Tippling Philosopher*) and then *OnlySky*. As an original member of the Tippling Philosophers, from which his blog title came (a friendly group of disparate believers and non-believers, and sort-of believers based in Hampshire), he is a big advocate of casual philosophy groups meeting over pints of good ale. He lives with his partner (and wonders how she puts up with him) and their twin boys (and struggles to put up with them...). Being diagnosed with primary progressive multiple sclerosis, he would like to personally thank God for that gift, though rely more reasonably on science to find a treatment (which he did do, with successful stem cell therapy that has kept him stable for several years—a resurrection of sorts, but certainly not a miracle).

Dr. Aaron Adair is a physicist, data scientist, educator, and biblical scholar. He earned his PhD from the Ohio State University (2013), taught physics as a professor at Merrimack College and Babson College as well as at the high school level, and now he is a research affiliate at MIT performing physics education research using machine learning techniques. His current profession is as a lead data scientist working in the defense sector. Among the places he has performed research include CERN (helping prepare optical equipment for the ATLAS particle detector) and SETI, where he mapped neutral hydrogen while scanning for intelligence among the stars. In biblical studies, he is the author of *The Star of Bethlehem: A Skeptical View* as well as several articles and book chapters on the intersection of science and religion. He is also a member of the Society of Biblical Literature (SBL), having presented on euhemerization in antiquity and the interpretation of the Book of Revelation. Additionally, he is an area chair for the Southwest Popular/American Culture Association (SWPACA) on the topic of science, technology, and culture, and he is on the editorial board for the journal, *Dialogue: The Interdisciplinary Journal of Popular Culture and Pedagogy*. Currently, he lives in Boston with his wife, enjoys going on long runs, and spends his free time learning languages.

Other books by Jonathan MS Pearce:

30 Arguments Against the Existence of "God", Heaven, Hell, Satan, and Divine Design

Why I Am Atheist and Not a Theist: How To Do Knowledge, Meaning, and Morality in a Godless World

The Resurrection: A Critical Examination of the Easter Story

Not Seeing God: Atheism in the 21ˢᵗ Century (ed.)

Did God Create the Universe from Nothing? Countering William Lane Craig's Kalam Cosmological Argument

Beyond an Absence of Faith: Stories About the Loss of Faith and the Discovery of Self

The Problem with "God": Classical Theism under the Spotlight

13 Reasons to Doubt (ed.)

Filling the Void: A Selection of Humanist and Atheist Poetry

13 Reasons to Doubt

The Nativity: A Critical Examination

The Little Book of Unholy Questions

Free Will? An Investigation into Whether We Have Free Will Or Whether I Was Always Going to Write This Book

As Johnny Pearce:

Twins: A Survival Guide for Dads

Survival of the Fittest: Metamorphosis

The Curse of the Maya

Greece Lightning

The Family Book of Word of the Day. For a Year. Every Year. Forever.

Other books by Dr. Aaron Adair:

The Star of Bethlehem: A Skeptical View

A note about the book:

A few words about acronyms may be warranted here. ETI stands for extraterrestrial intelligence. An alien—extraterrestrial life—may or may not have intelligence. A microbe living on Mars might be an alien, but it won't be shooting lasers at you any time soon. We don't think. We do use ETI in a number of different ways, though. ETIs will be shorthand for extraterrestrial intelligent lifeforms, or intelligent aliens, and thus (for most intents and purposes), aliens. We might also use ETs or ET life to mean broadly the same thing—(intelligent) alien or aliens. For example, we might say "ET is sending out a signal" where, technically, this equates to "an extraterrestrial lifeform or civilization is sending out a signal." Some of these terms will be used interchangeably and the reader shouldn't have an issue deciphering what we mean.

Acknowledgements:

So many people have helped to shape this book. In particular, we are grateful for the indefatigable and consistent proofing of Jörg Fehlmann. Not for the first time, he deserves and gets huge appreciation. James Fodor and Ed Buckner deserve our gratitude for their contributions and endorsements of this project.

Thanks are due to the many people who have read over the manuscript and endorsed the book.

Additional input from Prof. David Pritchard led to several useful additions to this book, including the chapter on the possibility of humans worshipping aliens. Lastly, Aaron was given so much encouragement from his wife, Xiao Di "Janice" Tong. 我非常爱你，小滴. Likewise, Jonathan was very grateful (and, again, not for the first time) for the focus his families have afforded him in the process of writing this book.

*For seekers: don't so much as endlessly look
for what is not there,
but discover what is.*

"I'm frequently asked, "Do you believe there's extraterrestrial intelligence?" I give the standard arguments—there are a lot of places out there, the molecules of life are everywhere, I use the word billions, and so on. Then I say it would be astonishing to me if there weren't extraterrestrial intelligence, but of course there is as yet no compelling evidence for it.

Often, I'm asked next, "What do you really think?"

I say, "I just told you what I really think."

"Yes, but what's your gut feeling?"

But I try not to think with my gut. If I'm serious about understanding the world, thinking with anything besides my brain, as tempting as that might be, is likely to get me into trouble. Really, it's okay to reserve judgment until the evidence is in."

- Carl Sagan, *The Demon-Haunted World: Science as a Candle in the Dark*

CONTENTS

FOREWORD 1

INTRODUCTION 5

PART I: ALIENS 9

1 – ARE WE ALONE? 9

THE COPERNICAN PRINCIPLE 9

THE KARDASHEV SCALE AND SPACE TRAVELERS 11

THE DRAKE EQUATION 16

THE FERMI PARADOX 25

SORRY, WRONG NUMBER 26

HIDDEN CIVILIZATIONS AND THE TEENAGER PROBLEM 28

ARE THEY AFRAID? THE DARK FOREST AND ITS LIMITATIONS 29

FIRST! 35

SO LONELY 38

THE COSMIC SCALE 43

2 – WHAT DOES EXTRATERRESTRIAL INTELLIGENCE LOOK LIKE? 47

BIOLOGY OR ELECTRICITY? 47

DETECTING INTELLIGENCE 50

3 – HAVE THEY ALREADY FOUND US? 57

LISTENING TO EARTH 57

UFOS/UAPS 60

MODERN ABDUCTIONS 63

AN ALIEN PROBE? 64

RELIGION FROM THE STARS? 69

4 – WHAT MIGHT AN ALIEN RELIGION LOOK LIKE? 75

PART II: REACTING TO ETI CONTACT 81

5 – RATIONAL HOPE AND FEAR 81

FEAR AND LOATHING 81

A NEW HOPE 86

6 – COGNITIVE DISSONANCE AND WHAT LEON FESTINGER CAN TELL US 91

7 – THE POLITICS OF POWER: A HUMAN REACTION 95

PART III: CHRISTIANITY: 6 CHALLENGES **101**

PROBABILITY AND ABDUCTIVE ARGUMENTS.. 105
8 – CHALLENGING THEISM **109**
THE FINE-TUNING ARGUMENT 109
THE POOR DESIGN ARGUMENT 112
9 – CHALLENGING CHRISTIAN SCRIPTURES...**117**
BIBLICAL COSMOLOGY.. 117
PROPHESYING ETI 122
SCRIPTURAL TUMBLEWEED AND THE ARGUMENT FROM SILENCE...................... 123
DIVINE DECEPTION 126
THE PROBLEM OF PARTICULARITY 131
10 – CHALLENGING CHRISTIAN DOCTRINES..**137**
JUST ONE INSTANCE OF *IMAGO DEI*? 139
ALIENS AS FALLEN MORAL CREATURES 141
ALIEN ATONEMENT 148
DIVINE INCARNATION IN ALIEN FORM.. 150
ONE JESUS TO RULE THEM ALL, ONE JESUS TO FIND THEM 153
A TRILLION PLANET-HOPPING OR SIMULTANEOUS JESUSES 164
A CONCLUSION ON DOCTRINE 173
11 – CHALLENGING CHRISTIAN TRADITION **175**
12 – EXACERBATING THE PROBLEM OF EVIL...**181**
ABDUCTIVE ARGUMENTS AND PREDICTING UNIVERSES................................... 182
RUHMKORFF AND THE COPERNICAN PRINCIPLE 184
THE CHRISTIAN RESPONSE 192
PHENOMENAL CONSERVATISM 194
13 – CHALLENGING THE CHRISTIAN NARRATIVE...................................**199**
NARRATIVE TENSION 199
OUT OF TIME... 205

PART IV: FROM ALLAH TO VISHNU – HOW OTHER RELIGIONS MIGHT BE AFFECTED **211**

14 – JUDAISM...**211**
15 – ISLAM **215**
16 – HINDUISM AND OTHER EASTERN SPIRITUALITIES **219**
17 – OTHER RELIGIONS OF NOTE...**223**
UFO RELIGIONS 223
BATTLESTAR MORMONICA.. 224
18 – ATHEISM: WHAT IF ALIENS ARE CHRISTIANS? **227**

19 – ALIEN APOTHEOSIS **231**
The Remote Religion 232
Gods Among Us 235
Ad Astra, Inter Pares 238

CONCLUSION **241**

20 – WRAPPING THINGS UP **241**
Cognitive Dissonance and Christian Coping Mechanisms 241
The Threat to Theology 244
The Existence of Extraterrestrial Intelligent Life Makes the Existence of God
Less Likely 245
AFTERWORD: The Life ∴ Religion? **247**
APPENDIX 1 .... **249**
Estimating the Population of Current ETIs in the Milky Way 249
APPENDIX 2 .... **259**
Updating Probabilities in the Problem of Evil and ETIs 259
BIBLIOGRAPHY **267**

FOREWORD

The idea that people's religious beliefs affect their views on social, political, and scientific matters is hardly novel. Indeed, in recent years we have seen just how potently theological beliefs can impact seemingly disconnected issues, such as the teaching of evolution in schools, efforts to combat climate change, and public acceptance of vaccinations. While these and other connections between religion, society, and science have been discussed at length by many authors, one area that has long interested me, but has received relatively little attention, is the relationship between religious belief and the existence of extraterrestrials.

There are many interesting parallels between God and aliens. Both are widely believed to exist despite the absence of any clear or direct evidence. Both are widely claimed by their adherents to be active in the contemporary world (albeit in different ways!). Both are thought to have such exceptional capabilities that their intervention is invoked to explain any and all sorts of strange or puzzling phenomena. Indeed, it is a common trope in science fiction for technologically advanced species to be worshipped as gods by less developed species.

And yet, skeptics are typically unimpressed by appeals to deities or aliens. After all, we know that humans have a tendency to anthropomorphize and seek for answers in terms of the intentional actions of agents. We see faces in clouds, in the stars, and on pieces of toast. Historically, we believed that the weather, the motion of the stars, and growth of plants, and the course of diseases were the result of invisible spirits or deities beyond our comprehension. As such, attempts to resolve our ignorance by appeal to powerful, unobservable agents is much more informative about our own psychology than anything else. Indeed, despite both invoking super-human agency in their explanations, the sorts of people who believe in alien abductions are liable to scoff at divine miracles, while those who invoke God's intervention as an explanation often do not take the idea of alien interference seriously.

I have encountered the latter situation on several occasions during debates with Christian apologists about the evidence for the resurrection of Jesus. Apologists typically claim that only a bodily resurrection can explain the empty tomb and the reports of Jesus appearing to groups of his followers. In response, I have sometimes argued that even if no purely psychological or sociological explanation of such facts were available, it is nevertheless more plausible that the resurrection was a hoax perpetrated by technologically advanced aliens with a perverse sense of humor, than a man actually came back from the dead. After all, we know that humans sometimes go to great lengths to perpetrate hoaxes, often for little purpose other than amusement and to gain attention from peers. Might not aliens be similar? Is it so implausible that aliens millions of years more technologically advanced than us would have the ability to make a corpse disappear, or produce some kind of apparition of Jesus sufficient to fool his disciples? Is it instead more plausible that the God of the universe came to Earth, lived for thirty years, got himself arrested and killed to atone for the corruption inherent in a world that he created, then resurrected himself and visited to his friends for a few days, before disappearing from history entirely?

I have never met anyone who found this appeal to aliens convincing. Nonetheless, in addition to the issue of how we assess the quality of explanations, it also highlights the close connection between extraterrestrial intelligence and Christian theology, and indeed theism more generally. For if humanity is but one species in a universe full of intelligent beings, can Christians really still claim that Jesus performed a universally atoning sacrifice? Is it just a coincidence that Earth, of all the other planets in the universe, was the lucky planet where God chose to incarnate? Or perhaps each world has its own incarnation, each of which is a single part or aspect of a God's atoning work in its entirety? Similar issues arise for the status of the Qu'ran as the allegedly perfect and final revelation of God. Given such examples, it is clear enough that the existence of extraterrestrials would pose significant problems for theologians, and draw into question many fundamental assumptions and doctrines of traditional religions.

In this volume, Pearce and Adair do a masterful job of examining these and other difficult questions that defenders of traditional religions are beginning to grapple with. One theme that emerges is the immense anthropocentrism of traditional religious doctrines. The writers of the Old Testament knew nothing of the world beyond parts of the Eastern Mediterranean and the Near East. Over the centuries, Christians had to grapple with the discovery of whole new peoples, fitting them into a theological framework which seemed to have little place for them. After all, if Christianity was so important, why had Jesus never visited China or the Americas? Later came the discovery that Earth was not in fact the center of the universe, despite what appeared (at least to many people) to be claimed in the Bible. Early in the twentieth century it was discovered that our galaxy is but one of billions of galaxies in a universe that was immensely larger than we had previously imagined.

Today, hundreds of new exoplanets are discovered every year, many with the possibility of harboring some kind of alien life. Viewed in this context, a God of all creation who busies himself with the attire and diet of a small group of Semitic-speaking tribes in the early Iron Age seems faintly ridiculous, at least to me. This book develops on this theme, highlighting the absurdity of literalistic readings of traditional religious teachings in light of contemporary cosmology and astrobiology.

Another highlight of this work is the importance of evidence and the scientific method. Confronted with the seemingly impossible question of estimating the number of intelligent alien civilizations in the galaxy, Pearce and Adair show how in the guise of the Drake equation, scientists have broken the question down into meaningful components, and then used the best available evidence and inferences to arrive at informed estimates for the various components. While we are not yet able to confidently answer the question with any precision using this approach, we learn a great deal along the way, using a framework which allows us to keep track of our uncertainties. The contrast to the approach of many traditional religions, where in response to difficult questions one is given an ancient book and told to read it until you find an answer, was not lost on me.

Of course, I realize this is somewhat of a caricature as not all religious believers or traditions respond in this way. Nevertheless, in my own experience, and I suspect in the experience of many readers, religion is often hostile (whether overtly or tacitly) to critical, out-of-the-box thinking.

This book provides a fun and informative exploration of one aspect of this tension between religion and critical thought. While readers may be disappointed at the lack of definitive answers, this highlights all the more the importance of carefully evaluating the evidence we have, without dogmatically holding on to preconceptions or reasoning to desired conclusions. I hope this book will be a guide and a resource to those interested in thinking about the big questions in life, such as "why are we here," and "are we alone?" Science and philosophy may not have all the answers, but they are the best guides we have.

James Fodor, author of *Unreasonable Faith: How William Lane Craig Overstates the Case for Christianity*

4

INTRODUCTION

A long time ago in a galaxy far, far away....

If you're anything like us (and if you are reading this book, we probably have some common interests), anything alien will pique your interest. Not in an "Ancient Aliens" *History Channel* explanation of almost anything sort of way, but from a cultural scenario of being surrounded by movies and TV shows, books and comics, in which sci-fi is the canvas onto which all sorts of fun and exciting stories are painted.

With those pictures hung about our lives, it means that we often end up thinking about the possibilities of the universe, wondering about those long-ago and far-away galaxies.

What if?

What if questions are fun but they are also useful. They can help us set up hypotheticals that allow us to understand certain situations or plan for various eventualities.

In the context in question, for many people, this isn't a "what *if*", but a "what *now*?" There are those who believe that we have already been in contact with aliens, with extraterrestrial intelligence (ETI). We don't, and we will be discussing why not in the proceeding chapters. Yet it pays to be prepared. Oh, and thinking about this is fun. So it's a win-win.

Many of us have grown up alongside or flying over a science fiction landscape full of characters and machines, lifeforms and catchphrases. "A long time ago in a galaxy far, far away..." is pitted against "To boldly go where no man has gone before" in a tribal culture war among the stars. Carl Sagan's torch has been passed to Neil deGrasse Tyson, and no doubt will be to someone else, as the rich tapestry of the universe in *Cosmos* evokes awe and wonder in the mind of the viewer. And that awe and wonder is only given eternal inflation by the bookshelves and comic stores, movie theaters and streaming services, chock-full of great (and not-so-great) sci-fi creativity.

Science fiction is a wellspring of philosophical stimulation. Legendary science fiction writer Ray Bradbury agrees:[1] "Science fiction is the most important literature in the history of the world, because it's the history of ideas, the history of our civilization birthing itself." Indeed, the genre itself may be preparatory in nature. After all, as actor Dwight Schultz once said, "With

---

[1] Lesnick (1990), p. 40.

science fiction I think we are preparing ourselves for contact with them, whoever they may be."[1]

However, this is not just the frivolous project or one about planning for a very low probability hypothetical scenario. If, and we will later argue that this is something of a certainty, intelligent alien life exists out there, then such a situation could well present a huge challenge to religious belief in the world, and certain religions in particular. For those adherents, such hypotheticals may not be so much fun and may render their beliefs or doctrines untenable.

This book is, then, a brief analysis of the following: what intelligent alien life might look like; the likelihood of it existing out there; whether we have already made contact with it; and, granting its existence, what ramifications this has on the different religions of the world (with an emphasis on Christianity).

We are unapologetic in looking in more detail at Christianity since that is the context that we have, as authors, found ourselves ensconced in or more knowledgeable of. But, later in the book, we will be looking at potential impact on other religions, though with less theological detail and discussion.

Part 1 of the book will deal with defining and discussing ETI before looking at whether we are alone in this rather sizeable universe. We discuss what alien life might look like. Then we pose the question, if we are not alone, have aliens already found us and *can* they even find us? Last, we ruminate on what an alien religion might look like. And yes, we do detail the Drake Equation and the Fermi Paradox for those interested.

Next, Part 2 discusses the psychology of hope and fear in finding out about or making contact with alien life. We also look at the vitally important psychological phenomenon of cognitive dissonance since this will become important later in our analysis. In the third chapter in this section, we deliberate, absent of considering religion, how we might react to alien contact (on a larger scale than a single drunk person being kidnapped from a Midwest farming community and then probed).

Part 3 deals with the potential challenges that ETI discovery present for Christianity. We structure this in a way that reflects work done by two Christian philosophers in a paper discussing the subject. The paper "Houston, Do We Have A Problem? Extraterrestrial Intelligent Life and Christian Belief," by C.A. McIntosh and T.D McNabb forms the skeleton of this section by dividing the challenges up into 6 areas. We first look at whether ETI is a threat to general theism. Then we see if ETI challenges Christian scriptures, doctrines, tradition, and narrative. We finish by investigating whether ETI exacerbates the problem of evil.

---

[1] From an interview *UFO Magazine* (U.S. Edition, 1994, 9:4), by Dean Lamanna, "It's No Act—He's into UFOs."

Part 4 of the book looks at some of the other major religions of the world in a similar fashion, if in a little less depth.

Lastly, we wrap up the project by clearly laying out our conclusions on the subject.

The book is aimed at the general audience and is written in an accessible style, but will also be dealing with some substantive philosophy and science. Hopefully, we will be flying on a trajectory that navigates a fine line between being accessible and impactful.

With that in mind, let us linger no more and get stuck into probing these ideas and seeing if there is anything surprising or even predictable that comes about if and when the world of extraterrestrial intelligence collides with the world of earthly religion.

8

PART I: ALIENS

1 — ARE WE ALONE?

One of the quintessential tropes of science fiction is the meeting with a member of an alien civilization. Sometimes it is a peaceful event, sometimes it is the beginning of the conquest of Earth. Never should it be a dull day. The possibility of ETI was first seriously considered centuries ago, and it is actively pursued today by both members of the SETI Institute as well as UFO enthusiasts. But we all are waiting for the undeniable signs that there is someone else out there among the stars.

Is there anyone out there to look for? Is there just one other civilization in the Milky Way galaxy, or in the entire observable universe, or is our galaxy teeming with life? And if so, are they looking at us now? Do we have the technology to find ETI, or conversely, is it possible for them to detect us? Like much else in science, we will try to answer the questions with observations and math, noting our uncertainties and how time may close our knowledge gap within our own lifetimes.

The Copernican Principle

In 1543, lying on his deathbed, a Polish astronomer finally saw his life's work in bounded form, a book with the Latin title, *De revolutionibus orbium coelestium*. It was the first major work in centuries written by a Western astronomer positing a model of the solar system with the Sun, not the Earth, at its center. While the thesis would become controversial in the decades to come, it did not lead to the immediate realization that a revolution not of the heavens, but of science itself, was underway. The modern version of the solar system is not the same as the original thesis, but we nonetheless name our new view of the universe after Nicolaus Copernicus.

When Copernicus's model was introduced, a new vision of the universe entered into the imaginations of people across Europe. Thomas Digges in England described this new cosmos, but with an amazing addition. In Copernicus's solar system, the stars were inert and centered around the Sun instead of revolving around the Earth. Digges, in his illustration, said the stars were not just far away, but went off into infinite space, and those stars were even more brilliant than our Sun.

With innumerable stars, could there not be innumerable worlds and inhabitants? This was a view supposed by the ancient philosopher, Epicurus, and it would be posited again when this new universe emerged in Europe in the 16th century. It was supposed that that there were infinitely many intelligent lifeforms by Giordano Bruno, and the notion was championed by philosophers and scientists for centuries, including Christiaan Huygens, John Locke, Benjamin Franklin, and countless scientists in the 20th and 21st centuries.

But if there are countless worlds, where is our place in it? Are we the pinnacle of creation or the muck in the corner? To place ourselves, we have two ways of knowing: prior knowledge and new observations. Our prior knowledge is guided by a principle named after the astronomer that helped start the ball rolling: the Copernican principle.

What the principle says is that, unless you have good reason to think otherwise, based on good evidence, you should expect to be in an unprivileged, average place in the grand scheme of things. In effect, we are a random draw of some distribution of the way things could be, and a random draw to a normal distribution means you should expect to be somewhere in the middle of the pack. Literally, the farther you are away from the average, the more unlikely to be there by random chance.

The Copernican principle is the result of us no longer finding the Earth as the preferred location in the universe, or special in that it is the center of the cosmos. Instead, our planet is one of several orbiting a star, and that star appears to be fairly average compared to the hundreds of billions of stars in our Milky Way galaxy, which itself is one of hundreds of billions of galaxies. We do not seem to be in a special place, but instead we are living in the suburbs of just one galactic city.

But we feel special because we are alive and intelligent. We are intelligent enough to do science and know our place in the universe. But the Copernican principle suggests that if we are a random draw from a bag of lifeforms, we are most likely close to average. What that means is that we should expect to find forms of life less intelligent that us, and that is clearly so. We don't need to device a new IQ test to confirm a human is smarter than a fern. However, if we are in the middle of the pack, doesn't that suggest there are not only intelligent species out there, but that some are more intelligent than us?[1]

The principle suggests, at least as an initial intuition, that we are not special: There are other intelligent lifeforms to find out there, either on Earth

---

[1] One of the first uses of the Copernican principle to analyze the distribution of alien civilizations, as well as calculate our future, is Gott (1993).

or among the stars. We seem to have done a sufficient investigation of the Earth, and no other beings on land, in the sea, or in the air appear to intellectually challenge us. So, to the stars we must search.

But if we are of middling intelligence, what is the range? If a bacterium or plant is the bottom of that range, imagine how far up it goes! We could expect there to not only be some sort of ETI, but it would seem reasonable to suppose forms of ETI *phenomenally* more advanced than us. Consider the Copernican principle in terms of technology. Again, perhaps we are in the midpoint of where machines and computers can take us; on the tails of the distribution of technological ability, there may be civilizations that are so powerful that they might as well be gods.

However, we must temper our intuitions with evidence. We could look at ourselves on Earth and say: We are the only intelligent species on this planet, so we are on the tail-end. Supposing we found no direct evidence for alien life so that we intuit our specialness. This may be so, but our lack of observations of ETI can be a product of numerous factors that make detection difficult, from technological hurdles to sociological grounds, which will be explored in the following sections. Basically, there are reasons to think the argument from alien silence is too weak to suggest we are special in terms of intellectual prowess.

We should consider one more feature when thinking about random sampling and ETI. It may well be that there are forms of ETI that are our equals or inferiors. There may be many worlds where the most intelligent species is the equivalent of our extinct relative, *Homo erectus*. However, these are not the forms of ETI we are going to find anytime soon. If we ever discover ETI with our radio telescopes or if they visit us, then their technological powers would be beyond ours. If we find ETI, there would be a strong selection bias toward ETI that has technology at least as advanced as our 20th century machines and electronics. Given that we will only see ETI with technology from our 20th century and beyond, then what we should expect to find, if it is out there, would be wonderous.

The Kardashev Scale and Space Travelers

The general expectation of ETI is that they will be far beyond us in technological prowess (given the probability of us not being the most advanced of species in the universe). We have no reason to think we have already reached the pinnacle of science and engineering, so the possible future discoveries and potential industries are vast. In the next few decades, we may finally land people on Mars, have sustainable fusion reactors, and computational power that makes our smart phones look as pathetic as the room-sized ENIAC computer does to us today.[1] Imagining that future technology is not so easy, and the history of futuristic predictions has a poor track record. Science fiction author Isaac Asimov imagined planet-sized computers instead of the Internet, for example. So we ought not to predict our own future technologies, lest this book age as poorly as the Victorians who imagined we would today ride on steam-powered horses.

However, if we were to think of a scale to weigh technological prowess and ability, physics does provide a universalizing strategy: energy. The first simple machines, such as levers and pulleys, were manipulators of energy inputs into more useful forms. With the Industrial Revolution, we then had machines that could output levels of energy far beyond what humans had been capable of before with only animal muscles. With steam power, humanity was no longer limited to the strength of men and mules. Electricity then made the flow of energy more efficient and efficacious. Today, we release energy from the atom and from the weather, in the form of solar and wind energy. Our expectations are that we will need more energy in the future, as well as more efficient use of that energy.

But then consider this: What are the limits of our energy use? There is only so much uranium in the Earth's crust, and only so much light from the Sun shines on the Earth. If we want to grow, then we would need to find other sources. The reason is not just to fulfill our economic desires; if we ever want to be space travelers, we need new power sources.

To get a sense of the energy needs for space travel, consider the problem with space itself: it's big. Very, very big. Our current methods of launching probes to the planets and beyond requires chemical rockets, which can achieve impressive speeds (or at least impressive compared to that which a bicycle can achieve). But in the context of space, those speeds are laughable if you have the goal of getting places within a human lifetime. The fastest probe sent into deep space still remains *Voyager 1.* Launched in 1977, it has a speed of almost 17 km/s. At such a speed, one could travel like a plane from London to Los Angeles in under 9 minutes. But if you just wanted to travel, say, to the nearest star to our solar system, Proxima Centauri, the trip would literally be ten times longer than the history of all of written civilization.

If the goal were to reach the stars, we had best get there while the crew is still alive, rather than their great × 20,000 grandchildren. And with the return trip, if even possible, the amount of time passed is on par with the age of *Homo sapiens.* By the time the crew were to return to Earth, humanity would have been gone, having possibly evolved into a new species genetically related to, but incompatible with, the returning Earthlings.

---

[1] ENIAC—the first digital, programmable, electronic computer—had an operating speed of about 500 floating point operations per second (FLOPS). An iPhone X's main chip is literally a billion times faster (roughly 600 GFLOPS).

Current rockets are just too slow.

If we want to reach the stars, we would need to go much, much faster. Our current physics indicates that the fastest we could potentially travel is just under the speed of light. However, the faster you travel, the more energy you need to get up to speed—not to mention more energy being needed to stop.

No matter what technology we ever develop, physics indicates an important grounding fact: If you want to go faster, you need more energy (literally the energy of motion, called kinetic energy). Chemical reactions can provide that energy, but this is a severely limited approach. Just to get 1 kg of mass into orbit requires at least 8 kg of chemical propellant,[1] and much more if the goal is to travel to the planets, let alone the stars. Worse still, because a rocket has to carry its fuel, putting more fuel in the rocket makes it more massive, thus requiring more fuel to lift the fuel. The mass of the rocket increases exponentially just for doubling its speed. Exponential requirements of a substance we need too much of—just for local trips—leads to absurd conclusions.

The problem is even worse if the hope is to reach an appreciable fraction the speed of light. In fact, it is physically impossible for chemical rockets to achieve such velocities, because chemical reactions simply do not have the energy to get their mass to such speeds.[2] If there is an advanced civilization out there that has an interstellar presence, they must have ways of utilizing much, much more energy than is available in chemical form.

To get a sense of the scale of the energies needed, suppose we wanted to get to Proxima Centauri in a reasonable amount of time. We would need to go an appreciable fraction the speed of light, and the faster we were to go, the more the effects of special relativity would come into play. If the craft went at about 87% the speed of light, the trip would last a bit less than five years, according to someone on Earth; but at such speeds, the effects of time dilation would mean the travelers would only measure about 2.5 years of travel time.[3] While we might not personally like to be trapped in a space tube for two and a half years, it's at least better than dying of old age before leaving the edge of the solar system.

---

[1] Using the rocket equation and a fuel exhaust speed of 5 km/s (comparable to the space shuttle's engines), to get to orbit about the earth (about 11 km/s), 8 kg of fuel is needed.

[2] A good rule of thumb is the energy needed for a mass to reach relativistic speeds (around half the speed of light) is at least the equivalent mass-energy of the mass. Chemical reactions have a billionth this much energy; if you think we just need a billion times more fuel, remember you must also accelerate that fuel, so even the biggest chemical rocket will fail. Only non-chemical means, such as mass-antimatter reactions, can even possibly get a rocket to relativistic speeds. Nonetheless, if one were to naively use the relativistic rocket equation for a chemical thruster with the same exhaust speed noted earlier (5 km/s), then more mass than is available in the observable universe is required. In other words, it will never happen.

[3] These results ignore the time needed for acceleration.

As for the craft itself, let us suppose it would be the same mass as the now-retired space shuttle, which is frankly a severe underestimate of what will be needed for the trip. If the craft could be externally pushed by some force, the energy needed to get the space shuttle up to 87% the speed of light would be about 7×10^{21} Joules (J). This is the equivalent of over 20,000 of the most powerful hydrogen bombs ever detonated. It's about half of all the solar energy that lands on Earth in a day. It's ten times all the energy produced by all of our power plants in a year combined.

This is just for one ship getting up to speed. It doesn't consider the needs of stopping the ship, let alone building a fleet to travel back-and-forth. The energy needs would simply be beyond what we have, let alone what could be economical.

But, if ETI can travel the stars in reasonable amounts of time, then they must have insane amounts of energy. And this energy criterion can be used to categorize civilizations. This was already done in 1964 by the Soviet astronomer Nikolai Kardashev. He developed the eponymous Kardashev scale. The scale has levels that correspond to the energy production or control of that civilization. A Kardashev 1 civilization can harness all the power available on a planet, while a Kardashev 2 civilization can harness the power of an entire star. The difference between just these two civilizations is about ten orders of magnitude in power consumption. That is, a Kardashev 2 civilization has access to roughly 10,000,000,000 times the energy of a Kardashev 1 civilization. The astronomer Carl Sagan projected the scale down to see where we humans currently were, and we would be about a Kardashev 0.73. However, this figure is deceptive, because this is on a logarithmic scale, so we are not even close to level 1. To reach Kardashev 1, we would need nearly a thousand times more electrical power plants—not a thousand more, but a thousand more for every plant on earth.

However, we should not look merely at the our own place on this scale, but what else might come next. The scale can be projected forward. A Kardashev 3 civilization, for example, would be able to control an entire galaxy's worth of stars, making it at least ten orders of magnitude more powerful than a Kardashev 2 civilization. Beyond that, there would be intergalactic forces at play in the universe.

With these scales, it would seem that Kardashev 2 civilizations could produce ships moving at relativistic speeds without it becoming cost-prohibitive. Accelerating our space shuttle at a rate of 3g (three times that of Earth's gravity) would require a power output of 8×10^{13} Watts (W) for most of the trip; by comparison, the output of the Sun is 4×10^{26} W, which is 5 trillion times greater. So, level-2 aliens would have plenty of energy to get about the galaxy at speeds that actually become plausibly useful. For a level-1 species, this may be within their output abilities, but it's likely not economical—

spending nearly 1% of one's energy output to launch *one* ship seems prohibitive. In which case, if we ever encountered ETI traveling between the stars, they must be at least at Kardashev level-1, and more likely 1.1 and above.[1] In other words, and perhaps more important for comparison, they would be at least a million times more powerful than us.

At this point, one might object to these calculations because they are assuming the laws of physics as we know them. Perhaps there is some way to go faster than light, for example, and in such a way that doesn't require multiple planet's worth of energy production. For decades, there have been stories about a way to actually go faster than light with a warp drive right out of *Star Trek*. This is actually real, and it is known as an Alcubierre drive, named after Miguel Alcubierre, the physicist who solved the equations of general relativity to get the needed reshaping of space-time to make super-fast space travel possible.

There has been a considerable amount of research that has gone into Alcubierre's original proposal, noting its issues but also ways to improve the original design to get something more likely to be engineered. However, the most recent papers on the subject have noted that reaching speeds beyond that of light (called *superluminal velocities*) require something known as *negative mass*.[2] You have never seen negative mass, and there is good reason to think it doesn't and can't exist. Among the things it could do is create perpetual motion, and it is a substance that if it were moving and you pushed it in the direction it was going, it would slow down. These requirements are sure signs that something is wrong with one's physics. Perhaps a very fast ship could be made to travel without these impossible forms of matter, but superluminal speeds appear to remain out of reach.

Similarly, the idea of creating tears in space-time to link different parts of the universe runs into the same difficulty. For example, a wormhole (or Einstein-Rosen bridge) can connect two points in space and provide near-instantaneous travel between them. However, not only do they seem to have gravitational fields of such severity that you are jumping into a black hole, which will have you torn apart atom by atom thanks to its severe gravity, but to keep the wormhole from collapsing requires that same sort of exotic negative mass.

We cannot dismiss the possibility that future insights into physics will reveal ways to circumvent the limitations of light speed. Perhaps quantum

---

[1] A Kardashev 1.1 species would have the energy output of about 10^{17} W, which means launching this one ship would require one ten-thousandth of the power output of the civilization. Assuming a very tight correlation between energy production and gross domestic product (GDP), it seems uneconomical to have a larger fraction of an entire world's GDP go into one voyage. This fraction is on par with the cost of a space shuttle launch compared to US GDP.
[2] Bobrick & Martire (2021); Santiago, Schuster, & Visser (2022).

gravity will have some detail that is missed in the theory of general relativity. On the other hand, what we seem to find is that any superluminal travel allows for time travel into the past, and, with that, the sorts of paradoxes that lead to unexpected and deadly encounters with our grandparents. The best we can tell, future physics will disallow superluminal space travel. However, special relativity does tell us that, as we approach light speed, the apparent travel time between points becomes smaller, so that even multi-million light-year trips can appear to happen in days, supposing you traveled close enough to light speed. The limitation will be in energy production and implementation, so once again the Kardashev scale is our best way to judge technological and space-faring abilities.

At this point, we have some ideas about the capabilities of civilizations that might travel the stars. If they can galivant about the galaxy at speeds close to that of light, they are millions of times more powerful than we currently are, and their technological prowess would be hard to estimate. However, we might also consider ETIs who are willing to travel the grand distances between the stars at a much slower pace, requiring either generations of travel or otherwise shutting themselves down for centuries, perhaps by cryogenic sleep for biological entities or low-power mode for electro-mechanical probes. But even these abilities are way ahead of what we can do, so *any* ETI that can travel here will be our technological superior.

So, when will ET come? That depends on how many ETs there are, how capable they are at travelling or communicating, and their willingness to do so. Can we estimate this? That is something scientists have been doing for decades.

The Drake Equation

How many piano tuners are there in Chicago?

This was the sort of question that Italian physicist Enrico Fermi would ask his students. Why such an odd and obscure question? There was no source to find an answer, especially in the 1930s when Google wasn't yet a word, let alone a search engine. However, to answer this sort of question, one can make educated guesses. In order to make those guesses reasonable, Fermi's students would have needed to consider what factors would influence the answer and apply reasonable estimates to those values. For example, the number of tuners must be related to the number of households in Chicago. Some fraction of those homes would have a piano. Perhaps one in twenty? We may not know for sure, but we can have reasonable guesses based on our background knowledge—not everyone has a piano, but we have been to plenty of homes with one, so it cannot be zero homes, nor can it be all of them. Consider also how fast a tuner can work, how many pianos they might need to service to make it a viable business, or if they have a larger business of tuning instruments, so they don't need to rely on just one instrument repair line to make ends meet. You may not come to an exact figure, and you should say that your result is perhaps within an order of magnitude (a factor of 10) of the true answer. But, with such reasoning, you can get much closer than a random number pulled out of a hat.

When Fermi was composing such questions for his students, he did so to make them think about how to estimate unknown quantities. It was something he needed when he was involved in the Manhattan Project, when he estimating the power of the first atomic weapon. When it comes to estimating unknown quantities, his approach was taken up by an astronomer to answer the big question: Are we alone?

One of the most famous equations in modern science is the result of answering a Fermi question. This is the Drake Equation, postulated in 1961 by Frank Drake. The equation is a simple combination of independent factors that together provide an estimate for how populated the Milky Way galaxy is with ETIs.

$$N = R_* \times f_p \times n_e \times f_l \times f_i \times f_c \times L$$

This equation[1] says that you can get an estimate of the N, number of civilizations in the Milky Way right now, by multiplying together all of the factors on the right-hand side of the equation. R_* is the rate of star formation; f_p is the fraction of those stars that have planets; n_e is the number of those planets that are earth-like or otherwise habitable around each planet-bearing star; f_l is the fraction of those habitable worlds that actually has life emerge; f_i is the fraction of those life-bearing worlds that become intelligent; f_c is the fraction of those intelligent worlds that can communicate to other planets, and L is the longevity of those communicating civilizations.

Let's consider what these values really mean. The rate of star formation is how often new stars are formed in the Milky Way galaxy. With each new star, forming by the gravitational collapse of gas and dust, a new solar system emerges as well, with its comets, asteroids, and planets. How often new stars form can be deduced directly from astronomical observations with

---

[1] There exist other, equivalent formulations, such as using the number of stars in the galaxy and using the fraction of a planet's habitable timespan as variables. However, the version above is more commonly found in the literature and seems a bit more intuitive.

telescopes, so it is one of the best-known and well-constrained values in the Drake Equation.[1]

However, do all stars have planets? Up until the 1980s, there were no confirmed observations of planets in another solar system. But new telescopes have been developed, some on Earth, some in space, that have found thousands of exoplanets (that is, planets orbiting stars other than our Sun). We are now confident that most stars have planets, but not all of the planets are likely great places to visit. That is, if your goal is to step outside and walk around. Many of the first planets discovered were more massive than Jupiter, were gaseous rather than solid, and were so close to their star that the surface of these gas worlds would boil you away. The techniques for finding planets were biased towards finding these so-called "hot Jupiters," since massive planets near their star were the easiest to detect. They either blocked the most amount of light from their star or caused it to wobble strongly and quickly; smaller, less massive planets farther from their star will block less light or budge the star less noticeably, as well as over longer periods of time because of their longer orbit. So, it was easy to find hot Jupiters and very difficult to find anything like a planet we would want to live on.

So, how many planets like Earth are out there? That is still being explored by astronomers with more technology, data collection, and physics simulations to constrain how likely such worlds are to orbit the stars in the sky. But how would we even know a planet is Earth-like? Why would we think a given exoplanet could have life? For that, the main considerations are twofold: Is there a surface to stand on, and could there be liquid water on its surface? Water is essential for life as we know it, so it is the major criterion for life to even be possible. For liquid water to exist, the surface of that world would need to be warm *enough* but not *too* warm. Since that surface temperature primarily comes from the planet's star, the main factors to look for are the brightness of that star and how far away the planet is from said star. If the star is brighter than our Sun, then for a planet around that star to have liquid water would require being farther away than we are from our Sun. If the star is dimmer, then the planet needs to be closer. The regions where this is possible are called the either the habitable zone or the Goldilocks zone. Not too hot, not too cold. Just right.

---

[1] However, there is serious issues with using the rate of star formation in the galaxy today. In the past, it was much higher, and current star formation has little to say about life forms that took billions of years to evolve. We might then use the star formation rate in the past, but how far into the past requires assumptions: How long does life take to emerge? Is it the same around stars generally? Do we use the rate of star formation for Sun-like stars or all types of stars? The R_* factor is not so certain given these issues, and it is a significant flaw with the Drake Equation as is. However, to be consistent with much of the scientific literature, we will continue to use it, though noting the caveats that come with the values for R_*.

The size of orbits within a Goldilocks zone is complicated by factors related to the atmosphere of a planet. If it has more greenhouse gases in its atmosphere, then it might become too warm, as it is on Venus; conversely, with more greenhouse gases, then it could be further away from its host star and still be a balmy, watery world. Additionally, the thickness of the atmosphere matters because atmospheric pressure effects the freezing and boiling point of water; if the pressure is too low, then liquid water cannot exist. On Mars, for example, if you took a bucket of water outside of your habitat, it would begin to freeze and boil at the same time, because the Martian atmosphere is less than 1% of Earth's atmosphere and does not provide enough pressure to make liquid water even possible. And how thick the atmosphere is depends, at least in part, on the mass of the planet.

Now, how could we possibly know the masses and distances of planets around another star? This can be determined from the detection methods and the laws of planetary motion first articulated by Johannes Kepler in the 17th century. The period of a planet's orbit can be determined by multiple means, and the orbital period is directly related to the size of the orbit via Kepler's third law. Astronomers can figure out the mass of the planet by seeing how hard it accelerates its host star (and there have already been methods of figuring out the mass of stars for decades) or more indirectly based on how much light the planet blocks from its star, giving an indication of its size.

With instruments such as the *Kepler* space telescope, we now have several known Earth-like planets in our galaxy. In fact, there is one orbiting the closest star in the night sky! Proxima Centauri b (or just Proxima b) is just over 4 lightyears from our solar system, and it orbits within the habitable zone of its relatively small star. There are other features about the planet that may not make it a great tourist destination (for example, it is likely tidally locked and one half of the planet is in permanent day and the other permanent night), its existence already tells us that habitable planets exist and are likely very plentiful in the galaxy.

In fact, our methods of saying what is a habitable planet may be too limited. Perhaps in the watery moons of Europa and Enceladus, which are heated not by the Sun but by tidal forces due to their orbits around their planets (Jupiter and Saturn, respectively), they could allow for life to emerge. There are also likely to be many, many rogue planets, worlds that no longer orbit any star after having been flung out by gravitational interactions during the formation of their solar systems. If those planets have Europa-like moons, or if they retain enough heat to warm water, not unlike the thermal vents on the bottom of Earth's oceans, then there could be many more places where water flows.

In any case, either with the more restrictive notion of Goldilocks zones or less restrictive, there will be some number of planets in a solar system that could allow for life to begin. So, again with good astronomical knowledge, we can limit the range of the values f_p and n_e.

Then comes life itself. Even if a planet has water, that does not necessarily mean anything is growing and evolving there. Only some fraction of planets that have the ingredients for life will actually have it emerge. That fraction, f_l, is very uncertain since we only know of one world in which life has emerged, and we still don't know how it came about. We are not totally ignorant concerning *abiogenesis*, the emergence of life from a previous non-biological state, but it is still incomplete and subject to considerable amounts of research. The main focus has been on the steps required to get to a self-replicating molecule, such as RNA. Once there is self-replication, then there can be descent with modification from an ancestor molecule, and then evolution by natural selection can begin.

There is the open question as to how much the emergence of life is a chance process or if there is something that drives chemistry into biology. If the first self-replicating molecules happen by chance interactions between simpler, non-replicating molecules, then life may never emerge even on a world where all the ingredients are in the oceans. After all, a million people can buy a lottery ticket in a scenario where no one wins; so having the ingredients does not necessarily mean the needed organic molecules will form. On the other hand, it seems as if life emerged on Earth as soon as it was possible. Early in the formation of the Earth was a period of millions of years called the Late Heavy Bombardment, wherein our planet was struck by asteroids at a precipitous rate. The craters on the Moon are some of the primary forms of evidence for this bombardment, and it is hard to imagine what could have stayed alive if Earth suffered a dinosaur-killing asteroid impact every few years. However, after the bombardment ended roughly 3.8 billion years ago, the first undisputed signs of life were found relatively soon after, about 3.5 billion years ago. There is even some evidence of biology from the period of the Late Heavy Bombardment.[1] In other words, even when hell was raining down from the skies, the oceans may have already been giving rise to life.

If life can emerge so quickly after the formation of the oceans or after the Late Heavy Bombardment, then that suggests the emergence of life is more probable than not, as if there were underlying physio-chemical reasons that self-replicating molecules arise beyond just random mixing.

Physics gives some hints that self-replication may be expected, because of the laws of thermodynamics. It is said a closed system, where no energy can enter or leave, tends to disorder, but an open system has the dynamics that can lead to spontaneous organization. Life is very good at finding places where there are different concentrations of energy and using that

---

[1] Dodd et al. (2017), p. 60-64.

energy difference to power itself. In general, energy flows from high- to low-concentrations, and every machine and living thing puts that flow of energy to work, either to accomplish a task or just to stay alive. Life, or more generally self-replicators, seem to enhance this energy flow. This energy flow seems to thermodynamically prefer self-replicators, meaning self-replicating molecules are more likely to be expected to form than not form where there is a non-equilibrium state.

The burgeoning field of non-equilibrium thermodynamics is giving us hints that self-assembly and self-replicators are not so improbable, given there is an open system like there is with the Earth.[1] Still, no one has gotten life to just pop out of a test tube given the chemical and thermodynamic conditions we think existed long ago, so we still do not know just how likely it is for life to emerge under the conditions of the early Earth, let alone on another habitable planet.

But even if life emerges on some watery world, that life may well be just microbes until the planet's star burns out, or worse, burns up the planet, as is the fate of Earth in a bit more than a billion years from now. To go from just simple life to thinking life that we would call intelligent is a long journey. There was about 4 billion years between the first lifeforms and the first tool-using hominids, and it is only this lineage that we have been willing to call "intelligent." After turning on the news, however, you might even think the term is ill-suited to describe us.

How could we even begin to estimate the fraction of life-bearing worlds that develop something we would call an intelligence? At this point, there are issues of definitions and criteria. What is *intelligence*, after all, and how can we tell if something is intelligent? There is a considerable amount of philosophical literature discussing the nature of intelligence and consciousness, which we cannot enter into here in any satisfactory way. Moreover, some of the classical ideas to test for intelligence, such as the Turing test for a computer, cannot be done with a species on a planet orbiting a thousand lightyears away.

Here we will adopt a functional definition of intelligence that is used in artificial intelligence (AI) research.[2] First, the subject of interest is considered to be an agent if it has agency—if it has the ability to observe its environment and autonomously use that data to take actions in order to achieve some instrumental goal. A simplistic agent would be a home thermostat: It detects room temperature, and if it is above or below some threshold, it turns on the climate controls to achieve the instrumental goal of keeping the temperature of the room within the pre-defined temperature range. However, a thermostat is not intelligent because it cannot evaluate its efforts to achieve its goals;

---

[1] England (2015); Dill & Agozzino (2021).
[2] Russell & Norvig (2016).

if the wires were connected incorrectly and the heat came on instead of the air conditioner, the thermostat would not change its behavior; it detects the room is warm, and then flips the switch it erroneously "thinks" is for cooling and continues to run the heater, making the room hotter instead of cooler. Never does the thermostat notice it's making things worse, let alone change its behavior. A rational agent could evaluate its performance and change its conduct in order to perform better than before. If this ability to evaluate a situation, autonomously act to achieve a goal based on the perception of the situation, and evaluate performance were generalized across many domains, then we would have an example of a general intelligence. Humans would fit this definition, since we have various instrumental goals, we act on our own to achieve them, and we can reflect on what we could do differently.

This definition of intelligence allows for the possibility of other animals fulfilling these criteria as well as the artificial intelligences that programmers and engineers have been developing. The functional definition of intelligence also suggests that life has an evolutionary pressure to become intelligent. A creature that is an agent can achieve its goal of reproducing more successfully than one that cannot. And an intelligent creature can improve its chances of success even more so, including fulfilling other instrumental goals. For example, birds can build elaborate nests in order to attract mates; the instrumental goal of being a better builder leads to fulfilling a terminal goal of reproducing. A species that can figure out numerous other instrumental goals to achieve their primary or terminal goal is more likely to be selected for than a less capable one. This isn't to say that life necessarily leads to intelligence, but it seems like intelligence and increased intelligence would be selected for by nature rather than against.

Returning to our working definition of intelligence, it also gives us a way to say if there is an intelligence on another planet. If there is an alien group that is achieving some instrumental goal and improving its abilities to achieve that goal, then it should count as intelligent. For example, if these aliens are figuring out how to grow more food or generate more energy, then they should be considered an extraterrestrial intelligence (ETI). The way we can tell if the aliens are doing these intelligent things is by comparing them to what would otherwise naturally happen. In nature, electricity does not generate itself more and more efficiently, but it is instead sporadic and controlled by non-intelligent factors. A lightning strike lacks agency. But if an alien builds a Tesla coil and makes their own lightning, even for the instrumental goal of "fun," it requires doing things that do not happen without agency. And those "artificial"[1] things such agents do could be detected by us.

---

[1] We put "artificial" in quote marks because there is an argument to say nothing is artificial that is carried out by a natural agent. Would we say a bird's nest is artificial? Then why a Tesla coil? This is, though, a philosophical rabbit hole.

If there is an ETI on a world, they will be developing their abilities to generate energy, since energy allows for the vast majority of other instrumental goals from food production and preservation to climate control, from entertainment to travel and trade. Agents need to communicate with each other to achieve their goals, and the easiest way that we know how to do this over long distances quickly and without wasting energy is the use of radio waves.

Radio is a form of light, and we have been using such light waves to send signals to each other for over a century. And those radio signals can do more than go between two agents living on an alien world. Those signals can travel between the stars and be detected by our radio telescopes. If we actually picked up a radio source that had the tell-tale signs of artificiality, that would instantly indicate that the source behind the signal had built something that nature would not have otherwise produced, and it was for some goal—the signal was made for a reason. That goal might just be facilitating bickering between Lrr and Ndnd on the planet Omicron Persei 8; or the goal might be to tell the universe that someone is living on that planet, and they want to say "hi" to anyone listening. In any case, the signs of intelligence can be noted by scientific means. In the next chapter, we will discuss what we would look for in a radio signal to identify it as intelligent in origin.

So, we have some sense of what the fractions f_l and f_i entail. The fraction of habitable worlds with life is unknown, and the fraction of those worlds that develop generally intelligent agents is also uncertain. It is unclear if life would normally generate some intelligent species after billions of years of evolution, but given the success an intelligent species can have in achieving its goals of survival and reproduction compared to non-intelligent species, it seems like they would stick around for a while. And if they are chatty, then we might notice their existence.

This brings us to the last two terms in the Drake Equation. Of the intelligent worlds, some fraction will have developed far enough to communicate, especially by means of radio. The fraction of ETIs that start to build and use radio transmitters is f_c, and again, we don't know what that fraction might be. It is going to be affected by how long that ETI has been around so it could even develop the technology in the first place. There is a million-year timespan between our ancestors first use of fire and our first use of antennas. If there are thousands of ETIs around the galaxy, and they are still in the Stone Age, then they are not going to be noticed by us any time soon. And even if they started to build radio antennas, our best chance of hearing from them is if they talk for a long, long time. The final parameter, L, concerns how long such a communicative species exists, either because they go extinct or because they don't send radio signals into space anymore. The hope is that they do not self-destruct and they actively pump radio waves into the interstellar void. If they also travel among the stars, then they should be sending signals back-and-forth, since lightspeed is still the fastest way to communicate. If an ETI becomes a galactic civilization, then they might well exist for millions of years. The value for L could indeed be very high, at least for some civilizations.

Thus, we have looked at the seven factors in the Drake Equation. Intuitively, one can see that the larger those factors are, the more civilizations there will be in the galaxy. One just needs to know the numbers.

At this point, the uncertainties involved become apparent. The fraction of habitable worlds that actually develop life is an even harder question than asking how many pianos a piano tuner must fix in a year to make for a viable business. Considering we have only one world that we know life has emerged on, our dataset is very limited, and our intuitions are not viable guides. The other factors are similarly dubious. How long can a communicating civilization last? At least a century it seems, given that we humans have been creating radio signals for that much time. But can we last for a thousand years? A million? Maybe, but it's speculative. And what fraction of life-bearing worlds become planets with intelligent creatures on them? Again, we have only one example, though we also note our planet only developed one kind of creature lineage that led to intelligent-enough tool users to start making cities, machines, and eventually radio transmitters.

So, what numbers end up going into the equation? Astronomers have tested out their intuitions and assumptions to end up with a wide variety of answers. When Carl Sagan presented the equation to those tuning in to watch the TV series *Cosmos* in 1980, he gave estimates from ten civilizations to millions. Others have been less optimistic than Sagan, suggesting we might be alone. With a range of zero to tens of millions in the scientific and popular literature, it seems easy to get the result one wants.

However, the equation can still be used in a scientific rather than merely speculative manner. In particular, given an estimate of N, it will have other potentially observable effects. The converse of the number of such civilizations is how far away they are. Given a random placement of these ETIs in the Milky Way, the more of them there are, the more likely they are close. If there are hundreds of millions of ETIs, then perhaps the nearest one is only a few lightyears away. The closer they are, the more likely we will notice them and pick up a signal. This means the value of N in the equation gives a time to find ET. The larger N is, the shorter the amount of time it will take for us to intercept their radio signals. It would also mean it would be easier to meet them physically and not just with electromagnetic signals.

The sorts of ETIs the Drake Equation considers are ones at least as advanced as us. The aliens must be able to create signals we can detect, or they have some sort of infrastructure that leaves a lasting mark in the galaxy. If they are building megastructures, then they may be blocking the light of stars, harnessing it to their own ends. If they are sending vehicles around the galaxy, they may also leave behind signs of their comings and goings. If they are in interstellar wars, the blasts from their weapons may light up our telescopes. The Drake Equation only cares about the period of time they have something detectable, though it was conceived as the length of time an ETI would transmit radio signals into space; Drake was a radio astronomer, after all.

The types of civilizations that Drake was considering would then fit across the Kardashev scale, from the minimum level of us (0.73) to inter-galactic superpowers (3.0 and above). If there were a type 3 civilization in the Milky Way, they would completely surround us, and every star system would have signs of their technology. If there are many type 2 civilizations, we might find at least the occasional star blotted out by alien megastructures. If there are numerous type 1 or below civilizations, they would nonetheless have the communication abilities we have with radios.

And so, we listen for them. At this time, our radio dishes and telescopes have found...nothing.

The Fermi Paradox

The Drake Equation is the product of asking a Fermi question. But, in 1950, Enrico Fermi also created another question. He asked a simple question about those aliens: Where are they? At the time, UFOs were entering into the public consciousness, and the physicist, conversing with his colleagues, blurted out a question that no one has yet definitely answered. His question leads to his namesake of a paradox: If we believe there are intelligent lifeforms in our galaxy, then why haven't we found them yet? If we predict that they are likely to exist, then not seeing them is very weird. If we predict that there are millions of alien civilizations, then this is even weirder.

The Fermi Paradox has its parallel with religious belief: If we are so certain that the supernatural OmniGod[1] exists, then why don't we find Him doing miracles? If God is not healing amputees is a particularly troublesome theological problem, then why isn't the Great Gazoo failing to perform technological marvels any less of an issue?

While we are here not looking into attempted answers to explain the hiddenness of the Almighty,[2] the similar need to reconcile the more scientifically-grounded search for ETI and our lack of detection can feel

---

[1] This is the term we will most commonly use to discuss God—the classical theistic notion of an omniscient, omnipotent, omnibenevolent god.
[2] Although, please feel free to peruse Pearce's back catalog of books; this is definitely something he analyzes!

Aliens and Religion: Where Two Worlds Collide

disconcerting. Scientists and other thinkers have come up with solutions to Alien Hiddenness, from intergalactic treaties to the extreme rarity of ETI. Some of these ideas have come from fiction, while others have a larger academic pedigree.

We will not be able to dive into all such explanations for why we have not yet found ETI,[1] but the major proposals have similar issues in resolving the paradox, and the weaknesses of all solutions thus far provide guidance on how to address the paradox and move forward in conducting new scientific experiments and observations.

Here, we will consider four major categories of explanations:

(1) They are not able to talk to us.
(2) They want to stay hidden.
(3) We are first.
(4) Life is rare.

Sorry, Wrong Number

The main assumption of SETI is that ETI are sending out radio signals that we can listen for with our correctly-tuned receivers. Of course, if they are calling us on 101.5 MHz, and our radio is tuned to 103.3, then we will get static rather than a message. While SETI researchers theorize which are the best radio bands to listen to in order to detect an artificial signal (see chapter 2 for discussion), there is nonetheless the need to listen to as many bands as possible. The larger the range, the more likely it is that a signal will be found.

But even this assumes ET is sending out radio signals to our benefit. We could imagine they have developed technologies they find far better at transmitting information across the galaxy. Radio signals can be blocked or at least interfered with by gas and dust, so perhaps they use a means we have not yet discovered. While it is currently very difficult for us to register, we now have gravity wave detectors; a gravity wave is less affected by the same materials that obscure radio. Our current instruments can only detect the largest of waves, such as when black holes orbit each other and finally collide. Perhaps an advanced form of detector can be built to find the subtler signals that also carry information at the speed of light, all while traveling under our noses.

There may be more technologies we have yet to learn of because of our incomplete knowledge of physics. This is, of course, speculation on top of speculation so it is not a great way to explain the paradox. It may be logically possible, but that does not make it probable.

---

[1] For many proposed solutions to the Fermi Paradox, see Webb (2015).

Even within the limits of what we currently know and can do with radio signals, it might be that the aliens just aren't trying to make a lot of noise. In the past, when we did not have satellites or had poor aim between ground stations and satellites, we had to transmit radio signals with much more power to make sure the message would be received. With time, we have become more and more efficient in signaling between ourselves, which means for an outside observer, our signals into space are getting quieter and quieter. While 1950s television broadcasts would have been sent by the most powerful radio antennas that corporations could afford in order to reach viewers, now with cable and direct satellite TV, that need has greatly diminished. Presumably, the same would be true for ETI, and for the same reasons (economics and signal quality). So the stray signals from ET should become harder to detect as time goes on.

The best way for us to actually get any indication that ETI exists by means of radio would be for them to openly broadcast their existence to us. They would have to actively choose to make themselves known and easy to detect, using a considerable amount of energy to transmit into the cosmos, just hoping someone might notice decades or centuries later. Consider the economics of doing the same on Earth: renting a powerful set of radio transmitters, constantly running, but with no listeners. It will be hard to get sponsorships for that, let alone a deal with Spotify.

So, perhaps ET is out there, but it's not worth the effort on their part to let us know.

Taking this hypothesis seriously, it must at least put a limit on how many civilizations are in the Milky Way. If there were millions of inhabited worlds, then one of them would be close by and even their weakening, leaking radio signals would have been noticed. If there were fewer civilizations, then a failure to detect them would be expected. ETI would be randomly placed around the galaxy, and if they are rare then they are likely farther away from us and their signals extremely weak or haven't crossed the gulf between worlds. Few alien civilizations could account for the lack of incidental signals being picked up by SETI. Thus, the null observation of ETI should put limits on how many civilizations there are in our galaxy.

However, the null observation and this hypothesis are consistent with there being no such civilizations at all. As the number of galactic civilizations decreases, it better fits our observation of no detected radio transmissions. For there to be more than zero ETIs in the Milky Way and be consistent with the missing signals of their technology, then additional suppositions must be added that are not in evidence. The quiet alien hypothesis has numerous variables, which are untethered to observations, and the simpler null hypothesis conforms to our observations just as well, if not better. Thus, the quiet alien hypothesis does not provide much insight into why we are not finding ET. Instead, a hypothesis that has more predictive power should be considered.

Hidden Civilizations and the Teenager Problem

Sometimes, it is suggested that we are not seeing aliens for a number of other reasons:

- Perhaps they are hiding their existence from us, either to treat us as a natural, undisturbed habitat...
- or maybe they *do* live among us but in hiding to study us (or with government knowledge in the style of *The X-Files*)...
- or they do not interfere in underdeveloped cultures *a la* the Prime Directive in *Star Trek*.[1]

These can be categorized into what is called the "Zoo hypothesis"—for one reason or another, the aliens are letting humans go about not knowing that there are other inhabited worlds.

While fictions provide all sorts of scenarios as to why the aliens are out there but not announcing themselves to us, there is a fundamental issue with the approach. If there are many civilizations, then they need to coordinate with each other to mutually not interfere in Earth's development. This may already be a suspicious scenario; if the civilizations are on uneasy terms, such as the US and former USSR were, agreements not to interfere in another country would be broached the moment it became possible to find an excuse. International peacekeeping is difficult as it is; magnifying the problem across star systems only makes it worse. But even assuming that the ETI nations of the galaxy could maintain such agreements, they would also need to internally police themselves. It becomes an issue of not just getting two powers to agree to not interfere with life on a planet, but it is also getting all of the citizens of those galactic civilizations to do the same.

---

[1] In *Star Trek*, the primary governing body demands that all members of their fleet of ships to not interfere with developing cultures, because they likely would be altering their natural development, leading to undesired results. Perhaps the most egregious failure to follow the Prime Directive in the original series is when a historian tries to help a struggling world rebuild their economy by providing the model of Nazi Germany, which Captain Kirk and the crew of the *Enterprise* must try to rectify. See *Star Trek*, "Patterns of Force", Season 2, Episode 21 (aired Feb 16, 1968).

28

This results in what we should call the "Teenager Problem": tell a teenager not to do something or create a rule for them to follow, and they will be prone to break the rule on purpose. Even if the equivalent teenagers of another planet are not the same as human teens, there is still the conceptual issue that it should be expected that not all members of a civilization will follow any given rule, and some will break it for their own, personal reasons. Or out of stupidity.

At the zoo, there is occasionally someone foolish enough to jump into the lion den. Sometimes alcohol is a factor, but people can be stupid without chemical assistance. Are there really no inebriated ET-teens in the entire galaxy who would jump the fence into Zoo Earth? To assume that all ETIs would perfectly conform to a galactic law for eons without fail is a very shaky foundation to set as a resolution to the Fermi Paradox.[1]

This sort of galaxy-scale policing has been subject to scientific analysis as well, and the studies generally find that it is hard to maintain such conformity unless the number of civilizations is very small or there was one early and dominating civilization that brought other, later civilizations into its galactic club.[2] However, even these analyses look at a "global" level, considering each civilization as a unit; to make matters worse, individuals *within* those civilizations would *also* be possible breakers of interstellar law. The Teenager Problem will be an issue so long as there is a galaxy of young whippersnappers not listening to their elders. So, the only way for the Zoo hypothesis to be viable is to have the number of civilizations in the Milky Way be very low, which is also consistent with there being no one else out there.

Are They Afraid? The Dark Forest and Its Limitations

The lack of sightings or detections of ETI could be the result of decisions on the part of the aliens themselves. If the aliens do not want to be seen, and they have the means to do so, then they would not be detected. If so, then the galaxy could be swarming with civilizations, but we would be none the wiser because of the decisions of countless worlds to keep quiet from the earthlings.

Why would one civilization do such a thing, let alone many? As mentioned, there is also the issue of maintaining the perfect black-out condition across the galaxy, requiring great cooperation and policing between civilizations and enforcement to keep those pesky teens from making a mess of

---

[1] Similarly argued by Wilkinson (2013), p. 103.
[2] Forgan (2011) and (2017).

interstellar law. What story could they tell each other to convince everyone to keep us humans in the dark?

Given the issues of coordination and policing, perhaps the assumption of some sort of federation of planets in our galaxy is faulty. Each civilization needs their own, internal reason to be quiet. In fact, there may be a better question to ask: Are the aliens hiding from each other? In other words, perhaps the reason we do not see ETI with our radio telescopes and other instruments is because the aliens do not want to be known by any other civilization out there.

This scenario is laid out most vividly in the novel *The Dark Forest* by Liu Cixin. Here is a simplified description of how he sets up the Dark Forest solution to the Fermi Paradox. Suppose a civilization made itself known to another. It is very likely that the two worlds would be at very different levels of technological ability. The detecting civilization could try to make contact themselves, but this would make their presence known, and with that the risk of how the detected civilization would react. If the detecting civilization is belligerent, then it would rather wipe out the detected civilization than let it grow more powerful. If the detected civilization is belligerent, the detecting civilization takes the risk of letting it grow and wipe *them* out. In either case, being detected makes other civilizations see you as a risk. Expanding this notion universally, your civilization should be quiet to avoid being discovered by an either belligerent or scared (yet more powerful) civilization; and because each civilization has the same fear (that there is another, more powerful civilization also hiding itself), everyone does their best to be quiet.

Every civilization is a hunter in a dark forest. Any noise might be someone there, ready to make the kill. The only way to be safe is to stay hidden, wait, and only when possible and certain, strike at the others you see.

Let us suppose the truth of this bit of galactic game theory. If so, then it puts necessary limits on the sorts of civilizations that could exist in the Milky Way right now. Anything from Kardashev 3 and beyond is disconfirmed by our very existence; the civilization in control of the entire galaxy would already reign over the solar system and would have wiped us out. Thus, we can be confident there are no such powers in our galaxy.

But what about less powerful civilizations, such as Kardashev 2? For such aliens to have the power of an entire star, they would have had to build a Dyson sphere or, more likely, a Dyson swarm (a gargantuan structure that would wrap around the entire star, collecting all the radiation from the star and using it for their own energy needs). But having such a massive energy drain would be notable. Even a super-advanced civilization must follow the second law of thermodynamics, which tells us that there must be waste heat from their consumption of an entire sun of power. That waste heat can be understood in the following way: In a combustion engine, the explosive energy in gasoline pushes the pistons to generate the torque to turn the wheels of a car; but in the process, the engine itself becomes warm, not to mention creating a warm exhaust. Not all of the energy goes into useful work, but much into waste heat. This is a consequence of fundamental physics, and it is also the case with electric engines and any other source of energy that is used for work. There will be waste heat, no matter what. And this can be detected, such as a strange source of infrared radiation in the sky with no star apparent. Efforts to detect Dyson spheres have been performed, but they have so far found no such worlds.[1]

But if the very use of Dyson spheres/swarms is detectable to another civilization looking for anomalous radiation sources, then they will not stay hidden; they would be a campfire in the dark forest, visible to any hunter looking in their direction. To remain hidden, they would need to somehow change the radiation output of their Dyson swarm. They still need to make it look like they have a normal star, not blocked by millions of reflectors. They would need to mimic a star of lower luminosity, so they can still have a net energy gain from their star. If their world orbited a star like our Sun, perhaps they could create a radiation profile similar to a lower-mass red dwarf. This would mean that a significant fraction of the energy of their star would go into cloaking their solar system, but the sacrifice for safety is hard to exaggerate, given the Dark Forest solution.

However, such masking would fail because their star would still have the same large mass, pulling on other nearby stars in a way that would indicate to observing civilizations how much mass that star really had. This gives away the expected luminosity of the star, given that stars tend to follow a trend line on what is called the Hertzsprung-Russell diagram. Most stars follow along the "main sequence," where more massive stars are brighter. This is a function of their greater mass causing more fusion in their cores, thus outputting more energy and glowing brighter.

The relationship between mass and luminosity is fundamental astrophysics, so deviations from that trend would become interesting targets for investigation by ET astrophysicists. There would be no clear, natural reason why the star looked dimmer than expected. That would mean if a high-mass star was giving off lower-than-expected energy, it would be a sign of artificiality. If a civilization hid their star as a lower-mass star, their higher-mass sun would give away the illusion by its gravitational effect.

Could the Kardashev 2 civilization know a way of hiding the mass of their star as well? Current physics doesn't show any way to do this, but even so, let us suppose they could create a balancing anti-gravitational field so that the average gravitational field of their star, from a distance, looked like a low-

---

[1] Wright et al. (2014).

mass star. This has to be a massive energy cost, because gravity is caused by mass, and even a small mass requires a large amount of energy. To balance out the gravitational effects of the star, the needed negating mass must be a significant fraction the mass of that star. For example, if we wanted our Sun to appear to have half its mass, a negative mass of half the Sun is required. From energy-mass equivalence, this is absurdly high; the Sun has a mass of about 2×10^{30} kg, so half that mass turned into energy would be nearly 8×10^{46} J. Such a number is too large to comprehend, but by comparison, the Sun's output is 3.8×10^{26} J per second. All the stars in the galaxy combined do not have this energy output in one second, not even in one year, to get to the required energy. To acquire the needed energy, assuming perfect energy conversion, the Sun would need to shine at its current rate for nearly 15 trillion years (15×10^{12})—over a thousand times the age of the universe! Moreover, the Sun never will output this much energy, since fusion only releases a small fraction of the mass-energy of atoms, and most of the gas of the Sun will never undergo fusion anyway. In other words, there is no way to use the energy of a star to mask the mass of that star.

The conclusion at this point is that a Kardashev 2 civilization cannot hide their technology from other civilizations. So, either the Dark Forest hypothesis is false or there are no such advanced civilizations and never can be, given the need to hide. We are thus limited to advanced Kardashev 1 civilizations. In which case, the civilizations in our cosmic neighborhood are not so advanced in terms of energy to gallivant about the galaxy.

It seems the Dark Forest must either be incorrect, or any advanced ETI will be much more limited than sci-fi scenarios imagine. The Dark Forest would also suggest we would never know until the very last possible moment. That is because this solution also suggests our own doom, because we have already broadcasted ourselves to the universe with our radio transmissions since the mid-twentieth century. While those signals are still local in galactic terms, it means we have already given away our existence and cannot undo what we have done. If the galaxy is filled with terrified ETIs, then the expected signal from them will not be a message of greetings, but more likely a relativistic impactor, carrying enough kinetic energy to wipe out all life on Earth in a single shot.

The universe is now a very frightening place. The silence of the skies makes us lonely, but a quiet stalker leads to far more anxiety. Perhaps the only thing scarier than being alone in a dark room, is not being alone.

But even with the clear, internal motivation of each civilization to hide itself from others, including humans, there is still the Teenager Problem. It only takes one rogue member of a planet to send out the signal that dooms everyone. It would be as if every ET had a button to launch nuclear weapons. Accidents and idiots would guarantee untold mayhem.

This is already the scenario in Liu Cixin's prequel novel to *The Dark Forest* (*The Three-Body Problem*). One lone, misanthropic scientist with a radio transmitter found a way to send a boosted signal into space, and this was picked up by the Trisolarins, the main ETI species of the novel series. So, even if there is a planet-wide embargo on transmitting into space, there is still the risk of earthlings being discovered, either by random people on the planet sending out signals, or by alien probes scouting out possible enemies. In terms of costs and benefits, there is still the possibility of complete destruction even when we try to hide.

There is another cost of staying hidden. Even at a macroscopic or civilization level, there are issues in assuming that ETI would remain perfectly hidden. The effort to hide costs that civilization its own development. As noted above, the Dark Forest scenario prevents the emergence of a Kardashev 2 civilization from forming. For a civilization to cap itself from growth based on fear is unstable, as it also induces a huge cost on its future. Worse still, such capped growth could itself be a death sentence; after all, if ETIs do not develop the ability to harness an entire star, might they then fall behind in the technological arms race and be destroyed by stagnating while the other civilizations outwit them?

In some ways, the Dark Forest hypothesis is a modified version of Pascal's Wager: Assuming infinite punishment for not believing in God (if he is real) and infinite reward for believing in God (if he exists), and the costs for believing in God are small, then one should believe in God.[1] With the Dark Forest, the scenario is analogous: Assuming our assured destruction if ETI finds us (given that ETI exists) and assured safety from ETI if we hide (if hostile ETI is real), and the cost for hiding is small, then we should hide while in the Dark Forest. But much like the Wager on God, the grounding assumptions are questionable. There is indeed a significant cost to hiding, if it means we have to forever remain in our solar system and not harness the true potential of the Sun; not expanding humanity is the effective cost of trillions of future lives. Even if we hide, our safety is not guaranteed, so there is still the possibility of what may as well be an infinite cost—total destruction of humanity. Lastly, if we do grow, we still have the chance of growing fast enough to outwit any hostile ETIs, so our destruction is not assured given we are detected. Stagnating may well be what actually gets us all killed.

To perform the game theory analysis more completely, we would need to calculate the expected utility of each decision we might make. Expected utility of a decision is the sum of the probability of each possible outcome of a decision multiplied by the cost/benefit of that outcome. This is a non-trivial and subjective calculation to make. For example, if humans must remain

---

[1] Yes, we are well aware of the many issues of Pascal's Wager from a theological and philosophical viewpoint.

below a Kardashev 2 civilization and make reaching for the stars too difficult, then the trillions of future humans cannot exist among the rest of the galaxy. How do we account for that? What rate do we use to discount the value of future humans? Also, the cost of being destroyed by ETI might be high, but is it infinite, like in Pascal's Wager? Infinities in such calculations make it nearly impossible to weigh options, and they lead to irrational conclusions. Complete destruction is possible on all hypotheses, even if the probability of doom is smaller in some cases; even a small probability multiplied by an infinite cost leads to an infinitely negative expected utility, and if every decision has an infinitely negative expected utility, then it is not possible to decide among them, as they are all calculated to be infinitely bad.

In other words, the Dark Forest scenario does not necessarily lead to the conclusion to do everything one can do to hide, because we can still lose while almost guaranteeing we won't win.

There is also a way to win in the Dark Forest: communication. In the novels by Liu, he argues that because of the immense distance between worlds, communication is too difficult to ameliorate hostilities. If a "phone call" between worlds took years for each message to be sent and received, this would be a problem, but it assumes we remain in place. A solution that an advanced ETI could have and which we might be able to achieve in the next few decades is create probes with a sufficiently advanced AI to talk to ETIs. We could literally send probes with an ambassador AI, which seeks out new civilizations, learns how to communicate with them, and begins a process of negotiating peaceful coexistence (or otherwise determining where hostilities reside). Sending such probes at high velocities would allow ETIs to learn about each other without risking lives. They can also hide their launch point if the probes hopped around from star system to star system in a random walk, changing direction to the next star at random. This way, the direction a probe came into a new star system could not tell that civilization where they originated. If such probes can remain with a ETI for years and have instantaneous communication, then the chain of suspicion that Liu imagined can be broken. These probes would need to make themselves appear safe, such as not having weapons, but also not transmit their coordinates to their sender, since that could be interpreted by the discovered ETI as a warning flare. Assuming the on-board AI is sufficiently advanced and diplomatic, peace is possible.

There are good reasons to think that the game theory assumptions of the Dark Forest solution are shaky, so it may not be the best solution to the Fermi Paradox.

First!

In any race, someone has to come in first. What if we are the first civilization in our galaxy? The immediate reaction to this would be such a case is unlikely and contrary to the Copernican principle—we should be average, not exceptional. However, our observation of a lack of evidence for other civilizations must come into our analysis, even if we have excuses for the observation. A runner who sees no one ahead and never saw anyone pass by her has good reason to think she came in first place when she crosses the finish line, even if she believes she wasn't the best runner; similarly, we might need to consider the possibility we really did arise to civilization-level first in our galaxy.

This possibility is given a plausible rationale by the idea of "great filters," first proposed by the economist Robin Hanson. In order for an advanced civilization to arise, several important steps had to have happened in the history of a planet. Not only did that world have to be in the right place to even allow for life to emerge (i.e., exist in the habitable zone around a star), but life needed to actually form, and then complex multicellular life, then species that could manipulate tools, and so on, until there is a civilization that then must not self-destruct in a short period of time, long enough so that it can actually leave its planet and explore the galaxy.

Consider just one of the steps to intelligent life, a step that happened at the cellular level: the jump from prokaryotic to eukaryotic cells. Among the differences between these types of cells is the mitochondria, found in eukaryotes but not prokaryotes. The mitochondrion far more efficiently generates energy for a cell than cells that lack it; in particular, the mitochondrion performs aerobic respiration, using oxygen to "burn" fuel and release more energy in the process. Current theories on the origins of the mitochondria (the endosymbiotic theory in particular) say that they were originally their own cell, and one was eaten by another prokaryote. However, in one instance, the mitochondrion was not devoured but entered into a symbiotic relationship with the cell that consumed it. Think about that; what are the chances that you would eat a cheeseburger, and instead of turning it into a trip to the bathroom, the two of you combined into a more powerful organism? The probability would surely be small. We can see how obscenely unlikely this is to have happened because the Earth had only procaryotic cells for over a billion years. In all the lakes, seas, and oceans, with trillions and trillions of cells in every cubic meter of water, it took over a billion years for this fateful event to finally happen, and it only happened once.

Without the mitochondria, larger structures (built up from many cells working together, specializing in their tasks) would have been all but impossible. There also would not be the available energy to run the organ of most interest to intelligence researchers—the brain. If this one, amazing endosymbiotic accident didn't happen, or otherwise took a billion more years to happen, intelligence on Earth would never have come about. If it came too late in Earth's history, there wouldn't be time for life to develop before the Sun roasted the planet. Yes, the Sun is getting bigger and hotter over time, and in another billion years or so, Earth won't be a pleasant place to stay. Our existence as intelligent beings depended on this one, freak accident long enough in the past for us to even be a biological possibility.

The endosymbiotic event was just one of numerous difficult or near-impossible steps for intelligent life to emerge. Modeling these sorts of "hard steps" that might halt the progress towards space-faring civilizations along with the time limit of a planet's period of habitability, Hanson and other researchers argue it is very likely that human civilization is expected to be one of the first such powers in the galaxy.[1] In the race to intelligence, we won, and that is because it was so very easy to get tripped up on the obstacles on the race course. Some worlds may never have had the first usable RNA molecule form; others may never have formed multicellular organisms; still others may have just never had a sophisticated tool-using species. We are the lucky ones who managed to get past all of those barriers, or so Hanson's model suggests.

When it comes to modeling, there is the question of the number of hard steps and the "hardness" of those steps. While Hansen *et al.* considered a significant range of possible numbers of hard steps, there is still the uncertainty of how much higher *or lower* that number should be. Perhaps there are no hard steps at all! After all, the time between Earth becoming habitable and life emerging is rather small; if abiogenesis were a hard step, then the Earth passed it very easily. It may also be that the steps are not so "hard", and a planet's biosphere can slowly push through those alleged barriers. For example, the endosymbiosis of the early eukaryotic cell with the mitochondrion may not have been seem unfathomably unlikely, since there are numerous examples of endosymbiosis in nature, including other pathways to get the same sorts of functions that mitochondria provide.[2] There is a lot of uncertainty in determining how "hard" these pathways to interstellar civilizations are. It is not even clear what the hard steps are supposed to be, let alone how many. We currently lack the data to make that a concrete, scientific conclusion, though it is a useful model to see what conclusions can be derived.

However, the most significant limitation with the model is that Hanson *et al.* consider "loud" aliens, ones that expand fast, are long-lasting, and make changes to the volumes of space they control that are visible. If the aliens are slow-expanding, or perhaps non-expansionist, then not seeing them is perfectly expected. If such civilizations cannot last for eons (perhaps across multiple star systems it's not possible to have cultural cohesion), then their

---

[1] Hanson et al. (2021).
[2] Bains & Schulze-Makuch (2016).

silence is not surprising. If most civilizations find the interstellar trips not worth the cost, then they won't expand. Again, the non-observation doesn't necessarily show we are first; it is just as consistent with aliens that like to stay home.

This isn't to say the expansionist ETI hypothesis has no justification. Hanson *et al.* suggest that the most likely kinds of aliens to be found would be "grabby," who expand and would dominate space against "non-grabby" or non-expansionist aliens. In a Petrie dish, it is expected that an aggressive-growing cell culture will dominate the space against a non-aggressive culture. However, the consumption of resources at such a rate may well make the "grabby" aliens unstable. Such a civilization, with power consumption across star systems, could well have a supply chain disruption that could end colonization efforts or bring their entire civilization to its knees (assuming that the aliens have knees). Consider the Bronze Age collapse in the 12th century BCE, wherein a combination of climatic changes and disruptions to the raw materials needed to produce bronze led to disaster and the fall of numerous ancient civilizations.[1] Supply chains across lightyears will almost certainly be more vulnerable to disruptions than modern or ancient land-based routes, and for expansionist, "grabby" cultures, the demands on fragile lines will lead to breaks. Similarly, in British history, the 18th and 19th centuries were very expansionistic, but with internal strife and war in Europe, the empire could not maintain its holdings across the world. Analogously, if ETI expanded but then had its own internal struggles, those colony ships would likely be deprived of the resources they would need to survive and thrive. Expansion does not mean success, and it can cause its own collapse.

Other scientists have also modeled issues with continuous rates of expansion and the speed limits of space travel. Those studies also show that there is an economic limitation to expanding when the rate of travel is limited by light speed, let alone limits of rocketry. For economic growth at any continuous rate, the expansion of our available resources must also grow at a faster and faster rate, but because there is a maximum speed of expansion, that continuous growth is not possible. Research on this limitation on growth suggests not only are there no galaxy-sized civilizations, but that our own interstellar empire would peak in size within a few dozen lightyears.[2] This is in stark contradiction to Hanson's grabby alien model; literally, his aliens' reach exceeds their grasp.

Ultimately, Hanson's model is dependent on what might be reasonable yet questionable assumptions and on one major non-observation—not seeing evidence of ETI. A singular observation is a shaky basis for deriving the expansion rate of unseen and future civilizations. Hanson *et al.* predictions are

---

[1] Cline (2014).
[2] McInnes (2002); Haqq-Misra & Baum (2009).

almost data-free, such as when we should encounter other grabby civilizations. The non-observation might well tell us something about the inability of grabby civilizations from existing. Instead of arguing that ETI will expand at near the speed of light, as Hanson deduces, perhaps the failure to observe grabby aliens means it was and never will be viable to cross the distances between stars at such speeds. There is not any strong, a priori reason to reject these deductions based on the same evidence, and certainly not with enough confidence to argue for our uniqueness in the galaxy.

In fact, another group modeled the sorts of civilizations we might detect, and they came to the opposite conclusions. Instead of us being first or advanced compared to any other civilizations that might exist now, this group found that we would much more likely than not detect civilizations significantly more advanced than us.[1] They do not make assumptions about the probability of life or intelligence arriving, nor do they assume anything about the psychology or expansionist policies of ETI. Rather, they take a Bayesian approach, model our ignorance of the lifetime of civilizations, and they realized that it is much more likely to be the case that any ET we encounter is significantly more advanced than us. Because their model makes fewer assumptions, their conclusions should be given more weight.

It would seem reasonable to suppose then that the limitations in Hanson's model are enough to leave open other solutions to Fermi's question about where the aliens are. Other models suggest that if ET life is out there, then it's going to be well-beyond us in technology, and thus the question of "where are they" persists.

So Lonely

Perhaps the simplest solution to the Fermi paradox is the quietest one: Life is simply not abundant in the universe, let alone the galaxy. Perhaps we are alone, and always will be. While there are hundreds of billions of stars in our galaxy alone, it could still be the case that the probability of life emerging on a world is low, and that that life would become intelligent is unlikely, and that that life would form a long-lasting civilization is implausible. It is indeed possible that all of those combined low probabilities counter the billions of opportunities that a civilization could have emerged.

Currently, there is significant uncertainty about the origins of life on Earth and just how abundantly those conditions exist elsewhere in the Milky Way. To get a sense of the results of those uncertainties, along with the current null results from SETI, a sampling of the ranges of possible values for the

---

[1] Kipping, Frank, & Scharf (2020).

Drake Equation was performed by a group of researchers at Oxford.[1] Some of the factors in the Equation were well-constrained, such as the rate of star formation. On the other hand, the probability of life emerging is extremely uncertain. This means that it could be the case that the fraction of planets with life is extremely small, even much less than one in a trillion. Some steps along the way to getting life to emerge might happen in seconds, or those steps might require many times the age of the universe. There is just that much uncertainty in the scientific literature.

Given the large error ranges, it means it is quite possible that life on Earth is truly special. The estimates from the sampling found that most simulations of the Drake Equation indicate that Earth is the only place in the Milky Way with a civilization. In fact, there may be no ETIs in the entire observable universe, according to their results. We really might be alone and never detect another intelligent kind of life even with the greatest telescopes the future may offer.

However, as noted, this is a result of the gigantic uncertainties in the parameter of life emerging. In fact, in the supplemental articles with the main paper, the authors note that the uncertainty in just one factor, namely the probability of life emerging, drives the simulations to suggesting we are alone in the Milky Way. There are reasons to suspect that perhaps the margins of uncertainty of life emerging are too wide, so we ourselves have repeated the analysis of the paper.[2] We leave the details of that analysis in Appendix 1, but the main points and results will be discussed here. Yes, in writing this book, we have done some science!

The way the Oxford researchers created their simulation was to set up a set of random number generators following certain probability distributions. Most of the distributions were uniform[3] between the upper and lower bounds. However, for the probability of life parameter, their system was more complex mathematically in such a way that it allowed for the probability of life emerging to be lower than 10^{-200}. These are odds of 1 against a 1 followed by two hundred zeros! It's the equivalent of the same 25 people winning the lottery ten billion times in a row. Now, the paper only allows this lower limit to happen rarely, while the mean value of the probability of life was close to 1 in 2. However, that lower limit still means that there are many simulations where the probability of life is much less than 1 in a trillion.

---

[1] Sandberg, Drexler, & Ord (2018). The method of taking samples for probability distributions and looking at the resulting, posterior distribution, is known as a Markov Chain Monte Carlo (MCMC) method or algorithm.

[2] The code for this analysis is found at https://github.com/adairaar/drake_simulator. You can run the code yourself in your web browser for free here: https://colab.re-search.google.com/drive/1g2ejjH0oZL94dj5Tt6M9qm8gzlvlpXvm?usp=sharing.

[3] More accurately, the magnitude of the values were in a uniform distribution; this became a log-uniform distribution when converting the magnitude to a countable number.

Looking at a recent paper on a minimally assembled, random string of functional RNA,[1] the pessimistic estimate for abiogenesis is nowhere near as low as the original simulations allowed. Rerunning the analysis of the Oxford researchers, keeping everything the same except the lower bound on life emerging, the prior probability of us being alone in the Milky Way drops from 52% (slightly likely) to closer to 43% (a bit unlikely). The median of the simulations' number of civilizations also grew from less than 1 to about 8. So, even with significant uncertainty in how life might have emerged and even allowing for it to be very improbable, we might still expect ETI to have emerged more than once in the Milky Way.

However, the next step in the analysis is accounting for the lack of success when searching for ETI (SETI is discussed in more detail in the next chapter). When the original researchers included this evidence, it was very model dependent, such as how the aliens were spread about the galaxy, if they were colonizers, and so on. With different models, the researchers got different posterior probabilities for us being alone in the galaxy. The least assumption-laden approach (that the aliens are randomly spaced around in the galaxy) provided some evidence for us being alone, but not much. When we repeated the analysis, we similarly found that the probability of us being alone increased, but more strongly since our model suggests ETI signals are more probable, and thus the missing signals are more unexpected. So, our posterior results, the probability that anyone else is in the Milky Way, is a toss-up, with a probability of just about 50% for us being alone, while the median number of civilizations was barely more than 1. These ambivalent results are still largely because of our uncertainties about life starting on other worlds.

In fact, there are several technical issues with the modeling done in the first place. For the probability of life emerging and the probability of life eventually becoming intelligent, the distribution of these values is not the sort of thing one should use when we are extremely uncertain about the values. Instead, the distribution should be something that indicates that we don't really know what the values should be (an uninformative prior probability), which is often done in Bayesian statistical analyses.[2] In fact, the probability distributions used do not rely on actual observations, but opinions, so the entire process of predicting the number of civilizations is strictly not empirical.

Again, we try to deal with these issues as can be seen in the finer details in Appendix 1.

So, once again, we incorporate the appropriate probability distributions based on both data and statistically justified reasoning. In particular, we use the data on planets from the survey conducted by the *Kepler* space

---

[1] Totani (2020).
[2] Kipping (2020).

telescope, which was built specifically for detecting and characterizing planets and their orbits.

The results again suggest basically a 50-50 chance that we are alone in the Milky Way, but the simulations were not so widely spread. This would suggest that if we are not alone in this galaxy, there are still unlikely to be even 100 ETIs. This low number means that the lack of success of SETI has almost no evidentiary value—the aliens are too far away for us to have noticed them.

Taking this altogether, the best we can say is we don't know whether we are alone in the Milky Way or not. We are likely not alone in the *universe*, but it is basically 50-50 whether there is anyone we are likely to meet while gallivanting around the *galaxy*. This is due largely to our ignorance of life emerging on other possible worlds and the uncertainty in the longevity of alien civilizations.

Nonetheless, large uncertainties are susceptible to large changes with just a few observations, or even just one good observation. If you had a jar of marbles, but you had no idea what color marbles were inside, you would have great uncertainty as to what shades you would see and in what quantities. What fraction of the marbles are yellow? Maybe all of them? Perhaps none? You have no idea. But pulling out just one yellow marble can indicate that there are at least some yellow marbles, and randomly pulling a handful more can already give you reasonable guesses as to the relative abundances of the marble colors. All from just a few observations. With more pulls, the uncertainty goes down, but even with one yellow marble you can move from not knowing if there were any yellows to being absolutely sure there was at least one and plausibly many more.

Similarly, new results in the study of abiogenesis would almost certainly shrink those margins of uncertainty, but those results may also indicate life is even more improbable. Instead of finding a hoped-for yellow marble, we might only get reds. We cannot know what the experiments on abiogenesis will say; if we did, then we wouldn't need the experiments. However, there could be some dramatic observations done by NASA in our own solar system in the next few decades that might tip the scales right over.

Various bodies in the solar system have been proposed locations to see if life had emerged, existing either in the past or even right now. Mars is a common target for these explorations, and chemical tests on the Red Planet have been conducted since the *Viking* landers scooped up Martian soil in the 1970s. While those results proved inconclusive, many still look for new ways to determine one way or the other if there ever were (or are) Martians.

However, in recent decades, different candidates have been proposed as even more likely to have life now. No, it's not the other planets, but the moons of Jupiter and Saturn. The ones with the greatest attention have been the Jovian satellite, Europa, and the Saturnian moon, Enceladus. Observations by the *Galileo* probe in the 1990s began to indicate that Europa has an ocean of liquid water underneath its frozen surface. The squishing and squashing of Europa by the gravity of Jupiter as the moon orbits actually heats the interior of that small world, providing the heat to keep more water than on the entire surface of the Earth in a liquid state.

In more recent years, it was discovered that another world filled with liquid water existed far from the Sun. The moon Enceladus not only has water, but it is shooting water geysers (also called cryovolcano) into space. This was only discovered by the most recent probe sent to Saturn, the *Cassini* mission. These cryovolcanoes have only been known for about a decade. Another Saturnian world, Titan, is also of interest to astrobiologists. This moon is too cold for liquid water, but instead it has another sort of hydrological cycle, but with methane and ethane, flammable gasses here on Earth under normal conditions. These are not just gasses in liquid form, but hydrocarbons, the main sorts of chemicals in life, and they form literal rain, rivers, and lakes on the surface, seen by both orbital scans and a lander. Perhaps there is nothing swimming in those extremely cold lakes of methane on Titan, but then again...

All of these tantalizing possibilities mean that NASA or other space agencies could answer the question of how plausible it is life will emerge by seeing whether life has emerged elsewhere in our solar system. There has been serious discussion of building a lander that would go to Europa, for example, drill down through its icy crust, and then examine the ocean beneath. If, in the upcoming years, a camera was deployed into those waters, turned on a light, and then suddenly something came up and licked the lens, the world of biology would change in that moment. While this might be the most fantastic observation, a more plausible one might be a sampling of the water looking for signs of organic materials, perhaps even proteins or RNA. If they were found, then the chances for life elsewhere in the universe would explode.

Finding just one other place in our solar system with life would mean that life emerging is not impossibly rare. If it can happen around one average star multiple times, then the floor on the probability of it happening around another is raised. In fact, the number of worlds known to be capable of having life in the first place will increase. A world like Europa is not warmed by being close enough to the Sun in a Goldilocks zone as Earth is, but by another means altogether. The definition of "habitable world" would be expanded. But most importantly, an "improbable" event happening twice in the solar system means that this event cannot be that improbable. It cannot be sextillions to one. And so, the numerical sampling method described above of possible results from the Drake Equation would be profoundly altered. Instead of credibly sampling for a probability as low as the current analysis does, only the higher probable cases would be simulated. Instead of it being probable that we are the only civilization in the observable universe, we would expect a galaxy of living, thriving worlds.

But we don't know that. Not yet. One discovery could change everything, but before that discovery happens, we must humble ourselves with our uncertainty. Given our ignorance, the best we can say is there are probably very few ETIs in our galaxy, and perhaps none at all. In the observable universe, there is more likely to be someone, but we might never know they are there. On the other hand, this uncertainty means we have the thrill of discovery. Strange new worlds may be in our planetary backyard. It will be the next generation of astronauts and mechanical voyagers that will explore deep space, and perhaps, just perhaps, within the next few years, the discovery of the century will be found under European glaciers, in a Titanic lake, or among the faint whispers from the stars.

For now, given our significant ignorance about important factors in the emergence of life, intelligence, and technological civilizations, we will be cautious in our conclusions. But even when modeling that caution, it still gives us reason to believe that we are not alone in the universe. There are probably some intelligent life forms in the cosmos. Less certain is if we are alone in the Milky Way or not. At this point, we find it is about a 50% chance that anyone else is in the galaxy with us. If we hope to find ETI, we might need to look for a while, given how much space there is to investigate.

The Cosmic Scale

How far do we need look to find ETI? How far is too far to matter? When looking for extra-terrestrial intelligence, it doesn't need to be outside of our solar system, but our scientific knowledge has advanced enough to say we are likely the only species orbiting the Sun that can direct its travel into outer space. If there is life on Mars, it's not in a form that can build machines and invade the Earth as depicted in *War of the Worlds*. If there are ETIs, they are farther away than Mars. But the scale of such a search is hard to fathom, where the distances become too large to be meaningful. The nearest star to our Sun is a gobsmacking distance away, unattainable by all current human technologies if the hope is to get there before our great-great-great grandchildren have died. But that nearest star is only one of hundreds of billions of galactic stars. We need a way to compare the huge, to the humongous, to the impossibly gargantuan.

The main unit used in astronomy to describe the distance between stars, nebulae, and galaxies is the lightyear. This is the distance traversed by any form of light moving through empty space over the course of one Earth-year. As lightspeed is a universal constant,[1] it is a great measuring rod. To the nearest star, Proxima Centauri, light takes about 4.2 years to travel there, so it is at a distance of 4.2 lightyears. But this is one of over 100 billion stars in the Milky Way galaxy, which is a spirally body with the majority of its mass of stars in the plane of a disk. To travel from one edge of that disk to the other, one would need to cover over 180,000 lightyears. That is, even at the fastest speed allowed in the universe, a stationary observer watching a light-speed ship would have to wait over 180,000 years for the journey to be completed.

But even this distance pales into insignificance in comparison to the scales of intergalactic space. The Milky Way has a few small satellite galaxies, such as the Large Magellanic Cloud about 160,000 lightyears away. The next *major* galaxy is the Andromeda galaxy. Even though it is incredibly far away, you can see it on a clear, dark night in the constellation of Andromeda. It is visible ultimately because it is, like the Milky Way, composed of hundreds of billions of suns, located about 2.5 *million* lightyears away. Another way to think of this distance is in terms of time. If there were aliens living right now in the Andromeda galaxy looking at Earth with their incredible telescopes, they would not see us with our cityscapes and farms; instead, they would see our ancestors in Africa who had somewhat recently started to walk upright. That is, Earth 2.5 million years into the past is the present observation for an Andromedin.

And yet, Andromeda is the closest major galaxy. Something residing in the beautiful galaxy, NGC 4414, would have just seen the end of the non-avian dinosaurs, as it is approximately 62 million lightyears away. An alien living in the Great Attractor, a culmination of gravitational bodies accelerating the Milky Way, will never have seen a *Tyrannosaurus rex*, while GN-11, the current record-holding galaxy for distance from us, has yet to see the Sun and Earth form, as it is billions of lightyears from us.

But this is not yet the edge of what we can see. The current estimate is that the part of the universe that can be recorded with telescopes, what we call the *observable universe*, is about 46 *billion* lightyears away. From beyond that point, space is expanding away from us at a rate greater than the speed of light itself, hiding perhaps a universe that goes on without end. We do not know if the universe is infinite, but it is consistent with current measurements of the geometry of space-time itself. Now, if the universe is infinite in size, then it is almost a logical necessity that there is another civilization out there; in fact, given infinite planets, the number of civilizations will be infinite. But if they are so far away as to be impossible to observe in the "small" bubble of space-time we call the observable universe, then they might as well not exist.

---

[1] In vacuum. In a medium, light travels slower, sometimes substantially slower. For example, light travels at 75% its normal speed when in water.

Even if they exist in another galaxy, they might be so far away that they do not matter and may well be forever beyond our technology to sense.

In this book, we will focus on two categories of regions to find ETI: the Milky Way galaxy and the observable universe. If life exists within the Milky Way, we have some possibility to meet it within hundreds of thousands of years, and communication would not require geological times to have passed before we know if they got our text message. The most relevant question will arguably be whether life exists elsewhere in our galaxy (what was once called an "island universe" as it is so separate from any other large galaxy). But the question of theological importance is whether life exists *anywhere* in the universe. If we discovered intelligent life anywhere in the universe, that would have theological impacts, even if ETI is stupendously far away. So, we will consider both the existence of ETI anywhere in the observable universe as well as more locally, even though it is only in our galaxy that we might have relevant congress with ETI.

2 – WHAT DOES EXTRATERRESTRIAL INTELLIGENCE LOOK LIKE?

If we asked you to imagine an alien standing in front of you, the first images that would likely come to mind would be a gray-skinned, large-eyed, lanky creature, On the other hand, if you were one of the first well-known contactees from the 1950s, you would have said the aliens look like tall, white, stereotypical Scandinavian-looking humanoids.

The image of the alien has been one of the human imagination for the simple reason that we have not yet seen one. Often, they are humanoid, though occasionally they are almost truly alien. But even the most imaginative forms of ET still show traces of analogy with the known world. For example, H.P. Lovecraft's monstrous creatures from the stars were described as combinations of earthly things, especially related to the sea. His most famous creation, the "indescribable" Cthuhlu, had bat-like wings and an octopus-like face with a mass of feelers. So... describable.

It seems unlikely that actual ETI will be like us in form, or any chimeras we might concoct. Perhaps we'll only know an alien when we see one. But it prompts the following question: How will we know when we see it? How will we know if ETI is out there?

Biology or Electricity?

When we talk of ETI (extraterrestrial *intelligence*), we are considering a subset of ETL: extraterrestrial *life*. Will these be carbon-copies of life on Earth, or can you get life without carbon? And when looking for ETI, what are we assuming about the intelligences? Do they think in symbols and metaphors, or is their cognition so alien to our own that we could only marvel at it? Perhaps we also wish to be so expansive in what we mean by "life" that we might include robotic or electronic versions of intelligence.

Some of these questions might well be answered in the coming decades as more and more is learned about computer technology and chemistry. There have been speculations about silicon-based lifeforms for a long time, since silicon has properties in chemical bonding similar to carbon, the premiere element in life on Earth. Future research may provide more firm conclusions about the possibility of non-carbon-based life, but for now we must consider the probabilities. If we were to encounter an intelligent entity, and it were supposed to be anything we could ever call biological, it would more likely than not use two chemicals we find in terrestrial life: water and carbon. Water is the second-most abundant chemical in the universe, so a living organism not using water is putting itself at a severe disadvantage. Water seems to be wonderfully useful to most any sort of chemical reaction, as it suspends the reactants so that they can mix and form the products. And, fundamentally, the core of life *is* chemical reactions.

Every cell in our bodies is sustained by a long series of physico-chemical events, from the bonding of H_2O with adenosine triphosphate (ATP) to provide energy to the cell, to the folding of proteins that are the building blocks of cellular machinery. Life is the beautifully complex parallel processing of chemical reactions, and water makes it possible.

Carbon, on the other hand, is so fundamental in the strong chains of bonds holding together the foundations of every cell that it is hard to find another element that can provide the same versatility. Carbon is all the more ideal that it is quite abundant in the universe.

While other chemistries are not conceptually impossible for life, they may not be common. We may also consider carbon-based life that uses another liquid as the solvent for its chemical reactions, but again water seems so wonderfully abundant that it should be the expected chemical in any discovered living organism. Thus, NASA's current dictum for finding life is to "follow the water."

But what if those organisms decide that there is a better way to live? As we know from our own flesh, there are many disadvantages to being chemistry bags filled mostly with water. A cut to the skin and too much fluid leaves our bodies, and functioning ceases. A microscopic protein shell with some RNA stored inside—what we call a retrovirus—can turn one's body against itself as the virus uses our cells to multiply. Even if we avoid the worst of viruses and injuries, the very dividing of our cells has set a limit on itself and leads to aging and mortality.

Perhaps these reasons (and more) pushed an intelligent, organic lifeform to create another form of life, one that can be easily copied, modified, updated, and saved. An electronic version of intelligence could be the future. This may well be the future of humanity, too, as our attempts at artificial intelligence (AI) become better and better, so that a general artificial intelligence may exist by the end of the century. We may well do more and more to integrate with such a system, giving ourselves a virtual world to live in, one without end or at least on the order of millennia instead of years. We may be technically capable of this in the decades to come; perhaps ETI will have had the same evolution on their worlds, or otherwise have created an AI that can travel the stars without the constraints of biology.

If we detect an alien species, we may be detecting a signal created by one of those ugly bags of mostly water. Or they may be of another kind of chemistry we have not theorized yet. Or we might come across one of their intelligent machines, probing the crevasses of space and time. If we were to consider, in terms of probability, which source of intelligence we would likely encounter first, we first suggest that it would be some sort of machine intelligence. Humans are striving to build that sort of general AI, and even non-AI machines can provide signals that would indicate intelligence, such as our deep space probes. Machines can also fly between the stars with fewer restrictions than biological ETIs.

What's more, we may also compare alien space exploration to what humans do with space travel. First, we send probes, then humans can follow. Before Neil Armstrong stepped on the Moon, probes had landed on the lunar surface in the years prior. Numerous robots and satellites have gathered data on Mars, and it is likely still years away for the first human to step on that planet. Probes have also gone to the outer solar system and now into deep space, such as with *Voyagers 1* and *2* as well as *New Horizons* (after it visited Pluto). If ETI is anything like us in exploring the cosmos, machines will go first.

Next in probability of what kind of ET we might discover would be some sort of carbon-based lifeform, again because the chemistry involved requires more abundant elements than other proposed biologies. That does not mean that they will look like us, though. Since they will have had unrelated evolutionary histories, there is every reason to expect such alien life to come out looking different from us. Humans are a modified monkey, which itself is a modified four-legged creature, which itself came from creatures that found it useful to occasionally stroll outside of the water in the Devonian period, hundreds of millions of years ago. Of the few constraints on alien physical form that may be in common with us, they might have some sort of body symmetry. Humans have bilateral symmetry: our left-side looks like our right-side in the mirror. Sea stars have another symmetry (radial). Symmetry seems to be efficient in terms of energy use, so there is a Darwinian selection pressure to have some sort of symmetry. Thus, the aliens are not so likely to be anomalous piles of goo. Then again, if alien life were goo, and talking goo at that, we would still try to shake its "hand" when greeting them at a soiree.

In a sense, it is not so much what alien intelligence might be, but what alien intelligence might look like for us trying to detect ETI lifeforms. An alien might have this characteristic or that trait of intelligence, but for the purposes of this section, it is how that intelligence manifests itself to us. From a long, long way away.

Detecting Intelligence

Whether ETI is on the lower or higher ends of the Kardashev scale, whether they are biological or electro-mechanical, the question for this book and its impact on religion is how would we would know if it were even to exist (there's no point worrying about the philosophy and theology of something if we are certain that it does not exist). What sorts of observations should we expect to succeed, and what should we fund in order to know if we are alone in the Universe?

There is an obvious answer, though the least likely one: wait for them to come and land. This is the classic Hollywood scenario—a flying saucer descends to the ground, gently lands on the lawn in front of the White House or the United Nations, the loading ramp is lowered, and our first visitor from outer space descends and greets us. Or it takes out a ray gun and starts the conquest. However, this scene is unlikely for the same reason that we haven't gone visiting them: Simply put, space is big! Even if a civilization can travel between the stars, it is still a massive investment of time and energy (as discussed), and we cannot simply hope they drop by as if it were a friend visiting us because they were in the neighborhood.

If we hope to detect ETI without them making the arduous journey, we need to consider what are the easy ways for them to become known to us. This would require a technology that can travel through the galaxy as fast as possible. It must be relatively easy to be created and received, so that not only are they more likely to have sent that sort of detectable signal but we could have noticed it with our own technology. Fortunately, the best candidate seems to have already been discovered over a century ago: radio.

Radio waves are a form of light, just with a much, much lower frequency than the light our eyes can see. Just like visible light, radio signals travel at the cosmic speed limit, so they will cross the galaxy faster than any possible space vessel. They are also easy to make; humans have had the ability to create radio signals, which have traveled well into space, for many decades now. In fact, in the Drake Equation (as discussed previously), the major feature in the calculation of the number of civilizations is how long the ETIs have been transmitting radio signals.

This approach of looking for radio signals from the stars has been the key component in the Search for Extraterrestrial Intelligence or SETI. While there had been precursors to the idea of detecting aliens, the focus on looking for ET's radio signals began in the 1960s. Since then, the sensitivity of our telescopes has increased by leap and bounds, but there are physical limitations to such a direct approach.

First, there is the issue of distance. If ET life is far away, it will still take a long time for their signals to reach us. If an ETI exists on the opposite side of the galaxy, a radio signal would take 100,000 years to reach us. By the time we picked up such a signal, they may have been long gone; similarly, by the time they picked up our old transmissions of *I Love Lucy* and *Doctor Who*, we will likely have forgotten these shows ever existed. Imagine having a phone call where each speaker had to wait 100,000 years before hearing what the other had to say, and they ask you about a show only your great-times-500 grandparents watched!

However, this assumes that we even can pick up the signal. Radio waves are also like visible light in another important aspect: The farther you are away from the source, the dimmer it will look. A bright torch in your hand will be hardly noticeable to someone a few kilometers away, even with no obstructions. This is because light spreads out, and its intensity drops off as the inverse square of distance; if you double the distance between yourself and a light source, the intensity of the light will go down by a factor of four. If you are ten times farther away, the intensity is reduced by a factor of 100. So the signals from ET life will become weaker and weaker as they pass through the space between the stars, eventually becoming so weak that they may not be noticed in the noise of all the other light sources.

Folks at places such as the SETI Institute are aware of this, and they are particularly aware that if the transmitting civilization is far away, even if we can detect their signal, we might not have the tools to read the message. Just as someone yelling from far away is hard not just to hear, but they are hard to understand, the same is true for radio. If we received a transmission from an ETI, we would almost certainly not be able to know what the message meant. It could be information about their location in the galaxy, or their version of the Nigerian Prince scam.

So, how can we possibly tell if we receive a signal from an extraterrestrial intelligence if we can't find anything...intelligent in the message? The method is not to look for interpretability, but another feature altogether. When we send a signal with our radio transmitters for our household radios to pick up, it is at a particular frequency called the carrier frequency. In a small band of frequencies above and below the carrier frequency, our radio signal is modulated, changing between higher and lower frequencies very quickly. Those changes in frequency are how we encode information, and we tune our radios to that carrier frequency to pick up the modulations and thus the information. The key thing to note is that all of the transmitted radio wave frequencies are found in a tight band. This is very different from natural sources of radio waves. A hot body, for example, gives off light across the electromagnetic spectrum, and other natural radio sources are also creating radio waves over a wide range of frequencies. No natural phenomenon creates narrow-band radio transmissions,[1] with one exception: intelligent life forms with an antenna.

When SETI points their radio telescopes into the sky, they are looking for any narrow-band signals. They might be coming from a local source, such as a satellite, but such signals cannot come from a star. Unless that star has someone there with an antenna. Currently the SETI Institute has an array of 42 telescopes in Northern California, and the radio wave data it picks up are copied and split; one copy goes towards general astronomical research, and the other copy goes into a bank of computers to look for any narrow-band signals, all in real-time. As the computers advance, they can search through more and more channels and through more and more telescopes to look for those possible signals.

The search requires very sensitive instruments, so these telescopes are usually far away from anyone else. A nearby cell phone would completely blind the telescopes, not unlike if you were listening to music on a tiny speaker while your brother screams into a megaphone next to your ears. In a similar way, a cell phone is "loud." If you took your phone and placed it at the distance of the Moon, it would be the third-brightest source of radio signals in the heavens, only behind the Sun and the supernova remnant in Cassiopeia. Since SETI is trying to find a "cell phone signal" (not on the Moon) from a hundred lightyears away, clearly the instrument must be very sensitive and can accept no distractions.[2]

But even if we look for artificial, narrow-band transmissions, we have to have some idea where to look. Just like with a home or car radio, you need to know what frequency to tune to in order to pick up the signal. If you are off a little bit, you will only get static. The signal also has to be noticeable above all the background sources of radio waves. Similar to being at a noisy party, it's much easier to notice someone calling to you if you are in a quieter part of the room.

So, where is the "quiet" part of space? In terms of radio, there is a fascinating region of the electromagnetic spectrum called the water hole. In the wavelength range of 18 to 23 cm (or frequencies of 1420 to 1662 MHz), the universe is relatively quiet. In that band, there are two interesting frequencies

---

[1] To see what narrow-band looks like, compare this to standards in modern technology. In the US, FM radio transmissions have a frequency deviation around the carrier frequency of 75 kHz. The carrier frequencies ranges between 87.5 and 108 MHz, so the deviation is less than a tenth of a percent.

[2] As an anecdote, one of the current authors was taking measurements of the Milky Way with a telescope run by the SETI Institute, and a documentary crew from the BBC was visiting to get B-roll. However, one or more of the team members had their cell phones on, and the few minutes it was on completely ruined the observations by the telescope. The record of that incident has been used as a visual example of why everyone needs to keep their phones off while at the telescope array.

corresponding to the frequency of certain chemicals, and they appear in our measurements of the radio spectrum as thin lines. These emission lines are for hydrogen (H) and the hydroxyl radical (OH)—in other words, in combination, water (H_2O). A water hole is also a place to chat in old parlance, so the double-meaning makes for a great name for this radio band.

Most of radio-based SETI focuses on this radio band, but only in part because of its ease of observation. There is also the assumption that if the aliens are transmitting in such a way for us to detect, they want to be detected. Either ETI is transmitting to members across the galaxy, and so they need to use the quiet space in the radio spectrum, or they are deliberately trying to let civilizations such as ours know that they are there. In other words, if SETI succeeds, it is likely because the aliens are trying to make it easy for us.

Another way the aliens could indicate their existence to us without radio signals is by another form of light sent by lasers. In what is called Optical SETI or LaserSETI, the idea is that ET could send pulses of laser light to particular worlds that they believe are likely life-bearing. This would be much more energy-efficient than a general broadcast with radio signals, as all of the energy of the laser is directed to its target instead of through all of space. Moreover, by using laser light at much higher frequencies than used in radio, the rate of data transfer can be much higher.

The laser-transmitting civilization can indicate the artificiality of their signal by having the laser pulse last for a nanosecond, much shorter than any natural light source. If the energy of the laser light is brighter in the light frequency than the star the aliens reside around, our technology could notice a strange increase of brightness at one narrow band of light with a pulsation rate unlike anything produced in nature. Basically, if the alien home star is red, then a blue laser pulse could be noticeable, and the shortness of the pulse would indicate a technology was behind the light source.

We currently have the technology to both produce and to detect such signals, so Optical SETI is a good addition to the search along with radio SETI. However, like radio SETI, Optical SETI has also come up with no positive identifications of ETI signals.[1]

There are perhaps other ways we could detect signs of alien technology. One recent proposal is to look for signs of relativistic spacecrafts—vessels that move at a significant fraction of the speed of light.[2] The ambient light in space reflecting off of a vessel moving would shift the frequency of that light as well as amplify the light by concentrating it due to relativistic effects. This sort of effect, known as Doppler beaming, could be detected with current technology, let alone future space telescopes, since it is already similar to

---

[1] Marcy et al. (2022).
[2] Garcia-Escartin & Chamorro-Posada (2013).

another effect used to detect very fast-moving, very hot gases (called plasmas) in extremely distant astronomical bodies called quasars.

The main advantage of this approach is it does not require the aliens to actively signal to us. Our detection of their existence is much more passive. ET goes on about its interstellar business, and we basically see them "drive by." No need for them to "wave" as they go by—we'll notice them because of their very fancy car, to continue to stretch the metaphor.

This isn't the only passive means of finding aliens, either. Another sort of technology we might detect is the energy collection of an advanced civilization. As mentioned in the previous chapter, an ETI that is trying to achieve type-2 status will harness the energy of an entire star. While other forms of energy production are possible, the raw power of a star, available for free, would be the economic way to generate so much energy. The collection of such power is similar to the use of solar panels on Earth or on satellites, but the grander engineering adventure is to capture the energy of *all* the light of the Sun, not the mere, pathetic fraction that lands on the Earth's surface. The idea of building a giant sphere around the Sun (or rather, around the orbit of Earth, so we don't have eternal night!) was first articulated by physicist Freeman Dyson, and so the megastructure is named the Dyson sphere.[1] This would clearly require not just a gigantic amount of material, but it would require intelligence. The proposal has some major issues in making it a stable structure or even building it in the first place, so the more plausible version a civilization could have is a Dyson swarm. Instead of a solid sphere, there can be millions and millions of reflectors that beam the energy they receive to where it is needed. Given enough such reflectors, it could return nearly all of a star's light.

To an observer in another solar system, the star would no longer be bright and have a spectrum of a hot ball of plasma. Instead, the observer would detect the waste heat of all of those reflectors. We could be those observers, looking for signs of warm reflectors, giving off infrared radiation. Surveys of the sky looking for Dyson-like structures have been performed, and as of yet have not found any such mega-projects.[2] However, future surveys may gather the record of either a current or a past civilization's grand construction.

Both of these techniques for finding life out among the stars—detecting relativistic vessels and Dyson swarms—would succeed only if there is a very advanced civilization, reaching at least to type-1 on the Kardashev scale; indeed, the Dyson swarm would be a sign of a type-2 civilization. But if we wanted to extend the search to a less advanced but intelligent species, we could learn about their existence by looking at what a similar species has

---

[1] As mentioned in the previous chapter.
[2] Wright et al. (2014).

done. Here on Earth, our industrialization process fundamentally changed the contents of the atmosphere. Over the centuries, we have added a considerable amount of carbon dioxide into the air, due to the burning of fossil fuels. Industrial processes also release other pollutants, such as nitrogen oxide. If another civilization is building its own machines and has a technology level somewhere between what we had in 1800 and now, they would similarly change their atmosphere. The particulars of that change may be harder to predict, but there are enough possible signs of industrialization to investigate.

To know what is in the atmosphere of a planet, we could train a powerful telescope towards that world and look at the light that passes through the air of that world. The absorption of light from that atmosphere leaves telltale signs of the chemistry, so even at a distance we can know the composition of such an atmosphere. In fact, the recently-deployed James Webb Space Telescope is tasked in particular to read the spectrum of light from exoplanets, so it might be the case that it could provide evidence of both biospheres and industrialization. It is hard to say how successful this space telescope, or a future innovation, will be in finding such signs of technology, but this is one more avenue for finding ETI, even ETI that is less, rather than more, advanced than us.

At this point, we have several astronomical methods for detecting intelligent life in our galaxy, if it exists. The closer it is to us, the sooner we will have clear detection of technosignatures, be that radio transmissions, scattering of light by relativistic spacecraft, massive structures, or atmospheric alterations. Which of these methods has the greatest chance of success is unclear, so it is best to seek out ETI by *all* these methods, and perhaps new ones as well. It is also unclear when we will make the fateful discovery, if ever. But there are scientific ways of conducting the search, and that gives us reason to think that science will progress towards the answer to the questioner "Are we alone?"

Or the aliens can just show up tomorrow and save us the trouble.

56

3 – HAVE THEY ALREADY FOUND US?

Every citizen of the world must become a sentinel watching the skies. Keep looking for the next flying saucer - watch the skies, watch everything over your head - throw a ring of watch towers around the earth. Keep looking - looking - looking...

The Thing from Another World (1951)

Listening to Earth

If humans can have SETI, why can't ET have their own STI—the Search for *Terrestrial* Intelligence? Since we have already developed the technology to detect radio transmissions, certainly ETI would have done the same, and they could have developed it beyond our own capacities. If their technology is only a century more advanced than ours, that would already make their abilities of detection very powerful. This goes beyond just radio detection, but to other astrophysical methods.

When we are looking to justify the parameters in the Drake Equation, one of the things we try to determine is the number of habitable planets in the galaxy. What makes a planet "habitable" is largely the determination of the potential to have liquid water on its surface. However, our telescopes are nowhere near capable of looking at the surface of planets in other solar systems (worlds called exoplanets). However, our methods for finding planets are not so much direct observations of the planets themselves but other ways they give telltale signs of their presence.

As a planet orbits a star, it will pull on the star by the force of gravity. This will accelerate the star a little bit, and as the planet moves in a broadly circular path, the star is also pulled around in a small circle. The movements of the star can actually be seen with our telescopes, because the velocity of the star changes the light from the star; in particular, it shifts the star's emission lines. As the star is pulled away from the Earth, it makes the light of the star redder and shifts its emission lines to lower frequencies (this is the Doppler effect); similarly, if the star is pulled towards Earth, the light is made bluer. By measuring the light from the star, the magnitude of those changes, and the period of that change, it's relatively easy to calculate the mass and orbital parameters of the planet swinging the star about.

Another way a planet can make its presence known is by blocking the light from its orbiting star. This requires the orbit of a planet to perfectly line up with our view from Earth, but when this happens our telescopes can see as the star dims when the planet is in the way. The period of the blocking of starlight and the star's dimming gives the period of the planet, which also gives the distance of the planet from the star using the laws of planetary motion. Knowing the distance from the star and how much light is blocked, it is also possible to estimate the size of the planet. This transit method was used by the *Kepler* space telescope for several years, and it helped discover hundreds of new worlds. Since the method can only detect a planet with the accidental alignment between our point of view and the orbit of the planet, we must statistically extrapolate to the other stars, helping us realize there are many, many worlds in our galaxy.

There is a significant limitation with this method, and it makes projecting out the habitable worlds more difficult. Both the Doppler method as well as the transit method most easily detect larger, Jupiter-like planets. A larger planet tugs on their star and makes its effects on the star more pronounced. Similarly, a larger planet blocks more light from its star if it lines up with Earth. Conversely, smaller planets are harder to notice, and the smaller planets are the ones more like Earth. There is an inherent bias in the ways we find planets that can only be compensated by creating statistical models and must allow for significant uncertainties.

Other methods exist as well for detecting planets, but it is worth noting our ability to find these worlds is relatively recent. The first confirmed exoplanet was declared in 1992, so the technology is barely thirty years old. But imagine an ETI that has developed the same or other techniques for a century more. Their ability to detect planets would allow them to notice even tiny planets. Even if Earth never made a single radio transmitter, ETI could determine that our planet is orbiting the Sun with orbital parameters that puts it close enough to the Sun to warm up and have liquid water without that water boiling away. They would have strong grounds to suspect ours is a habitable world. With their better measurement tools, they might also be able to learn about our atmosphere, something we are just starting to be able to do. Light that travels through the air will have some wavelengths absorbed based on the chemicals the light passes through. Looking at these absorption lines, one can figure out what gasses were in the planet's atmosphere. Perhaps then, ETI can confirm the existence of oxygen and methane—strong signs of life. If they could measure our atmosphere over time, maybe they can see the signs of a buildup of carbon dioxide and nitrogen oxide, clear indications of industrialization and environmental change on a large scale.

One of the tools they could use is one we have also theorized building, which would provide astounding image resolution across the universe, but without constructing a megastructure. The Sun's gravity bends the path of light, analogously as a lens does, meaning that gravitational lensing could create images, but with an aperture the size of a star. This would allow us to see a planet 100 lightyears away with the same resolution as the Apollo missions saw the Earth from the Moon. However, for this to work, a probe would need to move to the focal point of the Sun's lensing, and then maintain the correct position over time. This and many other technical issues mean we will not have such a gravity telescope for a long time. But ET might have a head-start, and they may have their own solar system-size telescope visualizing our seas and continents now.

Astronomers from another planet could have figured out we exist without traveling the vast distances across the galaxy, without noting any of our artificial transmissions. They could also know of our environmental degradation without being able to directly image the Earth. They would not need to see the clearing of the rainforest or strip mining for rare earth elements, processes that leave tremendous scars on the ground. Just the light of our air gives away what we are doing to ourselves.

But they also could be knowing this by watching the nightly news. For decades, terrestrial radio towers have been sending signals into space, spreading out at the speed of light. These signals are not just audio but also video, as television signals are transmitted in much the same way. If ETI could notice radio signals from us, they might invest in the hardware to read our messages with enough fidelity to actually watch and listen to our shows. Depending on how far away they are, they might be watching *The Dick Van Dyke Show*, or maybe they saw the inauguration of politicians like Reagan and Thatcher. Perhaps they could see the original transmissions of the first man walking on the Moon.

However, these possibilities are not likely, because those radio signals weaken over distance. Not only will civilizations far away have a larger delay in what signals they will receive from Earth, but the harder it will be to detect and read them in the first place. The chances of ET detecting us increases is if we actively signaled them in the quieter radio spectrum, such as the water hole described earlier. This has been done multiple times in a classification of SETI known as Active SETI or METI: Messaging Extraterrestrial Intelligence. For example, in 1974, the Arecibo radio telescope in Puerto Rico transmitted a 1679-bit message to a star cluster, M13, which will arrive in tens of thousands of years. More recently, in 2008, the Beatles song *Across the Universe* was aimed at the star, Polaris. It will only take about 400 years for anyone at the pole star to hear John Lennon. However, if anyone is there, they should have the technology to know who was the walrus.

Again, the size of space is a confounding factor for not just us to find ETI, but for ETI to find us. Our radio signals have only been able to go into deep space for less than a century, meaning the bubble of radio noise around the Earth is only about 200 lightyears in diameter. On the scale of the Milky Way, this is about a tenth of a percent of the distance across. In other words, unless ET is close by, they have not even noticed us. And even if they are within our radio bubble, unless they are carefully listening, they may not have realized we are here. Both for SETI and METI to succeed, ETI needs to be nearby.

UFOs/UAPs

Many watching news coverage in recent years might think that, well, of course ETIs are nearby—they're flying through our skies! Whether it is the allegations of a crashed ship in 1947 in New Mexico, the abduction story of Barney and Betty Hill from 1961, or recent videos and oral reports from the US military, there are many claiming to have seen the ships of aliens, or the aliens themselves.

Before we get into the analysis, there has been some recent changes to the nomenclature for these sorts of flying craft. The designation of "flying saucer" was originally a misunderstanding, as the pilot who first said he saw these strange craft said they moved "like saucers skipping across the water," not that this was their actual shape. Even so, a flying saucer is a popular term for an alleged ET's craft. More commonly, the term Unidentified Flying Object (UFO) is used, even though it is ambiguous. The term does not mean the object was alien, only that it was in the air and could not be identified; a bird could be a UFO, since it is based on uncertainty or ignorance. The more recent term, Unidentified Aerial Phenomenon (UAP), has the same issues, but it has become a new way of referring to alleged alien craft without the branding issues that UFO has garnered over the years. It can also be used to refer to phenomena of light, as an example of the scope of the umbrella term. This way, a military pilot can record an incidence with a UAP without the same baggage as claiming he encountered creatures that wanted to probe him. So, there is one less pain in the ass for those suggesting they saw something weird.

The books by proponents of alien encounters, government coverups, and telepathic messages from the stars would fill a modest home library, so it will be impossible to analyze every alleged case.[1] Instead, we must rely on

---

[1] Many older cases are critically analyzed in Frazier et al. (1997). See also Prothero & Callahan (2017), Scheaffer (1998) and (2015).

what have been the trends in this sort of research and then focus on the best alleged cases.[1]

First off, there are many examples of people misidentifying well-known objects in the sky. Because more and more people live in well-lit cities, they do not see the night sky as it was seen by our ancestors for thousands of years. A bright star like Vega may be invisible to a city-slicker until they finally go into the country, and as the clouds part, a mysterious new, bright light appears to them. There have been numerous examples of people reporting a planet or even the Moon as a UFO. There is even a documented case of police officers "chasing" a UFO at over 100 mph, but the records indicate they were chasing Venus.[2] This particular case was originally touted as an excellent example of an alien craft, so it is not a silly outlier but was a serious contender for advanced technology; that it could be explained as largely a misidentification of a bright planet shows the issues with witness testimony.

Second, when we move to the more recent videos of strange craft that have been plastered across the *New York Times*, TV news, and congressional hearings in the United States, we ought to come in as skeptical, even of the video evidence, because the videos of such craft are all blurry. They are unidentified aerial phenomena, not because they are alien tech, but because we can't clearly see what they are.

In the most recent congressional hearing, the strength of these UAPs as evidence for alien technology was found to be wanting. At that hearing, Scott Bray of Naval Intelligence said the following:[3]

> When it comes to material that we have, we have no material, we have detected no emanations within the UAP task force that would suggest it's anything non-terrestrial in origin. So when I say unexplained, I mean everything from too little data to the data that we have doesn't point us towards an explanation.

In other words, these UAPs are unidentified, not because they are from another world, but because there isn't sufficient data to say what they are.

It's also worth noting investigations that have come up with plausible descriptions of these UAPs in terms of well-known terrestrial objects, from aircraft to balloons. Mick West in particular has provided plausible hypotheses for a number of the most-recently alleged alien craft. For example, the video known as "Gimbal" gives the appearance of a craft that was unknown to

---

[1] Klass (1983) noted not only the weaknesses in various reports of alien craft but also the willingness for journalists and other officials to promote the strange instead of the real story.
[2] Sheaffer (1998). p. 235-54.
[3] Transcript of the hearing is found at: https://www.rev.com/blog/transcripts/congress-holds-historic-open-hearing-on-ufos-5-17-22-transcript. The comment from Bray starts at 51:43.

Aliens and Religion: Where Two Worlds Collide

the F/A-18 pilot tracking it, and at one moment the craft appears to rotate itself without visible thrusters or movements of wing flaperons. It's like a flying saucer moving without well-understood aeronautics. However, the "Gimbal" object's strange motions are likely caused by the automatic rotation of the infrared sensor used by the jet following the UAP. This is a sensor that needed to make adjustments to track the object while preventing the mechanism from getting stuck into what is called "gimbal lock," wherein the sensor can no longer rotate further along its axis. Another rotation avoids the gimbal getting stuck—perhaps this is why the video was called "Gimbal" in the first place, as if it was known this was a camera rotation artifact. The craft is not rotating, but the glare around the object is rotated because of the rotation of the infrared sensor; the remainder of the objects on the screen remain in place because of software that de-rotates the background.[1] Overall, this UAP is likely a sensor artifact around a normal aircraft, which fits the data well, rather than the extraordinary claim that it is the ship of an ETI, which oddly only seems to rotate as to stay in sensor-lock for one aircraft.

This explanation for this UAP by Mick West is now the accepted solution by the Pentagon, and now most of the UAPs seem to be explainable as various terrestrial objects. Some are balloons, while others are surveillance drones, perhaps from China.[2] Other objects still cannot be explained, but for the same reasons given by Scott Bray above—not enough data to say one way or the other. Strange sensor data can sometimes be attributed to operator or equipment error.[3] So now, expect UFO proponents to ignore these videos as old news and move on to the next unexplained video, leaving us to forget the last time an alleged alien spaceship was caught on camera, but it turned out to be a weather balloon, a satellite, a bird, or a drone. So far, it hasn't been Superman or any other Kryptonian, nor any other extraterrestrial. The scoreboard is not looking good for UFO hunters.

However, this book cannot investigate all of the recent UAPs, and it is likely that more information will come out to further solidify what objects are being detected, just as we have seen in updates to the US government's investigation into these sightings. Instead, we turn to an important observation about the reporting of these stories. Already decades ago, UFO skeptic Phillip Klass noted that there was a general bias within the media to promote UFOs to the public. Such articles and news bulletins were profitable and thus ripe for lax journalistic standards. It was also more prone to hype by those with media savvy and political connections. That seems to be happening again with this round of UAPs that have been pushed by former political advisors, such as John Podesta of the Obama Administration, by celebrities like Tom Delonge

---

[1] West (2022).
[2] Barnes (2022); Office of the Director of National Intelligence (2023).
[3] Office of the Director of National Intelligence (2023).

of the band *Blink-182*, and by questionable scientists like John Puthoff who worked in parapsychology.[1]

While it would be a well-poisoning fallacy to say that these individuals cannot show that UAPs are extraterrestrial in origin, it is fair to note that this sort of promotion of UFOs by social and political elites is nothing new, and given the low quality of the evidence, we have every reason to be skeptical unless and until strong evidence becomes public. Until then, it looks like another dog and pony show.

Modern Abductions

Instead of detecting an individual space craft with our technology, there are numerous stories of people meeting the aliens themselves, sometimes even being taken aboard their vessel for reasons of revelation or experimentation. UFO religions, such as that founded by Claude Vorilhon (now named Raël), are often established by their leaders who claimed to have met with aliens, but there are also people who report other-worldly experiences without trying to become a modern messiah. The foundational and perhaps most famous abduction story, that of Barney and Betty Hill from the 1960s, has not been found to be convincing, especially as their testimonies were not all that concordant, and numerous details seem to be derivative of episodes of the show *The Outer Limits* (1963-1965) that premiered in the weeks before they underwent hypnosis to get their allegedly suppressed memories.[2]

There have been more abduction stories since then, and many such testimonies were presented at a conference in 1992 at MIT.[3] While the early psychological analyses of these folks suggested they were credible,[4] the lack of physical evidence was certainly a hamper on accepting what was alleged. This wasn't for a lack of trying to find such evidence. Examples of supposedly

---

[1] Colavito (2021).
[2] Adair (2012).
[3] Pritchard et al., eds., (1994); Bryan (1995).
[4] However, the abduction experiences may have more in common with sleep paralysis than with actual ET encounters. Blackmore (1998) conducted an experiment with adults and children who reported these sorts of experiences and found there was no good correspondence with sightings of aliens and their descriptions, in particular as "grey" aliens, which undermines the quality of the witnesses' memories. On the other hand, seeing "greys" among adults seemed to increase with exposure to television, suggesting a cultural influence. Nickell (1996) argues that the subjects investigated by John Mack, the Harvard psychologist who helped put together the MIT conference, were fantasy prone, so their testimony is more likely to be invented from cultural products.

alien materials left behind turned out to be from local sources.[1] Thus, the best possible evidence for abduction stories did not pass examination, which also suggests that experiencers of ET abductions are looking for things to justify what they believe in. The stories are very emotive, even if they ultimately are not due to true alien encounters.

Perhaps these emotional experiences are more like another phenomenon that also seemed to be consistent across multiple experiencers, but ultimately the events never happened as imagined. One observer at the conference noted that the stories were reminiscent of what was alleged at conferences of people who suffered during satanic rituals.[2] That comparison does not bode well for those saying they encountered ETI, since the satanic rituals were never real, but were part of a larger moral panic happening in the United States at that time. If the testimonies of the abduction experiencers are more like what was said of victims of satanic rituals, then we will need to consider these experiences as a cultural phenomenon rather than a physical one involving men from the stars. For now, the testimony consists of extraordinary claims, but we still do not have the extraordinary evidence that needs to be commensurate to those claims.

However, if you do find yourself aboard any sort of space craft run by beings from another world, just grab anything! The material science of such an ETI would be beyond ours, so even the simplest items will show signs of not being the sorts of stuff you find at Walmart. We await your findings.

An Alien Probe?

In 2017, the solar system got its first confirmed interstellar visitor. Hurtling through space, its trajectory was impossible for a body normally orbiting the Sun; its kinetic energy required us to deduce the object came from another star system, and though its trajectory points back in the general direction of the 25 lightyears-distant star Vega, it is still impossible to know its origins. The visiting body was given the formal designation 1I/2017 U1, but it's better-known as 'Oumuamua (OU-MUWA-MUWA), a Hawaiian word meaning "scout."

The scout from the stars was moving so fast that there was no way we could have launched a probe to intercept it. Our only chance of getting to see and observe it was with the telescopes we had. Telescopic readings of the

---

[1] Bryan (1995), p. 232, n. (see p. 50 for an earlier mention of the artifact); Pritchard et al., eds., (1994), p. 295.

[2] This was mentioned in conversation with one of the conference organizers, David Pritchard, and the comparison is also found in Bryan (1995), p. 142.

body suggested it had a reddish hue and perhaps had the shape of a cigar. The oscillations of reflected light, becoming light and dark and light again, suggested that the body was tumbling as it was zipping through the solar system. Several radio telescopes were also pointed at 'Oumuamua that were listening for narrow-band transmissions.

The idea of listening for transmissions from a space rock seemed in part to be inspired by science fiction. The story of an elongated vessel on a hyperbolic orbit through the solar system was the main plot device of Arthur C. Clarke's award-winning novel, *Rendezvous with Rama* (1973). But to say that there were only imaginative reasons to suggest 'Oumuamua was an alien vessel would be unfair. There have been a few other features cited by astronomer Avi Loeb to indicate that the interstellar traveler was really *sent* by an ETI. The most important feature is the oddity of the orbit of the body, namely it was moving as if it had another force acting on it besides gravity. While Loeb was not proposing there was an engine on 'Oumuamua, he did suggest that it was acting as a solar sail. By using the light of the Sun as a force on the reflective surface, it provided an extra bit of umph to accelerate and perhaps direct the "craft" through its travels.

To evaluate this argument, we need to consider the probability that this is how an alien craft would move through the galaxy, how probable is it that the same results can be gotten with a non-technological source of acceleration, and what problems remain on either hypothesis.

The first[1] detail that makes it hard to accept that 'Oumuamua is using a solar sail is that it is tumbling. It is hard to direct yourself if you are constantly flipping around. If 'Oumuamua had some sort of deployable solar sail, that sail would have damped such oscillations and made it steadier. Secondly, if the purpose of 'Oumuamua was to perform reconnaissance—if it were a probe—then it was a crummy attempt. If you want to collect data on an astronomical body, you need to either remain in orbit or land on it. A flyby that lasted a matter of months severely limits the time to gather data. This also leads to the third issue: At the speed 'Oumuamua was traveling, at best the object took hundreds of thousands of years to get here. That is one heck of a long mission! Moreover, if the purpose were to send the probe to gather one-years' worth of flyby data, but it took 100,000 years to get it, that is rather inefficient (to put it mildly), and again raises the question of why not make it an orbiter rather than a flyby mission? Lastly, it sure seems like 'Oumuamua wasn't sending anything back to its sender, since we picked up no radio transmissions from it. It's also not heading back the way it came, so if it's not sending radio signals to home, its data is simply never going to be used by its alleged creators. All of this makes for a very strange case for a space probe.

---

[1] These points were first made by Katz (2021).

Conversely, these issues do not exist for the hypothesis that 'Oumuamua was something more like a comet or one of the distant Kuiper Belt objects. A tumbling, non-radio-transmitting rock flying by is completely consistent with observational expectations.

But what of its anomalous acceleration? This has required more effort to explain, and the leading hypothesis is that 'Oumuamua has comet-like features, including having a fair amount of nitrogen ice. While the heated nitrogen ice could sublimate to a gas and provide a push, the more important feature is the ice's reflectiveness (or albedo), not unlike the solar sail proposed by Loeb but without the need for alien tech.[1] This explains the acceleration of the object quite well, so the probe hypothesis cannot be favored based on the motions of 'Oumuamua.

However, Loeb has countered that, given our models of the formation of asteroid or comet bodies in deep space, the amount of nitrogen needed to make an object like 'Oumuamua is too high. Loeb and his student, Amir Siraj, argue that for 'Oumuamua to be a nitrogen iceberg, chipped off of a body like Pluto, there would need to be far more collisions than is allowable on current models.[2] This does appear to be a problem, and while there seems to be no consensus on what is the best mechanism for explaining the small anomaly in the acceleration of this newly-discovered body, new scientific efforts exist to avoid even this problem, such as the gas is hydrogen instead of just nitrogen.[3] However, even the papers saying there is no consensus still suggests the evidence is against the alien probe hypothesis, for much the same reasons we noted earlier.[4]

If we compared the nitrogen/water ice to the alien hypothesis, we would start with the prior odds. While Siraj and Loeb suggests that the required nitrogen ice was beyond what models allowed at the 95% confidence level, and so we should have lower prior odds for the nitrogen ice hypothesis (perhaps 1 in 19 or less?), their reasoning is not sufficient to make this evidence *for* ETI artifacts. While it might seem unlikely that the ice model is correct, one needs to ask how does that compare to ETI existing and sending probes like 'Oumuamua? As noted in the first chapter, the evidence is ambivalent as to whether ETI exists in the Milky Way at all, let alone close-by enough to send slow probes to Earth. For ETI to be within 25 lightyears based on our current modelling of the Drake Equation, that probability is much, much lower than 5%. Add to that the strange features of the object if it were a probe, features that are completely consistent with it being an icy rock, and the prior odds that this is an alien artifact are much worse than that of the nitrogen ice

---

[1] Jackson & Desch (2021).
[2] Siraj & Loeb (2022).
[3] Bergner & Seligman (2023).
[4] Levine et al. (2021).

model. The problems with the nitrogen ice model are not worse than the alien probe model, so the likelihood favors even the problematic ice hypothesis. Adding in the evidence of no detectable radio signals from 'Oumuamua, then the posterior probability, accounting for prior knowledge and evidence, strongly suggests that there is no plausible case for 'Oumuamua being an alien object.[1]

It's also worth noting why the probe hypothesis is not a great one to use because of inherent issues with suggesting a motivation for 'Oumuamua's existence. The reason the nitrogen ice thesis has problems is not because of its fit to the observations of 'Oumuamua, but other astrophysical constraints on the formation of bodies *like* 'Oumuamua. Basically, natural or non-artificial explanations for things are strongly constrained by our considerable background knowledge. The alien probe, on the other hand, can just do whatever it needs to fit observations and for people to just say "the aliens wanted it that way." Such a hypothesis has little precision, unlike any of the other astrophysical hypotheses.

This is comparable to the problem of the God-of-the-gaps argument. Just because something is currently unexplained doesn't mean the whims of a deity are the better explanation; if there is no good, a priori reason to believe OmniGod would want to do something a particular way, but instead we can make excuses for any outcome with the phrase "it's what God willed," then there is no predictive precision or predictive power in the explanation. Virtually any explanation can come along, fit the data, and actually improve our overall knowledge. Time and again, when God-of-the-gaps reasoning has been used, it has been overturned when new and better data made clear what was happening. We learn more about physical phenomena, but we don't learn more about God.

We can analogize with reference to horse racing, and a favorite analogy of the authors as set out by atheist scholar Richard Carrier:[2]

The cause of lightning was once thought to be God's wrath, but turned out to be the unintelligent outcome of mindless natural forces. We once thought an intelligent being must have arranged and maintained the amazingly ordered motions of the solar system, but now we know it's all the inevitable outcome of mindless natural forces. Disease was once thought to be the mischief of supernatural demons, but now we know that tiny, unintelligent organisms are the cause, which reproduce and infect us according to mindless natural forces. In case after case, without

---

[1] This is using what is known as a Bayesian analysis, which relies on the mathematical probability equation known as Bayes's Theorem, to compare the probability of two competing hypotheses.
[2] Carrier (2006).

Aliens and Religion: Where Two Worlds Collide

exception, the trend has been to find that purely natural causes underlie any phenomena. Not once has the cause of anything turned out to really be God's wrath or intelligent meddling, or demonic mischief, or anything supernatural at all. The collective weight of these observations is enormous: supernaturalism has been tested at least a million times and has always lost; naturalism has been tested at least a million times and has always won. A horse that runs a million races and never loses is about to run yet another race with a horse that has lost every single one of the million races it has run. Which horse should we bet on? The answer is obvious.

We suggest a similar thing is likely to happen (and is happening) with 'Oumuamua. So often, explanations that involved aliens are found to be wrong and are replaced by non-alien explanations, but this doesn't appear to happen in reverse. In other words, we have good inductive reason to think, initially, that such explanations are more likely to be non-alien-based.

We are just learning about a whole new class of objects from parts of the solar system and wider galactic region that we have never explored. Of course there will be things we don't know or have modeled incorrectly! In fact, just a bit more than a year after 'Oumuamua visited the inner solar system, we found another interstellar traveler, which was christened Borisov. This was much more clearly a comet, and our observations of Borisov allowed us to learn more about the outskirts of the stellar neighborhood. With time, we should expect to find more visitors, and with new observatories coming online, we will learn even more. On the other hand, unless the aliens signal us, we don't learn more about them, so we only have one plausible avenue for learning more about 'Oumuamua, and that is waiting to be told what it was by its alleged creators. We think a lot more science will be done in the interim.

So, at this time, we do not have good grounds to suggest 'Oumuamua was sent by an ETI. If anything, the evidence is against it, while other proposals can explain its anomalous kinematics.[1] The weaknesses with those hypotheses can be resolved with further research, but the alien probe hypothesis has little hope of being better understood with a new telescope. Our money is on science using non-artificial explanations for 'Oumuamua. And given the uncertainty that aliens even exist in this neck of the interstellar woods, 'Oumuamua cannot provide evidence anywhere near strong enough to convince us that we are being watched.

---

[1] Kinematics is branch of physics concerned with the geometrically possible motion of a body (or system of bodies) without consideration of the forces involved (for example, the causes and effects of the motions).

Religion from the Stars?

The best-selling book on the subject of archaeology came out in the late 1960s, but its writer was not a professional in the field. In fact, his day job was as a hotelier; yet Erich von Däniken's book *Chariots of the Gods?* has sold over twenty million copies. His thesis was not original to him, as many of his arguments and sources were derived from another, esoteric book called *Morning of the Magicians*, but his book was written with a more engaging style.

Before setting up his science fiction scenario, von Däniken opens his book with an argument for the plausibility of life in the universe. The next chapter makes for a gripping introduction to a novel. He imagines a technologically superior culture traveling for dozens of lightyears and coming to an earth-like world. Upon landing, the travelers notice the planet is inhabited by creatures who can use stones as basic tools for hunting and marking on cave walls.

For the astronauts, this is a quaint tribe; for the natives, the sight is beyond description. The craft from the sky carries beings in strange suits with metal rods sticking out, flashing lights, and perhaps soaring about with rocketry. These marvelous travelers will have left a mark just by being seen by the natives. But at some point, a member of the tribe wants to make contact with these travelers, providing them gifts because of their nobility. The spacemen might learn to understand the tribe's language with computer-assisted communication, and then the tribe members can learn even more from these "gods." The astronauts might well provide not just technological advice, but also moral precepts to these planetary natives. Thus, the spacemen have left their mark on the people and their stories, advancing the technological capabilities as well as their laws and mores. Von Däniken then supposes that this is what happened not on some distant world of the future but on ours in the past.

While the remainder of his book is used to justify this scenario, the claims of impossible technologies and the depictions of alleged aliens have been found to be wanting when careful researchers have looked into his claims.[1] From the pyramids of the Giza plateau in Egypt to the Nazca lines in Peru, there are allegedly signs of things that would not have been possible without outside help. Unfortunately for those speculating about ancient astronauts, these claims not only do not have good evidential foundations, they are often warmed-over racial and colonial narratives from the Victorian Era, where Europeans would travel to some "exotic" land, find the inhabitants too

---

[1] Numerous books have been written to debunk the Ancient Astronaut Hypothesis (more recently calling the travellers ancient aliens), but most recently we recommend the books and articles by Jason Colavito, such as Colavito (2005) and (2012).

"primitive" to have done some local wonder, and then suggested some great civilization must have done it instead—ostensibly one with white people.

It would be erroneous to lay the entirety of the idea of ancient aliens at the feet of racism pushed forward by dilettantes. Before von Däniken, the astronomer Carl Sagan was intrigued by the idea of past contact between early humans and ETI. While he said it was a tentative view, he suggested that the stories of the fish-man Oannes (who imparted knowledge to early humans) might be memories of such early contact.[1] On the other hand, Sagan stepped away from even his tentative approval of this idea a few years later, noting that a priest claiming their knowledge was of divine origin was just as able to explain the story as actual ETI-visitations.[2] In fact, Sagan would spend a considerable amount of time debunking ancient astronaut theories, including that of von Däniken.

But even though Sagan turned away from ancient ETI-visitors, there are many proponents who claim to find aliens in sacred texts in the Western cannon. When turning to the Bible, von Däniken and others have claimed to have found UFOs and ETs, but instead they were called God, the Throne of God, or one of his many angels. Genesis 6 is a key text suggesting the "sons of God" were aliens that impregnated the early humanoids so that our ancestors became the smarter, more capable *Homo sapiens* we see ourselves as today. Monkeys were just that sexy to aliens, so it seems. Less genetically problematic, the vision of "wheels within wheels" in the Book of Ezekiel are also likened to some sort of flying machine. One way or another, Clarke's Third Law is the claimed paradigm for what is happening: any sufficiently advanced technology is indistinguishable from magic.

However, as is often the case with these sorts of claims, they are done in the ignorance of so much cultural context that makes the strange stories understandable products of human creative powers—no need for aliens. The UFO of Ezekiel is a description not of a flying contraption but of a throne with wheels that moves along the vault of heaven (or firmament; Ezek 1:25-26; cf. 10:1). It fits within the mythological/religious depictions of other gods in the region.[3] When re-contextualized, the ancient myths are no longer weird tales but indigenous cultural artifacts.

So far, no plausible evidence of alien contact in our past can be deduced, though the claims are legion and cannot all be dealt with here.

However, in abstract, how would we even know if a given belief about supernatural powers was in fact a legendary account of aliens visiting us in

---

[1] Shklovskii & Sagan (1966), p. 461.
[2] Sagan (2000 [1973]), p. 204-05.
[3] For examples of the imagery in other Ancient Near Eastern cultures, see Keel (1977). For discussion of the vision, see Block (1998), p. 89-110.

our remote past? It seems the best way of telling if there was some cultural contact with ETI is the same way we can tell if there was cultural contact between human cultures. When a new belief or cultural norm enters into a region, there is a noticeable discontinuity between what was found before the contact and what was found after.

Sometimes the change is stark. Before Native Americans encountered the British, they did not have the same designed ceramic pots and plates as the Europeans had. When doing an archaeological excavation, the layer with white tea plates is a strong sign that a new people-group has made themselves at home. There is also linguistic evidence of such contact, as new words are exchanged; for example, coffee is a staple drink for many Westerners, but it was not a major beverage in mainland China, thus it had no native word. If you travel to Shanghai and do not know Mandarin, you can at least confidently say *kā fēi* and be understood, because there was significant contact between caffeinated Europeans and Chinese men and women. Conversely, Europeans were introduced to Chinese goods and words, and from different Chinese dialects, we English-speakers can refer to that tomato-based red sauce as *ketchup*, the bean curd as *tofu*, the fruit *cumquats*, and the elegant *qipao* dress.

Telltale signs such as these are used by archaeologists and linguists to identify cultural exchanges as well as distinguish between gradual evolution of a people-group and a sudden outside influence. So, we should be asking: Where is the alien pottery? What is an ET loan-word? Most importantly, what is an example of an ancient alien proponent showing they actually know the relevant cultural context so they can actually see there is a sudden change in the native population's ways of acting, talking, or building that cannot be explained by gradual change or influence from another human culture? We await a demonstrable example.

To see why we believe no such examples have been given, consider one of the structures that people have attributed to alien intervention for decades: the pyramids of Egypt. Specifically, the three large pyramids on the Giza plateau are said to have either been impossible to build with Bronze Age technology, or that the natives in Egypt started from nothing and then suddenly started building these monumental structures. These claims, however, collapse under scrutiny and go to show the inadequacy of the scholarship performed by ancient astronaut theorists.

Let's look at this greatest example of the difference between the claims of ancient people's construction methods by ancient alien proponents and the reality of what archaeologists and historians know. The Great Pyramid of Khufu, for example, was not the first pyramid for the Egyptian pharaohs, but the pinnacle of decades of development. The previous pharaoh and Khufu's father, Sneferu, had multiple pyramids built to act as his tomb, and the reason there were multiple structures is because the early ones failed. The first built at Meidum (south of Cairo) either collapsed during construction or was stopped when the workers realized it was unstable. The structure was not built in a good location and suffered from the ground settling, among other issues. The next attempt changed a few features of the design, but the designers realized half-way through that it was also unstable; at that point, they changed the angle of the sides of the pyramid. This bent-looking structure is thus called the Bent Pyramid. Lastly, the Red Pyramid, which stands to this day, was sufficient for the pharaoh.

As for Sneferu's pyramids, he was himself adopting an earlier pharaoh's structure. The Step Pyramid of Djoser is what it would look like if you stacked a series of smaller squares on top of each other. This is ultimately a result of the previous structure to the pyramid, the mastaba, which was one of the earliest burial structures in pharaonic history. A mastaba is a flat-roofed structure with sloped sides—basically a burial mound with a flattened top and sides. If you stack a set of smaller and smaller mastabas on top, then you get a step pyramid. Fill in the steps to make smoother sides, and you get the classical pyramids of the Egyptian Old Kingdom. Rather than being a sudden innovation out of nowhere, there are centuries of development, from older Egyptian burial practices, to elaborations on those structures, to aesthetic enhancements, to the magnificent form the pyramids took in Giza.[1]

As for how the pyramids were built, this has been an area of over a century of research, and there are issues that are not yet settled. However, no serious Egyptologist suggests the ancient used power tools or alien technology. That is bolstered not only by the considerable research into how the stones were excavated, shaped, and moved into place, the markings left by the builders, and archaeological context, but in some recently discovered papyri, we have detailed work reports from some of the original workers and administrators.[2]

And yet, with all this knowledge within Egyptology that explains the development and construction of the pyramids, these monuments and others are used on shows like *Ancient Aliens* as proof of extraterrestrial contact. At best, it's a dismissal of the evidence. At worst, it is a denial of the capabilities of the people who did build these structures—the allegation is that they were "too primitive" to have figured this out. If the only way these monuments and others can be used to discover ETI is by denying our own knowledge and intelligence, then we humans are diminishing ourselves to make these new gods of the imagination powerful enough for our worship.

Now, in the future, archaeologists may discover something that truly is out of place. If an archaeological team were digging into a burial mound in

---

[1] Watson (1987); Smith (2004).
[2] Tallet & Lehner (2021).

Illinois, and they found strange glyphs printed onto plastic circuit boards, then we would have an anomaly worth discussing. Until then, the mysteries of our past are likely to be of this world and not another.

4 – WHAT MIGHT AN ALIEN RELIGION LOOK LIKE?

It seems like an absurd task to imagine what the religious views of ETI might be. On Earth, there is a long history of diverse views, and even within any religious tradition, there is a cacophony of opinions and interpretations. With everything from pantheism to polytheism, from monotheism to non-theism, and with so many gradients of views into what gods are, if they exist, and what myths representing them mean, one must stand back and remember: This is all just for humans. And that entails humans also having a similar world of experiences and a fundamentally similar psychology. We are all, at least, the same species and live on the same rock with the same stars in the sky. How impossible would it then be to fathom what the stories would be on a civilization with a different evolutionary history and psychology, with a different world to attach meaning and symbolism to!

Religious stories are highly contingent things. If you grew up in India, you likely learned very different stories than someone else growing up in Brazil. A few thousand years ago, the mythic landscape and the gods in it were unlike what most anyone uses today to make sense of the world. Rather than something universal like the laws of nature, religion is a highly contingent thing. Even if there were a singular, universal, eternal, omnipresent God, we can see that not everyone on Earth gets the same (correct) message. Beliefs about God are not universal, but contingent on time and place. And if they are not universal, then there is every reason to expect it to be different to the people of another world.

To see just how much religious stories are a cosmic accident, consider how much stories in one world religion would have changed merely if the solar system were in a different part of the galaxy. If our planet and Sun were transported elsewhere, the patterns of the stars in the sky would be altered, perhaps beyond all recognition. The constellations as we see them are, in essence, 2D projections of the 3D locations of the stars in the sky. Leo the Lion, for example, is easy to see in the springtime because its head is composed of fairly bright stars in the shape of a backward question mark. However, the brightness of those stars is in part a product of our distance from those stars. From another part of the galaxy, what might be bright for us would there be impossibly dim. Moreover, the relative position of those stars would alter. If you could tour "around" the stars that make up Leo, their pattern would become unrecognizable until you returned to your starting position.

This would mean that the skies above an alien world would be vastly different from our own. What things the primitive aliens would see in those patterns is perhaps impossible to guess, if for no other reason than the environment of plants and animals would likely be so different from our own. If their world did not have lions and bears, then there would be no such constellations. There is every reason to think that the stories *these* strange beings told of their sky would be extremely different than the ones *our* ancestors told.

Why that matters can be seen by tracing the genealogy of the oldest story that can be constructed and how it affected the telling of one story in the Christian canon. The oldest myth that can be currently reconstructed is the Cosmic Hunt. Derivatives of this story can be found from Western Europe to South America. The details of the story, along with phylogenetic analyses (creating a family tree of sorts) of the related versions of the tale, place it at least 15,000 years (perhaps 30,000 years) into the past and likely have it originating in Siberia. The story is of a hunter, carrying a cooking pot and very hungry, who finds an elk drinking by a stream. The hunter chases the elk, which is running towards the setting sun and has the sun in its antlers. At the last moment, the hunter throws the spear, and the jumping elk is hit by the weapon just as the sun is setting. The elk is transformed into stars in heaven, and the hunter is also placed in the sky, where the hunt continues. The elk becomes stars in the Big Dipper, or, in the Greek system, the Great Bear (Ursa Major).[1]

Remnants of the story would come to Greece, and the elk or ungulate would become a bear as seen in other cultural and linguistic sources. The story of the barely saved animal in the stars is seen in the Greek myth of Calisto, one of Zeus' lovers, who was transformed into a bear; at just the last minute before the bear-woman was speared by her own child, Zeus tossed her into the night sky.[2] Stories about a hunt from tens of thousands of years in the past changed into another cosmic tale of how a bear came into the sky, especially one with a long tail. In other words, our human ancestors have the stories that we have today because of the stars they saw in the sky, something that would be completely different in another part of the galaxy or universe.

Of course, humans have more than just stories about the stars. The people who carried these tales to Greece also carried another mythotype that takes its name from the Homeric version of the story.[3] In the *Odyssey*, the giant cyclops, Polyphemus, is a brutish goatherd, and it traps the hero Odysseus and his men in its cave. Odysseus and company get the monster drunk, and, while it sleeps, they fashion a sharpened branch to blind the one-eyed

---

[1] d'Huy (2013).
[2] Lushnikova (2003).
[3] In folklore studies, the tale type is labelled ATU 1137.

monster. Then, hiding under the goats of the cyclops, the men escape, even after Polyphemus chases them, hurtling stones and praying for vengeance from its father, Poseidon. This story itself has numerous similar stories from around the world, and it also appears to have great antiquity and moved with the same people that told the Cosmic Hunt.[1]

The cyclops itself also appears to be the accident of the history of life on earth. While there were no actual giant, one-eyed men in either human history or in the paleontological record, there were animals that were likely seen as just that sort of monstrosity. Before human settlement of much of the Mediterranean region, there had been elephants, and their bones have been found by ancient people, people who often imagined them as the remains of giants. In particular, the skull of an elephant gives the impression of a grotesque face, and where its trunk had been, it becomes a singular, central hole where one might expect an eye to be found. It would seem that stories of cyclopes, and perhaps numerous other mythological creatures, were the results of people interpreting old bones.[2]

The sorts of stories these people told is important because one of those stories was taken up in the Gospel tale of the Gerasene demoniac (Mark 5:1-20). As argued by Professor of New Testament and Christian Origins Dennis R. MacDonald, the story of Jesus healing the man possessed by many demons was derived from the Homeric tale of Odysseus' escape from Polyphemus.[3] Among the interesting connections is the name of the possessed man—Legion, that is, having many demons; Polyphemus is also similar in this regard, as his name means "many-voiced," as if the monster, too, were a multitude in one body. That the figures are also extremely strong, live in caves/tombs, and their stories begin and end with a sea voyage by the protagonist (Odysseus/Jesus), are suggestive of the author of the Gospel of Mark emulating this older myth from the Homeric epic.

How contingent this Gospel story is, then! It depended on Homer, who was utilizing a common, archetypal myth or mythotype, which itself was likely influenced by ancient findings of animal bones. On a world without such animals, there may never have been a story of cyclopes, no story of Polyphemus, and then no story of the Gerasene demoniac. The ET's Gospel of Klaatu could not well have been the same as our Gospel of Mark.

More examples can be given to show how the world we view created our myths and canons. This might be a mixture of stories being overtly aetiological and stories simply using stimulus for the world around them. Aetiological stories are those created to explain phenomena—geographical

---

[1] d'Huy (2015).
[2] Mayer (2001).
[3] MacDonald (2000), p. 63-76.

features and landmarks, names, or even bones. These aetiological myths could not be the same on worlds that never had such features or phenomena.

An even more direct effect of astronomical observations on the Bible and its interpretation is the fact the Earth is the third planet from the Sun. This allows for the two planets closer to our star (Mercury and Venus) to appear only in the morning and evening skies, a straightforward result of the geometry of the planetary orbits. Venus in particular has the moniker of "morning star" in modern English, and that designation is very old. If Venus never existed, or if our world were in another solar system with the Earth closest to its star, then there would never be such a morning star.

The motions of Venus seem to be encoded into the very ancient Sumerian myth of the *Descent of Inanna*, where the goddess associated with Venus goes to the underworld, dies, but returns after several days, much like how the planet itself will set in the west, not having been seen for about three days, and then rise again in the east. While the motif of three-day death-and-resurrection may not come from this story, the connection to the Jesus narrative is uncanny.

But can one find the planet Venus even more directly in Scripture. In Isaiah 14, there is a dirge directed at the Babylonian king, and at one point he is referred to as "Morning Star, Son of the Dawn" (Isa 14:12). The reference seems to be part of a larger, mythic complex of a deity associated with Venus trying to take the place of the high god Baal on Mount Zaphon (Isa 14:13). The king/god is said to be thrown down, like the falling star. This imagery of the morning star falling from on high was a chief influence on the myth of the fall of Satan, also known as Lucifer—the Latin name for Venus as the morning star. In particular, ancient and medieval readers of Luke 10:18 saw this as the same story from Isa 14, so the fall of Satan/Lucifer is largely dependent on a mythic tradition that exists because of an accident of astronomy. On another world, the myth of Satan's fall may not even be a possibility, because the imagery it depended upon would not have existed.

What this all indicates is that the religious stories told have long antecedents, sometimes affected by the sorts of skies those earliest of hunters looked up to on a cold, clear night. It also means that just by changing the skies, one changes the stories these people told. This has the long-term effect of changing how the natural world and the stars would affect the stories told in modern religious communities. If a religious tale is so dependent on a history of accidents, then imagine just how different the stories of an alien people would be from our own, considering they will have different skies, different animals, and likely different psychologies. Extraterrestrial stories might be at least as alien to us as the ETs themselves.

This understanding of the genesis of our religious stories invariably calls into question the truth of those very claims. If we can see how Gospel authors mimicked, emulated, mirrored, and paid homage to earlier narratives, then it is natural to call into question the historical veracity of those later Gospel tales.

Such is the converse of what is imagined by the Christian novelist, C.S. Lewis. He wrote in his *Cosmic Trilogy* of aliens communing with Maleldil, their name for God, in peace and harmony. The Martians lived without strife, which was only interrupted by some Earthlings. On Venus, Lewis conceived of an Adam-and-Eve couple who remained in an Edenic world until those pesky humans showed up. But Lewis imagining Judeo-Christian aliens on Mars and Venus shows the limits of what he considered. Again, the theistic beliefs of aliens should be at least as different from those of humans as human theistic beliefs are from each other. For all we know, the ETs would have a theology closer to that of the Klingons—there used to be gods, but, when they became too much of a burden, they were eliminated.[1]

This, however, assumes the aliens even have such stories at all. The formation of myths is largely an exercise in analogical, metaphorical, and anthropomorphic thinking. Seeing the gods acting in the world is itself a plausible outcome of our evolutionary history, namely from hyperactive agency detection, the tendency to see agency in things and events beyond what is actually there.[2] If ETs didn't have this same evolutionary history, then their minds would have developed differently, perhaps in such a way that they would not see the mythological approach as valid. ETI may well move through the universe not in a biological form but an electronic or robotic form, which may not house analogical thinking any more than it would house a spleen. They may not even have a god concept.

If, tomorrow, aliens landed in front of the White House, and the President was giving them a tour of Washington DC, he would likely show them the Washington National Cathedral. If he tried to explain his, say, Catholic beliefs, such as bread and wine becoming flesh and blood, or that churches are places to commune with the invisible power of the universe, the aliens may simply have no more understanding of this than we would them explaining the intricacies of Vogon poetry—both would be similarly incomprehensible and painful to understand.[3] There may well be no vocabulary in the alien language for the many religious beliefs, doctrines, traditions, and customs that humans have invented over the millennia. It is possible that ETs are atheists in the same way human babies are—they don't even have the concept.

This is not a very outlandish possibility. Not only are there human atheists (gasp!), but some have claimed that people on the autism spectrum have

---

[1] *Star Trek: Deep Space Nine* Season 4, Episode 11.
[2] Atran (2004).
[3] One could say that Catholic doctrine is already painful enough to understand, but we wouldn't be so presumptuous.

a decreased likelihood in believing in a personal god because of issues with projecting a theory of mind onto others, especially onto invisible agents.[1] Belief in God has at least some dependence on the nature of the human brain. If the psychology of ETs does not have the same evolutionary history as human psychology, there is little reason, a priori, to think they would have the neurotypical beliefs most humans have, given the fact that people with that same history can lack such beliefs. And if the universe is infinite, then the existence of an atheistic civilization is a near certainty. While the aliens may not be atheists in the style of Christopher Hitchens, they could well be non-theists.

So, if you met an alien tomorrow, what might you assume about their beliefs on God? The best answer is that their beliefs will almost certainly be different to our conceptions, assuming ET has any such beliefs at all. Those beliefs will be the product of a series of cosmic, contingent accidents, just like humans mythologies. Told around the campfire, the tales told by aliens will seem just as strange, if not much more so, than the most diverse stories told by any human story weaver.

---

[1] Norenzayan et al. (2012).

PART II: REACTING TO ETI CONTACT

5 – RATIONAL HOPE AND FEAR

In the classic, 1950s scenario, strange flying, metallic disks zoom over the nation's capital, finally landing in front of the White House. The craft's gangplank extends, and one or two of the space-faring creatures exits down the ramp. It is not just our first time confirming the *existence* of extraterrestrials, but we are all able to see what one actually *looks like*. No longer are these creatures just in our imaginations, but they are standing right in front of our cameras, in front of our politicians.

This creature has two arms and wears something like ceremonial robes. It reaches into its robes. To pull out...

A ray gun?

Or perhaps...it's a form of greeting?

The cinematic scenario may not be the likely way we will know if ETI exists, and the assumed importance of the United States may be just as erroneous as the belief that the aliens will have a well-defined set of arms. But how we imagine that first moment reveals something about us: How will we react, and what are we worried about?

Given the higher likelihood that our first confirmation of the existence of ETI will be through detections of their technology rather than a close encounter, what might science have to say about what we should be scared of? In fact, might detecting ET even give us scientifically valid hope for our own future?

Fear and Loathing

"The only thing we have to fear is...fear itself." This famous line from President Franklin D. Roosevelt's first inaugural speech in 1933 was in the wake of the economic downturn known as the Great Depression. FDR spoke of "nameless, unreasoning, unjustified terror which paralyzes" any chance of progress. This was a terror due to a failing financial sector, the disappearance of life savings, and the combination of falling food prices and failing crops forcing many workers into hunger and homelessness. However, what makes something depressing, perhaps more than the pain or initial shock, is the feeling that it won't end, or that it will only get worse. Such depression can be self-defeating, which it why it was squarely addressed by the new president.

But what would a president or prime minister say to the country if the issue wasn't banking but battle, not with people but with powers beyond our comprehension? What if humanity received a declaration of war from the stars?

If an ET civilization were space-faring and able to move about the cosmos at relativistic speeds, as noted previously, they would be at least a million times more powerful than we are now. So, if their intentions were our destruction, our chances would not be all that great. If the mission were to simply eradicate humans, the aliens could use technology we already have. A few thousand atomic bombs, and our civilization is over. And this is if we assume they use our current weapon capabilities. Certainly a more powerful civilization could do better. Think the Death Star in *Star Wars*, but with better science.

One possibility is something straightforward yet surprisingly effective. It's not ray guns or black holes, nor is it some strange quantum oscillation or specialized virus. All ETI needs to do is punch us. Really, really hard. With a mass of merely one kilogram launched at Earth at over 99% the speed of light, such a mass impacting our planet would provide more energy than the most powerful bomb ever detonated by humans. The impactor could be a kilogram of anything: a rock, a slug of lead, copies of *Armageddon* on DVD—anything of such a minor mass, and in any shape. With a few such impactors, our world would be devastated, and with such velocities, we wouldn't even see it coming. If a civilization has the energy to create fast interstellar vessels, making a bunch of rocks fly at relativistic speeds would be easy. And there is nothing we could do.

This is one scenario where fear could paralyze us without us even knowing the sources of our fear actually exist. While we don't even have hard evidence aliens exist in our galaxy, let alone that they have such ill intentions for us, the very idea that they could wipe us out with both speed and ease, all in a fashion we could not even notice, let alone prepare for—it makes folks with home bunkers and months of MREs look quaintly optimistic. Just the mere possibility of near-instant death from the skies could make you into a sort of cosmic hypochondriac of the most existential proportions. But remember, this is merely being afraid of the merely possible. Better to focus on the fear of fear itself.

Now, we think it is more likely that our first awareness of ETI will be by means of SETI, where the mere existence of their radio transmitters would be what we sense. It would not be reading their messages, let alone their war declarations. If the scientists at SETI or another program that accidentally find a signal from an intelligent, far-away source, then what will be reported is the existence of ETI. Beyond that, it would be hard to say anything else. Perhaps by measuring the intensity of the signal and its approximated distance from us, we could determine the strength of their transmitter and perhaps extrapolate to their technological abilities. If they have a very powerful transmitter, then they have plentiful energy. More than that, and it would be hard to say anything about them, until we found a way to read anything they transmitted.

From that point, the news could only tell the general public that there is at least one ETI, how far away it seems to be, and some general statement on their technical prowess. From there, our imaginations will run wild. And very often, our imaginations run to fear.

On the less likely scenario that our first detection of ETI is when their ships or probes come here, then we could say more about them. If they were within a matter of light-hours from Earth, having entered our solar system, then communication would be plausible, as reading messages from vessels that close is easy for NASA, given that they already have receivers for our probes that have gone beyond the orbit of all of our planets. If they were close enough for our telescopes to register, then we could say even more about their capacities. And if they landed, their intentions could be known very, very quickly. Then again, our imaginations in these scenarios would probably race towards the idea of invasion.

If ETI made their presence known, would we have scientifically valid reasons to be afraid? On what grounds could we say that their presence would be either benign of malignant? As already noted, if they could get here at relativistic speeds, then they would be well beyond us in technology. Some, such as Carl Sagan, have supposed that to become so scientifically advanced, it would require ETI to have also advanced their morality to such a point that they could only be altruistic.

That seems difficult to justify. Do we think that just because our technology improves, that increases the moral worth of an ant to us humans? This is a non-sequitur, and no amount of energy production from fusion reactors is going to make this a logical deduction. What is or is not moral is not simply based on our knowledge, but far more on what we value. And just because the aliens might be more advanced than us doesn't mean that they will value us. In fact, the expectation should be these ETI believe in self-preservation and self-flourishing. If we are in the way of their growth, then why expect an advanced civilization to limit their expansion for our sakes?

On the other hand, if ETI were so advanced, they may simply not find us a threat and might ignore us. After all, we can barely build ships that leave the Earth, let alone travel to another planet, and we are a long, long ways from even traveling to another star. For ETI to think of us as a threat would be like you worrying about a Chihuahua that is behind a fence in another country.

But let us speculate about the psychology of these aliens for the sake of argument. These aliens might well be rational in searching for any risk, even at low probability, and then bringing that probability to zero. For ETI to consider us a threat, they would have to believe we would advance significantly and so fast that we could outpace them. On the order of thousands of years, maybe we could become a threat to them, and thus they might strike at us to prevent that from happening. Would they not take a preliminary strike to make sure we never even become a threat?

This is now a series of improbable events: We advance enough to reach ETIs, they do not advance fast enough and are overtaken, and we decide attacking is justified. Conversely, there is a lot of effort in striking the Earth; if the aliens are not that advanced, then the energy needs for relativistic bombardment are unavailable; in which case, they are not really all that advanced compared to us. There is an effective stand-off, not unlike the US and USSR during the Cold War; the main difference being that the separation between the countries isn't thousands of miles but trillions. On the other hand, if ETI is so far advanced that it could perform a first strike with ease, then there is a very low probability we could ever compete—the more advanced ETI is, the larger a gap we would have to cover. It seems then, in the abstract, there is no clear reason why ETI would perform a first strike, given the energy needs to do so combined with the probability we could actually harm them, given their energy-generation technology.

However, some have considered another, paranoid way in which aliens could doom us with little energy, and all we have to do is listen. Some have considered the possibility that the aliens transmit to us a message, but that message carries a virus.[1] If we had robots performing most of our tasks, and they could be infected by an interstellar computer virus, then the aliens could use our robots against us, all with the power of one radio transmitter and us willing to even search for ETI. In other words, SETI could doom us all! But even this passive method towards Armageddon is unlikely for the reason that SETI could not actually read an alien message. The needed resolution for such a message would either require a gigantic radio dish, the transmission rate would be painfully slow, or a combination of the two. By slow, we mean a data transmission rate of one bit per second in order to be discernable over the background noise of the cosmos. Given the typical virus is on the order of

---

[1] See Langston (2013), p. 131-140 for discussion. This is a plot point as well in *The Killing Star* (1995) by Charles R. Pellegrino and George Zebrowski.

kilobytes on the small end, megabytes more typically, and if it is carrying some form of artificial intelligence (i.e., a trained neural network), then it will be quite large. The transmission time would be very long, so long that someone looking at the incoming signal could grow suspicious of its contents. Moreover, for a virus to operate, the writer of the code must know something about the computer system it is infecting, and as far as we know, Microsoft has not sent any representatives to Sirius to sell licenses for PowerPoint. Other attacks on computer systems, such as SQL injections, which can try to break into databases to steal otherwise secure information, require back-and-forth transmission, so it is easy to avoid such a hazard by just not transmitting back whatever the aliens demand. Lastly, even if a computer were infected, it would just need to be disconnected from the network and thus the virus would be contained. It is also worth noting that if the aliens could reprogram our robots for terrible ends, then so could we, and more easily. We have already been able to prevent such infections from people with deep understanding of our computer software. If we can harden our hardware from local virus infections, then ETI has less chance of passively transmitting instructions that would doom us. Put another way: update your antivirus software already!

It seems that if the aliens want to harm us, they are first going to come here, which requires far more energy, and thus returns to the previous point about the war mission not being worthwhile.

But what about the aliens coming to take our resources, enslaving or destroying us in the process? This colonial scenario manifests many of our fears, grounded in human history. Imagining the fate of the natives of Australia with the arrival of European settlers, but on a grander scale, makes for a terrifying prospect. However, the evils of colonialism were motivated largely by *economic* exploitation. Is it really worthwhile coming all the way to our solar system to take raw materials from our world? The trip requires far more in terms of energy and time than any trans-Atlantic voyage by many orders of magnitude, so it would surely make little to no economic sense to come here for resources.

And even if this *were* the intention, why take it from the Earth in particular? If the aliens want water, why not go to the Saturnian moon, Enceladus? It has more water, and the aliens would not have to fly into and out of a deep gravity well. They could get access to more, and more easily. If they wanted iron, then they should go to Mercury, which has more that Earth; or, again avoiding the harsher gravity well, they can mine the asteroids, something that we are likely able to do in the not-too-distant future. And of course, there are all the other star systems and their planets. The Earth is not the only depository for important resources in the galaxy, so making us the important place for "capitalist space-pigs" to exploit is making our world one of inflated value.

Conversely, what the Earth has that is unique is life. So, if the aliens wanted biological samples, then destroying it would be very counter-productive. Raw materials can be gotten by mining any other world into pieces, but the thing that makes our planet unique requires keeping things intact.

For now, it seems hard to justify a reason to be afraid of the existence of aliens, and their discovery does not indicate in and of itself our doom. If their intentions are hostile, then we have reason to worry. Then again, if they want us gone, we won't even see it coming. We suggest not to worry about aliens being an existential threat. If we do discover the existence of ET, the only thing we have to fear is fear itself.

∧ New Hope

The biologist Richard Dawkins had a mathematical insight into why our finite lives were worthwhile.

> We are going to die, and that makes us the lucky ones. Most people are never going to die because they are never going to be born. The potential people who could have been here in my place but who will in fact never see the light of day outnumber the sand grains of Arabia. Certainly those unborn ghosts include greater poets than Keats, scientists greater than Newton. We know this because the set of possible people allowed by our DNA so massively exceeds the set of actual people. In the teeth of these stupefying odds it is you and I, in our ordinariness, that are here.[1]

We are so lucky indeed that life emerged in the first place. The Earth can be hostile to those living, and most of the universe is instantly hostile to life existing at all. So many fantastic coincidences had to have taken place for you to read these words.

But we *will* die. Each individual reading this book will hopefully be around for years to come, but not *all* the years. In fact, we have a fear of death not just for ourselves but for the collective entity we call humanity. In the 20th century, we developed technologies that could kill millions of us. By the middle of that century, we had the power to destroy all of us. Worse still, that technology is still sitting in silos around the world, and the possibility of an accidental apocalypse has not disappeared. A false alarm and a few phone

---

[1] Dawkins (1998), p. 1.

calls, and everyone in the world could be gone in a matter of hours. If some lucky few did survive the nuclear fires, the radioactive wasteland that remained would make them wish they had burned away.

Notwithstanding the possibilities of nuclear war, there are other ways that human progress may be so destructive that our future looks somewhat bleak. After decades of warning, the situation with global warming has become worse, and the efforts to reverse the trend may be too slow. Perhaps with a few elections of certain politicians, in the US in particular, the possibility of making any gains in reducing the production of carbon dioxide may be lost and preventing the worst of climate change might be beyond our technologies. Current modeling of business-as-usual production of CO_2 indicates, by the end of the 21st century, a 3-degree Celsius increase in global temperatures, at least a half-meter of sea-level rise, while in places like North America, the winters will be wetter and the summers drier. Food production will decrease, and people will be on the move looking for respite from the devastation of crops and the rising of the oceans. While it may not be the end of the world, the suffering of the human race will be horrendous and could well lead to even more violence and destruction.

We may not feel as lucky as Dawkins would have us believe. Worse still will our grandchildren feel. Optimism for the future could be suppressed by the sweltering heat.

But suppose, tomorrow, a strange radio emission is detected at the Allen Telescope Array run by the SETI institute. Just by chance, on that night, scanning the skies, a distant probe of another civilization broadcast itself to the galaxy, and one of those telescopes was pointing at the right place, at the right time, tuned to the right frequency to see it was there. Suppose nothing else could be determined about the signal other than it was artificial and it was from beyond our solar system.

If that day were to happen, with the mere detection of such a narrow-band radio source, we would know we were not alone in the universe, or even alone in the galaxy. And this would give us reason to believe we were not just lucky to be alive. The odds would be in our favor to survive and thrive.

Remember that in the Drake Equation, one of the factors that estimated the number of civilizations in the galaxy was the length of time an intelligent civilization was able to communicate outside of its planet, namely by radio transmissions. If such a civilization were short-lived, then there would be fewer such civilizations in the galaxy. Conversely, if a civilization could survive a long time, on the order of thousands or millions of years, then the galaxy could be teeming with intelligent life.[1]

---

[1] In our simulations discussed in chapter 1, the potential number of civilizations in the Milky Way was small, and perhaps we are alone. However, even if the probability of life, intelligence,

The determinants of the longevity of a civilization are numerous, but the key question is this: Does a technologically advanced civilization self-destruct, or does it find ways to push through its growing pains to last for millennia? Perhaps a civilization invents radio, discovers atomic fission, and soon annihilates itself over political differences. Conversely, an advanced civilization also has the tools—scientific, philosophical, and political—to avoid such a horrendous fate. If a civilization is smart enough to communicate with radio, perhaps they are smart enough to work through their issues. This is not to say that with sufficient technology there would be a utopia, but it might be enough to avoid the worst possible outcomes.

This relates again to the idea of "great filters." Perhaps technology is also a filter on long-lived, intelligent lifeforms. How difficult it is to pass through that filter suggests the longevity of communicative ETIs and how many would exist in the galaxy.

As noted in chapter 1 and worked out in Appendix 1, there are unlikely to be that many civilizations in the Milky Way. The only way for that number to be larger than we estimated, given our current astronomical data, would be if the longevity of ETI civilizations is very long, and much longer than the estimates we suggested. Moreover, if life or intelligence is rare, then seeing an alien civilization would mean the longevity is even greater—the more unlikely for a civilization to emerge, the longer-lived they need to be so the chances of us detecting them are high.

So, if we did detect such a signal from another world, it will tell us that it is possible to get past the technological great filter. The aliens would not need to tell us how they did it; just by existing, they would give us hope. If they could get past their growing pains, then so could we. The Copernican Principle says that we are not likely to be special. And not being special would be a blessing given a detection of ETI. If other civilizations can make it, then it is more likely that we can, too. And sometimes, all it takes is hope to make a better future possible. A hope built on scientifically-sound knowledge is a stronger hope than one built upon fables. Science allows our hopes to not be blind, and if science can find another world where technology has allowed ETI to grow and survive, then we have not a blind faith in our future but a hope with clear and powerful lenses, letting us peer into our future with something paralleling divine clarity. We could make it, and we must!

However, it would be unwise to allow our policies about climate to depend on finding radio signals from the unknown reaches of space. We have so many good, local reasons to make changes. So, let us not wait until it is too

---

and becoming communicative were higher (near 1), the value of N in the Drake Equation would not increase much, based on our knowledge of the astrophysical values in the Equation. The only other factor that had the greatest effect on N which is not well-known is L. Therefore, discovering ETI would most-directly change our estimates for L.

late. Nonetheless, knowing that we are not alone would tell us our grandchildren have a brighter future ahead.

This is all predicated on never even knowing anything about ETI, beyond their mere existence. We could imagine even more. Their radio transmissions to Earth could be more intentional, providing insights into the laws of the universe, for example. While sending *Encyclopedia Galactica* would be arduously slow, given that over cosmic distances the transmission rate would be greatly reduced—far worse than an old dial-up modem—it would be a boon to human knowledge.

Others have imagined even more. With such an advanced society, perhaps ETI would come to us with wisdom beyond anything we have imagined. They may be so gracious and helpful that they would save us from our own destructive ways. In other words, ET could be the real Messiah.[1] Or even better, The Doctor![2]

However, we should not expect aliens to be savior figures. They do not have a necessary reason to save other civilizations; in fact, with few other civilizations around, there are fewer competitors. We might hope they are supremely altruistic, but that desire seems no more grounded than the belief that any alien species would immediately initiate a genocidal mission against Earth. In science fiction scenarios, it is imagined that more advanced civilizations wait to interfere in our affairs until a point when we have sufficiently developed, and then we can join the larger federation of ETs. This is, of course, imagination and not scientific data, and the scenario requires numerous civilizations in the Milky Way, something we cannot justify given our estimates from the Drake Equation in chapter 1.

The main point is that we do not have reason to think the aliens are out there waiting to tell us everything they know. Maybe they would, or at least some members of those worlds might. Still, we cannot wait for the aliens to tell us about quantum gravity, cancer biology, or new forms of democratic economics. We need to figure that out for ourselves. After all, we don't even know if ET exists, let alone would they be willing to help us. Perhaps even more literally than the old saying intended, heaven helps those who help themselves.

---

[1] Cf. Klaatu from *The Day the Earth Stood Still* (1951).
[2] The Doctor of *Doctor Who* tends to save humanity every other week, not just once, in an abstract sense, 2000 years ago.

06

6 – COGNITIVE DISSONANCE AND WHAT LEON FESTINGER CAN TELL US

When being concerned with how a *religion* might react to a given situation, it is wise to understand first how *people* act because religions are, of course, made up of people. While on the one hand we will be dealing with how doctrine and theology might be affected or what they might say, the actual reactions of people will more likely be defined by their psychology rather than the theological content of their holy books, though one might well be a tool utilized to conform to the other.

Primarily, we will be discussing cognitive dissonance and cognitive dissonance reduction. And there is a certain amount of irony at play here since the development of the psychological theory of cognitive dissonance was, decades ago, born out of infiltration into a quasi-religious UFO cult.

The theory looks at how disharmony can be generated when two competing attitudes, beliefs, or behaviors exist together in conflict. The mind doesn't like conflict and much prefers harmony, and so, through cognitive mechanisms (heuristics), sets about reducing the disharmony.

Let us go back to the 1950s to paint a picture of how this theory came to be. Social psychologists Leon Festinger, Henry Riecken, and Stanley Schachter were already studying the effects that disconfirmation of prophecies had on groups of believers when they saw a story in a local paper that detailed a prophecy claiming much of the world would be consumed by a flood on December 21st of 1954. The prophecy believers actively recruited others to this cult, a cult who predicted that they and only they would be rescued by a flying saucer before the flood.

Festinger, his colleagues, and some paid observers infiltrated the cult in two separate locations to study them from the inside as trusted members of the groups.

Such was the strength of the belief of the genuine members that they left their jobs (or even lost them) and academic studies, relationships and friendships. With the impending end of the world, they gave away or got rid of money and possessions. As with many religious adherents the world over, these cult members were *invested* in their belief, a belief that was at the core of their identity and behavior.

The date came and the date went, but the flood didn't come, and the UFO didn't arrive to rescue them.

Aliens and Religion: Where Two Worlds Collide

If one was to consider the outcomes of this from a rational point of view only, one might be tempted to think that the cult would recognize that their main prophet, Dorothy Martin (who also claimed to have the psychic ability of automatic writing), was wrong. That their belief was bunkum.

But humans are an irrational lot at the best of times.

What happened was surprising to the team analyzing the cultists' behavior. The members had a core belief that their prophecy was true and that there would be some kind of UFO rapture to save them from a flood. But now they had evidence that this belief was not true: The flood did not occur, and the UFO did not appear. Either one of these two bits of data must be wrong, or there is a third option that the data can be harmonized.

The strength of the core belief, together with the perhaps subconscious recognition of how much time, effort, and money had been invested in this belief (as well as social and psychological costs being incurred), meant that this held primacy over disconfirming evidence. While they couldn't deny the non-event, it could be that they just got the date wrong or some other notion that allowed the core belief and the disconfirming evidence to harmoniously coexist.

Festinger and his colleagues broadly found that those members with the highest levels of belief retained their belief and committed to it *even more strongly*. While also explicitly trying to gain more media attention, the members produced rationalizations to account for the absence of the flood and accompanying space craft. Conversely, some of those with much lower levels of group investment and adherence ended up leaving the group or reduced their levels of adherence and involvement.

There were various ways that the cult members' minds achieved cognitive harmony by reducing the dissonance, and this most commonly involved adapting the new evidence or elements of their belief to maintain the *core belief.* Either they got the date wrong, or the Earth was actually saved as a result of their dedication and faithfulness.

This phenomenon is ubiquitous. It happens everywhere and in every context, whether it be chasing bad money with good on the stock markets, or continuing to believe that a bad relationship will work, despite the evidence.

We have seen this happen, too, in more overtly "religious" scenarios—such as with the rapture prediction of Harold Camping and his Family Radio Christian movement (cult) in 2011 (May 21st, to be precise)—these could just as easily have served as the subject for Festinger's work. Harold Camping's reaction to a lack of end days rapture, and by extension the beliefs of his followers, was to adjust a small part of his core belief: the date.

Camping had actually previously predicted the rapture to occur on September 4th or 6th of 1994. Nothing like hedging your bets. In the readjusted 2011 version, one that garnered a huge amount of media attention, 3,000 billboards were erected, a five-car caravan toured the States, and a massive amount of money was raised, at one point being about $100 million. One transportation agency worker donated $150,000 of savings to the cause.

May 21st became *October* 21st, and that became a *further* let down. Eventually, the Family Radio network was broken up and sold as donations failed to meet costs. In 2013, after *13 failed predictions* and countless follow- ers enduring countless manifestations of cognitive dissonance reduction, Harold Camping died.

The overarching lesson to be learned here is that people will go to ex- traordinary lengths to maintain a core belief. This might mean experiencing one of the following:

- Adapting the core belief marginally.
- Ignoring the contrary data.
- Compartmentalizing the contrary data and core belief.
- Adapting the contrary data.
- Denying the contrary data.
- Delegitimizing the source of the new data.
- Reducing the importance or value of either the contrary data or the core belief.

More rarely, and if there is less personal investment, the believer will give up on the original core belief as being incorrect or untrue.

We might all look at the data concerning both Festinger's original UFO cult or Harold Camping's rapture claims and agree—regardless of geography or creed—that all the data, from any point in time, shows that those core be- liefs were irrational and unjustified.

And yet, as a result of the above heuristics, people still believed them. Really vehemently.

The point is that despite the amount of challenge that the theological and religious analysis will present to religious believers, they will no doubt continue to believe. If religious believers either find out the ETI exists, or de- duce that it must, and discover there is a theological crisis to deal with, any of the above is likely to take place. But the value of the contrary data, or the value of the theological crisis, is likely to play second fiddle to the value of the core religious belief.

Such belief systems are tough nuts to crack. There is every chance that the discovery of ETI in the universe will cause religious believers to entrench even more fervently in their core religious belief in spite of the theological ramifications that such a situation might entail.

Aliens and Religion: Where Two Worlds Collide

Alas, we will continue to explore for the joy of such exploration. For, as Lucian of Samosata, a satirist of some renown from the Roman province of Syria, wrote in the second century CE:[1]

> Once upon a time, setting out from the Pillars of Hercules and heading for the western ocean with a fair wind, I went a-voyaging. The motive and purpose of my journey lay in my intellectual activity and desire for adventure, and in my wish to find out what the end of the ocean was, and who the people were that lived on the other side...

Which is apt. And all the more so, considering that Lucian then told of his meeting with aliens on the Moon.

---

[1] Lucian, *A True Story*, Book I.

7 – THE POLITICS OF POWER: A HUMAN REACTION

The project of searching for ET life splits those who have thought about it into two camps: those who think it is a bad idea and its discovery a threat to our own existence, and those who are excited by the prospect. Indeed, two scales have been created to attempt to quantify the social impact of extraterrestrial intelligence (the Rio Scale) and extraterrestrial life (the London Scale).[1] Furthermore, there have been a number of different protocols drawn up to detail courses of actions for experts and governments, but these are not legally binding and are likely to be ignored.[2]

Predictions of hypotheticals are arguably spurious pastimes. We have to resort to generalizations, while being careful of the analogies we draw, as space science researchers Martin Dominik and John Zarnecki point out:[3]

> If data are absent or ambiguous, we tend to argue by retreating to analogies or theories about universalities. Historical examples, however, need to be well understood before these can serve as a guide, which is demonstrated by the fact that history is full of misinterpretations and misconceptions of itself.

For humanity in general, the discovery of intelligent alien lifeforms (or, indeed, them discovering us) will probably not be a religious affair, though this will depend on the type of discovery. We doubt that news channels will devote a huge amount of time, initially at least, to the considerations that alien life might have on the atonement. To begin with, impact on our own religious ideology will play second fiddle to other, more pertinent and potentially practical concerns.

To answer the question of how we would react to these sorts of considerations is largely dependent on what the intentions and actions of the alien life that discovered us would be, assuming that contact would happen this way round. It would be safe to say that chaos, panic, and hysteria would meet the news of an alien military armada amassing behind Saturn. But we may be far more welcoming if contact was initiated in a completely different manner.

---

[1] Vidal (2015).
[2] Dominik & Zarnecki (2011).
[3] Ibid.

The question of how we might react can be asked in differing contexts, too. For example, do we discover microbial life in our Solar System? Do we discover some long-range radio-wave evidence of life from millions of lightyears away? Or is there a fleet of galactic war ships hiding behind some planet in our solar system? Each of these scenarios will entail entirely different reactions.

The first two options would allow us to cogitate, over time, on the meaning and implications of such. The latter would invoke an immediate reaction and might look something like a panicked emergency. And yet a long-distance contact is thought to be a far more likely event. Dr. Michael Schetsche (political scientist and sociologist from the Institute for Frontier Areas of Psychology and Mental Health) states the following in a paper reflecting on extraterrestrial confrontation:[1]

> The hopes of almost all SETI-researchers today concentrate on a long-distance-contact through radio waves - perhaps also because that would have quite likely less far-reaching consequences for mankind than a close contact. The further away we know the aliens to be, the less threatening their existence appears to be....
>
> It is my thesis that in this case too the spatial distances are of great importance: the closer to earth such a physical contact occurs, the more negative will be the psychological and social consequences. One can substantiate this thesis first with our sociological and psychological knowledge about the short-term consequences of unexpected meetings with strangers and secondly with the historical experiences of long-term consequences of symmetrical cultural contacts here on earth.

There is no doubt that geographical proximity is a huge factor in reactions to anything. People are more worried about crime in their own neighborhood than a city across the state, or in another country. This leads Schetsche to add, "We can conclude from this that the eruption of mass panics is most likely when the contact occurs on Earth itself, in the 'living room' of mankind so to speak."

We should also look at the diversity of reactions based on the diversity of the people who might learn the news of ET discoveries. One piece of research looking at the attitudes of American and Chinese undergraduate students produced some interesting results.[2] The paper illustrated many interesting data points, including these below:

---

[1] Shetsche (2005), p. 3-4.
[2] Vakoch & Lee (2000).

- There were differences in attitudes across the two cultures.
- With a lack of data, interpretations said more about the respondents than the reality (such that there was a lot of psychological projection).
- Projecting again, alienated people were more likely to see ET life as hostile just as their own world seemed hostile—a negative view of the world meant a negative view of aliens.
- The more anthropocentric the respondents (who saw humanity as the center of the world, the apex), the more unsettling they would find alien signal detection. The discovery of ETI could pose "a significant challenge to the world view of strongly anthropocentric individuals."[1]

There will obviously be a lot of cultural influence and psychology at play in how humans react to the discovery of ET life.

Besides these more clinical studies, there are the real-world reactions of people who have thought the aliens had arrived. Most famously, in 1938, a radio play of H.G. Wells's famous The War of the Worlds novel caused panic as some people took it to be actual news, spreading rumors to yet others. Some thought the end of the world was nigh, others that the Germans were attacking, and people were reported to take to the streets or to get out of Dodge. The reality was probably somewhat less hysterically widespread than the newspapers ended up reporting, but we can certainly learn something from this episode. With 24-hour rolling news and internet media, there is no doubt that alien contact would be at the center of everyone's lives for a long time.

Primarily, if we discovered that we were being visited or clandestinely checked out by ETI, humanity would feel threatened. It would be nice to think that we would react morally neutrally, welcoming our newfound universe-inhabitants with both curiosity and decency. Realistically, though, there would be a power imbalance. Any lifeform that could find their way to us before we have got past our own Moon would no doubt have technology and abilities that far outstripped our own, as we have previously discussed.

There is no doubt that such power inequality would breed feelings of threat. Straight away, humans would have an inclination toward a defensive position full of suspicion and worry. The calculations made would be political, involving an understanding of who held what power.

Historically speaking, this same situation has happened a number of times. Take the Aztecs (among many other colonial scenarios). Spanish conquistadors made contact with the indigenous Aztecs who had never seen horses or muskets before. This situation invoked Arthur C. Clarke's Third Law: "Any sufficiently advanced technology is indistinguishable from magic." And

---

[1] Ibid., p. 743.

people get generally freaked out when they see something they interpret as magic; dangerous magic, one would imagine, even more so.

When a civilization that has never before glimpsed a tall ship, a musket, a man riding a horse, or metal armor, sees all of these things, one can excuse them for thinking magic is at play. Or that the gods were visiting. Or both.

This power imbalance, and certainly the initial shock in these historical scenarios, would have caused terror (after amazed excitement), and would have been cause for caution. Eventually, appeasement in the face of the futility of resistance became the order of the day. Of course, we are simplifying a complex situation, but one can learn from history and apply it to a potential future; it has some predictive value.

The sum of this interchange of resources, food, ideas, diseases, populations, and so on, in the New World of the Americas in the 15th century and thereafter is known as the "Columbian exchange" after Christopher Columbus. The bad news is that this exchange led to an 80-95% reduction in indigenous populations. Broadly speaking, the destruction of human civilizations has resulted from these interactions with "contactors." This is something of a warning to humans awaiting a galactic exchange.

The supply routes and harsh environment for the new invaders in South America was somewhat levelling, and similar challenges may or may not meet intelligent alien visitors. We will, like the Aztecs and Incas, be left wondering what reason aliens might have for coming to our lands.

Indeed, as SETI Institute senior astronomer Seth Shostak told *Space.com*, "Personally, I would leave town. I would get a rocket and get out of the way. I have no idea what they are here for."[1]

If we were to learn from history, we would punt to ideas of resources and enrichment (or necessity), which might come at a price for us indigenous earthlings. Interstellar travel, again as discussed previously, would probably be an awful lot of effort to go to in order to merely be interested in meeting or investigating other civilizations. That said, we have spent much money ourselves creating wildlife documentaries that seek to understand our own planet. Perhaps, once harnessed, the necessary technology for such travel might not be that resource expensive. It might even be relatively easy to jump into hyperspace and visit another galaxy for a civilization that has had a million years of technological and scientific head start on us. But according to our previous reasoning, this seems rather unlikely.

There is nothing to say that contact could not be benign, and we could even end up vastly benefitting. With a million years of technological advance over us, there is every chance that alien technology, if shared, could help us

---

[1] Howell (2018).

out of our global predicaments, from global warming to food shortages, from sufficiently powerful clean and renewable energy to health.

This will initially cause some bruising to our collective sense of self in the same way your clever, top-of-the-class child might feel when they qualify for a leading university only to end up being surrounded by vastly more intelligent people than them.

There are many unknown variables when playing with such hypothetical ideas. But what can be sure is that we, as humanity, would be wary and scared. We would likely have one hand outstretched for a handshake (or whatever appendage, if indeed it had appendages, our new neighbors would have), with the other hand hovering over a red button.

How we might react is something that many have sought to understand before. Of course, this is fertile ground for science fiction creative content (which often portrays such scenarios negatively), but it is also something that the American Association for the Advancement of Science have reported on.[1] There is very little original research on this topic (which may have something to do with us having not yet come across many aliens in reality), but the accompanying AAAS research paper, "How Will We React to the Discovery of Extraterrestrial Life?"[2] found that humans would be far more positive about the discovery of alien life than previously imagined. However, before we prepare our welcome placards, it is rather important to note that this study looked at the discovery of alien microbes, so there is still work to be done.

Science fiction, when done well, however, is probably as good as a guess at what might happen as one could find, though there are so many versions given the myriad variables that might, as mentioned, be at play. As Duncan Forgan, research fellow at the University of St. Andrews in Scotland said in interview to *Space.com*, "If you pick the right science fiction—the hard science fiction—it's placed in the best possible educated guesses about what will happen,"[3] where "hard science fiction" is the type most interested in accuracy and realism.

The main gist of our case here is that human considerations would be political and practical first, thinking of our own safety given this vastly superior technological entity. Morality, and questions of religious impact, would only be of concern once we were very sure that we were not under immediate threat. Although many people might pray to their respective gods for answers or assistance, theology would certainly be further down the list of relevant concerns than safety and politics.

Various science fiction movies and books that have fired up our imaginations have dealt with these ideas of the collisions of worlds. We would

---

[1] See Drake (2018).
[2] Kwon et al (2018).
[3] Howell (2018).

suggest that we learn from the history of our own planet here in understanding how we would react, and history would suggest a process of appeasement and subjugation.

After all, we own a history of invaders and the invaded, conquerors and the conquered, colonizers and colonials.

And perhaps, for Christians and Jews, theology might move from explaining the legend of Hebrews in bondage to the Egyptians to humanity enslaved to alien colonizers.

Invariably, there will be a huge denting to our global pride. No one likes to be knocked off the Number 1 Apex Species spot.

PΛRT III: CHRISTIΛNITY: 6 CHΛLLENGES

This section represents the beating heart of the book in many ways and is what drew us to embark upon this journey into the unknown. There are, as you may find obvious, two distinct positions: (1) The discovery of intelligent alien life will have no impact upon religious belief and theology, and (2) The discovery will fundamentally challenge known beliefs (in this case Christianity) such that something has to go or change.

Various (if not too many) thinkers have considered how ETI might affect Christianity and Christian thought. For example, Theologian Ted Peters wrote a paper ("The implications of the discovery of extra-terrestrial life for religion") that sought to answer five questions:[1]

- Will confirmation of extra-terrestrial intelligence (ETI) cause terrestrial religion to collapse?
- What is the scope of God's creation?
- What can we expect regarding the moral character of ETI?
- Is one earthly incarnation in Jesus Christ enough for the entire cosmos, or should we expect multiple incarnations on multiple planets?
- Will contact with more advanced ETIs diminish human dignity?

Ted Peters is still a theologian so one can imagine what his answers to these questions were. ETI discovery, it appears to him, will not signal the collapse of religion. He concludes:

Despite the conventional wisdom, it is not reasonable to forecast that any of Earth's major religious traditions will confront a crisis let alone a collapse should we confirm an encounter with extra-terrestrial intelligence. Theologians will not find themselves out of a job. In fact, theologians might relish the new challenges to reformulate classical religious commitments in light of the new and wider vision of God's creation. Traditional theologians must then become astrotheologians. Perhaps in preparation of this eventuality, the time now is ripe for some speculation. What I forecast is this: contact with extra-terrestrial

---

[1] Peters (2011), p. 644.

intelligence will expand the existing religious vision that all of creation—including the 13.7 billion year history of the universe replete with all of God's creatures—is the gift of a loving and gracious God.

Christianity faces no problem at all, then! If that really were the case, however, this would be a much shorter book. We are, as you might imagine, not in agreement with Peters and will seek to explain why.

Peters, as part of the paper reports, conducted a survey among people of the major religions and the non-religious, finding that "religious adherents overwhelmingly registered confidence that neither they as individuals nor their religious tradition would suffer anything like a collapse."[1]

However, this does run contrary to other data. George Pettinico, Professor at Plymouth State University, has found that the more religious Americans are, the far less likely they are to believe in ETI.[2] This may have a lot to do with how such people believe that humans are the center of the universe (or more accurately, creation) in some way.[3] This is a topic that we will dwell on in greater depth later in the book.

Of course, even given this survey were methodologically sound and the sample representative, this only tells you what adherents think from a layperson's quantitative position. It doesn't tell us whether religions would *actually* be confronted with theological quandaries and incoherencies. And, anyway, what people say in the quiet comfort of a hypothetical compared to the real event of alien discovery may be two very different things. We must keep this in mind as we examine other surveys.

In the early 90s, for his paper "Not the sons of Adam," behavioral scientist Michael Ashkenazi interviewed twenty-one theologians and clerics from various religions. While many responded with amusement to the questioning, ETI existence "did not create a theological or religious problem for any of the respondents."[4] But this must be taken in the context that "none of those questioned was prepared to suggest ways of dealing as a religion with ETI, nor were they aware of any such suggestions in their respective churches."[5] At that time, at least, religious thinkers hadn't given the problem too much thought and appeared, perhaps, to be blithely confident that ETI discovery would cause no theological problems.

---

[1] Ibid., p. 645.

[2] George Pettinico's chapter "American Attitudes about Life beyond Earth" in Vakoch & Harrison (2011), p. 104.

[3] Although Ted Peters challenges Pettinico on drawing the causal line from Conservative Christian anthropocentricity to lack of belief in ETI see Peters (2018), p. 185.

[4] Ashkenazi (1992), p. 343.

[5] Ibid.

One aspect of contact with ETI is worth bringing up here. In light of the notion that any such aliens may have their own religious beliefs, the religions of Earth will have to be on the defensive. Ashkenazi talks about there being three classes of religion in general: (I) proselytizing religions, (II) isolationist religions, and (III) agglutinative religions (that do not proselytize, not rejecting other religions and often assimilating their beliefs into their own).[1] When the Christian conquistadors with their class I religion met the class III Aztecs, Christianity prevailed rather easily, though painfully, helped along by the large power imbalance. Imagine the technologically advanced aliens coming to us with a class I religion, confronting humans with their own class I religions, but who might be on the wrong side of that power imbalance.

Christianity and Islam would certainly be in a defensive mode, but we can't imagine most of them would be willing to assimilate. Such a situation could be one of religious conflict the likes of which we have seen throughout our own history.

What we are interested in for this section is whether the core tenets of Christian theology and belief really are threatened by the discovery of ETI. We will be returning to the rest of Peters's questions listed above as we evaluate further, but the discussion will be structured not by the Peters questions but by another paper. We will be relying on Christian philosophers of religion C.A McIntosh and T.D. McNabb's paper "Houston, Do We Have A Problem? Extraterrestrial Intelligent Life and Christian Belief" to provide the scaffold against which we will be pushing. As far as we are concerned, this paper gives the best treatment to the subject matter that we are concerning ourselves with.

While the two philosophers remain, as Peters, believers who feel unchallenged by the discovery of ETI, we, as we do with Peters, disagree with their conclusions. The proceeding chapters follow the same subject structure as McIntosh and McNabb's paper, looking at the following areas of potential conflict and threat:

- Challenges to theism (i.e., a general belief in God).
- Challenges to scripture.
- Challenges to doctrine.
- Challenges to tradition.
- ETI and the problem of evil.
- Challenges to narrative.

When answering the question "Would the existence of extraterrestrial intelligent life (ETI) conflict in any way with Christian belief?", the two rightly

---

[11] Ibid., p. 347-48.

introduce their discussion by defining what they mean by "Christian belief." With the simple fact that there are tens of thousands of different Christian denominations, "Christianity" represents a broad church. Thus, "Christian belief" may not be as easy to pin down as one might think.

We will, for the sake of argument, agree with their definition:

> By ["Christian belief"] we mean the broad intersection of beliefs that, traditionally, Catholic, Orthodox, and Protestant believers alike would be happy to raise their glass to: that there is one God who is three persons, perfect in knowledge, power, and goodness, responsible for the creation and sustenance of the universe, and whose character is specially revealed in the Bible, in particular, the New Testament, the central message of which is that while we are estranged from God by sin, God, in His grace and mercy, extended to us an offer of forgiveness and reconciliation through the sacrificial death and triumphant resurrection of Jesus Christ, the incarnate second person of the Trinity, and all those who accept that offer will enter eternal communion with God, readied in this life by the sanctifying work of the Holy Spirit, the third person of the Trinity.

We feel it is important to lay their definition out in full in order to avoid presenting a straw man of their position. Their belief entails something that we will henceforth call OmniGod where relevant. This idea of God is the one of *classical theism* that establishes God as omniscient, omnipotent, and omnibenevolent (though with added Christian bells and whistles). God is all-knowing, all-powerful, and all-loving, and this will include omnipresence (God is everywhere) and full divine foreknowledge (God knows the outcome of all future events and decisions, even in hypothetical worlds that might not exist) arguably as subsets of omniscience.

Sin, grace, mercy, and God's incarnation to work with these ideas are also pivotal components.

So, without further ado, let us see whether there really is "conflict between the existence of ETI and Christian belief."[1] Before we look at Christian theism in light of ETI discovery, let us look a little at probability and abductive arguments.

---

[1] McIntosh & McNabb (2021), p. 16.

Probability and Abductive Arguments

It is worth spending some time discussing the different types of arguments we will be dealing with in the coming chapters.

In terms of the forms of logical arguments we generally see in the philosophy of religion, we have *deductive, inductive,* and *abductive* arguments. Deductive arguments are at first glance very attractive in that they seek to outright prove or disprove something in a "once and for all" manner. Unfortunately, more often than not, they suffer from the principle of "garbage in, garbage out," meaning that their strength depends on their premises. Though the logic may be valid, and the conclusion follows indubitably from the premises, if the premises themselves are unsound, then the argument terminally suffers.

The classic example is as follows:

1) All men are mortal.
2) Socrates is a man.
C) Therefore, Socrates is mortal.

We might argue that much of the work in such arguments is definitional to the point that deductive arguments could be accused of being tautologous. While some thinkers might see them as a sort of gold standard of argument, they have done relatively little to disabuse people of the notion of God or cement God's existence into philosophical thought. People just don't agree on the premises.

Alternatively, inductive arguments look more at observational data and probability. For example, we might use this argument:

1) Earth has orbited the sun for at least a million years.
C) Therefore, Earth will continue to orbit the sun next year.

Or, famously in philosophy:

1) All swans we have seen have been white.
C) Therefore, all swans are white.

The conclusions for either of these do not *necessarily* follow; it's just a case of probability. The idea of a black swan used to be that it was an impossible thing. That was, indeed, until we came across a black swan.

However, inductive arguments are very useful in a data-driven world where we see it as rational to believe in the most probable thing, all other things remaining equal (*ceteris paribus*).

105

Abductive arguments are similar to or even are a form of induction whereby one has an incomplete set of data and uses inference to argue toward the best explanation for that data. Inference to the best explanation can be a very useful mechanic when we don't have all the information to hand. It can be used as a source for Bayesian probability analyses.

For example, we might look at religious pluralism. We know that there is a huge number of religions in the world entailing a huge number of religious adherents believing many different things. In fact, given that we could argue that most religions are mutually exclusive, we can surmise that most people believe the wrong religion. Is this data better supported by a particular strain of OmniGod theism (where God is all-loving and presumably wants all people to come into a loving relationship with him) or naturalistic atheism? *Ceteris paribus*, it seems (to us, at any rate[1]) that this data is better supported by naturalistic atheism.

Furthermore, we can switch the perspective and talk about the predictive qualities of both hypotheses (OmniGod theism of a particular strain and naturalistic atheism). What would we predict of people in a world created by, say, Christian OmniGod? Would we think that an all-loving Christian God would create a world in which most people not only *didn't* come to love him (believing in a vast plurality of religions over time and place), but where many were simply *unable* to come to love him (being born in times and places where access to Christian belief was impossible, even if they would and could have been Christian believers[2])?

On the contrary, what would we predict given naturalistic atheism? Given how we understand evolution working, where *patternicity*[3] and *agenticity*[4] (among many other mechanisms) give us understanding of the natural reasons why religious belief manifests,[5] we would *expect* religious pluralism.

The data of the huge plurality of religions and religious belief not only better *supports* the hypothesis of naturalistic atheism, but it is also *predicted* on naturalistic atheism.

This is, of course, an illustrative example and one we don't want to bog ourselves down in arguing for in any great depth. However, it is these sorts of abductive arguments that will be pervasive throughout this text. What would we *expect* about the universe, and ETI, given OmniGod? If we start with the assumption of OmniGod, what universe would we *predict?* Would we expect the universe to be teeming with life willing to enter into a loving relationship

---

[1] Though you could use the same data to argue for a very general theism or spirituality.
[2] In philosophical literature, these people are called *nonresistant nonbelievers* and are components of a school of arguments known as *divine hiddenness arguments*.
[3] The ease with which we find patterns in the world around us.
[4] The over-determination of agency in the world around us.
[5] See Shermer (2012) and Wathey (2019), for example.

with God, or a vast space where death is far easier than life, and where most intelligent entities believe in the wrong deities?

Therefore, the questions we ask you, the reader, to consider when contemplating anything concerning ETI life are, "Which hypothesis does this data or idea better support?" and, "What would I predict of the universe in this context, given OmniGod or naturalistic atheism?"

8 – CHALLENGING THEISM

We tend to agree with the many theists who declare that theism in general won't be challenged by the discovery of ETI, though there is still room for some interesting discussion here.

Essentially, one can treat the discovery of ETI not too dissimilarly to the earthly discovery of the New World or new civilizations, or even old civilizations and extinct hominids. Take, for example, that it is now thought that homo sapiens existed alongside eight other now-extinct human species.[1] With each new discovery and expansion of knowledge of our world, religions have had to adapt (or not, simply being unaffected by new discoveries) to such new data. One can create an analogue of Earth with its constituent populations and civilizations with the universe and its constituent planetary populations and civilizations.

In short, the discovery of ETI would not disprove the existence of God, certainly as some abstract entity such as OmniGod. When people start bolting add-ons and plug-ins to the god-concept (whichever one they have), they start narrowing down their options. If new data is thrown into play, then something with many more constraints (extra revelations, scripture, theology, and doctrines) will be more likely to struggle to accommodate the new information. OmniGod, for example, is a fairly simple, abstract concept that will not suffer from human contact with ETI.

As McIntosh and McNabb observe, they are not aware of any logical syllogism that has "ETI exists" as one of its premises and "God does not exist" as its conclusion.[2] We are similarly unaware of such an argument, and we present no such syllogism. In fact, to reverse expectations, McIntosh and McNabb suggest it might be the case ETI is *more likely* on theism than atheism. That will depend on the arguments they and other astrotheologians have to offer. That said, there are also potential hurdles for the OmniGod theist.

The Fine-Tuning Argument

One famous argument that is often used by apologists arguing for the existence of God is the fine-tuning argument (FTA). The existence of ETI may well show that this argument is rather incoherent: It is a case of theists having their

---

[1] Plackett (2021).
[2] McIntosh & McNabb (2021), p. 2.

cake end eating it too. Now, unless they have access to Schrödinger's cake, this really shouldn't happen.[1]

The FTA usually looks something like this: Our laws of nature and our cosmos appear to be delicately fine-tuned for life to emerge, in a way that seems hard to attribute to chance.[2] Given the very precise nature of the physical parameters (laws and constants) governing the universe and underwriting existence, theists argue that this can only reasonably be explained by a god who has fine-tuned the parameters in order that life might exist.

We won't go into the many objections to this argument, such as the anthropic principle, the multiverse, the fact that the universe looks more fine-tuned for death (for example, black holes, but also all the ways it is very easy for life to cease to exist...), probability arguments, and cherry picking, We will look instead at the Goldilocks principle and the idea that, for theists, the FTA ends up being heads they win, tails the skeptic loses.

The Goldilocks zone is a narrow range only within which life-supporting planets *seem* to exist. The conditions are just right—not too hot and not too cold, containing plenty of water and perhaps oxygen. The theist uses an argument from incredulity (they just can't imagine that it could be so, or that something with a low probability might happen),[3] arguing that the conditions are so *specific* that only God can explain life.

However, notice that the intuition that drives the Goldilocks zone argument for planets runs in the opposite direction for the FTA. The FTA suggests we live in the best-possible universe for finding life in it, that almost any other universe would make life nigh-on impossible. But the Goldilocks zone suggests that life can only exist in perfect conditions and in very rare occasions, even in the same universe *ideally built for life*. So, is the universe designed for life, or can life only exist in the most exceptional of locales? If these arguments are not in contradiction with each other, then they are certainly in tension.

One would expect that if something was "fine-tuned" to some end or purpose, it would be really very good at producing said thing. The FTA would suggest that life is abundant. The Goldilocks principle suggests that life (and intelligence in particular) is rare. These two arguments for theism predict rather different outcomes for the paucity or abundance of life in the cosmos.

---

[1] Of course, if one were to observe Schrödinger's cake, it would turn into one of two cake states, thus collapsing the cake function.

[2] Landsman (2016).

[3] This is an extreme version of the Rare Earth Hypothesis, derivative of Ward & Brownlee (2000).

What we will be showing is that no matter how much life there might be or could be in the universe, the theist would use *that* to argue for a fine-tuned universe.

Imagine that a given universe (U1) was teeming with life. Life was everywhere. Every planet and every moon provided habitation for endless ecosystems. Everywhere you looked, there would be different forms of life with differing levels of intelligence. This smorgasbord of life would be, for the theist, evidence that God exists, since U1 amply provided habitation for life. U1 really was designed—fine-tuned—such that life was abundant.

Now imagine another universe, U2. Bono notwithstanding, in this universe there is Earth—a planet so particular that it can just barely sustain life—as the only planet that does, indeed, sustain life. The vast universe out there has no life in it. Extending out perhaps infinitely, there is a void of life; there are planets and black holes, suns and nebulae, but no other life. Theists would argue (and have done since this has been one conception of our universe) that this universe is evidence of fine-tuning, since everything had to be *just right* so that in this *one* place, the apex of creation managed to form, clinging on tenuously to life.

Then imagine yet another universe, U3. In this set up, life has started and evolved on Earth, but also in a vast number of other solar systems and galaxies stretching out into a perhaps infinite universe. Here, it's a game of numbers. Life starting may be improbable, like winning the lottery, but if you buy enough lottery tickets, you'll win the gamble. Potentially a lot.

Probability is all about frequency. If life starting is only a one in a billion chance, we look and think, *prima facie*, that this is an unlikely event. But if there are a googolplex number of worlds, then the chance of life is now a relative certainty and will no doubt be widespread.

In U3, then, life is rare when we look into an individual solar system or two, but in the whole scheme of things, life is plentiful (even if it isn't everywhere, and is separated by huge numbers of lightyears). Theists, again, would (and again do) argue that U3 would also provide evidence of God through the fine-tuning argument.

Heads the theist wins, tail the skeptic loses.

No matter what amount of life there is in the universe, from one instantiation to infinite cases of life, the theist argues that this is some kind of proof (or at least evidence) for the existence of a fine-tuning, designer-creator god.

The discovery of ETI would help to exemplify this. As it stands, and for a hundred years or so since the chemist Lawrence Joseph Henderson wrote *The Fitness of the Environment*, people have argued for fine-tuning in the context of really only humanity existing as the apex of divine design and creation. Each subsequent discovery of ET life would seem to devalue this argument to the point of rendering it useless, because the theist would continue, at each point, to argue that *now this* understanding of life in the universe was evidence of God.

While McIntosh and McNabb may or may not agree with our criticisms of the FTA as set out above (and the many other criticisms that can be levelled at it), they do understand that some theistic arguments may not survive ETI discovery. Of course, in their defense, it is one thing to say ET alien life *debunks an argument for* the existence of God, but quite another to say it *disproves* God's existence in and of itself. As they state:[1]

> We conclude that the existence of ETI poses no special problem for theism generally, although we may have to revise or abandon a theistic argument or two.

They also admit that such contact might well psychologically and sociologically impact religious belief on Earth, as already discussed:[2]

> What interests us, however, is not whether discovering ETI would have a psychological impact inimical to religious belief generally, but whether such an impact would be philosophically and theologically *justified* for Christians. We argue not.

We shall see.

The Poor Design Argument

Where a *teleological* argument is one that is an argument from design that implies the universe around us is (perfectly) designed by a (perfect) designer-creator god, *dysteleological* arguments seek to show the God hypothesis as likely false given the multitude of design flaws.

For the naturalist, there is no such thing as poor design in the prevailing mechanism of evolution. Simply put, there is no agency involved in evolution, and the process works in bringing things about using only pre-existing building blocks with no eye on the big picture. Evolution is only "interested in" (it has no agency and so is not really interested in anything—this is a turn of phrase to help explain) successfully getting an organism to reproductive age and to then reproduce.

---

[1] McIntosh & McNabb (2021), p. 3.
[2] Ibid., p. 4.

A teleological argument for a designing God looks like this:

(1) Living things are too well-designed to have originated by chance.
(2) Therefore, life must have been created by an intelligent creator.
(3) This creator is God.

On the other hand, a dysteleological argument might run as follows:

(1) An omnipotent, omniscient, omnibenevolent creator God would create organisms that have optimal design.
(2) Organisms have features that are suboptimal.
  (1) Therefore, either no deity created these organisms or they are not omnipotent, omniscient, and omnibenevolent.

In the human body alone, bad design is rampant. Let me list some examples of poor design that simply wouldn't or shouldn't be the case in the event of OmniGod designing us: narrow birth canals in women, often leading to death at childbirth; ectopic pregnancies; our digestive canals sharing the same pathways with our breathing system easily leading to choking; hernias; wisdom teeth; the appendix and appendicitis; some muscles and nerves having no use (e.g., ear twitching), being evolutionary "leftovers"; the poorly designed back; the gene for synthesizing vitamin C being defective, potentially leading to scurvy; human (and many animal) eyes being poorly "designed" with an inverted retina that has caused many adaptations, resulting in blind spots; and so on.

We would posit that this is a good enough argument in and of itself to invalidate the creative and tinkering OmniGod. Seriously, this alone is powerful enough to invalidate bothering to write this entire book. Yet here we are, gluttons for punishment.

While we have no idea as to what the design status of ET life would be, it would be problematic to special plead that all other alien life is perfectly designed in a way that Earthly life simply is not. This would render God unfair in dealing one set of creatures a bad hand while giving all others special (or at least decent) treatment. If life on Earth is (or appears to be) poorly designed, then all ET life must be (or appear to be) poorly designed.

On the other hand, if our own understanding of the wealth of poor design throughout Planet Earth is anything to go by, then multiplying this almost infinitely further compounds issues of God's poor design, or makes it somewhat harder to explain.

Take into account carnivorousness (and this is something to which we will return in the Problem of Evil chapter). The sheer volume of pain and suffering experienced on Earth alone over its entire history due to natural predation is vast. Each and every instance of one organism killing and consuming another just to merely survive—to exist—causes the theologian some serious intellectual suffering.

If it happens here, and God is fair, then it happens everywhere. But if we assume this kind of mechanism exists throughout the universe to a gargantuan level, then the dysteleological argument is given even greater ammunition.

Of course, the theist will state that if the argument can be solved once, it can be solved an infinite number of times.

The problem for the theist is that this argument *hasn't* been solved once.

There really seem to be only two viable options here:

(2) Skeptical theism: God moves in mysterious ways.
(3) The Fall. Everything was perfect before then.

These are both deeply unsatisfying answers. Skeptical theism is the ubiquitous get-out-of-jail-free card that we *don't* or *can't* know the mind of God. There *could* be a reason why God might design us poorly, a greater good that comes about from predation, cancer, mothers dying in childbirth, masses dying from malaria, and Uncle Bernard choking to death in the restaurant. There could also be unicorns, but I am not prepared to believe in them until given good enough reason to do so. Just positing a conceptual *possibility* is not good enough, especially given the myriad other arguments that the OmniGod idea appears to fail to overcome.[1]

And if God is designing poorly to bring about a greater good, and this poor design manifests itself in a cruel cornucopia of ways with resultant voluminous instances of suffering, then one would imagine an all-loving being to at least communicate *why* this was happening, why God created it so. To be met with nothing other than the mantra of "Don't question God" is nothing short of another example of a valid argument from silence. In other words, not being given any evidence *for* there being a good reason for all the poor design and suffering is *disconfirming* evidence *that there is* a good reason for it,

In the absence of such defeaters for this dysteleological argument, according to a number of philosophers (*epistemologists* who deal with truth and knowledge), we are rational in adopting something called *phenomenal conservatism*:[2]

---

[1] See Pearce (2022).
[2] Huemer (n.d.). See also Huemer (2007).

Phenomenal Conservatism is a theory in epistemology that seeks, roughly, to ground justified beliefs in the way things "appear" or "seem" to the subject who holds a belief. The theory fits with an internalistic form of foundationalism—that is, the view that some beliefs are justified non-inferentially (not on the basis of other beliefs), and that the justification or lack of justification for a belief depends entirely upon the believer's internal mental states. The intuitive idea is that it makes sense to assume that things are the way they seem, unless and until one has reasons for doubting this.

The approach of skeptical theism is one of *externalism* where the believer is relying on external arguments, in some sense, to justify belief that there *is* a reason for such poor design.

Putting aside these asserted appeals to there simply having to be a reason without doing anything to provide that reason, we turn to the second option of everything being perfect before the Fall.[1]

The first thing to say about the Fall is that it never happened. Historically speaking, the Fall, as described in the Bible, never took place. While it is claimed that before the Fall there was no death in the world (Rom 5:12), we know from evolutionary history that there was predation long before there were primates. There was no historical Adam and Eve since the whole idea contravenes known science and the evolution of man. Indeed, we can see the biblical story of an amalgam of pre-existing myths from in and around Mesopotamia and the Fertile Crescent.

The issue with using the Fall as an explainer here, irrespective of the fact that it never historically happened, is that it seems to fail as an idea when considering carnivorousness alone. To punish the animal kingdom with untold amounts of pain and suffering as a result of the decisions of one serpent-tempted human (in the literalistic sense) or of a faultily designed humanity is prima facie morally reprehensible.

Imagine having a child and a dog. Your child is rude to you, so you go to the dog and punch it, or take away its food for a day, or whatever punishment you can imagine. In no moral framework would it be morally acceptable to punish an animal for the wrongs of an unconnected human action. You would be charged for that and prohibited from keeping pets. Now imagine punishing *all* such animals for *the rest of time*.

But with ET life, imagine this is happening all over the universe, all of the time, for billions of years.

An analogy might be this. One could, as a child, forgive a friend giving them a single dead arm. Maybe two or three in quick succession. But a billion

---

[1] We will consider issues with the Fall and the design of Adam and Eve in the chapter on doctrine.

dead arms?[1] That appears horrific, and absent of any defeaters, it shows the dead-arm giver to be morally abhorrent.

---

[1] Yes, analogies become stretched. Children *usually* only have two arms and a finite time to receive such punches. But you get the point.

9 – CHALLENGING CHRISTIAN SCRIPTURES

Remember, we are playing with a hypothetical here. It *might be* that there is no intelligent life in the universe outside of Earth's atmosphere. But, as discussed, this is incredibly unlikely. Given that there is more and more of a recognition that life really must exist out there, and likely in huge quantities, what do Christian scriptures say?

Nothing. Absolutely nothing.

But does the fact that the Bible says nothing about intelligent alien life out there actually have any real consequences?

Most probably, yes.

The question, certainly as far as McIntosh and McNabb are concerned, is arguably one of deception. By *not* saying anything about this state of affairs, if we are confident that ETI exists, is God *deceiving* us?

Before we answer that question, there are a few other relevant ideas to discuss. There is arguably a more fundamental issue than alien existence concerning biblical coherence with the universe; namely, whether the Bible gets the cosmology of the universe completely and utterly wrong.

Biblical Cosmology

Sometimes we have to remind ourselves just how strange and backward the cosmology is as found in the Bible, not to mention other religious writings, when compared to the modern view. Though it may not be accurate to speak of "biblical cosmology" as if it were a singular thing, there are threads of a worldview that appear to be consistent across the sources. When Genesis (or *Geneses*) was compiled sometime after the Babylonian captivity, the view suggesting the shape and position of the Earth was primitive and derivative of the cosmographies of other Ancient Near Eastern cultures, though it was theologically innovative. Much the same can be said for the other books of the Bible, both in the Old Testament and even in the New.

First, consider the land in the biblical schema. As indicated by various passages, the earth was believed to have been a flat body, surrounded by water. Our inhabited world is compared to a footstool (Isa 66:1), which of course is a flat body—no one has found a spherical footstool in antiquity, and they are not readily available at the furniture store today. A footstool also requires legs for support, and the earth is also said to be held up by pillars (1 Sam 2:8,

Aliens and Religion: Where Two Worlds Collide

Ps 75:3, Job 9:6), or otherwise the earth is said to have foundations (Mic 6:2, Isa 51:13, 16, Job 38:4-7). The flat but round body of the land is expressed in numerous places, including mention of the "circle of the earth", something that is "inscribed" on the "surface of the deep" (Prov 8:27, Isa 40:22, Job 26:10). It must be said, a circle is not a sphere, and because the circle of the earth is 'drawn' on the surface of the waters, the shape must be 2-dimensional—one cannot draw a 3D body on a 2D surface.

The earth of the biblical cosmos also shows its extent with passages that indicate that, from a high-enough mountain, one can see all the kingdoms of the world (Matt 4:1-11, Luke 4:1-13); conversely, it is possible to imagine a tree large enough to be seen to all the ends of the earth (Dan 4:11). This is only imaginable with a flat earth, not a spherical world. This view of the land also explains why the Bible frequently talks about the "ends of the earth" (Job 28:24, Ps 67:7, Isa 45:22, Mk 13:27), the most distant points one can go, all of which is inscribed within the circle drawn by God on the deep at creation.

But this is all just said of the land we live on. When the Bible turns to the heavens, again the view of the cosmos is consistent with the ancient worldview, not the modern one. For example, in Genesis we are told of the firmament (1:6-8), that is, a hard dome to separate the waters, which is telling us that there are two waters: the seas and the ones above the sky. This firmament, according to the writers, contains the Sun, Moon, and stars (1:14-17), above which is water (1:7).

What is a firmament? It seems to be a hard material like metal or glass (Job 37:18)—a literal sky dome into which the stars were placed.[1] This is in common with the cosmologies of the other major cultures of the time, so the Hebrew Bible is not out of place as a piece of literature, though it is wildly out of place as a claim to scientific fact.

However, if one were to read this theological story and find room for our potential alien brethren, where could one put any ostensible ETIs? The stars are not places with orbiting worlds, because they are lights in the dome, little more than an afterthought in the Genesis account ("he made the stars also"—Gen 1:16). No one could plausibly read this and say it provides any space for another kind of creature to exist. Those stars were created after the earth and seas (Gen 1:9-13 vs 1:14-19) to provide a calendar system for humans (1:14); they exist for no independent reason other than to tell humans when to farm and feast. Moreover, there is an "above" to this world, where God and his angels reside. The throne of Yahweh is described in strange details in the vision of Ezekiel 1, all of which suggests that above the dome are divine powers, and not more entities of flesh and blood.

---

1 Edward T. Babinski, "The Cosmology of the Bible" in Loftus (2016), p. 109-147.

The vision of the heavens filled with supernatural entities expanded in the centuries after Genesis, as the Greco-Roman cosmology became a layered universe of spheres. The lowest was that of the Moon, then the Sun and planets, and finally the sphere of the stars. Journeys through these celestial spheres were found in the works of the Roman senator Cicero (in the "Dream of Scipio"), as well as multiple Jewish and Christian sources.[1] For example, in *1 Enoch* 18, beyond the great mountains at the ends of heaven and earth, there is a prison for stars that did not follow God's orders; later in chapter 71, Enoch is taken up to heaven and visits the realm of the stars and the angels who obeyed God. In *2 Enoch* 8, the titular character comes to the third heaven and sees Paradise; the remainder of *2 Enoch* consists of many more visions of the heavens.

The most biblical of these kind of ascension stories is 2 Corinthians 12, wherein Paul describes someone (probably himself) rising up to the third heaven, which was the heavenly location of Eden or Paradise. Of a similar note, in the Book of Revelation, John is taken up to heaven as well to see the heavenly Jerusalem and the throne of God. In the future, the heavenly city will descend to Earth. Revelation 12 also depicts the vault of heaven as a place of war between Satan and his minions against the archangel Michael and the rest of the heavenly host, from which Satan would be thrown down to the land and sea.

Heaven is a place of God, not of flesh and blood. This is in fact something Paul is adamant about in 1 Corinthians 15, where he says that the new, spiritual body of the resurrected will be heavenly rather than earthly (1 Cor 15:48-49), and flesh and blood cannot inherit the kingdom of God (1 Cor 15:50). The only blood in those higher realms is the blood that Jesus used in the eternal sacrifice of himself, according to the Epistle to the Hebrews. In heaven is the true tabernacle of God, of which Moses made a copy (Heb 8:5), and it was there that Jesus used his blood to establish the new covenant (Heb 9:12-14, 28). So, it was no issue for imagining Jesus ascending to heaven as a place above the Earth (Acts 1:9-11).

Reading such passages, how does one find a place to put other, naturalistic intelligences? Everything above the firmament is the realm of God and his angels, and this would continue to be believed into the early modern era, influencing the *Divine Comedy* of Dante, among other works, authors, and thinkers. Only with changes in astronomical knowledge did this view of heaven as the unchanging divine realm disappear. Space was no longer filled with angelic armies and cities, nor temples and prisons as found in numerous Judeo-Christian sources.

---

[1] I.e., *Testament of Levi, Ascension of Isaiah*, Babylonian Talmud *Hagigah* 12a, *1 Enoch, 2 Enoch, Apocalypse of Moses*.

The only way to have ETI in such a realm is to say the Bible's view of the cosmos is fundamentally *wrong*. Heaven is not a place God lives in, because we can now travel there and see it for what it is.

A modern theologian has room to reinterpret scripture in numerous ways, the most obvious one being that God accommodated his message to fit the understanding of his primitive readers, who viewed the world as flat with a sky dome, derivative of other Ancient Near Eastern cosmologies. One can question why God would fit his message to humanity's erroneous views, but even if the text has some other referent beyond the physical nature of the cosmos, it is clear that the Bible provides no grounds to expect other sentient creatures living among the stars. Quite the opposite! The heavens were always depicted as a world of angels and supernatural powers. No Christian or Jew before the Copernican revolution argued there were advanced ETIs, not because the Bible explicitly denied it, but because the book could not even conceive it as a possibility. ET could not exist in or around the stars, nor the planets, as those were either mere lights or angels, but not places. The heavens of the Bible allows for ETI just as well as the Bible allows for a square circle—it says nothing about them, because it doesn't even register them as possible.

Really, this alone is enough to invalidate the Bible as any kind of truthful guide. But it is also an example, one could argue, of outright divine deception if one is to actually believe that the Bible is divinely inspired and divinely sanctioned. If God signed this book off, then he signed off a very shoddy representation of reality.

However, let's further consider the idea of God accommodating his message to his audience. The assumption in this approach is that, had God not placed his ideas in a package that was understandable to his followers, then they could not have understood and followed his commands, let alone achieve deep communion. From the viewpoint of a teacher, this sensible— one does not introduce complex mathematical operators to a group of 8-year-olds when discussing the basics of motion. A good teacher, or even a good public speaker, must reach their audience at their level.

But this approach will not work well in the case of the Bible's explanation of the cosmos and its extraterrestrial inhabitants. God could have accommodated his message to the world using a model that did include the possibility of life elsewhere in the universe. That model already existed for centuries before the time of Jesus. In particular, the atomists of the classical world speculated on an infinity of worlds like our own. While the atomic theory of the ancient philosophers, such as Democritus and Epicurus, is not the same as the modern scientific theory, it has in common the notion of natural laws and random motions giving rise to so much of what we see around us and above us. Epicurus in particular argued that there is an infinity of worlds with life on them, as documented in his letter to Herodotus.[1] In less philosophical literature, the satirist Lucian depicted the Moon and other planetary bodies as inhabited,[2] thus indicating the notion of biological life outside of the Earth was not outside of the range of possibilities to ancient Greeks and Romans. Thus, *the conceptual framework for ETI existed at the time of the composition of the New Testament.* God could have mentioned aliens in such a way that was accommodating to the Greco-Roman philosophies of the time, but instead *rejected even this moderately correct view.* In other words, God's message was backwards even in its own time, let alone today.

No wonder then that the Church Fathers such as St. Augustine directly *rejected* the notion of the infinity of worlds, not to mention the creatures of those worlds, declaring it heresy.[3] The century after Augustine, a deacon named Rusticus Diaconus stated that there cannot be many worlds, because there are not many Christs, apparently seeing the issue of multiple worlds requiring multiple incarnations (something to which we will return in depth).[4] One medieval pope, Zacharias, said a priest expositing such views should lose his position.[5] Thomas Aquinas also stated that, using both scripture and philosophy, there is only one world.[6] Even early Protestants denied the possibility of other words, in part because of their views of Christ and the incarnation.[7]

So it cannot be that God did not include the possibility of other worlds in his sacred book because his audience wouldn't have understood; it was a viable option to his audience. Nonetheless, the Bible spoke of a world without the possibility of ETIs, in such a way that the idea was actively avoided by the biblical authors and rejected by those reading these sacred texts. No one could derive the existence of aliens from the Bible, and instead every pre-modern reader interpreted scripture to mean there were no such beings.

Such interpretations of the Bible also played a role in the condemnation of Giordano Bruno (a 16th-century Italian Dominican friar, philosopher, mathematician, poet, and cosmological theorist to boot), who was famously found guilty of numerous heresies. It was almost certainly the case that his

---

[1] Diogenes Laertius, *Lives of the Eminent Philosophers* 10.74. Note that Epicurus and the followers of his philosophical school were not ancient empirical scientists, as exemplified by their confusing statements suggesting that the Earth could be flat, even after the proofs of the Earth's sphericity that existed in antiquity. See Bakker (2016).
[2] Lucian, *True History.* This book is sometimes called the first work of science fiction.
[3] Augustine, *City of God* 11.5; *De Haeresibus ad Quodvultdeum* 77. For more discussion see McColley (1936), esp. p. 392-95.
[4] Martinez (2016), p. 355; see Diaconus (1528), p. 248 reverso.
[5] Martinez (2016), p. 354.
[6] Ibid., p. 355; in particular, see Thomas Aquinas, *In Quatuor Libros Arsitotelis De Coelo, & Mundo Commentaria*, Book 1, Lesson 16.
[7] Martinez (2016), p. 372.

alleged denial of the Trinity, the Virgin Birth, and the divinity of Christ were of utmost importance to the Catholic Church, but the documents we have of his interrogations indicate that he was also questioned about his beliefs in the infinity of worlds and their inhabitants,[1] a view also condemned by another cardinal the same year Bruno was executed.[2] Perhaps if Bruno did not indulge in other serious heretical views, he may not have been bothered, but the fact that life on other worlds was a charge against him indicates that it was the orthodox view that the Holy Scriptures were read as denying this possibility.[3]

A historical reading of the Bible and its traditional interpretation sees a rejection of biological life beyond our small world. The main theological approach to avoid the conflict—*accomodationism*—seems to be at best dubious, and at worst undermined by the history of ideas and early Christian tradition. To say that there is no conflict between Scripture and ETI is to reject this history.

Prophesying ETI

Before we move onto the idea of divine deception, we have an opportunity to have a little dig at biblical prophecies.

Prophecies, usually delivered by prophets (as one might expect), are predictions of the future. The problem that skeptics have with biblical (or any holy book) prophecy is that they never *actually* predict something significantly in the future as to be evidence of the prophecies being true.

What we mean by this is that prophecies are invariably *ex eventu*, which is to say they are made after the event, and then predated to apparently be from before the event. Technically, these are called *vaticinium ex eventu*. For example, Jesus' foretelling of the destruction of the Temple in Jerusalem in Mark 13:14 and Luke 21:20 were written by the Gospel authors *after* the destruction of the Temple. Skeptics see these as examples of the authors putting words in Jesus' mouth to make it look like he was making prophetic claims.

The same can be said of the so-called prophecies found in the Book of Daniel, some of them very vague and certainly *ex eventu*. And we could name many others.

---

[1] Firpo (1993), p. 247-304. For important contextualization and the mythology around Bruno's so-called martyrdom, see Shackelford (2009).
[2] Martinez (2016), p. 367.
[3] McIntosh & McNabb (2021), p. 10 n. 29 deny that Bruno was condemned for his views on the plurality of worlds, but their source (from 1908) is out of date; we have documentation rediscovered in the 1940s that indicates he was accused of believing in the plurality of inhabited worlds.

The problem is that any such prophecies (when they clearly *are* prophecies and not just random verses taken out of context, applied to Jesus, and then *claimed* to be prophecy[1]) only ever seem to involve making very obvious predictions set squarely within the historical and cultural context that they are written. This is most clearly a product of being, indeed, *ex eventu*. It is no surprise that a biblical prophet makes a vague future prediction of some great battle victory here, or of the destruction of a building there. We might shrug our shoulders with a lack of excitement. *Even if* these predictions *really were* written before the events they purport to predict, such predictions would be pretty easy to make. We could probably make some predictions now of future political scenarios and there is good chance, if we were sufficiently vague, that such "prophecies" would come true.

What would *really* show prophetic prowess would be if biblical prophets predicted hospitals, the germ theory, nuclear power, inter-continental ballistic missiles, cars, and airplanes.

But no, they are silent on really useful or truly provable prophetic claims.

What is pertinent to point out here, for the purposes of this book, is that biblical prophecies make no claims as to flying to the moon, let alone ETI.

If the universe has millions or billions of instantiations of alien life, and our eventual discovery of only one of them, the Bible's prophecies are silent. Instead, they only predict a war being won within a few hundred years of the original "prophecy" apparently being uttered.

Scriptural Tumbleweed and the Argument from Silence

It's not just prophecy that is silent, but God is completely close-mouthed about these matters—aliens, space travel, astronauts, radio messaging. There is not even a hint of such talk from the divinely foreknowledgeable deity.

Writing for the BioLogos Foundation,[2] Deborah Haarsma discusses the lack of mention:[3]

> Many parts of the Bible are provincial, and intentionally so. Scripture focuses on the work of God in one small geographic region of our planet,

---

[1] See Chapter 11 in Pearce (2021).
[2] A Christian advocacy group, founded by biologist Francis Collins, that supports the view that God created the world using evolution of different species as the mechanism. It attempts to harmonize Christianity, scripture, and doctrine, with science.
[3] Haarsma (2019).

centered on the descendants of one family. The Bible does not attempt to be comprehensive about the entire Earth or people living on other continents. Rather, God revealed himself in a way suitable for the first audience in the ancient middle east, leaving out information that would not make sense to them.

And yet the Bible's claims are also cosmic in scope.

How Haarsma justifies this scope is to reference biblical quotes that make claim to God creating *everything*. It's that easy. We think this is somewhat disingenuous since this scope can justify anything in human conception from unicorns to evil demigods. Of course, "everything" to contemporaneous biblical authors and thinkers really only entailed a stunted view of the cosmos, as already set out.

Christians could hide behind the defense that there is *no conflict* with ETI discovery and Christianity since the Bible simply doesn't mention ETI. If there is no discussion of other life in the scriptures, then there can't really be any theology concerning it, and thus no resultant doctrine and narrative. No scriptural conflict means no other conflict; Christianity is then completely open to further intelligent life being part of the cosmic landscape.

We're not nearly so sure.

The reader might well be cognizant of the oft-mentioned Argument from Silence. Christian apologists often claim that this is not a sound argument to use when criticizing the Bible for its claims. As an example of this, we may look at the claims Paul makes in the New Testament epistles about the life of Jesus, about the details of his crucifixion, the empty tomb, the details of the appearances and the ascension. The astute reader will know that Paul details none of this. There is a black hole of detail concerning the end of Jesus' life other than the vaguest claims of dying and being resurrected again.

Skeptics claim that the fact that Paul makes no detailed claims about these events, that he doesn't even mention the empty tomb (for example), is evidence that they didn't happen as claimed, or at all (in the case of the empty tomb). Paul's silence is *evidence against* the Gospel claims.[1]

Christian apologists, conversely, claim that just because Paul doesn't mention these things—the empty tomb, the earthquake, the three hours of darkness, dead saints parading around Jerusalem for everyone to see—doesn't mean they didn't happen. Just because you can't see any evidence of your neighbor's house being broken into doesn't mean it didn't happen.

The key to this argument is understanding that if we *fully expect* there to be something reported about an incident, and we have silence, then this silence is indeed *disconfirmation* of the claim that the event took place.

---

[1] There is a lot more detail to be expressed on this example. See the chapter 9 in Pearce (2021).

Let us analogize. Imagine that Polly makes a claim that there was an earthquake in your town last night while you were asleep of a 7.5 measurement on the Richter Scale. If you look out of the window and see no evidence of this, walk around town and see no broken windows or cracked walls, speak to people and they know of no earthquake, listen to the news and hear nothing, then you would be justified in believing that the silence on the matter of the earthquake *is* disconfirming evidence of the claim Polly made that there was an earthquake of 7.5 on the Richter Scale in your town.

If there is no evidence of an event where you would fully expect there to be evidence, then this lack of evidence is disconfirming evidence that the event took place.

That is a sound formulation of the Argument from Silence.[1] Let us now take this idea and apply it to the Bible. The question is, then, if ETI exists, to what extent would we *expect* it to be mentioned in the Bible?

Imagine the universe was teeming with life. There was life on all the planets in our solar system and God knew full-well that we earthlings would discover it. Would we expect this detail to be revealed in the Bible, particularly in the Book of Genesis? We think so. What if there were more intelligent creatures in such close proximity to us? We definitely think so.

How much would we expect the Bible to have mention of ETI if it was in a remote solar system that we may not discover for hundreds or thousands of years from now? And if the life was even more distant? What if ETI was vastly more intelligent than human life such that we could not conceivably be seen as the apex of creation as many theists see us, being the most intelligent creature on Earth (leaving aside issues of defining intelligence, and there being different categories of intelligence that we may be lacking in comparison to other animals).

We are of the opinion that, all things remaining equal, if we are not the apex of creation, and if there is ETI in the universe, God would have made mention of this. And yet, God failed to mention our actual place in the cosmos. As discussed, the Bible can't even get cosmology correct, so there is little hope of it getting a firm grip on ETI existence.

The Bible over and over suggests that humanity is in a superior position, only lower than the angels (Ps 8:4-5, cf. Heb 2:9), but otherwise superior to all other creatures on the Earth (Gen 1:25-26; Job 35:10-11, Ps 8:6). In the future, righteous humans would even be made superior to some of the angels (1 Cor 6:3). The only plausible take-away a reader would have from such passages is that humans have a uniquely special place in God's eye, to the point that the faithful will have power in heaven. The failure to mention even the possibility that humans would have to share that special place with intelligent squids is

---

[1] Carrier (2012), p. 117-119.

for one simple reason: the Bible and its authors knew nothing about such creatures or thought them as even possible.

There is absolutely no evidence of divine revelation pertaining to the existence of intelligent lifeforms that are at least as intelligent as us, and arguably (as will later be discussed) also made in the image of God,

Finally, if ETI exists, and we would *expect* such mention in the Bible but it is *not* referenced, there are two options: (1) The Bible is not true (i.e., there is scriptural conflict with the existence of ETI), or (2) God is purposefully not telling us.[1]

Divine Deception

We say "purposefully" because we are not sure that the idea of God merely *omitting* telling humanity about ETI makes any sense. OmniGod certainly couldn't *forget* to tell us such that it "slipped his mind." And if "it just wasn't important enough or worth it" to tell us, then that is a conscious decision not to reveal to humans ETI existence by God *registering* that it was not worth it. It appears, when God has the full suite of omni characteristics, that nothing can "slip his mind" and all things that happen, happen for a reason (as we are so often told by theists!). Unlike humans, who cannot think of, or remember, or cogitate on all things at all times, God has perfect knowledge and cognition. God *would* be able to think of, or remember, or cogitate on all things at all times. God would have the knowledge of all things important and unimportant—God's knowledge is infinite—and he would be able to consider all things at any one time in order to know what to reveal and what not to reveal.

In other words, if ETI exists, and this is not revealed to us by God, then this lack of revelation is *intentional*. If this is intentional, it is in some sense deceptive. But is it necessarily so? For example, God doesn't reveal in the Bible who the President of the US will be in 2050, or that a blue flower is growing in a certain garden in Austria at a certain time. These are deemed not important enough to reveal and not part of the remit of the Bible. It is hardly fair to call God "deceptive" for not telling us about the Austrian flower in the Bible.

This prompts the question, again, as to whether we would *expect* ETI revelation to be in the Bible. Is ETI life, in its potential abundance, worthy of mention? Is it analogous to the Austrian flower? Or is it important enough to be seen alongside all other biblically revealed claims and ideas.

---

[1] Divine deception is dealt with in the Christian narrative section of the McIntosh & McNabb paper, but we feel it fits very well within the context of the revelation of the Christians scriptures, so we have included it here.

Again, this is a hard question to be able to answer because it depends on a whole host of things that we, at this present moment, just don't know. Is it intuitively surprising? Yes, we think it is, all things remaining equal. The fact that the creation story in the Bible is so stunted in its scope, and so...parochial *is* surprising—or, at least, it is surprising if OmniGod really exists, thus having some impact on the most optimal revelatory text. It is *not at all surprising* on a naturalistic, anthropological understanding of the development of man and religions, where religions reflect the communities and societies from whence they arise, and the imaginations and knowledge (or lack thereof) of those people.

Remember our discussion of abductive arguments.

This, in turn, becomes a problem of evil scenario (and we will return to this subject in a later chapter). The problem of evil looks at why suffering or evil can happen given that God is apparently knowledgeable enough to know *what* to do about it, powerful enough to be *able* to do something about it, and loving enough to *care* or *want* to do something about it. No suffering or evil, given OmniGod's omnibenevolence, can be unnecessary or gratuitous.

What this then means is that, if God has decided not to tell us about alien life, there must be a net benefit. In essence, deceiving is not always morally bad or blameworthy (using this sort off consequentialist ethic[1]). It must be noted that the previous statement is very much dependent upon what moral framework one adopts.[2]

McIntosh and McNabb do importantly admit the following:[3]

Would the surprise of an S5-like [physical interaction with a hostile ETI causing an existential threat] event make God, the author of life as we understood it up to that point, a deceiver? The God of the Christian scriptures is full of surprises. There are even instances where God acts in deceitful ways, and praises acts of deception. That may sound damning, but a moment's reflection reveals that not all deception is morally wrong. Many jokes rely on deception, as do many sports and, indeed, good stories. The question before us, then, is whether an S5-like event would constitute a morally blameworthy act of deception on God's part. And we don't think so.

The examples that they give for God's acts of deception or praise for deception are numerous: Exodus 1:15-21; 1 Samuel 16:1-5; Judges 4:17-21; Luke 24:28; 1 Corinthians 2:8, 2; Thessalonians 2:11, and so on. McIntosh and

---

[1] Although, problematically, this is a moral value system that Christians generally eschew.
[2] Someone who believes in absolutist morality or deontology such as that of Immanuel Kant would maintain it is still bad to lie, no matter the consequence.
[3] McIntosh & McNabb (2021), p. 16.

McNabb claim that God can only be morally blameworthy for deception if God intentionally distorts our understanding of reality. Here, for them, not telling us something is *not distorting reality*.

And there is the nub of it for them.

We have two areas of potential debate here:

(1) Does God distort our understanding of reality? They think no, we think yes.
(2) Can you only be morally blameworthy by distorting something rather than not saying something? They think yes, we think no.

What we are faced with here is either analogous with or directly reflective of the idea of *omissive will* and *active will*. This is the difference between *letting* something happen and *making* something happen. But, especially when it comes to God, there is actually very little meaningful difference. For instance, if Jake was walking past someone drowning in a pond and could easily have saved them but didn't, and just walked on, then he is worthy of moral blame on most moral systems (discussions of free will notwithstanding). Likewise, if Jake pushed someone into the pond on purpose such that they drowned, he would also be held accountable. In both cases, all things remaining equal, Jake intended that the person drowned, even though in one case he didn't actively make the person drown by pushing them in.[1]

In God's case, of course, it is a little different due to his design and creation powers and general omnipotence. Let us take the example of the 2004 tsunami that killed 230,000 people and countless animals in mass ecosystem destruction. Was God the Jake walking by, able to help the drowning person, or the Jake pushing the person in the pond?

Both.

With the tsunami, God did nothing to help those people and animals when it was well within his power, but he also knowingly designed and created this universe, with its plate tectonics, in such a way that the tsunami would happen. God pushed the victim in the pond *and* walked away. (But it must also be noted that God also designed and created the victim, ponds, and the very process of death by drowning.)

Now we need to apply this logic to the scenario of potential deception. But before we do that, it is important to understand the difference between lying and deception. Lying is actually quite difficult to define and has some

---

[1] It might be worth noting here that causality is notoriously difficult to both define and pin down. One might say that Jake did not "cause" the person to drown—water filling the lungs did. We might talk about *causal circumstances* whereby a whole host (all?) factors concerned (perhaps the whole universe at a prior point in time) contribute to a given event. However, we digress. Suffice to say that causality, as a philosophical subject, is a tricky one.

nuance. The traditional view is that it is an intention to deceive in making an untruthful statement (adherents are known as *deceptionists*),[1] though there are some who believe it is just making an untruthful statement (*non-deceptionists*), and there are variations of both positions.[2] Lying is a manner in which one can deceive.

And deception itself? General dictionary definitions look something like "To cause to believe what is false." More refined definitions, though, might look like this: "To intentionally cause to have a false belief that is known or believed to be false." And, to be even more philosophically particular, the following appears to be more robust still:[3]

> [T]o cause another person to acquire a false belief, or to continue to have a false belief, or to cease to have a true belief, or be prevented from acquiring a true belief, or to allow another person to acquire a false belief, or to continue to have a false belief, or to cease to have a true belief, or be prevented from acquiring a true belief.

With that in mind, imagine five different scenarios:

(S1) Lyra is sitting talking to Ali, Ali is petrified of spiders and there is a spider behind Ali's head. Lyra doesn't tell Ali about the spider. He comfortably carries on the conversation, gets up, and leaves.

(S2) Callum works for NASA and knows indubitably that a meteor will hit Earth next week and completely destroy the world. He neglects to tell anyone about this (you could even argue he did this for the psychological benefit and for other benefits to humanity for that time).

(S3) Charlene sees her friend Lukas on the other side of the road, looks and sees nothing coming, and, believing the road is clear, starts to cross the road. Lukas sees an oncoming truck that looks to be on course to hit Charlene but does not tell her that it is coming.

(S4) Gita is an only child. Alex knows that Gita actually has four siblings that she doesn't know about. Alex does not tell Gita.

(S5) Ramos knows that Benny loves the movie *Grease*—it's Benny' favorite. Ramos knows it is on television that night but purposefully doesn't tell Benny.

In no scenario here is an agent actively lying. We have selected these for their various analogies to the ETI scenario. S1 is not telling someone about something that they perceive as a threat because it would cause the other

---

[1] It can be a lot more complex than this and also involve ideas of breaking trust of faith.
[2] See Mahon (2015).
[3] Ibid., cf. Chisholm & Feehan (1975), p. 145.

person unnecessary psychological harm. S2 represents a scenario where there actually will be existential harm but communicating that would cause excessive psychological harm to others. S3 represents harm happening by withholding true information about reality. S4 represents withholding truth about reality that the other person would like to know so that the agent continues to have a false impression, which will vastly alter future actions and beliefs. Finally, S5 represents a situation whereby an agent is deprived of something they would like to know absent of any larger narrative.

All of these scenarios are instances of agents not telling other agents some kind of information about reality. Is it fair to claim that they are not "distorting reality"? Whose reality and from whose point of view? What about an agent's right to know about things that may affect them? These take into account the desires of one agent but not those of the other.

There is no doubt to us that deception is involved in all of these scenarios as per the robust definition above. Which is to say that if ETI exists, as it stands, God is deceiving us, because each of these carefully invented scenarios could map onto God not telling us about ETI existence.

Much of the evaluation of these scenarios depends upon the intentions of the agents involved, the outcome, and the *perceived* outcome.

Given that we believe that if ETI exists, it should be somehow referenced in the Bible, if God is deceiving us, then he must be doing it for a good reason.

This is again the purview of skeptical theism in the context of the aforementioned problem of evil. God is committing/allowing a moral evil in light of a greater good. But, as skeptical theists assert, we don't or can't know the mind of God, so we just don't know why this might be so. Since God is perfect, there *must* be a good reason for God deceiving us. This all looks rather circular and can be used to account for any level of evil and suffering in the universe (as we shall see in a later chapter) under the assertion that *there could be a good reason for it.*

Therefore, all the usual criticisms of skeptical theism will also apply here.[1] Moreover, the justification of deception lies in consequentialist ethics such that the ends justifies the means. A theist can only really appeal to this kind of reasoning if they are happy to adhere to this secular understanding of morality. And yet, more often than not, they eschew moral consequentialism.

To add another potential defense for the theist here, it could be that God *will* communicate to us, but just hasn't yet done so—now is not the right time. God was happy to deceive us with how the cosmology was formed and what its existence actually looks like, and God is happy to deceive us about ETI, but he *will* let us know when the time is right.

---

[1] See Green (2021), for example.

Because, as you can see, we like a good list, here is one to document what we think are the range of options here:

(1) God does not exist and neither does ETI.
(2) God does not exist. ETI does. It would make no real difference as to whether or not we knew of its existence (it might be too far away with communication impossible, and our knowledge of its existence might not significantly affect our lives).
(3) God does not exist. ETI does. It would be nice and/or beneficial or useful to know of its existence but we don't yet know.
(4) God does not exist. ETI does. It would actually be harmful to know about its existence, either psychologically, socially, or existentially.
(5) God does exist and ETI does not. God has not yet confirmed this though might still.
(6) Both God and ETI exist. It would make no real difference as to whether we knew of its existence, theologically, psychologically, or existentially (it might be too far away with communication impossible, and our knowledge of its existence might not significantly affect our lives). God has decided we should not know.
(7) Both God and ETI exist. It would be nice and/or beneficial or useful to know of its existence but God has decided we should not know.
(8) Both God and ETI exist. It would actually be harmful to know about its existence, either psychologically, socially, or existentially. God has decided we should not know.

The theist is only concerned with options (6)–(8) but cannot know which is true, only that we should not know. For, if we *should* know, we *would* know. This works in the same way that a theist has to pick up the pieces if their partner dies horribly from cancer over a prolonged period, and though they might think up some reason as to why OmniGod would let this happen, they wouldn't really *know* and would have to rely on faith *that there was a good reason*.

The Problem of Particularity

The creation story as set out in the Book of Genesis is not a creation story of Earth, but apparently of *everything*. However, it appears to be very Earth-centric and anthropocentric. As the story continues, we hear about breathing into the dust of the Earth to make two humans, made in God's image, who invariably let him down and fall from grace. We follow the trials and tribulations of their children in establishing mankind.

Next, we have a flood that allows OmniGod to start again as if he hadn't already (foreknowingly) realized the shortcomings of what he had created. After Noah's family, the focus shifts to Abraham, in particular, who makes a covenant with God to contractually bind his descendants to their creator. These are the *chosen people*. They become enslaved and are freed by their leader Moses, to whom God reveals himself and his name.

This all seems very parochial, very *particular* as opposed to being universal— particular to certain places and people on Earth, let alone to Earth in relation to the rest of the cosmos. Theologians Andreas Losch and Andreas Krebs are aware of this:[1]

> Interestingly, Jewish and Christian religious traditions have always been conscious of the particularity of those events and memories through which they belie[ve] actually to know of God—among them the revelation of God's name to Moses, the Exodus from Egypt, the giving of the Torah, and for Christians certainly the life, death and resurrection of Jesus Christ...

> We would maintain that the "scandal of particularity" is even more scandalous than Polkinghorne holds. Jesus was a Jewish man of Galilee, and Christians believe him to be the promised Messiah (anointed one, Greek "Christ"—hence "Christians" are those who believe Jesus to have been the Messiah), which in fact is inseparable not only from his being human, but also from his being a Jew. Thus, when we say that Jesus Christ is the center of Christian belief, we must not forget that we can neither understand the historical person Jesus nor the complex of ideas connected with "Christ" independently from a very specific historical, social and religious context.

One cannot help but think that the reams of moral codes in the Bible, from the Ten Commandments to the Covenant Code—all 613 *mitzvot*—are very particular. Absent of the discussion about ETI, this is a criticism that has been levelled at the Bible. It has a lack of universal moral dimension in much of its teaching. Is circumcision universal, or just something that was expected of a particular group of men in a particular time and place? If aliens are in the picture, the situation is all the stranger. How can circumcision apply to aliens? Do they even have comparable genitalia? Some of the ceremonial laws in the Bible are virtually unintelligible to modern men and women, so we shudder to think what aliens would think of them.

Take Deuteronomy 22:11 that decrees that people shouldn't wear clothing made from wool and linen together. This makes very little sense to us

---

[1] Losch & Krebs (2015), p. 236, 240.

now, but for aliens? One theory, and this coheres with why Abel was the keeper of sheep and Cain the tiller of the ground, is that this possibly represents the hill communities versus the urban lowland, where the shepherding (i.e., wool) Judahites of the hills are given primacy in the narrative over the Canaanites (i.e., linen) of the arable lowlands.[1] The two shouldn't mix, This is why Cain is the cursed bad boy. But so much of this symbolism is all but lost to us modern readers as we try to pick up the pieces from the clues left in this revelatory text. Does this have any relevance to us nowadays? And to aliens from across the universe who do not have wool or linen or Judah or Canaan?

The late Irish theologian and philosopher, Ernan McMullin, stated of this issue:[2]

> The particularity of this account is striking. ... The tension in this complex narrative between the more and more explicit affirmation of the God of Israel as the Creator of all that is and the expectation that this same Creator will favor one particular people over all others in war and in peace leaps from the page. ...
>
> Independently, then, of the questions raised by ETI, the religions of the Book have always had to face the difficult issue of particularity. It was inherent in the very idea of God's choosing particular individuals through whom to communicate and a content of that communication that would mark off one human group in a way that seemed to privilege it in God's sight. Is salvation possible outside the chosen group? What of those who have never heard of Moses, of Christ, of Mohammed? If they too are eligible for salvation, what then was the function of the Book? Theologians of all three faiths have struggled with this knot of questions for centuries, softening the harshness of exclusions in a variety of ways, interpreting fidelity to the Book as indispensable witness, for example, instead of as some sort of inside track.

He admits this this is not a new issue, but one that is expanded in scope when considering ETI. His tentative solution is that "the Creator of a galactic universe may well choose to relate to creatures made in the Creator's own image in ways and on grounds as diverse as those creatures themselves."[3] Such a conclusion is only ever as strong as the modal possibilities upon which it rests ("may well"). The aforementioned Losch and Krebs maintain that "it is no other than the universal God who chooses to disclose himself exactly *in* that particularity."

---

[1] See Isbouts (2019) and Carlson (2004).
[2] From his chapter "Life And Intelligence Far from Earth: Formulating Theological Issues" in Dick (2000), p. 161-62.
[3] Ibid., p. 162.

That is all very convenient. Little more than "that's just the way it is."

The "Scandal of Particularity" is a somewhat connected idea that sees Jesus (as per the Gospels) as the *only way* to God. Christian philosopher of religion Ron Belgau explains:[1]

> God is the creator of everything, and is equally present in all places and all times. Yet Christians believe that He chose Abraham (and not any other person in the ancient world), and from Abraham's descendants (through Isaac, but not through Ishmael, and through Jacob, but not Esau), He created for Himself a covenant people. Then, at a particular time, in a particular place, among this particular covenant people, He became a particular man, Jesus Christ. Moreover, Christians believe that Christ is the only way for any human being, in any time or place, to be saved and brought to final communion with God....
>
> Theologians call this the scandal of particularity, because it seems that God favors some (His covenant people, those who have heard the Gospel) while neglecting others.

If Jesus is the only way to God and salvation, and if the core tenets of faith and salvation, moral behavior and theistic belief, are grounded in the particular events at particular times and places on Planet Earth, then what does this mean for ETI life? They would have no connection to or knowledge of these events, with the ritual and legal codes also being utterly irrelevant to them.

Circumcision, for example, has no use outside of the context of this planet, and, moreover, has no use for modern Christians *on this planet!* The problems of particularity in the context of ETI hammer home the issues as they pertain to humanity without the considerations of ETI existence. Indeed, modern anthropologists and religious scholars struggle to understand, for example, circumcision. It has no *intrinsic value* (the value of circumcision in and of itself as a physical state or procedure) and thus seems only about its *extrinsic value* (what it shows to be circumcised).

One of the better theories is that circumcision prevailed for a particular community and represents cultural identity that is accidentally left over from slave marking.[2] ETI's existence provides all the more reason to think that the procedure and state of circumcision is irrelevant, leading one to conclude that the extrinsic value of the marker could be achieved in some more benign (and universal) way.

---

[1] Belgau (2013).
[2] Avalos (2013), p. 71-72.

The same could be said for so much that we see in the Bible, in the particular moral laws. For example, the Bible...

- strictly forbids eating rabbit, shellfish, pork, weasels, scavengers, reptiles, and owls.
- claims all waterborne creatures that do not have fins and scales are an abomination.
- prohibits boiling a kid (of the goat variety) in its mother's milk.
- dictates that menstruating women are impure for seven days, as is anyone who touches her, and anything she lies or sits upon.
- clearly states that anyone whose testicles are crushed or whose penis is cut off cannot be admitted to the assembly of God.

So on and so forth. The list of bizarre proclamations is immensely long. You would be hard-pressed to find a modern Christian (though there are some orthodox Jews who do) who accepts these as universal laws, which is to already accept the problem of particularity.

Let us come to a brief conclusion to this chapter. First, there is nothing in the Bible about ETI when, if we are confident such intelligence exists, it should be in there. We claim that God, through the revelation of the Bible,[1] has deceived us with the claims as to biblical cosmology, especially in the knowledge that the Bible would be used to disseminate falsehoods and punish heretical (though true) beliefs thereafter.[2]

Not only this, but if ETI exists, God has deceived us concerning its existence (or has not confirmed its lack of existence).

But the theist, as ever, can rely on God moving in mysterious ways, and having an unknown but good reason for this deception. The strength of this defense rests upon the strength of skeptical theism as a concept. We would argue that it often looks like the *possibiliter ergo probabiliter* fallacy: It's possible; therefore, it's probable. Theists often present something that is conceptually possible as explaining something, jumping in unjustifiable fashion from "This *could conceptually* be what happened" to "This is *probably* what happened" to "This *is* what happened."

---

[1] If this is God's intended revelation or not makes little difference since he has not clearly communicated otherwise and presumably could.

[2] This is a similar argument as pertains to slavery and how, the actual verses notwithstanding, God would have moral culpability in knowing that the Bible (as a primary revelation) would be used to countenance and justify slavery for the next 1900 years, yet God did nothing to stop this.

Or, there *could be a reason*, therefore, there *is* a reason for God's deception. As mentioned, the theist would also have to adhere to moral consequentialism, a secular moral value system, in order to do this.[1]

The existence of ETI puts great emphasis on the rather self-evident truth that the Bible, as scriptural revelation, is largely parochial, contextualized to a people in a time and place that bear little relevance to modern man, let alone alien life forms.

If intelligent alien life existed, it would help to consign the Bible to the dusty shelves of human history: interesting, but not immediately relevant.

While ETI may not invalidate scripture (or, indeed, vice versa), there is enough to consider here to present a number of issues for the Christian (or Jew).

---

[1] Some theists, like William Lane Craig, try to get out of this by claiming that God is not a moral agent (which arguably permits him to do anything), and so these claims of God being a consequentialist do not hold. Pearce deals with this in his chapters "God Is a Consequentialist" and "Dealing with Objections to 'God Is a Consequentialist" in Pearce (2021b).

10 - CHALLENGING CHRISTIAN DOCTRINES

"Doctrine" can be understood as a set of beliefs or policies held by a church or religion, often as their stated principles. Of course, these principles don't just come from nowhere but have their basis in scripture and interpretive authorities. Though their basis is in scripture, there is a lot of spadework to do to get to doctrine, and this is carried out by theologians, the authorities of interpretation.

The skeptic might have a more cynical approach to the undertaking of theology, such as was described by the comparative religion specialist Lewis Browne in his 1931 book *Since Calvary: An Interpretation of Christian History*.[1]

> Someone has said that a philosopher looking for the ultimate truth is like a blind man on a dark night searching in a subterranean cave for a black cat that is not there. Those Gnostics, however, were theologians rather than philosophers, and so—they found the cat!

There is an element of looking for something that is not there and inventing it out of scripture, of conforming scripture to the needs of the theology. This could be argued to be the case for the doctrine of the Holy Trinity, which is scarcely evidenced in the Bible, but was developed as views about Jesus metamorphosed into a full-blown divinity while such interpreters also proclaimed monotheism. This, of course, needed to be sanctioned by the original revelatory texts. This kind of project starts with the authors themselves. Take the Gospel of John. Written much later than the other Gospels, and after decades of theological development among early Christian communities, the author rewrote some of the stories about Jesus to conform to a theological agenda.[2]

After the authors write to communicate their own theological agenda (and not to mention the input of later scribes, translators, and editors), the work of the theologian starts. In a manner of speaking, we see this from the quote in the previous chapter from BioLogos's Deborah Haarsma, who claims

---

[1] Browne (1931), p. 81-82.
[2] For example, see variously in Pearce (2021), where Pearce discusses how the author of John changes the day on which the Last Supper and the crucifixion take place in order to make Jesus the *actual* paschal lamb of the Last Supper.

Aliens and Religion: Where Two Worlds Collide

that *this* biblical chapter and *that* verse can be theologically attuned to ETI existence in this particular way.

After decades and centuries and millennia, we get doctrine. But it must be remembered, different Christian communities and denominations will often disagree on the finer details, or perhaps the larger ones. For Catholics, Mass takes on something of a different flavor given the doctrine of transubstantiation. This entails that the wafer and the wine *actually turn into* the flesh and the blood of Jesus Christ upon consumption. Protestants, on the other hand, protest about such theological cannibalism and see it more as a symbolic partaking. Atheists complain about the taste.

The point is that we will get disagreement about theology among Christians themselves. This is no more so the case, we could argue, than with one of the central components of Christianity: the atonement. Still, after almost 2,000 years, Christian apologists, theologians, and philosophers cannot agree on how the atonement works. There are a host of varying theories, with most of them being mutually exclusive.

Put another way, on atonement alone, most Christians have it wrong.

McIntosh and McNabb admit that "[p]articular focus has been on the interconnected doctrines of the Fall, incarnation, and atonement, those being the central events of the gospel."[1] It is on these ideas that we will focus.

Doctrinal conflict with ETI is, surprisingly, something that philosopher and revolutionary Thomas Paine described back in the 1790s and early 1800s in *The Age of Reason* (partly written in prison!):[2]

> [T]o believe that God created a plurality of worlds, at least as numerous as what we call stars, renders the Christian system of faith at once little and ridiculous and scatters it in the mind like feathers in the air. The two beliefs cannot be held together in the same mind, and he who thinks that he believes both, has thought but little of either. ... From whence, then, could arise the solitary and strange conceit that the Almighty, who had millions of worlds equally dependent on his protection, should quit the care of all the rest, and come to die in our world, because, they say, one man and one woman had eaten an apple? And, on the other hand, are we to suppose that every world in the boundless creation had an Eve, an apple, a serpent, and a redeemer? In this case, the person who is irreverently called the Son of God, and sometimes God himself, would have nothing else to do than to travel from world to world, in an endless succession of deaths, with scarcely a momentary interval of life.

---

[1] Mcintosh & McNabb (2021), p. 4.
[2] Paine (A), p. 42, 49, cf. McIntosh & McNabb (2021), p. 5.

This looks to be impressively ahead of its time. Here, Paine talks about Eve, the apple, the serpent, and a redeemer. In other words, to talk about ETI and how it might conflict with Christian doctrine is to wonder whether alien life would be moral, whether it would "fall from grace," and whether it would require redemption.

Moreover, as we will see, there is a potential for our world to lose its unique place in the Christian cosmos such that we see "Christianity's fear over losing her absolute claim and possession of God evidenced in the belief in one God, one religion, one savior and one salvation for all."[1]

Just One Instance of *Imago Dei?*

We will be discussing, to some significant extent, whether humans on Earth are special and take up a privileged spot in God's universal creation.

One of the doctrinal concepts under the microscope is *imago Dei*—the idea that we are made in the image of God. There are, as one might predict, various theological interpretations as to what being made in the image of God means. In general, this is our special relation to God given our likeness—our similarity—to God. In antiquity, some noted the oddity of saying that gods and humans have the same image or form. As Xenophanes of Colophon (6th-5th century BCE) said:[2]

> But had the oxen or the lions hands, or could with hands depict a work like men, were beasts to draw the semblance of the gods, the horses would them like to horses sketch, to oxen, oxen, and their bodies make of such a shape as to themselves belongs.

Modern philosophical understandings and conceptions of an abstract style of god are now in tension with the biblical notions of a corporeal divinity (see the excellent *God: An Anatomy* by Dr. Francesca Stavrakapoulou). As such, *imago Dei* is seen more in terms of *rational* similarities—a likeness of personhood. We supposedly have conscious recognition of the reflection of God in our human nature.

There are, indeed, many philosophical problems with this. If, as many claim, our nature concerns human freedom, then it is difficult to conceive as to how God could have this nature himself as a being (OmniGod) constrained by his own nature in having to be all-loving, since he cannot do anything that

---

[1] As according to Dr. L.N. Okwuosa and colleagues at the Department of Religion and Cultural Studies, University of Nigeria. See Okwuosa et al. (2017), p. 161.

[2] Clement of Alexandria, *Stromata* 5.110.

Aliens and Religion: Where Two Worlds Collide

invalidates this highest of benchmarks. Moreover, with full divine fore-knowledge, God cannot act contrary to his own predictions and knowledge of himself and his own future actions. This is worth dwelling on. If God has infallible foreknowledge, this would entail knowledge of his own future beliefs and actions. If God knows he will infallibly do X and Y at two future moments, then God can do nothing other than fulfil that foreknowledge in doing X and Y. Any deviation of action from his own internal predictions will mean his foreknowledge would be fallible. In other words, God is a divine automaton, a trolley hurtling down a track on an unwavering, immutable journey.

If we had freedom to sin and turn away from God, as theologians tend to believe, then it is hard to see this as any sort of reflection of the unchanging and unchangeable God.[1]

It can be seen that Adam was made in the image of God and that the incarnated Jesus is, in a sense, a second Adam (1 Cor 15:45) with a renewed conception of *imago Dei*. If this is the case, what would this say of multiple incarnations?

One particular problem for the image-of-God thesis is that it may suffer in light of the discovery of ETI that is (far) superior to us. There is a very real chance that we will forfeit the mantle of *imago Dei* in this event. Philosopher and theologian Joshua M. Moritz disagrees in asserting, "Regardless of which traits ET possess—superior or inferior—this will not affect either our election or our responsibilities."[2] Unfortunately, he fails to offer any defense of this position.

Let's do our due diligence and see the implications more fully. In our estimate, there is a deep issue in need of debate.

Consider the implications of identifying *imago Dei* as defined by rationality, and then imagine we discovered ETI. More likely than not, an ETI we can find will be more technologically advanced than us. The average alien civilization will be older than ours, and they will have had perhaps centuries of more technological development. Given the diversity of possible aliens, we should expect that some aliens will be more intellectually capable than humans. Similarly, if *imago Dei* relates to morality, there should be the expectation that some ETIs will exhibit more advanced morality that humans. If aliens can be more rational and moral than humans, that suggests that some creatures conform to the image of God better than others.

Everything exists on a spectrum, we often hear. Rationality and personhood are arguably characteristics of other species (and humans, and extinct hominids) to differing degrees. Are dogs, dolphins, chimpanzees, and

---

[1] See Pearce (2022) for expansion of these ideas and further arguments along these lines.
[2] Moritz in Peters (2018), p. 343.

bonobos made in the image of God? Because to claim that we and no other species are looks very much like special pleading in a just-so scenario.

That special pleading begins to look like crying if ETI exhibit all of the great qualities of the *imago Dei*, and more so than us lowly humans. Such chest-pounding self-importance is an example of egotism and speciesism, which...doesn't look rational. Perhaps it's the aliens that have the *imago Dei* and not us... At the very least, more rational, more moral aliens undermine the idea that humans have a special relationship with God, if those aliens clearly exhibit godliness more than us.

There are two avenues a theist might be considering: Only humans have the *imago Dei* or both ETIs and humans have it. If the former, it becomes special pleading when the aliens are smarter and kinder than us; if the latter, then why do the aliens seem to have the closer relationship to God? On the first horn, humans deny the "humanity" of ETI; on the second horn, humans are not special.

As you might imagine, the most prevalent of such divine characteristics is morality, and to this we shall now turn.

Λliens Λs Fallen Moral Creatures

Morality appears to be a foundation stone for theology, so let's use this as a springboard to the potential doctrinal conflicts.

Humans are a broken creation. This alone is a powerful argument against OmniGod since we did not design and create ourselves (with full knowledge of our own outcomes); that was OmniGod's job. God, as the perfect designer and creator that he must be, designed and created us such that we are moral creatures, but that we are fallen. This is a problem that Christians try very hard to evade. As professor of theology and science, Robert John Russell believes:[1]

> [W]e live in a paradox. On the one hand, sin is not an intrinsic part of human nature, and yet on the other hand it is an inevitable component of fallen human behavior. Sin by definition, then, is unnecessary; but without grace it is inevitable. Indeed its only remedy is the grace of God. In sum, each of us is created in the *imago Dei*, the image of God, and each of us inherits the inevitability of sin, i.e., sin as original.

---

[1] Robert John Russell's chapter "Many Incarnations or One?" in Peters (2018), p. 304.

Christians can and do interpret the Bible's claims to this in different ways. Adam and Eve's antics in the Garden of Eden can be seen either as a literalistic account of what actually happened or a symbolic representation of the imperfect nature of humanity in the moral sphere.

If we are to believe the story of Adam and Eve literally, then one of the most prominent questions is whether all aliens in the cosmos are affected by Adam and Eve's sin. It's difficult enough to argue coherently for sin and punishment being passed down from generation to generation (though this is what the Bible clearly advances in several places). Yet, to think that Adam's sin has some causal efficacy for the outcome of intelligent life on the other side of the universe is quite the claim.

The suggestion is also atemporal. More likely than not, advanced civilizations came into existence well before the first humans came into existence, let alone received the alleged *imago Dei*. The antiquity of the cosmos suggests that some of these civilizations have also passed away. If these civilizations came and went before the time of humanity, how could they have felt any effect from the sins of humans? Unless Adam's sin can time-travel, there are billions of entities that Adams' actions could not have influenced. It is bad enough to suggest that we must suffer for the sins of the past; how much worse would it be to say we must suffer because of sins *that haven't even happened yet?*

In this way, we think that a literal interpretation of Adam and Eve's story is problematic in light of ETIs unless one believes that there are extraterrestrial parallels—alien Adams and Eves—committing the same sorts of misdemeanors on various planets around various galaxies.

In the Genesis story, Adam and Eve disobeyed God and ate from the tree of knowledge of good and evil (incidentally, disobeying *before* they had eaten the fruit and so not knowing what was good and evil at this point, and still being punished for it!). They were given extra temptation by the serpent and were afterwards banished from the Garden as now morally imperfect beings.[1]

As we have already discussed, where *teleological* arguments concern themselves with how the universe looks as if it has been designed, or has meaning or purpose, *dysteleological* ones are a family of arguments that provide evidence that God does not exist because there are clear-cut examples of very poor design. It is unlikely that a perfect being would be in the business of suboptimal design.

In the context of Adam and Eve, we have the problem that the design of intelligent humanity falls victim to a two-horned dilemma. The pioneering

---

[1] There are many problems with the theology of this account, especially when read literally. See Pearce (2022).

pair appear to have been given a test, a command not to eat fruit from the tree of knowledge of good and evil in the Garden of Eden.

God knows in advance the result of the test, but he picks (by design and creation) Adam and Eve to do it anyway. And he knows they are going to fail. This is the consequence of divine foreknowledge where God supposedly knows the outcome of all future freely willed events.[1] God knows he has a glaring design flaw on his hands and yet still sets the test up. Further, if Adam and Eve are representative of humanity, then any human taking that test would have failed, and *we are all inherently faulty.* You, us, anyone and everyone we know—we would all have done the same thing as Adam and Eve. This throws perfect design and creation down the proverbial drain.

If, however, Adam and Eve are not representative of humanity, then God has chosen non-representative people to take a test and fail. This is to say that had God ordered you, me, or other people that we know, not to eat from the tree, then there is every chance in this scenario that we would have followed orders. We would have passed the test. On account of this failure of a non-representative Adam and Eve, all other people, given the Fall and Original Sin as commonly understood, are punished.

Analogously, imagine you have studied for a math test, and you were ready to pass it with flying colors. However, you get to the classroom, and you see another student has taken the test for you, failed it miserably, and you are thrown into prison for that student failing your test. Any conception of justice would render this situation ludicrous. But this is what the non-representative interpretation of the Genesis story suggests.

It's bad enough to know we are being punished for the choice of Adam and Eve because we are all equally as shoddy as they are, but it's possibly worse to think we are being punished for the wrongdoing of this test when we could have passed the test ourselves! Adding aliens into the mix only further accentuates the problem.

One option for theologically differentiating between aliens and humans is to consider whether alien life might develop morally differently. Perhaps you could imagine alien life having perfect moral existence such that Jesus only needed to die for *our* sins and not *theirs*—because aliens have no sins, with us being the bad boys and girls of the universe.

This would be hard to believe, though it is *conceptually possible.* In fact, it is something that author and famous apologist C.S. Lewis included in his fictional space trilogy (*The Cosmic Trilogy*). In this series, the aliens of Venus and Mars had not fallen into sin, perhaps reflecting our world if Adam and

---

[1] The huge and terminal issues of libertarian free will notwithstanding. Also, there are some theists who deny divine foreknowledge for all the problems it entails such as this one. The most common position is "open theism," which sees divine foreknowledge as incompatible with libertarian free will. But this position has as many problems of its own as it seeks to solve.

Eve hadn't eaten the fruit.[1] It could be that their paradigm is so different that our own concepts of sin and redemption are...alien to them.

The problem for this approach is that if God has designed and created intelligent life that could freely act morally perfectly or decently, then why has he not done that with us? This scenario looks (given God's foreknowledge of what would come to pass, and his design abilities) like creating some entities *so that* they fail, while others get the free pass. And if it is good or right for *us* to be fallen, why is this not the case with other sentient life? Remember, God has ultimate power and sovereignty over what is created. What *he*[2] has created.

This approach of seeing humanity as somewhat special compared to ETI life has been labelled variously *planetary chauvinism, geocentrism* or *anthropocentrism,* and even *pre-Copernican chauvinism.* We haughty humans (that being all of humanity, not just the authors, though we might well be particularly haughty; you can decide) think we sit at the top of the ladder, but this denies personhood and thus moral agency to these outsiders. Theologian Andrew Burgess weighs in on this:[3]

> As long as someone is thinking in terms of a geocentric universe and an earth-deity, the story has a certain plausibility. . . As soon as astronomy changes theories, however, the whole Christian story loses the only setting within which it would make sense. With the solar system no longer the center of anything, imagining that what happens here forms the center of a universal drama becomes simply silly.

If aliens were to be seen as persons—to have personhood—then they would be moral agents.[4] Personhood is a notoriously tricky notion to accurately pin down for a whole host of reasons (including whether we have the required properties when we were children, asleep, in a coma, or lacking in mental faculties). Many philosophers over many years have had a stab at the properties that constitute personhood. It is perhaps somewhat academic since it depends upon what ontological philosophy one adopts. We (authors), as conceptual nominalists[5] adopt a position whereby such labels as abstract ideas exist only in the minds of conceivers and not *out there* in the objective,

---

[1] Joke's on the aliens there. Venus and Mars are hot and cold as Hell, respectively.

[2] Though God would have no gender.

[3] Burgess (1976), p. 1098.

[4] See Puccetti (1969), p. 106, for example, who says that "despite secondary biological differences [ETIs] would certainly qualify for person-status, since they would be both capable of assimilating a conceptual scheme and the sort of entity to which one can quite reasonably ascribe feelings. Since these are…the essential requirements for constituting a moral agent, to say that extraterrestrials are persons is the same as to say they are moral agents."

[5] See Pearce (2021b).

mind-independent aether. We often agree on certain ideas, and when we do, we write dictionaries, encyclopedias, and laws; but if all sentient life was to die, those ideas would die with it.

But, even if we don't agree on what these properties are, there is some intuition to be had such that we can recognize properties in other intelligent life that mirrors our own that would entitle them to the title of "persons" under most definitions, and for pragmatic reasons. Variously, such properties as the following have been suggested in differing combinations:

- Rationality
- Genetics
- Continuous consciousness over time
- The ability to consider the future; to formulate plans
- Free will[1]
- Being alive
- Being aware
- Positive and negative sensations—the ability to feel pleasure and pain
- Emotions
- Sense of self (self-aware)
- The ability to controls one's own behavior
- The ability to recognize other persons
- Intentional action
- Language or verbal behavior

And no doubt more still. Again, if humans as intelligent life forms have the qualifying properties of all or some of the above list as persons, then we can assume that other ETIs would have these properties too. Because what we are really doing is saying that the above list in some sense at least partially denotes intelligence. Thus, intelligent extraterrestrial life is extraterrestrial life that has (some or all of) those properties.

If we are imagining aliens visiting us, then it would be rather naïve to think that they wouldn't have these properties in order to work together to harness the necessary technology for interstellar travel! Theologian Robert John Russell agrees:[2]

I assume that since the underlying laws of physics are the same everywhere, the processes of evolutionary and molecular biology most likely

---

[1] Noting that many philosophers deny the libertarian version of free will.
[2] Russell in Peters (2018), p. 305.

will be quite similar. Accordingly, ETI would be gifted by God with an *imago Dei* similar to ours (the image is, after all, that of the one universal God). Moreover, as I suggested in chapter 4, ETI would be tragically flawed by an ambiguous ethical and moral character even as we are. In essence I expect that ETI will experience a kind of moral dilemma that in many ways resembles the moral quagmire of terrestrial human experience, though obviously differing in the moral morphology of personal and social ethics.

In reality, it is hard to get away from the idea that intelligent alien lifeforms would be rather similar to us in that they would operate within the same moral dimensions. As theologians Andreas Losch and Andreas Krebs state, "let us assume that 'all have sinned and fall short of the glory of God' (Romans 3:23), including potential extraterrestrial fellows."[1]

This is what theologians and apologists would think if they were fans of the "Reformed Epistemology" of Christian philosopher Alvin Plantinga, who adheres to something he called "transworld depravity." In the case of aliens, as opposed to merely other sets of humans, the term would merely be translated to "transuniverse depravity." There would be at least one bad apple in every bunch of apples found in the cosmic orchard.

Which is to say that if humans are moral, aliens are likely moral, and if humans are fallen, aliens are likely fallen, too.

McIntosh and McNabb agree since they "think most Christians would be inclined to accept" that "[a]ll intelligent life would fall and need to be saved."[2] They use Paul's writing in Romans 8 to support their position:

> [20] For the creation was subjected to futility, not willingly, but because of Him who subjected it, in hope [21] that the creation itself also will be set free from its slavery to corruption into the freedom of the glory of the children of God. [22] For we know that the whole creation groans and suffers the pains of childbirth together until now.

We might wonder here what God's intentions are. It seems somewhat odd to design and create a universe in which countless creations—billions upon billions of them—are made in such a way that they would fail a key test (literalistic interpretation) or were just naturally plain faulty such that they deserved punishment for it (symbolic interpretation). But then again, if the whole creation of a broken entity (Adam, Eve, humanity at large) creates a net benefit (using moral consequentialism), then we might expect God to create

---

[1] Losch & Krebs (2015), p. 239.
[2] McIntosh & McNabb (2021), p. 6.

as many broken entities around the universe as possible to maximize the net eventual good or benefit.

But here is one of the main criticisms of the account that McIntosh and McNabb give. On the one hand, they argue (as we saw in the previous chapter) that the scope of the biblical text concerns humanity only:[1]

> The reality is that the Christian scriptures are chiefly about God's relationship to man. At the risk of taking a popular analogy too far, it would be odd, to say the least, for a man to mention that there have been others in a love letter to his wife. That is a conversation for a different time and forum. Scripture's silence on the matter should therefore be expected.

And yet on the other hand, they are claiming that the Bible *does* have scope to cover a universal scale that includes ETIs, that Paul's letters are universal enough to cover ETIs. This is a classic case of having one's cake and eating it too. Conflating biblical teachings as both parochial and universal undermines those deductions from the texts.

A further nuanced question might also arise here: Are we more or less moral than ETIs, as far as we might be able to work out? Or does that question not even make sense? How would we calculate this? By intrinsic moral value, or by measuring the outcomes of moral activities? Of course, this will depend on what moral framework one is adopting.

If this question does make sense, then perhaps we need to revisit the Copernican principle.

As we have mentioned in the book previously, if there is a huge number of ETIs out there, we would expect ourselves to be somewhere in the middle of a distribution of any given characteristic. If you recall, this is called the Copernican principle "which claims that when we have no information about our position along a given dimension among a group of observers, we should consider ourselves to be randomly located among those observers in respect to that dimension."[2]

In this particular context, the Copernican principle implies that, morally speaking, we would expect there to be far "more moral" (however you might understand this) alien life out there than us (and far less). Philosopher Samuel Ruhmkorff observes:

> [S]uppose we discover that *noetism*...is true: there are many intelligent extraterrestrials. We should then think it very likely that there are intelligent extraterrestrials whose most unfortunate members experience

---

[1] Ibid., p. 4.
[2] Ruhmkorff (2019), p. 297.

more evil than the most unfortunate humans, whose most fortunate members experience more good than the most fortunate humans, and whose worst calamities are worse than the worst human calamities.

Ruhmkorff does a probability analysis of the sort (in his case "evil") that might be useful for what we are saying here in terms of morality. He concludes:

This is a sobering result. Take whatever you consider to be the worst evil in human history. There is an 80% chance that something at least one quarter again as bad has happened somewhere else in the universe, and a 50% chance that something at least twice as bad has happened.

We will return to this paper and its reasoning in the chapter on the problem of evil. Suffice to say that, for our purposes here, we can use this approach to argue that we are unlikely to be the most intelligent or moral lifeform in the universe. Given this, if we are fallen because we are moral creatures—and this is some kind of collateral for being intelligent and moral (in the image of God)—then there will be ETI out there who will arguably be *more* fallen than us.

The scenario presented here gives the skeptic cause for critical concern. There are issues explaining sin for aliens on any conceivable scenario. If aliens are infected by human sin, then this is punishing others for the actions of one, including punishing people in the past for things not yet done by anyone; if some aliens fall and others do not (God made some species better than others), then there is injustice in God's acts of creation; and if every alien falls, there is massive faulty design in God's creation. The Copernican Principle suggests that because we are fallen, then other ETIs will also be fallen, but some will be in more or less fallen states that us. What's a god to do to make this right?

Alien Atonement

Talking about how one solves the problem of God creating a fallen creature, we must discuss the solution presented in Christianity: the atonement. The issue here, to begin with, is that there are a number of mutually exclusive ways in which atonement is supposed to work, and theologians to this day disagree on how it does, indeed, work.

This mystery has been circulating for quite some years. At the Council of Chalcedon in 451 CE, while also ratifying the Nicene Creed, the ecumenical leaders also proposed the Chalcedonian Creed (or Definition) that is still binding for most denominations. It determined the famous phrase according to which Jesus Christ was "truly God and truly man" but as "one and the same" person, though also "in two natures." Theologians since that time have been trying very hard to show how that might work, and, in our opinion, without an awful lot of success.

Jesus is the human incarnation of God whose death somehow atones for the sins of humanity. Atonement, therefore, depends upon (or *supervenes* on) the existence of the Holy Trinity. The prevailing theory of the Holy Trinity is *mysterianism,* which declares that "It's a mystery! We don't know *how* it works, only having faith *that* it works..." We maintain that the Trinity is plainly incoherent, so it is no wonder that mysterianism prevails.

To simplify, the main reason Jesus even "exists" or "existed" (historically speaking at least) is for atonement. And yet no one can agree on how that works! This, to us, seems rather bizarre. But the mystery of atonement is built on a foundation stone of a further mystery: that of the Holy Trinity. The very core elements of Christianity, then, are a mystery at very best, and downright incoherent at realistic worst.

The principal challenge for atonement is why a god with divine foreknowledge would knowingly design and create so many of these faulty entities that sin to the point that the god (ultimately responsible for the faulty creation) demands payment (to himself) for those faults. This same god is supposed to represent divine simplicity and oneness, and yet doesn't (and can't, in terms of necessity) exist in the simplest form, but exists in three "forms" or aspects. Why not two, or twenty-two?

Furthermore, if our sin is such an affront to God, and the incarnation and atonement so costly to God to reflect our huge misstep—something that could be argued from a single instance of a fallen species—then what does it say if there are millions, billions, or perhaps infinite examples of such a fall? This devalues the Christian story and doctrine, the redemption and the glory. It's hardly the *ultimate* sacrifice if God has to do this billions of times.

Although we do not have the time and space for an exposition of these problems[1] in this book, we cannot emphasize enough the fundamental and terminal nature of the issues for the Holy Trinity and atonement doctrines. For example, we could spend a great deal of time analyzing the faults with OmniGod utilizing penal substitution theory, ransom theory, Christus Victor, or satisfaction theory of atonement, among others.

In partial conclusion. These ideas don't work once (on Earth). Multiplying them endlessly in countless instances across the universe doesn't make them any more workable.

---

[1] Although it is discussed to some length in Pearce (2021).

Divine Incarnation in Alien Form

If we can agree that, in the event of multiple instances of ETI existence, these ETIs would be as equally fallen as humans, then it is reasonable to think the same method of redeeming humanity would be chosen.

This is largely due to the constraints of fairness in God's all-loving character. If there is a single best (most loving, most moral) way of humanity's sins being redeemed,[1] then God would have it this way so not to have gratuitous suffering or gratuitous lack of benefit. If ETIs, somewhere else in the universe, got access to a better, or even worse, form of redemptive mechanism, then this would create some unfairness. The only way around this (libertarian free will being a broken concept notwithstanding) is for certain ETIs to be more or less deserving of the best form of redemption. We both deny the existence of libertarian free will, though, again, we are unable to give the necessary time and space to the argument here.[2] Such a lack of praiseworthiness and blameworthiness renders the just desserts of the varied apportioning of redemptive mechanisms truly unfair.

That aside, if God is truly forgiving, and loves all of his creations equally, it would seem most appropriate for God to use the most optimal mechanism for redemption, which would appear to be atonement through the sacrifice of himself incarnated in human form.

11th/12th-century philosopher and theologian Anselm argued that only a human could atone for human sin and only God could make that payment; hence, the "God-Man" of Jesus. With this thinking, all other incarnations would take the form of "God-particular intelligent alien lifeform." This would require multiple incarnations of Jesus (and, by "multiple," we mean "millions, or billions, or trillions, or infinite numbers of").

On the other hand, perhaps not. Aquinas (for example, though he was not the only one) says incarnation was not a *necessary* mechanism...and yet it was. In *Summa Theologica*, in the Third Part, answering the question "Whether it was necessary for the restoration of the human race that the Word of God should become incarnate?", he states:[3]

> I answer that, A thing is said to be necessary for a certain end in two ways. First, when the end cannot be without it; as food is necessary for

---

[1] Don't forget that those sins were foreknown at the design stage and created into humans, fully knowing what the outcomes would be and that they would need atoning. Nothing is a surprise here to God since humanity has been designed and created *such that* this would come to pass. See Pearce (2022).

[2] Pearce has written variously on the topic, with the most recent large-scale iteration being his chapter "Free Will" in Loftus (2016).

[3] Aquinas, Thomas, *Summa Theologica*, Part 3, Question 1, Article 2.

the preservation of human life. Secondly, when the end is attained better and more conveniently, as a horse is necessary for a journey. In the
first way it was not necessary that God should become incarnate for the
restoration of human nature. For God with His omnipotent power could
have restored human nature in many other ways. But in the second way
it was necessary that God should become incarnate for the restoration
of human nature. Hence Augustine says (De Trin. xii, 10): "We shall also
show that other ways were not wanting to God, to Whose power all
things are equally subject; but that there was not a more fitting way of
healing our misery."

Which is to say that it wasn't necessary for God to become incarnate...but it was. He had the raw ability to achieve redemption in other ways,
but not the desire to. In other words, it appears that God had the "ability" to
achieve redemption through another method, but this was the one he wanted
on account of it being the best method. His nature constrained his ability,
which leads one to question whether that really is an ability if it is something
that one could never *actually* do, because one was *unable* to *desire* it.

Therefore, it really seems that God's sacrifice of himself to himself to
pay for the sins of man that he himself had knowingly designed into the system and actualized is most likely (for the theist) the best payment for sins
there is, even though those sins still keep getting committed.[1] Jesus as God
incarnated in human form was the best method and the one that God wanted,
presumably dictated by his necessary nature.

Incarnation, at least as far as the eventual sacrificial element is concerned, is predicated upon the coherence of atonement, as mentioned. But
there are some theologians who think incarnation is independent of the Fall
of Man. It is for fundamental disagreements like this that we shake our heads,
as skeptics. That there is still, after several thousand years, a widespread lack
of clarity at a base theological level is a good sign that Christianity is not true,
that it is not the best explanation of reality. Theologians can't agree on original
sin, the Holy Trinity, the incarnation, the atonement, and any number of other
subjects, and that's a real problem. In discussing two differing views of incarnation, astrotheologian Ted Peters sets out the views of German-American
existentialist theologian Paul Tillich and German theologian Wolfhart Pannenberg, both Lutherans:[2]

[W]hat warrants incarnation? Had Adam and Eve not fallen, would the
Son of God still appear in incarnate form? Is it the world's fallen state
that creates the need for redemption? Or, might the incarnation be due

---

[1] Arguably more so, given rampant population growth.
[2] Peters (2018), p. 286-87.

Aliens and Religion: Where Two Worlds Collide

strictly to God's self-communication in creation? Tillich seems to assume that our situation of estrangement calls out for an incarnate visitor from the ground of being. Fallenness calls out for redemption. Pannenberg, in contrast, makes the incarnation independent of the fall. "The incarnation cannot be an external appendix to creation nor a mere reaction of the Creator to Adam's sin." God's presence in Jesus Christ adds the grace of redemption upon the grace of creation. The latter is a completion of the former. Tillich and Pannenberg represent two models of terrestrial incarnation, the fix-a-broken-creation model and the divine-self-bestowal or incarnation-anyway model.

Although, as Peters goes on to admit, most theologians would place incarnation and atonement together, it is not a given. It is what Peters calls above the "fix-a-broken-creation model," which is a model that brings into serious question the design and creation abilities of OmniGod, as we have already discussed.

Alternatively, Eastern Christianity proposes something called *theosis*— a deification of humanity that Jesus sends us toward after healing us. Peters says of this that "[s]uch deification had been God's original plan in creation, to be sure; but because of human sinfulness God found it necessary to take redemptive action."[1] Yet again, we are left questioning OmniGod's foreknowledge and forethought, wondering why OmniGod would create a humanity that he had plans for, but had to "readjust" to cope with sinfulness that surely he knew about because he designed it into the system and foreknowingly created! If this bizarre design and creation process is to be repeated ad infinitum across the universe, then it just seems all the more bizarre.

It is theology like this that leaves us shaking our heads in both confusion and disbelief.

Because theologians love to disagree as a result of the distinct lack of revelatory clarity, there is this second model of incarnation (*self-bestowal*) that warrants a moment of thought. This is the idea that, irrespective of whether humanity would sin (the Fall), the incarnation would happen anyway. 13th-century thinkers Bonaventure and John Duns Scotus advocated this approach. Incarnation wasn't some kind of afterthought or Band Aid to fix a broken creation, but instead had its intrinsic value, that of perfecting nature.

And the famous reformer John Calvin appears contradictory in straddling the fence between both positions.[2] Theology isn't easy.

This idea that incarnation is some kind of grace through divine self-communication, where we become part of God's own history, perhaps, is

---

[1] Ibid., p. 287.
[2] Ibid., p. 289.

tricky. If we as humans are afforded this form of grace, this privilege of God existing in our form, then why don't all other ETIs get this? If it is a historically and geographically contingent action, then why are we the ones so privileged? Surely, this approach shows God to be unfair, giving humanity the better access to God. Again, this shows an inequality of opportunity of access to God.

Or is it that this incarnation is not necessitated by moral creaturely sin, but by God's desire to be all-loving and thus fair? In this way, there could be multiple Jesuses across the universe but that all of these instantiations of God would not be *necessitated* by the sin of the ETIs but out of God's love and sense of fairness.

If Jesus as God incarnated wasn't exclusively about atonement, then there were other supposed benefits for Jesus' appearance on Earth. While he didn't communicate germ theory or that we should boil medical equipment before and after use, that slavery really was bad, or the principles of sustainability, he *did* communicate a number of moral lessons. His parables and teachings were a way for God to impart a certain amount of (moral) wisdom. There are other ways that God can achieve revelation, as we see in the Hebrew Bible with burning bushes and prophets, dreams and angelic visitations. We should also expect such methods to be employed elsewhere in the universe.

One Jesus to Rule Them All, One Jesus to Find Them

Could it be that one earthly Jesus as incarnated God can pay for all fallen creatures across the universe? This is what aforementioned astrotheologian Ted Peters calls "pre-Copernican Earth chauvinism"[1] and suggests a feeling we have that "it is all about us."

Imagine if we discovered an intelligent alien civilization, and one that didn't have our understanding of Christianity (or even theism in general). How would the most fervent of Christians, used to witnessing and proselytizing, react? One would assume they would take the good news of Jesus, atonement, and redemption to these newly found alien persons. This would be predicated upon the universal redemptive qualities of Jesus' sacrifice. It would be similar to scenarios that have taken place on Earth with the discovery of new cultures in newfound geographical locations.

---

[1] Ibid., p. 291.

Vatican II, the Second Vatican Council, expressed this kind of universalism, even if not with ETI in mind:[1]

> Christ in His boundless love freely underwent His passion and death because of the sins of all men, so that all might attain salvation. It is, therefore, the duty of the church's preaching to proclaim the cross of Christ as the sign of God's all-embracing love and as the fountain from which every grace flows.

Some theologians see Jesus' incarnation as "cosmic in scope"[2] and "do not believe Christian theology can posit a multiplicity of Christs and remain Christian theology."[3] Jesuit L.C. McHugh once stated in an interview on the subject that ETIs "would fall under the universal dominion of Christ the King, just as we and even the angels do."[4]

Lecturer of philosophical theology, Joshua M. Moritz argues (in a redemptive concern for other animals) that the incarnation will work from non-human creatures on Earth, and as such, one can extend the incarnation out to have cosmic scope.[5]

Another voice to add to this chorus of single-incarnation defenders, though there are plenty of others, is New Testament scholar J. Edgar Bruns, who claimed that "the significance of Jesus Christ extends beyond our global limits. He is the foundation stone and apex of the universe and not merely the Savior of Adam's progeny."[6]

For such thinkers, the universality derives from the Logos. The Gospel of John starts "In the beginning was the Word, and the Word was with God, and the Word was God." This is an understanding of creation that later developed into trinitarian theology.

Brother Guy Consolmagno makes some bold claims in his piece "Would You Baptize an Extraterrestrial?":[7]

> Just how this "Word" might be "spoken" to the rest of the intelligent universe, I don't know. But it will be in "words" (that is, events)

---

[1] "Declaration on the Relationship of the Church to the Non-Christian Religions", *Nostra Aetate* 4, p. 667.

[2] Marty (2014), p. 3.

[3] Theologian Mark W. Worthing in his chapter "The Possibility of Extraterrestrial Intelligence as Theological Thought Experiment" in Kelly & Regan (2002), p. 83.

[4] McHugh (1969), p. 29.

[5] See Moritz's chapter "One Imago Dei and the Incarnation of the Eschatological Adam" in Peters (2018).

[6] Bruns (1960), p. 286.

[7] Consolmagno (2000).

appropriate to those beings. In any event, good extraterrestrials (ETs), just like good humans, do not need to know about Christ for salvation; that's the tradition of "baptism by desire."

The point there is that, even though the life of Jesus occurred at a specific space-time point, on a particular world line (to put it in general relativity terms), it also was an event that John's Gospel describes as occurring in the beginning-the one point that is simultaneous in all world lines, and so present in all time and in all space. Thus, there can only be one Incarnation-though various ET civilizations may or may not have experienced that Incarnation in the same way that Earth did....

ETs may not be aware of the idea of an Incarnation, or they may have their own experience of the matter. Their experience may be so alien from ours that even though they have experienced God in their own way, it's an experience that we will never be able to share, nor they share in our experience.

We are not sure that this is particularly coherent since it appears that Jesus' spatio-temporal existence on Earth can be the only one, and ETIs may or may not be able to experience it in the same way. Perhaps the bodily sacrifice of Jesus as a human might be, well, an "alien" concept, both materially and conceptually. Thus, it seems that Consolmagno is asserting that God would have to miraculously communicate this event and its meaning to all other instances of ETI life because "there can be only one" Jesus (and I guess Immortal, since yes, that is a *Highlander* quote).

It seems a stretch to assume that a completely alien event, as Jesus' life and sacrifice would be to all other ETI life, would have the same power and purchase for those alien societies. Imagine the scenario the other way around: that we were to receive stories of an alien divine incarnation, perhaps of Qaternix of Dev'argon'ii. Would that story have as much meaning and evoke as much empathy for humans as that of Jesus of Nazareth? We already have discussed issues of Jesus universalism on Earth alone with regard to the Problem of Particularity. We already struggle with the idea of Jesus being "whitewashed" in art to become an Arian of sorts, and there is difficulty in taking particular images of Jesus (as either white or Middle Eastern) and trying to sell that to other cultures. The idea that there is only one incarnation means that there is already a cultural, perceptual barrier between that incarnation and all those entities that are required to derive meaning from it.

Consolmagno's claims above essentially run into the famous philosophical problem of what it feels like to be a bat. We, as humans, cannot access those experiences. We empathize with other humans precisely because we are humans ourselves. Aliens trying to understand human behavior and empathize with the predicament of Jesus in Jerusalem might well be like us trying to empathize with being a bat.

This presents the scenario of God being unfair in this geocentric chauvinism. All other ETI lifeforms are at some kind of experiential and thus psycho-theological disadvantage. We are of the mind that a single earthly incarnation shows a problematic geocentrism that favors us over any other ETI life.

Ted Peters discussed how incarnation might suggest more of the idea that God "became flesh" (John 1:14a), and that incarnation is about God reaching into *material* existence, and not so much *human* existence—Christ is part of God's omnipresence still. His earthly life wasn't a show to be watched. It is that "God becoming flesh stretches to every nook and cranny of the physical universe, including plants and animals who do not actively or consciously share communion with God."[1] It is like God's incarnation becomes a figure of speech—"an ontological synecdoche"[2]—for the material world. This is something Peters and some other theologians call "deep incarnation" and would entail a "deep resurrection" that has implications for the whole natural world.

The problem is, we're not sure we buy this, and we're not sure how to make sense of Peters[3] claiming "If God's presence in the cross on Earth applies to off-Earth creatures, then God will be present in their struggles and sufferings as well."[4] This sounds nice, but what does it really *mean*? What does it *actually* mean to be present in our and all alien life's struggles?

Moreover, it means nothing to us unless we know that this is the case. Most humans don't even know this is the case (indeed, most humans have been born in times and places where the evidence for Christianity has been impossible, either strictly or practically[5]). So, unless God is playing out the human Jesus story to all other ETI civilizations like a galactic drive-in cinema, this sort of incarnation can only be *for the benefit of God*, and not all his creations.

In the same breath, Peters quotes University of Oxford theologian Celia Deane-Drummond, who claims that the resurrection "gives stronger account of a God who acts decisively to raise Christ from the dead, and as such, it offers the possibility of hope for all creaturely beings, not just humankind."[6] But, unless we have either multiple incarnations that are known to each ETI civilization, or a galactic theological movie or similar divine revelation, how

---

[1] Peters (2018), p. 293.

[2] Ibid.

[3] Ted Peters doesn't actually adhere to such a multiple incarnation belief entailing deep incarnation, though he discusses it. It is advocated later in his collection by Robert John Russell in his chapter "Many Incarnations or One?"

[4] Ibid., p. 295.

[5] See Pearce (2022), Arguments 10 and 11, concerning *divine hiddenness*.

[6] Deane-Drummond (2009), p. 179.

are all ETIs to understand or *even know* about this "hope for all creaturely beings"?

To continue to critique Peters and this idea, we can't see how one can make sense of the claim that "If by single incarnation we refer to the entire history of Jesus that includes death and resurrection, then we can think of the incarnation as a divine work that establishes within the physical world a promise of redemption from sin, evil, suffering, and even death."[1] How can a promise have substantive meaning for both parties if one party is not aware of it? Harry can promise to give Jane a free holiday if she is kinder to her children, but if Jane has no knowledge of this promise, then it will give Jane no incentive to change her behavior. It is a promise that only *means something* to Harry. It has no causal efficacy outside of Harry himself. God's redemptive mechanism only has meaning for God himself and no causal efficacy unless clearly communicated to all necessary entities.

This could only work if we could even make sense of an atonement and incarnation. If redemptive incarnation were to happen without our knowledge and without having any effect on our behavior, it leaves us questioning what the point of the promise of redemption would be. It might make God, or Harry, feel better, but other than that?

We wonder whether such writing and claims act as comforting soundbites that feel good on the ear, but when one digs a little deeper, they appear to be vacuous in nature. As a further example of this, Peters quotes George Coyne, SJ, the former director of the Vatican Observatory:[2]

> How could he be God and leave extraterrestrials in their sin? After all he was good to us. Why should he not be good to them? God chose a very specific way to redeem human beings. He sent his only Son, Jesus, to them and Jesus gave up his life so that human beings would be saved from their sin. Did God do this for extraterrestrials?... There is deeply embedded in Christian theology...the notion of the universality of God's redemption and even the notion that all creation, even the inanimate, participates in some way in his redemption.

What does "participates in some way in his redemption" *actually* mean? Again, it sounds nice, but it's otherwise entirely vague and unsatisfying. Coyne goes on to conclude that knowing whether God does or would incarnate "fully a human being" on multiple planets is difficult to fathom; we cannot derive the answer from Scriptures or churches. Rather, we can perhaps learn it from the sciences about which God has written in the "Book of Nature."[3]

---

[1] Peters (2018), p. 295.
[2] Coyne (2002), p.187, referenced in Peters (2018), p. 295.
[3] Coyne (2002), p. 187-88.

Given all of the arguments for and against a multiple incarnation, Ted Peters defends a single *soteriological* event[1] that fixes a broken creation for the universe, but where we should not infer a geocentrism or superiority from our place as the single location for incarnation in the universe. Instead, we should comport ourselves with the sort of humility we are shown in the Scriptures where "the last shall be the first." However, this is not so much about the benefit for humanity but for the negative connotations for everyone else in the universe. Peters sees that we should accept this paradox: "It would fit hermeneutically, so to speak, for we earthlings to live with the paradox that the cosmic event of salvation took place on our marginal and humble planet."[2]

But if this is a logically self-contradictory state of affairs (paradox), as properly understood, then something has to give. Simply put, a problem cannot be solved with a paradox; a paradox is what happens when you *don't* have a solution—especially if the paradox is currently an antimony—an impossible paradox (using influential philosopher W.V.O. Quine's categories of paradoxes). Peters might as well just say "it's a mystery," which is the honest assessment. Peters recognizes the difficulties here:[3]

Defending this position risks some unhappy repercussions. For example, I might get accused of geocentrism; because it appears that I grant our planet a specially chosen status. However, I do not wish to defend geocentrism, because I deem salvation to be an eschatological gift of divine grace for all of creation, all of the galaxies. What happened on Earth in the Jesus event was a prolepsis, and an anticipation of the cosmic wide transformation which the Jesus event promises. This is the case for the cosmos, I think, whether conscious beings realize its truth or not.

For Peters, there could well be other revelatory events throughout the universe for ETIs to behold, but the soteriological intervention of divine incarnation is a once-only procedure. This is how he seeks to evade claims of geocentrism.

But, again, we can't get away from the feeling that this would, invariably, entail the sort of anthropocentrism he denies.

The scenario here is very similar to the problem of evil whereby God cannot allow a single unit of *gratuitous* pain or suffering in the universe given his OmniGod qualities. In this soteriological and revelatory situation, there can be not one single unit of unfairness, if you will, for any civilization in the universe. If one civilization is, through this single incarnation on Earth, given

---

[1] *Soteriology* concerns the doctrine of salvation.
[2] Peters (2018), p. 297.
[3] Ibid.

less access to (or opportunity to access) God, knowledge of God, union with God, understanding, experience, and comprehension of God, or salvation, then God is being unfair.

It is not good enough to say that it will all be alright in the end as any material world injustices will be balanced in heaven, since this is compensation and not moral justification. The only way the theist can get that calculation to work is to employ a strict form of consequentialism, but this creates more problems than it solves as consequentialism is not favored at all by theists. The moral value system requires no god or gods for it to function, and so theists decry it. For example, William Lane Craig, famous apologist, theologian, and Christian philosopher, calls it a "terrible ethic."[1]

Without employing moral consequentialism, the theist, such as Ted Peters here, has to plead that the incarnation on Earth isn't that special. Literally, it could have happened anywhere, such as to the aforementioned Qaternix of Dev'argon'ii, and this (or in combination with other revelations) would have to have the exact same power and effect on *us* as it does to the *Dev'argon'iians*, and vice versa.

This seems to render the incarnation, life, and earthly atonement of Jesus as nothing special. What we mean by this, is that it cannot confer any advantage to humanity. We should not wish for Jesus to be born here because it holds no extra value. However, most any Christian, when asked, would say that the incarnation was special, probably the most important thing to happen in human history. If it is an intrinsically (theologically, philosophically, etc.) valuable (or useful, etc.) thing for humanity, then this would render God unfair and afford humanity special privilege.

The theology to escape this problem seems something of a series of *post hoc* rationalizations. Peters might want to keep Jesus Christ as a uniquely human entity, but he has to show that there is no benefit, no overall good, of Jesus being incarnated on Earth that puts us earthlings at an advantage. But the flipside is that if you can generate all the same benefits to ETIs without a Jesus incarnation for *them*, then this arguably renders the *actual* incarnation unnecessary. If you can generate all the benefits of the incarnation for everyone else by proxy, then it can *all* be done by proxy, and Jesus didn't need to be sacrificed at all. If every ETI civilization can suffice equally with humanity without actually having to have a sacrificed incarnation of God, then you can just add Earth to that list. *Everyone* could be played the divine movie and we wouldn't really need to have the suffering and death of a god-man.

---

[1] As he did in his 2013 debate with physicist Laurence Krauss in Brisbane, Australia: "Life, the Universe, and Nothing (I): Has Science Buried God?". A transcript can be found online at https://www.reasonablefaith.org/media/debates/life-the-universe-and-nothing-i-has-science-buried-god (accessed 06/11/2021).

Aliens and Religion: Where Two Worlds Collide

It's rather like how modern communities anywhere in the world now learn about, through books, movies, songs, and rituals, the story and theology of Jesus of Nazareth from a different earthly time and place. Of course, the situation (of differing levels of evidence and access to the correct god) we supposedly have on Earth is already unfair without even consideringly alien life. That Thomas was given the luxury of first knowing Jesus, then denying him with that knowledge and witness of miracles, and then being given the opportunity to touch Jesus' holey hands, is a privilege not accorded to any modern potential convert. Thomas is arguably a saint in heaven, and certainly an eventual believer, but he was afforded more evidence for his belief than any modern person can hope to be given.[1]

In other words, the people of that time and place who knew Jesus and saw his miracles were given unfair levels of evidence. This is something Pearce has discussed in his book *30 Arguments Against the Existence of "God", Heaven Hell, Satan, and Divine Design* and is another thorn in the side of Christianity.[2] Not only does it seem that the evidence and access to God is stacked unfairly in a single-incarnation-on-Earth universe, but it is stacked unfairly on Earth itself. Doubting Worf on the planet Boreth[3] gets no special knowledge of Jesus as Doubting Thomas did.

"Well, it has to happen *somewhere*."

But does it? That sacrifice is all about making sense to *God*. And quite why God would need a payment to settle the score for his own designed and broken creation is quite beyond us. But if all other aliens (since we would be aliens ourselves to others) can get value and meaning *exactly equally to us*, then we can imagine a situation where every civilization can get the same meaning, value, and understanding by proxy (as already mentioned) and so an incarnation *doesn't* have to happen somewhere. The only reason it appears it must do is to settle the books *for God*. This atonement reasoning makes little sense at any rate. *We* don't feel the need for the books to be balanced. Atonement after incarnation is not something that was developed for a need that we had did not have. Rather, it was the other way around. Jesus died rather unexpectedly (for a Jewish Messiah), with the Godhead as Holy Trinity and atonement theory only being developed *post hoc* to make sense of this death.

"God *must* have needed to pay for our sins and that is how we can understand the Messiah dying, and this makes more sense if the Messiah was actually God." These were the sorts of ruminations that took place for hundreds of years after the death of Jesus until some kind of half-coherent conceptual understanding of the data could be theorized. But even now,

---

[1] This is, of course, if you accept the face-value truth of the Gospel of John…
[2] Pearce (2022), Argument 11.
[3] *Star Trek: The Next Generation*, "Rightful Heir", Season 6, Episode 23 (aired May 17, 1993).

theologians disagree on both how the Holy Trinity works and how atonement works. Meanwhile, humanity continues living and sinning, and being good and dying.

The point is, atonement is all about God's desires and needs, and not about humanity's, no matter how theologians will try to dress it up. If we all "experienced" (through some kind of revelation) an "alien" incarnation, then there can be no value difference between that scenario and an incarnation taking place *somewhere*. But that somewhere could end up being a *nowhere*, as long as all intelligent creatures in the universe could understand the same thing by it.

Moreover, it's not like God wouldn't be paying his debts by not committing to an actual sacrifice, because it is supposed to be a vicarious sacrifice so *we* are the ones that somehow owe our designer for our corrupted creation. God wouldn't be defaulting on owed payments, because those payments are due *to him*.

Or, in other words, Jesus' sacrifice (to himself?) was arguably unnecessary—an example of gratuitous suffering. Rather than saying that "Well it had to happen somewhere,"[1] we can say it *didn't* have to happen *anywhere*.

If a singular atoning death and resurrection of Jesus is going to cause all sorts of unresolvable theological paradoxes, then we should consider the antithesis. What if there was more than one incarnation? Then all of these particularity paradoxes resolve themselves.

The aforementioned Robert John Russell offers some other arguments against a single incarnation, as we shall set out.

First, there is no way that humanity could communicate any revelation to other civilizations due to the vastness of space, as previously discussed. The ontological view of incarnation that Russell claims is involved in single incarnation "emphasizes the significance of the incarnation as the act in which God redeems the world from such physical laws and regularities as to produce the universal facts as sin and death." This has physical ramifications, which are important given how the speed of light and other physical constants will govern how things happen in the universe. He continues, saying, "Instead God's act begins the transformation of the world into the New Creation, including the change in these laws and regularities."[2] Only a multiplicity of incarnations makes sense of this:[3]

---

[1] Peters (2018), p. 298, says, "for something to be real everywhere it must be real somewhere. So, we might argue, any redemptive event cannot escape particularity, even if it bears universal significance."

[2] Russell in Peters (2018), p. 306.

[3] Ibid., p. 306-07.

[H]ow would the changes in the laws of physics, initiated on earth by God, "spread" throughout such cosmic distances? Would the change be spread at the speed of light? Then in an ever expanding universe (indeed, one in which the rate of expansion is accelerating) it might never reach the farthest parts of the universe.[1] Or would the changes occur "simultaneously" throughout the universe? If so, how is this universal change connected to and dependent on the unique initiating event of the incarnation only here on earth?

Russell sees the multiple incarnation view as "essential" for answering these questions.

Second, Russell agrees with us that ETIs would lose something in having to access a human narrative and incarnation, claiming, "I believe they must have their own access to the revelation of the incarnation and the dispensation of the Spirit in their own histories in species-specific ways and on their own planets."[2]

Third, there is the issue of probability, or "concern about absence."[3] The more widespread ETIs are in the universe, the less likely it is that the single incarnation actually took place on Earth. If there are trillions of ETI civilizations, then this probability becomes vanishingly small.

His following two arguments are ones positively in favor of multiple incarnations, and so we will save them for the next section.

Does a single incarnation communicate some uniqueness and specialness to humanity? The following acts as criticism against the single, earthly incarnation hypothesis. As the previously referenced scholar Ernan McMullin observes and asks:[4]

It was easier to accept the idea of God's becoming man when humans and their abode both held a unique place in the universe. But is it any longer credible in the light of the new questions that the plurality of inhabited worlds poses?

Indeed, not all theologians are on board with incarnation maintaining its significance in light of the discovery of ETI life. Theologian Norman

---

[1] In actual fact, there are distances so far away that, because of the expansion of the universe, light can never reach them. This is known as the Hubble volume, and its radius is more than ten billion lightyears. The expansion rate is also increasing, due to the mysterious dark energy, so the limits to what light signals from Earth can reach is decreasing. Given a few billion years, many more galaxies will be beyond our sight even with any future telescope designs.

[2] Ibid., p. 309.

[3] Ibid., p. 310.

[4] McMullin in Dick (2000), p. 165.

Pittenger opined, "How can the Christian gospel, concerned with the salvation of men in this world, have any universal significance when we know that there may well be intelligent life on other planets?"[1] Oxford biologist and theologian Arthur Peacocke agreed, thinking this dangerously close to rendering the theology nonsensical:[2]

> What can the cosmic significance possibly be of the localized, terrestrial event of the existence of the historical Jesus? Does not the mere possibility of extraterrestrial life render nonsensical all the superlative claims made by the Christian church about his significance?

Professor of physics and writer Paul Davies pulled no punches in a 2002 piece for *SETI*:[3]

> Theologians and ministers of religion take a relaxed view of the possibility of extraterrestrials. They do not regard the prospect of contact as threatening to their belief systems. However, they are being dishonest. All the major world religions are strongly geocentric, indeed homocentric. Christianity is particularly vulnerable because of the unique position of Jesus Christ as God incarnate. Christians believe that Christ died specifically to save humankind. He did not die to save little green men.

Although in the same piece, Davies sees ETI existence as more unlikely under atheism such that some kind of deistic hypothesis would better support ET life, the greater ire is directed at the theist. Davies' claims are strong here: Theologians and ministers who find no doctrinal and theological issue with ETIs are being dishonest.

Theoretical physicist and theologian John Polkinghorne suggests that there are surely "many sites in the universe suitable for the development of some form of life" and that the Word would apply just as much to "little green men" as to us.[4]

We would tend to agree with the position such as Davies's above as opposed to those of thinkers like Peters, who claim to find no issue with a single incarnation-on-Earth hypothesis. A single incarnation event that just so happens to have taken place here, in some parochial Middle Eastern province, is difficult enough to believe in, given its own host of particularities and theological issues previously discussed. And that's not to mention all the various historical, archaeological, and exegetical problems concerned with the

---

[1] Pittenger (1959), p 248.
[2] Peacocke (2000), p. 103.
[3] Paul Davies's piece "Transformations in Spirituality and Religion" in Tough (2000), p. 51.
[4] Polkinghorne (2004), p. 177.

biblical claims.[1] Indeed, to think that this would suffice for the whole universe, and potentially billions of ET civilizations, is stretching credulity.

Λ Trillion Planet-Hopping or Simultaneous Jesuses

Whether it be that all aliens are fallen, or whether it be that incarnation is God's preferred way of interacting with sentient intelligent lifeforms irrespective of their moral situation, the alternative to the single incarnation theory is one of multiple Jesuses. And multiple means anywhere from two to potentially an infinite number.

That's a lot of Jesuses.

If God is love—infinite love—it would seem more reflective of this characteristic if God was bestowing his salvific generosity on as many ETI civilizations as divinely possible. To restrict an incarnated divinity to one instance alone seems incredibly miserly and runs contrary to what we might predict, at least intuitively so, since it perhaps runs contrary to God's omnipotence. As long as there is the right type of intelligence to grasp the Word of God, says Franciscan Sister and theologian Ilia Delio, there will be "[i]ncarnation on an extraterrestrial level.... Many incarnations but one Christ."[2]

If incarnation is, on balance, a good thing, then surely OmniGod would want as much of this good thing as possible: He would want as many incarnations as possible! As Catholic theologian Peter M.J. Hess asks, "If the purpose of creating is to bring that creation into communion with God, why would God not use any means at the divine disposal?"[3] However, taking this logic to its conclusion, this would oblige an all-loving God to create a world with infinite incarnations and/or infinite worlds with single, or indeed multiple incarnations.

After reflecting the theological split among and within various denominations in his chapter "Many Incarnations or One?" in *Astrotheology: Science and Theology Meet Extraterrestrial Life*, theologian Robert John Russell opines (referring to the work of other scholars):[4]

---

[1] See analysis of the Nativity in Pearce (2012) and the Resurrection in Pearce (2021), and a forthcoming volume historically, exegetically, and theologically critiquing the Exodus accounts.

[2] Delio (2012), p. 169.

[3] Hess in Peters (2018), p. 326.

[4] Russell in Peters (2018), p. 307-08.

On the one hand, I believe it would be incoherent to argue that we humans can encounter the Good News without there being any actual, ontological basis for it, and yet on the other hand I believe that this ontological basis must be manifest (Tillich), appear (Pannenberg), and be received (Niebuhr) by people reflecting the specificities and diversity of human history. I therefore start with the revelational view of the incarnation and argue that this lends support for multiple cosmic incarnations.

... In sum, if Peters is correct, then from an ontological perspective a single incarnation of the divine logos in Jesus of Nazareth is sufficient for the redemption of the universe. A unique incarnation on earth can ontologically alter the entire universe and the laws of nature which describe natural processes. But if I am correct, from a revelatory perspective our participation by faith in the Resurrection requires that this revelation be based on an ontological act of redemption by God and that it be known to all species needing redemption. In this perspective a single incarnation is insufficient for the redemption of the universe because the participation of ETI will be impossible. Instead multiple incarnations are required.

This position reflects what we have already discussed: That if human Christians derived some extra value (understanding, etc.) in the participation in the redemption process, then this would need to be accorded to all ETIs.

As well as setting out arguments against single incarnation, Russell presents a couple of further arguments for multiple incarnations.

First, he details an issue concerning the "grace of God as God's self-communication to God's creatures." This is essentially the approach that God would not want all of the ETIs in the universe to be without the knowledge and grace of God. God would want union and fellowship with all of his able creations. This leans into the previous argument about the idea that such interaction with humanity on Earth in this "particular fellowship through the Son of God"[1] would entail some kind of unfair privilege for humanity.

Whilst we have detailed other arguments in the previous section, Russell's final argument is one about eschatology (end times), whereby "life on every planet will experience this eschatological transformation." There are three aspects to how intelligent life is "blessed": incarnation, resurrection, eschaton. Russell sees multiple incarnations as theologically supporting these ideas.[2]

This is a pluralism of a different sort that we might see on Earth whereby one could see a sort of single religious truth in the plurality of the varied Earthly religions. Instead, this pluralism stretches the length and

---

[1] Ibid., p. 311.
[2] Russell in Peters (2018), p. 312.

Aliens and Religion: Where Two Worlds Collide

breadth of the universe (if such dimensions make physical sense here), encompassing all known ETIs.

But here we come across something of a problem, and one to which Russell somewhat alludes. In arguing that God must allow all ETI species across the universe access to such a salvific incarnation, then we should be able to apply that argument to Earth across history and geography.

Which leads us onto some atheistic arguments presented in Pearce (2022) in his chapters "The Accidents of Geography and History" and "Unfair Levels of Evidence." In a sense, these are forms of divine hiddenness arguments and ones that concern issues of God's lack of fairness.

The problem is that over time and place, humans have not had access to God (here, particularly, the "right god"). Amazonian tribespeople in 3000 BCE will not have had access to the incarnation and resurrection, and its connected theology, of which we have been talking. Some people on Earth have had precisely no evidence for, and access to, God, while others have been given confounding "evidence." We might think of someone born in Riyadh, Saudi Arabia in the 1990s who would almost certainly be destined to become a Muslim on account of the accident of their birth.

There are, and this is an indisputable fact, distinct geographical concentrations of religions around the world. Christians may be concentrated in Europe, South and North America, and other pockets of colonial history. Islam prevails in the Middle East, North Africa, parts of subcontinental India, and the islands of the Indian Ocean. Shintoists primarily exist in Japan, and Hindus in India. These and other statistical generalizations and trends exist.

If God were perfectly loving, then he would ensure that all people could participate in a relationship with him unless people had somehow deliberately excluded themselves through some kind of resistance. Some philosophers call the people who under other circumstances would have happily believed in the right god given the right context *nonresistant nonbelievers*. They don't believe in the right god not because they are wholly against the idea of that god but because they just weren't lucky enough to be in the right geographical and/or historical context.

Moreover, in a Judeo-Christian context, some people are afforded an unequal amount and access to evidence and thus unequal access to God. Take, for example, the account we have already discussed, found in the Gospel of John. God saw fit to convince Doubting Thomas, who—after all—knew Jesus and saw him do his miracles. He was a disciple—not just any disciple, but one in Jesus' inner circle. And yet *even he* didn't initially believe in the Resurrection, attested to by his friends and eyewitnesses, until he had Jesus standing in front of him and until Jesus made him touch the wounds.

As John 20 relays:

[24] But Thomas, one of the twelve, who was called Didymus, was not with them when Jesus came. [25] So the other disciples were saying to him, "We have seen the Lord!" But he said to them, "Unless I see in His hands the imprint of the nails, and put my finger into the place of the nails, and put my hand into His side, I will not believe."

[26] Eight days later His disciples were again inside, and Thomas was with them. Jesus came, the doors having been shut, and stood in their midst and said, "Peace be to you." [27] Then He said to Thomas, "Place your finger here, and see My hands; and take your hand and put it into My side; and do not continue in disbelief, but be a believer." [28] Thomas answered and said to Him, "My Lord and my God!" [29] Jesus said to him, "Because you have seen Me, have you now believed? Blessed are they who did not see, and yet believed."

Thomas got to touch and prod Jesus who was bodily resurrected in front of him. He got to feel the skin of the real and resurrected God, and only then did he believe.[1] And yet almost the entirety of the rest of humanity is not remotely afforded this level of evidence and is expected to believe, arguably on pain of hell.

These sorts of examples show how God is inherently unfair.

As such, there seems to be some disconnect in arguing that it is more likely that all intelligent life in the universe should have full access to union with God through a redemptive incarnation, while observing that this access has not been *and often is not possible* on Earth itself. This problem is accentuated by further, perhaps infinite, instances of it.

Added issues with multiple incarnations as an idea revolve around a single mind needing to be split over perhaps billions of entities simultaneously. In many understandings of the Holy Trinity, the human properties of Jesus are "assigned to and even given over to [Christ's] divine nature."[2]

In Catholic tradition, there is the explicit denial of this possibility by Thomas Aquinas. In his work refuting non-Christian philosophies, Aquinas states that the number of souls is equal to the number of physical bodies, and it is impossible for one soul to be united to many bodies.[3] This is quite the theological wall to run into, if one is hoping to remain consistent with Christian (let alone Catholic) teachings. To move forward with any proposal of Jesus incarnating into multiple bodies, then one would be considered a heretic by the world's largest Christian denomination.

---

[1] Made all the more controversial because Jesus declares that "Blessed are they who did not see, and yet believed" (John 20:29) – these people are to be congratulated for their belief.
[2] Russell in Peters (2018), p. 314.
[3] Thomas Aquinas, *Summa contra Gentiles* 2.83.37.

Aliens and Religion: Where Two Worlds Collide

And yet, there are efforts to try to square the circle. McIntosh and McNabb seem to stretch credulity somewhat in their appeal to multiple personality disorder (MPD—or DID, dissociative identity disorder) in explaining how Jesus could be incarnated in multiple (and simultaneous) times and places. MPD is sometimes used to explain "how Christ could be of two minds, divine and human, in the incarnation."[1] As they claim:[2]

> Alters have been said to have their own thoughts, sense of humor, beliefs, feelings, skills, memories, mannerisms, fears, voice quality, visual acuity, and even their own tolerance to medication and allergic responses. We find MPD just as serviceable, if not more, in illuminating how Christ could be of more than two minds, one divine and indefinitely many other non-divine, in the case of multiple incarnations. Of course, we hasten to highlight the obvious point of disanalogy, namely, that such a "condition" would not be for the omniscient creator a disorder, but wholly consistent with cognitive perfection. We thus agree with Robin Collins that "God's overall consciousness would not in any way be diminished even if God took on an infinite number of finite mental systems, from an infinite number of fallen races; in fact, if anything, it would enhance God's consciousness."[3]

That with this condition, humans can have functional "alters" (alternate personalities) seems at least intuitively implausible. We don't want this to be *an argument from incredulity* such that we are claiming the theory can't work just because we can't get our limited imaginations around it. However, the idea that Christ could be incarnated billions of times and have, in each separate instance, a functionally operating mind that is simultaneously both independent and one-and-the-same appears to be theists having their cake and eating it, too.

However, this is something of a false analogy or model.

In DID, the person goes back and forth between persistent personalities, but when going between them, there are gaps in the memories of the personalities. When personality A is out, personality B is not expressed and generates no memories. So, if Jesus is in some sort of DID, then only one personality can exist at a time. But in the multiple incarnation scenario, there are potentially billions of simultaneous Jesuses.

To say that Jesus can have DID but not like DID is to just say Jesus can have multiple locations and personalities at once just because McIntosh and McNabb say so! They are having their cake and eating it, too.

---

[1] McIntosh & McNabb (2021), p. 8.
[2] Ibid.
[3] Collins (2015), p. 218-219.

168

But what's better than one cake? A multiuniversal cake, of course!

McIntosh and McNabb, referring to Christian philosopher Alvin Plantinga, suggest that if the Gospel story and incarnation is good, as hinted above, then it will take place in various possible worlds.[1] The idea is that possible worlds with the Fall, incarnation, and atonement are better than possible worlds without them. Therefore, there would be multiple Gospel events and stories *across* possible worlds, and not just *within this one*.

So now God can not only have potentially billions of simultaneous minds operating across a single universe or world, but God can exist as separate-but-unified minds across multiple universes or worlds! And yet, as Catholic theologian Peter M.J. Hess emphasizes, "There is only one spirit and one Christ no matter how many times the Cosmic Christ might become incarnate throughout the universe."[2] This sounds nice, but really how does this work? Is this just divine logical, psychological, and theological *magic*?

The end result of such theologizing is that God ends up magically being able to be whatever the theist wants God to be, irrespective of how incoherent the idea or move is. If the theist wants God to have a billion separate minds that are, at the same time, one and not separate, then voilà, they can have that. God is not what is plausible but what is conceptually conceivable without necessarily trying hard to work out how it might not be workable.

Of course, the theist will argue (rather like with skeptical theism) that such a thing is *conceivable*, and just because it might seem *intuitively* implausible, it doesn't mean that it is *actually* impossible.

Nonetheless, let's put this mere possibility into previous generations of theological thought. Thomas Aquinas rejected[3] the idea that all humans could be God incarnate in some universal incarnation because it would result in the end of the human race but also that it would take away from the glory of God. If multiplying the instances of incarnation devalues the glory of God in this context, then we could propose that this devaluation of glory would happen in the scenario of multiple incarnations across the universe (trans-universe) or universes (or inter-universe).

If we return to the Council of Chalcedon in the 5th-century CE where the idea of the Trinity was ratified, Jesus of Nazareth is proposed to be personally identical with the second person of the Trinity. But this second person is thus also personally *identical* with all other incarnation entities. Christian philosopher Christian Weidemann outright rejects the position of McIntosh and McNabb (and by extension Plantinga) above:[4]

---

[1] McIntosh & McNabb (2021), p. 9 and Plantinga (2011), p. 59.
[2] Hess in Peters (2018), p. 326.
[3] In *Summa Theologica*, III, Question 4, Article 5.
[4] Weidemann (2016).

That members of completely different biological species ("truly man", "truly Alpha Centaurian") are supposed to stand in a relationship of *diachronic* personal identity is already something most people will find too hard to stomach. Even worse, if the number of sinful species in the universe exceeds a certain threshold, God (the second person of the Trinity) would be forced to incarnate himself simultaneously. (The purported difficulties that the Special Theory of Relativity poses to the notion of cosmic simultaneity will be ignored here). For instance, given, say, 100 billion extraterrestrial civilizations in need of salvation so far, an average incarnation time of 30 years, and roughly 12 billion years since the first emergence of habitable planets, God's incarnations would on average have had to live at 250 places simultaneously, in order to accommodate all sinful civilizations! If these numbers strike you as far-fetched, reflect on the more than 100 billion stars in our galaxy, on the more than 100 billion galaxies in the visible universe and on the idea of an enormously (perhaps infinitely) big multiverse, advocated by many prominent cosmologists.

It is obvious, I think, that no matter which reasonably serious position in the philosophy of mind one may adopt, the conclusion will always be the same. No single person who is an embodied being, a "truly" biological organism, can be more than one such being at the same time. The same corporeal being simply cannot exist at many distant places simultaneously

Weidemann, in his paper, argues that this leaves the Christian believer with a two-horned dilemma: they either have to drop some aspect of doctrine, or the pervasiveness of ETI. You cannot have widespread ETI *and* the received doctrinal theology. Perhaps Jesus is not "fully man" or not "fully God."

Theologian Robert John Russell, whom we have been discussing at length in this section, espouses the previously mentioned "deep incarnation" theology to rationally substantiate a multiple incarnation theology. This is derived from the Gospel of John and "the Word became flesh." Although we would make accusations that even though such thought might sound good, the ultimate meaning is unclear. Russell sees this phrase showing that the "incarnation embraces, and extends to, all life on Earth" whereby he seeks to extend this idea "to a higher level of generalization which allows me to include the flesh of all life in the universe."[1] This flesh that embodies the Word has commonalities that traverse the universe so that, he sees, there is a "'many-in-one' incarnation of the *Logos*."[2]

---

[1] Russell in Peters (2018), p 315.
[2] Ibid.

Again, apologists and theologians favor ideas that sound good but that are ultimately vague, lacking in theological clarity. Nonetheless, let's interrogate the idea that Christ's sacrifice extended to "all life on Earth" and thus extendable to the universe. By "all life," does Russell mean every sort of plant, animal, fungus, and single-celled organism? If so, then apparently the *E. coli* bacterium was destined to cell Hell and needed salvation. Moreover, if Jesus' incarnation could extend to "a higher level of generalization," then this prompts the question: Why did Jesus have to come in human form? Apparently having any sort of "flesh" is good enough for enacting the plan of salvation, so could Jesus not have incarnated into the simplest life form, die, and be resurrected without having the same sort of suffering he would have as a human? If the incarnation in lowly humans can extent to a "higher level," why would that not be true for single-celled Christs?

In response, one might cite Aquinas, who argued that that irrational creatures could not have been suitable for incarnation,[1] but that was premised on Aquinas arguing that such low creatures could not have union in God—the opposite of what Russell and his "deep incarnation" says. Moreover, Aquinas' argument for why Jesus had to take on human nature rather than any other is because it is the closest to the divine image, which will run counter to the Copernican principle. If there are aliens, some will be even closer to the image of God than humans, so *those creatures* should have had the single incarnation, and not a human. For Jesus to have incarnated as a human is another example of human chauvinism or theological geocentricism, which is what astrotheologians are trying to avoid.

Once again, trying to make sense of the single incarnation requires some previous theological conclusions to be left aside. "Deep incarnation" allows for Jesus to have incarnated into *any* living creature, so why human form is just a mystery, and a strange one—Jesus could have incarnated as a senseless fern or bacterium and not have needed to suffer so much while still achieving the exact same results. For "deep incarnation" to work, it follows that all living things can obtain salvation, and this is contrary to Aquinas' conclusions. Conversely, if one holds to Aquinas' analysis, then one would expect some other, higher alien species to have had the incarnation event instead of us.[2]

Many such dilemmas exist with the single incarnation atonement theory. Then again, there were serious problems for multiple incarnations. There is not just one dilemma a (single or multiple), but dilemmas within dilemmas.

The theological arguments for single incarnation atonement act as implicit criticisms of any multiple incarnation account.

---

[1] Thomas Aquinas, *Summa Theologica* 3.4.1, reply to objection 2.
[2] More issues with Deep Incarnation are discussed in Nesteruk (2023).

Aliens and Religion: Where Two Worlds Collide

A multiple-Jesus account of reality that runs into the trillions is no easy pill to swallow for theists and plays fast and loose with intuition.

Yet an unintuitive theology may be the least of the problems that the multi-incarnating Jesus would have. Multiplying the incarnation is really another way of saying Jesus is reincarnated (literally, made incarnate again). Reincarnation (or otherwise called the transmigration of souls) was explicitly called out as impossible by numerous early Christian figures, including Origen, who was denying that John the Baptist was the reincarnation of Elijah.[1] Similarly, reincarnation was also refuted by Aquinas.[2] In order to hold to the possibility of Jesus reincarnating billions of time, the theologian will need to rebuild a lot of pre-established theology, as she is running up against apposition, widely-repudiated both by older Church authorities as well as modern theologians, such as Paul Tillich.[3]

Besides the highly unintuitive and unorthodox idea of Jesus' personhood existing in more than one place at a time, there is also a problem with the atonement by God's Son if atonement is required on every alien world. The sacrifice of Jesus was supposed to be a unique and singular event to provide forgiveness of sins. This is a point indicated multiple times in the New Testament (Heb 9:26-28, 1 Pet 3:18). In Romans 6:9-10, Paul states that Jesus will never die again because he has conquered death. How, then, could Jesus have traveled to another galaxy, died, and needed to be resurrected again?[4] This would also undermine the power of Jesus' sacrifice; if Jesus can't save everyone with his death, then his suffering was not sufficient at Cavalry. This undermines the idea that even we humans are saved—if salvation is species-limited, then could it not be even more limited still?

Let us explain. Think of it another way: Is there a distance limit to salvation? Can Jesus only save within 1000 kilometers of the cross? We suspect most Christians would say no. If Jesus' sacrifice is supposed to have the power Christians allege, it must be for all people, since Jesus came to save everyone, not everyone up to some special barrier. With no distance limitation, then the entire universe can be included in the salvation act. Conversely, are there genetic limits on salvation? If aliens do not participate in the death of the human Jesus, even though they have the *imago Dei,* then the reason they are not saved is genealogical—they aren't "children of Adam," a restriction based on a mythological character. A genetic-based limitation on salvation seems just as arbitrary as a distance-based limitation. Such a view would also allow for

---

[1] i.e., Irenaeus, *Against Heresies* 2.33.1-2; Tertullian, *Apology* 48; Origen, *Commentary on Matthew* 10.20, 11.17, 13.1.

[2] George (1996).

[3] Tillich (1976 [1963]), p. 416-417.

[4] This seems to have been an issue with the idea of many inhabited worlds in the first place, as expressed in 1549 by Philipp Melanchthon. See Martinez (2016), p. 372.

terrible racist views to gain traction with theological accoutrements. If certain "kinds" of people are not saved by Jesus' sacrifice, then everyone has to ask how they know they are the "kind" of person that can be saved. Putting limits to who Jesus saves fundamentally undermines the universal gift of salvation.

Paul makes clear the finality and immensity of what Jesus was supposed to have accomplished, but the thesis of multiple incarnations and sacrifices does away with it all. If Jesus must die many times, he is not the master over death (contra Rom 6:9). If Jesus can die again, then he is not imperishable, and thus the promise of an imperishable body for us is seriously undermined (cf. 1 Cor 15:42, 50). If Jesus' death could not save everyone, then Jesus (and hence OmniGod as Jesus) is not all-powerful but strangely limited.

Λ Conclusion on Doctrine

The initial challenge for Christians when considering the discovery of ETI is the intrinsic theological coherence of the existing doctrine itself. The first hurdle is a teleological one involving the design and creation of a broken model that needs fixing. If that is an example of poor design and creation once in humanity, then multiplying that perhaps endlessly across the universe arguably shows God to be even less competent.

Arguments concerning the unfair levels of evidence and access to the knowledge of God and God's love also seem to be compounded given intelligent alien life—or at least equally problematic for those lifeforms.

We believe that the Holy Trinity and atonement make little philosophical sense and the discovery of ETI would only offer to compound these issues. Theologians *still* wrangle and disagree over these core Christian tenets; finding intelligent alien life would promote an even greater set of debates about these doctrinal issues.

But, for the sake of charity, if we were to grant that the Trinity and atonement work (even if we do the Christian move and punt to mysterianism), then the discovery of ETI still presents problems for doctrine. First, we must question whether the story of Adam is representative of all ETIs, and if so, this would then render God an even more incompetent designer than we would otherwise have thought, as mentioned.

Yet if God knowingly designed and created a faulty Adam for a *greater good* (thus employing moral consequentialism and using agents instrumentally, a big problem for Christian moral philosophy), then we could imagine that God would be obliged, as being all-loving, to extend that goodness across the universe. In other words, all ETI lifeforms would be made in the image of God, but also fallen moral creatures such that they could then bring about this greater good. The universe would then be full of instances of morally fallen intelligent life in need of salvation.

There are issues with what *imago Dei* really signifies and whether a much greater, more intelligent alien being (given the principle of mediocrity) would invalidate our having that label. If God's incarnation as a second Adam in human form is the most optimal way of achieving redemption for humanity, then it seems logically consistent to argue that this should be the primary (though equivalent) form of salvation for all ETI life.

However, such a scenario brings about perhaps billions or trillions of Jesuses, incarnated in a myriad different alien forms, and many simultaneously.

Furthermore, then can be no specialness afforded to humanity—no Earth chauvinism—that entails humans get greater value or greater access to God and God's love. If this were to be the case, God would be seen to be unfair.

Although a number of theologians favor a single incarnation event to work as a soteriological Band-Aid for all of the universe's ETIs, we think that this struggles to get away from Earth chauvinism and would show God to be unfair.

As such, if the Christian has jumped through all the hoops they need to in order to get this far, then they would have to take on the principle of multiple incarnation. And, as mentioned, this could entail trillions of Jesuses. Or more. This idea plays merry havoc with theories of mind, and we think that something like the psychological condition of multiple personality disorder is too much of a stretch to explain the problem away. To be honest, this is reminiscent of one of the problems with trinitarian theology in trying to explain how God can be made of three separate-yet-different entities.

Further problems for the multiple incarnation theory come from biblical excerpts and exegesis, meaning that both theories of incarnation fall victim to robust criticism in the light of the existence of ETI.

Essentially, we agree with all of the criticisms that single-incarnation theologians and philosophers level at multiple-incarnation advocates, as well as agreeing with all of the criticisms passing the other way.

The discovery of ETI in the universe, then, not only presents new problems for theology and theologians, but expands fissures that are already evident in the edifice of Christianity to create chasms.

11 – CHALLENGING CHRISTIAN TRADITION

On initial thoughts, this looked to be one of the shortest chapters in the book, reflecting the short shrift that McIntosh and McNabb give it in their paper. The question is whether the discovery of ETI would be in conflict with Christian tradition, and we found that there was more to take issue with in the source material than we had first imagined.

Let us start by defining our terms. There are two ways of looking at Christian tradition, with one being a very generalized notion of what the religious movement has taught or believed, and with the other being the collection of practices, rituals, beliefs, and catechisms developed by (a) religion as a whole or by individual churches and movements. We will adopt a more generalized understanding.

McIntosh and McNabb "find in the Christian tradition a remarkable openness, and in some cases outright enthusiasm, about the possibility that there be ETI."[1] They substantiate this belief in the following way:

> One of the earliest known examples is the fifteenth century Catholic Cardinal Nicholas of Cusa, who, in his treatise *De Docta Ignorantia*, entertains with no theological difficulty the idea that in every "stellar region" among the uncountably many stars there are "inhabitants, different in nature by rank and all owing their origin to God." Others, in the spirit of the Medieval dictum *bonum est diffusivum sui,* argued that God's goodness implies that He create an infinite number of worlds with creatures in them.

Of course, this entails an unhealthy dose of cherry picking. For instance, in 1600, the Italian philosopher Giordano Bruno was burned at the stake in Rome for being a heretic. Part of the reason he was executed concerned his rather more accurate beliefs about cosmology than those of his peers (indeed, than Copernicus and Kepler themselves). Bruno claimed that the universe had no center and that stars were actually suns that had planets and moons surrounding them. Yet many Christians deny that he was put to death on account of such beliefs.

Take the Catholic Encyclopedia, which claims "Bruno was not condemned for his defense of the Copernican system of astronomy, nor for his

---

[1] McIntosh & McNabb (2021), p. 10.

Aliens and Religion: Where Two Worlds Collide

doctrine of the plurality of inhabited worlds."[1] Then there is historian Frances Yates who stated that "the legend that Bruno was prosecuted as a philosophical thinker, was burned for his daring views on innumerable worlds or on the movement of the Earth, can no longer stand."[2] Furthermore, fellow historian Michael Crowe dismissed the "myth that Giordano Bruno was martyred for his pluralistic convictions."[3]

In *Scientific American*, however, Alberto A. Martinez (professor of history of science at the University of Texas at Austin) expertly debunks these claims in his article "Was Giordano Bruno Burned at the Stake for Believing in Exoplanets?":[4]

> But when I examined old treatises on heresies and canon law, I learned otherwise. In fact, in the 1590s Bruno's claim *was* considered heretical. Many authorities denounced it, including theologians, jurists, bishops, one emperor, three popes, five Church Fathers and nine saints. In 384 A.D. the belief in many worlds was categorized as heretical by Philaster, Bishop of Brescia, in his *Book on Heresies*. This condemnation was echoed by subsequent authorities, including Saints Jerome, Augustine and Isidore.
>
> Moreover, it was heretical according to the highest authority. In 1582 and 1591, Pope Gregory XIII's official *Corpus of Canon Law* included this heresy: "having the opinion of innumerable worlds." The *Canon* embodied the laws of the Catholic Church: all inquisitorial and church courts obeyed it.

Indeed, in analyzing all of the accusations against Bruno,[5] Martinez found that "the Inquisition's strongest case against Bruno was, in fact, and contrary to the conventional wisdom, his belief in many worlds. It was the most frequently recurring charge." The inquisitors even wrote, "Again, he posits many worlds, many suns, necessarily containing similar things in kind and in species as in this world, and even men." Some two decades later, the same inquisitor would go on to confront Galileo.

Martinez elsewhere provides a much fuller survey of pre-17th century Christian scholars who clearly condemned the idea of there being multiple worlds.[6] Over thirty such examples are recorded just in the Catholic tradition,

---

[1] "Giordano Bruno", *The Catholic Encyclopedia*, https://www.newadvent.org/cathen/03016a.htm (Accessed 09/06/2022).
[2] Yates (2001 [1964]), p. 355.
[3] Crowe (2011), p. 8.
[4] Martinez (2018); see also Martinez (2016).
[5] Firpo (1993), p. 247-304.
[6] Martinez (2016), p. 358.

while several Protestants also agreed in calling the view that there are other worlds with life on them either nonsense or full-blown heresies. Those like Bruno were the exception, and he paid dearly for being exceptional.

Stepping back to the cited author endorsing ET's existence, further context undermines the point that was being advanced by McIntosh and McNabb. The medieval philosopher Nicholas of Cusa did indeed suggest the possibility that there are inhabitants in the "stellar regions" that were unknown to us, but he *also* goes on to tell us that these will be creatures of less intellectual ability than humans, not more.[1]

> Now, even if inhabitants of another kind should exist in the other stars, it seems inconceivable that, in the line of nature, anything more noble and perfect could be found than the intellectual nature that exists here on this earth and its region. The fact is that man has no longing for any other nature but desires only to be perfect in his own.

While he does not directly state it, Nicholas seems to say that only humans have the *imago Dei* and not any creatures among the stars; there may be ET, but no real *ETI*. This then looks like McIntosh and McNabb are not being genuine or thorough with their source material when making such claims.

One must also note that Nicolas does not say that there *are* ETs, but just that it is *possible*. Saying something is possible, and asserting it is true, are two different things, especially given what Pope Gregory XIII would say: "having the opinion of innumerable worlds" was a heresy.[2] The same phrase was used by another Nicholas, the Inquisitor General Nicholas Eymerich, writing before the more famous Nicholas of Cusa, so we can understand why the philosopher couched his writing in possibilities rather than opinions. Bruno, unlike Nicholas, positively asserted there were countless worlds with ETIs, and he was executed for that belief.

So much for an alleged, time-honored tradition of supporting the existence of intelligent aliens. There is at most a few mealy-mouthed *possibiliters* for non-intelligent aliens, while there was strict condemnation of promotors of actual intelligent creatures in the heavens. The beliefs in an acentric universe with innumerable planets and suns, with the potential for some to resemble our own *inhabited world*, were in direct conflict with Christian tradition, and Bruno suffered the ultimate consequence for holding to those very beliefs.

---

[1] Nicolas of Cusa, *De Docta Ignorantia* 2.12. Translation by Heron (1954), p. 115.
[2] Martinez (2016), p. 357.

Aliens and Religion: Where Two Worlds Collide

Which is to say that the idea of ETI in the universe, as well as many other givens in today's understanding of cosmology, *definitely were* in conflict with Christian tradition.

Today, it is generally not Christian tradition to burn people at the stake for believing in what science quite clearly shows us. The Catholic Church and other Christian denominations have, in the main, adapted to scientific knowledge. That said, there are still young Earth creationists aplenty, doctrinally imposed homophobia, and nefarious persecutions around the world for people who do not adhere to the more "traditional" views of local or national churches.

It would be rather difficult for a geocentric or Earth chauvinistic religious group to argue with the actuality of an alien visit, or with uncompromising evidence for ETI life. Sure, humans are great at burying their heads in the sand, but reality has a habit of catching up with us. At least, that's what we, as skeptics, hope for!

There is enough plurality of belief—given the modern social-moral framework in which this book is set—to think that ETI discovery wouldn't too greatly affect Christian tradition in the long run. The traditions already survived the Copernican revolution, so another sort of Copernicanism isn't likely to be the end.

In the short term, however, there would almost certainly be a period of adjustment for everyone, but most certainly for those whose traditions have been built on a bedrock of human uniqueness, and the understanding of the problem of particularity as it pertains to the tradition of the Jesus story.

In fact, Pope Francis has been quite explicit on the matter:[1]

"That was unthinkable. If - for example - tomorrow an expedition of Martians came, and some of them came to us, here... Martians, right? Green, with that long nose and big ears, just like children paint them... And one says, 'But I want to be baptized!' What would happen?

"When the Lord shows us the way, who are we to say, 'No, Lord, it is not prudent! No, let's do it this way'... Who are we to close doors? In the early Church, even today, there is the ministry of the ostiary [usher]. And what did the ostiary do? He opened the door, received the people, allowed them to pass. But it was never the ministry of the closed door, never."

So rather than burn such thinkers at the stake, it appears that the Catholic Church is itself willing to elect a Pope who is one to take on such previously heretical thought.

---

[1] Ohlheiser (2014).

Given how much society had changed, and the massive proliferation of freethought and science fiction content, we would tend to agree with McIntosh and McNabb, though emphasize that there will be many in various fundamentalist communities who already find it easy enough to deny basic science (the age of the Earth, evolution, etc.) who would find their own Christian tradition in direct conflict with the discovery of ETI life.

To be fair, McIntosh and McNabb admit this themselves, saying "We speculate that this stance would be surprising only to those who have an unduly and often dogmatically narrow conception of Christianity, especially the relationship between faith and science."[1]

So the discussion probably revolves around the question of "Whose Christian tradition are we talking about?" While McIntosh and McNabb's iteration of Christian tradition may not be threatened, there will be many in the Christian world who will feel differently.

---

[1] McIntosh & McNabb (2021), p. 10.

12 – EXACERBATING THE PROBLEM OF EVIL

Perhaps the best-known and most forceful argument against OmniGod is the problem of evil, which might more accurately be called the problem of unnecessary suffering. Since at least Epicurus formulated the issue, this problem has vexed believers, and particularly those who adhere to some notion of OmniGod. The vast quantities of pain, suffering, and death, all without any clear and apparent good coming from it, has made it hard to believe all things are under the control of a potent force for good in the universe.

The first thing to note is that the theist who might be reading this should not fall into the *tu quoque* (you too) fallacy trap. Which is to say, they should not accuse the skeptic of not being able to justify suffering, evil, or even good. These problem of evil arguments stand or fall irrespective of the existence of skeptics and atheists. Imagine, theistic reader, that atheists simply do not exist. These problems will still stand. If God is all-knowing, all-powerful, and all-loving, then God should know what to do about suffering, should be able to do something, and should care enough to want to do something, So why the suffering?

There are numerous attempts to solve this theological problem and they are so well known in the philosophy of religion that the explanatory mechanism gets its own fancy term—a *theodicy*. These act as defenses of a good God in a world of evil. They usually amount to the position that there is some reason for such evils (suffering), such as to bring about some greater good. Under skeptical theism (the idea that we *do not* or *cannot* know the mind of God, that suffering is a mystery), we cannot know what that reason is; *but there is one*, believe us! More positive cases, such as developing souls or allowing for free will, are very common in the debates between atheists and theists, both online and in academic papers.

It is worth also noting that there are two different forms of the problem of evil argument: the *logical* problem of evil and the *evidential* problem of evil. The logical problem of evil seeks to disprove the existence of God outright by saying that the suffering we see is incompatible with OmniGod, and, therefore, OmniGod doesn't exist. Theists typically retort, as per skeptical theism that we have already mentioned, that OmniGod might have a reason for such suffering to exist, but that this reason is unknown to us or even beyond our ken. This is a valid move since OmniGod and suffering can be argued to be logically consistent (irrespective of how improbable this might be).

The ramifications here are many. A few are worth pointing out. First, it means that every unit of pain, from the largest heinous genocide or plague down to the reader stubbing their toe, needs to be explained in light of OmniGod. Both the genocide *and* the toe stubbing must be necessary—the optimal way—in bringing about whatever the greater good might be. Sometimes, focusing on the smallest units of suffering can be just as illuminating as the biggest.

Moreover, one can argue that this means that the present world (and for the purposes of this discussion, the universe at large) must be the best, most loving, optimal world or universe there can be. If God could have achieved his objectives by producing a universe with less suffering, then God would have done. Thus, every unit of pain and suffering in this universe is necessary for this universe to be the "most loving" universe there is, or to fit into God's remit as an all-loving creator.[1]

Every single unit of pain or suffering in every single inch of land, sea, sky, or gas cloud on every single piece of ground or sky on every single planet in every single solar system in every single galaxy all over this universe...must be necessary. Any *gratuitous* suffering—suffering without reason or pain that does not *need* to happen in terms of God's omnibenevolent objectives—invalidates OmniGod's existence. For much more on this topic, please see Pearce (2022).

So, what does ET have to do with the problem of evil? It doesn't seem like aliens existing is an evil thing in and of itself. What ETI might mean, however, is that the problem of evil is *magnified*. It becomes a *bigger* problem. The existence of a population of ETIs who experience sufferings and evils means there is a distribution of such suffering, and we can treat this statistically. It comes down to a numbers game, and it leads to some individuals winning the worst possible lottery, pulling from the lever of the Copernican principle slot machine.

Before we discuss the Copernican principle and how it might affect issues concerning the problem of evil, let us dwell for a short while on abductive arguments and what we might expect.

Abductive Arguments and Predicting Universes

We have previously discussed abductive arguments and how they are important components of our analysis. To remind the reader, abductive

---

[1] Some theists contest that such calculations can be made. See Pearce (2022b).

arguments are inferences to the best explanation. Given the observation of some data, what best explains the phenomena? If we have two competing hypotheses—OmniGod existing and OmniGod not existing—which of the hypotheses does the data best support? This is not a case of outright disproving either hypothesis, but ends up being a way of looking at what hypothesis is most probable, *ceteris paribus.*

To put this into context, given the sheer volume and variety of pain and suffering on this world alone, what hypothesis does this data better support—(Christian) OmniGod theism or naturalistic atheism?

There is another way of looking at this. We can also look at the predictive value of a hypothesis in light of any given framework or data. For example, in the case we have here where we are discussing evil and suffering, starting with either OmniGod or naturalistic atheism, what sort of universe would you predict?

Imagine that you commence this thought experiment understanding pain and suffering, and also knowing what love and goodness are. If you were to start off understanding or believing that God is love—infinite, unbridled love—what universe would you predict? Would you, for instance, predict a universe full of genocide, tsunamis, rape, black holes, empty and life-denying space, cancer, pillaging, loneliness, debilitating mental health and crippling physical health? The whole range of pain and suffering that we merely experience *on Earth*, let alone the greater range and instances of them *across the whole universe*, would be more supportive of naturalistic atheism than an all-loving god.

Oftentimes, when one questions the rationale for God creating at all,[1] the theist maintains that God has a superfluity of love: God has so much love that it spills over into creation (...of cancer, genocide, tsunamis, black holes, malaria, debilitating mental health conditions). This, for the previous implications, seems a thoroughly problematic tack to take.

But, as we will see discussed in the next section, the principle of mediocrity and the Copernican principle lead us to believe that the pain and suffering experienced on Earth is distinctly moderate in comparison to what most probably exists out there, in various galactic neighborhoods. This can be seen in the *amount* of suffering (evil), the *frequency* of suffering, or even the *type* or *class* of moral badness.

If our Earthly experiences of suffering in the context of the problem of evil are enough to lead us to think that God probably does not exist, then one can surely posit that finding (potentially infinitely) *more* and *worse* examples would at least shift the probability further in favor of naturalistic atheism as

---

[1] See the argument "Why Would God Create At All?" in Pearce (2022).

an explanation of the data and reality and away from Christian OmniGod theism.

Ruhmkorff and the Copernican Principle

As noted earlier, unless we have good reason to think otherwise, our first guess is that a random measurement will be close to the most likely value in a distribution. For humanity, we should expect to be an average civilization, unless we have reason to think otherwise. The standard result in statistics is, given a normal distribution,[1] that you will find your observation to be close to the middle (or average), and the further away from the middle, the more unlikely it is to get that observation. Most things that could be observed will be closer to the average value, few far away, and extremely few very far away. So, given an observation, you would expect it to be a more likely result than an unlikely one, let alone a very unlikely one.

When it comes to humanity on the cosmic scale, our first guess is that we would be closer to the middle than at the tails of a distribution of civilizations. We wouldn't be the smartest, nor the dimmest. We wouldn't be the most violent, nor the most peaceful.

But shouldn't this principle also mean that, when it comes to the suffering humanity has endured, we are also more likely to be on the average than at the extremes? This is a point made by philosopher Samuel Ruhmkorff.[2] His approach in his paper "The Copernican principle, intelligent extraterrestrials, and arguments from evil," is defined mathematically, using prior probabilities and then updating it with our experience of terrible things. We will repeat his strategy here.

To start, we will use something called a "Jeffreys prior." This is a way of describing a probability distribution before you have information about the subject of investigation. You only know what the *possibilities* are, but nothing else about how *likely* some possibilities are more than others. For example, with a coin-flip, there are two possibilities: a heads or a tails (ignoring the coin landing on its side, for simplicity's sake). But what is the probability that the coin always comes up heads, or always tails, or equally both, or just slightly more heads than tails, and so on? You know from experience that coins are closer to being fair than lopsided, but let us suppose you don't even know that. What is the likelihood of the coin being fair or any level of unfairness?

---

[1] By normal, we mean the particular sort of distribution also called a Gaussian distribution, and the form of the probability density function (PDF) looks like the shape of a bell—hence the nickname, the bell curve.
[2] Ruhmkorff (2019).

184

The Jeffreys prior is what we can use. In the example of a coin, if the probability that it flips and gives a heads is p, this value is between 0 (never comes up heads) and 1 (always comes up heads); a value of 0.5 would indicate a perfectly fair coin. For this Jeffreys prior,[1] the distribution of probabilities goes as $1/p\sqrt{1-p}$. What this suggests is that the probability is maximized for p nearing either 1 or 0 (completely biased coins), but with even a little bit of evidence, that distribution changes. If you flip the coin five times, and you get 1 heads and 4 tails, you know it cannot be a completely biased coin. Based on those results, you can then update the probability distribution and see just how likely or unlikely it is that the coin is fair.

This logic can apply to anything that can be described in a mathematical way. If what you are measuring is some sort of continuous quantity (for example, the height of a person), then it follows some sort of normal distribution;[2] if it is something counted over time, then this is a Poisson distribution (predicting the frequency of something happening); if you are measuring a probability and it takes some value between 0 and 1, then this is a binomial (or Beta) distribution (such as our coin flip). For various distributions, there is a Jeffreys prior, and then with the evidence we have, we can update. It doesn't even require a computer, though it helps when the problems get very hard.

Now, let's apply this method to the problem at hand. Or better still, let's get mathematical! Bear with us as we will try to explain our steps (or you can skim read the next few paragraphs and trust our conclusions!).

Suppose that the worst thing that ever happened on Earth (in terms of pain or suffering) is something we can measure, or at least put on a scale. Let's say that worst event ever (call it event X) had a value of 100 standard badness units (SBUs). So, stubbing one's toe is probably less than 1 SBU, while events such as trench warfare are closer to 100—at the very least, we suppose that toe stubbing is less than a hundredth as bad as the fighting that took place during World War I (or we could use the 2004 tsunami that killed 230,000 people and countless animals).[3] But is 100 SBUs the absolute worst thing in the universe? 100 SBUs is for the worst thing in *Earth* history, but now we must consider other worlds. Could there be an event X^* that has a value of more than 100 SBUs? The prior probability distribution that there exists X^* greater than 100 SBUs can be given by the Jeffreys prior. In other words, we can assign a probability to there being this X^* event without knowing anything particular about this potential event.

---

[1] This prior is also equivalent to a beta distribution (a binomial distribution between 0 and 1) with parameters of 0.5 and 0.5, written in mathematical notation as Beta(0.5, 0.5).
[2] Also known as a bell curve.
[3] We have covered our bases here by using both a human and a natural example of evil.

Aliens and Religion: Where Two Worlds Collide

We want to know how probable it is that there is an event worse than our worst human experience: How probable is there an X^* more evil than some value x, given X? Or, in mathematical notation, $P(X^*>x|X)$ for badness x.

For something that follows a continuous normal distribution, like badness, Jeffreys tells us the prior probability distribution is proportional to $1/x$, where x is the value of badness in this case. We can write the prior probability as $P(X)dx \propto dx/x$, where dx is a small interval of x which we would look at to see the chances that something of badness x is in that interval. This will be important later when we do a step of calculus, which requires calculating based on this tiny region dx.

The key takeaway at this point is this: as the value of badness for a given event becomes larger, the probability that that event took place becomes smaller. In other words, worse things are less probable. Intuitively this makes sense; if the heaviest thing you ever lifted is 200 kg, then lifting 201 kg is possible, but not certainly doable, and lifting something 300 kg is very unlikely. In other words, to be able to lift something heavier than the heaviest thing you have ever lifted is less than 100% likely, and a much heavier object is much less likely. Similarly for badness. So the math seems to match our intuitions.

We are next going to show that this entails Earth's worst event as not likely being the worst event in the universe.

Now, we earthlings know we have experienced the worst thing on Earth, with a badness of 100 SBUs, so we update our prior probability distribution (of how likely there is an event worse than X), to a new posterior distribution. This is where we now employ some serious mathematics in the form of a formula known as Bayes's Theorem. The objective for this formula is to be able to compare competing hypotheses to work out which is the more probable. In our case, we are comparing the hypothesis that there is worse evil in the universe than on Earth and the hypothesis that X as the event on Earth has the greatest amount of evil or suffering in the universe.

For our purposes, we are using a simplified version of Bayes's Theorem wherein we multiply the prior probability of there being greater suffering outside of Earth by the likelihood of observing the evidence of what evil we see on Earth.

That new, posterior probability is the prior probability (proportional to dx/x) multiplied by the likelihood of seeing such evidence, which is also going to follow a probability distribution of $1/x$. This posterior probability distribution is proportional to dx/x^2.

If you have struggled with the math so far, then be warned as the next few paragraphs will stretch your gray matter even further. Don't worry, there won't be a test at the end.

So, the probability distribution of the worst thing in the universe being that terrible thing we have experienced (X) is proportional to $1/x^2$, but where x must be at least 100 SBUs. This would look like $P(X^* > x|X)dx = (c/x^2)dx$ where c is what is called a proportionality constant that we don't yet know. To explain what this means, take a popular formula like Newton's 2nd Law, $F = ma$. Here, F is proportional to a, and m is a proportionality constant that lets us directly calculate F, given a. Proportionality here tells us the relationship that as a gets larger, F gets larger even if we don't directly calculate F.

We now work on performing the Bayesian updating. In our calculations, we need to exclude values of x less than 100 SBUs because we know they are not the worst events.

Our next step, then, is to figure out the probability of X^* existing. The basic axiom of probability theory is that all probabilities have to add up to 1, and we know that x cannot be lower than 100 SBUs. These two points and the probability distribution above will build up to our results.

First, we need to figure out how to get everything consistent with basic probability theory. To do that, we add up all possible values of x (badness of events) equal to and greater than 100 SBUs by using an integral from calculus. This takes all x values that are possible and performs a sort of weighted sum (i.e., integrates) over the function c/x^2 (posterior probability density) so that the total probability will add up to 1 and we can figure out the value of c. This allows us to directly calculate the probability of X^* existing.

In mathematical notation, the result looks like this:

$$\int P(X^* > x|X)dx = \int_{100}^{\infty} \frac{c}{x^2} dx = 1$$

Now, assuming you know how to take an integral (finding the area under a curve) with the limits of 100 and infinity, you will understand the following:

$$\int_{100}^{\infty} \frac{c}{x^2} dx = -\frac{c}{x}\Big|_{100}^{\infty} = \frac{c}{100}$$

Now we can see that $c/100 = 1$. With one step of algebra, we realize c is equal to 100. And so, we now come to our final equation:

$$P(X^* > x|X) = \frac{100}{x}$$

That's a lot of math, and you might just have to trust us if you are not mathematically inclined. But what does this tell us? Well, it means that for

Aliens and Religion: Where Two Worlds Collide

something just 25% worse than the worst thing ever experienced on Earth—that is, event X^* being 125 SBUs—the probability that that even worse thing has happened is 100/125 = 0.80 = 80%.

In other words, it is likely, given alien life, that some of these ETIs have experienced an evil significantly worse than here on Earth. If the worst thing in human history was the Holocaust with 11 million killed, then there is likely an ET-Holocaust that was worse still, with our example saying it could be closer to 14 million exterminated. The math also suggests that there is a 50% chance that ET life somewhere has experienced something twice as bad as anything ever seen on Earth ($X^* =$ 200 SBUs).

Up to this point, this is what Ruhmkorff concluded in his paper upon which we are drawing.

Now, to go beyond what Ruhmkorff calculated, let's consider not just the *badness* but the *frequency* of such badness. Might ET have experienced more than one of the worst things ever? This counting of events follows a different probability distribution (called a Poisson distribution), and its Jeffreys prior goes as $1/\sqrt{n}$, where n in the number of times the event has happened.[1] Now, the worst thing on Earth has happened once, so we can do a similar sort of update as previously, though the math is more difficult. Nonetheless, we have worked out the following. The posterior probability follows something called a gamma distribution,[2] which is not so easy to calculate or intuit, so we have tried a few values to see what it suggests.

If the number of the worst events ever is 2, the probability of two of those events happening in the universe is more than one in three (about 37%), so it is unlikely that an ET civilization will experience two Holocausts. It is even less likely that there will be 3 of the worst events ever (less than a 1 in 7 chance). More generally, the probability that more than 1 worst event in the universe will happen is a bit more than 50%. This all suggests that there is a coinflip chance that the worst thing ever can happen more than once.

We shouldn't also forget the converse. All of the same mathematical logic would also apply for goodness, and there is an 80% chance that ET life has experienced something 25% better than the best thing that has ever

---

[1] Technically, for a Poisson distribution, we should use the mean value of events, that is the number of times an event happens over some observational period. For simplicity, we say that the time frame is the age of the universe, so we only care about the number of times some event has happened in the entire history of the observable universe.

[2] Starting from relative ignorance, the prior distribution can be treated as the gamma distribution $\Gamma(1, 0)$. Forming a posterior distribution, we get a probability of the form $\Gamma(1, n)$, where n is the number of events. There must be at least 1 example of the worst thing ever, so at minimum n is 1. We then perform summation and normalization, and the probability of seeing n events of maximal badness is $2.718 \times \Gamma(1, n)$. For n=1, the probability is exactly 100%; for n>1, the probability monotonically decreases.

happened on Earth. There is also roughly a 50% chance that the best thing in the universe happens more than once.

Taking together these points, let's bring it back to the problem of evil. The issue is that we think we see terrible things happening on Earth that are not justified. We might argue that no greater good can be claimed to necessitate childhood brain cancer, for example. If there are gratuitous evils on Earth, then it is very likely ETIs have experienced even more gratuitous evils.

On theism, the number of gratuitous evils in the universe should be zero, since an all-good OmniGod would never allow for gratuitous suffering—by definition. While theists claim that all the evils we see are somehow justified, nonbelievers gawk at such assertions when looking at the tsunami in 2004 that washed away hundreds of thousands of lives, or the billions of animal lives burned up in Australia's 2019/2020 brushfire season, or the many children suffering and dying from starvation, cancer, or lethal birth ailments. How many examples of gratuitous suffering have existed in Earth's history? That is hard to argue, given that theists deny any of these examples of suffering are without purpose—though we'd rather not be the person that says billions of animals burning to death was for the greater good, or that childhood leukemia is a blessing in disguise. And yet, as terrible as these things are, it is likely that ET has suffered even more gratuitously, both in quality and quantity.[1]

All of this follows just from our prior knowledge of suffering in our world and looking at how that likely stacks up when suggesting that there are more examples of suffering out among the stars. It is not just imaginable that on a world in another galaxy there are crueler forms of slavery, harsher forms of torture, viler tyrants, and even more sickening wars, not to mention even more terrible diseases. It may well be the case that some aliens can experience evils more than humans can; it is possible that they have a physiology that allows for more and deeper pain given otherwise identical phenomena. The Copernican principle can be applied in each of these contexts.

But what of the good in the world? Theists generally, and Christians in particular, think the good in the world is possible because of the evil. However, looking at how we experience good and evil, there does not seem to be some great equalizer, in that someone who experiences a great evil in their life also experiences an equal or greater good. In psychology- and philosophy-speak, this is called *just world theory*, and for many people this is interpreted as *karma*. In fact, this is primarily why the ideas of heaven and hell (which did not exist in the Old Testament in any meaningful way) were developed in the intertestamental period. They were mechanisms that could

---

[1] If there have been 100 gratuitous examples of suffering on Earth, the chances that there is more than 100 such examples in the entire universe are close to 90%. If the number of gratuitous examples on Earth is larger, than this probability is also larger.

Aliens and Religion: Where Two Worlds Collide

be used to explain how bad things could happen to good people, and vice versa.

For example, some people only live as infants in pure pain before succumbing. Conversely, some live in luxury and avoid the sorts of suffering that many others feel. This is recognized by both the secular and the religious, including several authors in the Bible (Ps 73: 3-5, Jer 12:1, Job 21:7, Matt 5:45). While some claim that, in the afterlife, all is equalized, that is both a realm without evidence (that argument is covered in Pearce (2022)[1]) and it goes against the evidence of our eyes. If the universe is under OmniGod's dominion, and it doesn't balance things out before our eyes, to suggest that it will when our eyes are closed for the last time is wishful thinking in its purest form.

It is also a form of compensation *and not moral justification*. This, again, cannot be emphasized enough. To employ heaven or the afterlife as a way to balance the earthly books is to employ a strict form of moral consequentialism whereby humans (and animals) are used in a utilitarian and instrumental fashion to bring about some greater good, and moral value is derived from the eventual consequences—for example, heaven—for the agent. If the theist is not a moral consequentialist (and very few are), then the afterlife can only act as compensation and not moral justification.

Bringing this back to aliens, we have every reason to suspect that the ETIs experiencing the greatest evils in the universe are not also necessarily experiencing the greatest goods.[2] In all probability, one ETI is getting the greatest good in the universe and avoiding the worst evil, while another ETI is getting the worst evil in the universe but not having the greatest good. In other words, the injustice we see on Earth is almost certainly worse on another planet. If we can see something that looks like gratuitous suffering here on Earth, then there is (therefore) very, very likely more gratuitous suffering than we have ever seen on Earth taking place on another world.

Today, most philosophers pondering the problem of evil consider it as an evidential rather than a logical problem, as we have discussed. Perhaps there is some way that OmniGod allows for gratuitous suffering, so it cannot be a logical contradiction as is the claim that there are four-sided triangles. On the other hand, it is at least *unexpected* that there are such evils, so it is evidence against the existence of OmniGod. The surprise that there is so

---

[1] Pearce's book [Pearce (2022)] *30 Arguments Against the Existence of "God", Heaven, Hell, Satan, and Divine Design* deals at length with the issues and inconsistencies regarding the ideas of an afterlife, heaven, and hell.

[2] Contra McIntosh & McNabb (2021), who suggests that evil is not a problem "so long as [good and evil are] distributed equally." There is no evidence that good and evil are distributed evenly here on Earth and plenty against, so to suggest that boons and burdens are equally distributed on another planet is blind faith.

much evil gives rise to a whole theological tradition dedicated to explaining quite why this should be.

It may well be debatable as to how strong that evidence is, but we can consider the trend. The more gratuitous suffering in the world, the more evidence there is against OmniGod theism. If aliens exist, then the gratuitous suffering in the universe very, very likely increases. Therefore, aliens would likely decrease the probability that OmniGod exists.

Just to exemplify this last point (and we don't need to quantify this explicitly), if there were only rare instances of evil or suffering, then these would be more accessible to explanation. We would find a world with only the occasional murder, or the common cold as the worst ailment, to be a world much easier to explain in light of OmniGod. But a world with malaria and cancer, tsunamis and genocides, is much more difficult to explain in terms of theodicies than the former world. The greater the evil in the world, the harder theologians have to work to solve the problem. This seems intuitively unequivocal. It is no surprise that problem of evil arguments defer so quickly to the most poignant and terrible instances of mass suffering, because they are simply harder to explain away. Thus, postulating evils and suffering that potentially cause such human instances to pale into insignificance by comparison, is to present data that is even harder to explain, making God at least somewhat less plausible.

This is perhaps one of the most important points in this book: If the problem of evil (POE) is a thorn in the side of theists, then the discovery of ETI presents the POE to be at least a larger thorn, and quite possibly a very sharp battering ram. At the very minimum, the POE in light of ETI makes OmniGod if not DOA then certainly less probable.

In Appendix 2, we formally demonstrate how we should update the probability of seeing evil in the universe, given how unlikely we expect the worst evil ever experienced on Earth, if God exists, to be. We then update that probability on the assumptions that there are ETIs. As an example, if we suppose that the expected likelihood of seeing the worst event in human history on theism is 50% (that is, we think there is only a 50% chance that OmniGod would have prevented something like the Holocaust or the 2004 tsunami), then this is evidence against OmniGod's existence. Updating this number based on the mere existence of aliens, we find that this likelihood to about 23.8%, because of the expected greater sufferings that ETIs would face. In other words, the POE becomes about twice as bad for theism.

More generally, we can say that, given the history of suffering on Earth and the existence of ETIs, the probability of theism is lower than on aliens not existing. In mathematical terms, $P(T|X, ET) \leq P(T|X)$, where T is theism, X is the great suffering on earth, and ET is the existence of aliens. We also find that these probabilities are equal only when suffering is impossible on theism $(P(T|X) = 0)$ or suffering is completely expected on theism $(P(T) = P(T|X))$.

It is this latter option that the theist is likely to take up.

The Christian Response

Essentially, Christians beg the question by arguing in a circular fashion to the existence of God in this context. The critical move that they take is to assume that whatever situation we are in *must* provide a net good, and to be more precise, the optimal amount of good.

In other words, whatever amount of suffering there is in this world or this universe, they already believe that God exists, and so there can be no net evil, and with God's all-loving characteristic that is assumed, they argue that the world's amount of pain and evil is optimal. It just has to be that way.

Another way of putting this is that they argue *from* God rather than *to* God. They start with the assertion that OmniGod exists and then post hoc rationalize all the data that they see in the universe to fit this assumption. The conclusion is that, sure, God exists because the pain and suffering is not disconfirming and *can conceptually* fit their presupposed conclusion. This somewhat circularly derived conclusion can then be used to support other ancillary arguments.

If we use the *reductio ad absurdum* technique, though, we can see that no matter what level of pain and suffering in the universe, the theist can still invoke skeptical theism and the idea that there *could* be a reason. Take the ridiculous scenario that you, the reader, are the last sentient creature in the universe. You have seen all other sentient lifeforms across the universe tortured by a monster in front of your eyes. Last up are your family and friends who are tortured with terrible sadism until only you are left with this monster for company, painfully almost alone. And then the torture starts on you. Even in this scenario (or any worse one you could cook up), you, as a skeptical theistic Christian, could *still conceptually argue* that your omnibenevolent God still exists because there just *could be* a reason why such abundant, terrible suffering might (necessarily) exist for some greater good.

So, here, we need to tease apart something that is conceptually possible from something that is probable. As discussed previously, unicorns *could* exist, but they probably don't, and just thinking that they *could* does not justify thinking that they *actually do.*

There is a spectrum of pain and suffering, and we would suggest that everyone has their own level (*ceteris paribus*) at which the amount of suffering eventually triggers a disbelief in OmniGod. For most atheists, that threshold is lower than for most theists (there will be other variables involved,

192

such as motivated reasoning). But, at some point, one will surely say, "Okay, enough's enough; this is not the sort of data that supports my OmniGod hypothesis."

Whatever suffering Christians not only experience but think there is (and it is interesting to note that this will be different for every Christian on Earth) must be optimal, in line with their OmniGod. If a Christian is blissfully unaware of all ETI life existing and thinks that the total suffering in the universe is optimal, and is then introduced to the notion of vast, vast numbers of ETIs existing together with the baggage of innumerable quantities of suffering, then this new *hugely increased* amount of suffering must now, also, be optimal.

Whatever evidence there is just happens to conform with their God-belief—rather reminiscent of the heads I win, tails you lose approach explored earlier with fine-tuning for life in the universe.

For some Christians, psychologically speaking, this new placement along the spectrum of suffering may trigger a deconversion moment. Or the power of belief may keep this shift played down.

For us, to suddenly expand the volume of suffering in this way is a problem for OmniGod theism. However, just taking into account Earth existing, theists already claim that suffering is optimal and that *there must be* a net good. Even though there are tsunamis and cancer, rape and Ebola, these are worth it and even necessary for (so theodicies tell us) the greater, net good.

The move that the Christian then utilizes is to merely extrapolate this mechanic out across the universe.

For instance, the Christian could point at a single village and observe the philosophizing that could be done by those villagers thinking they are the only people on Earth; the Christian could then claim that it is not unreasonable to expect village theodicies to continue to work now in a larger region of here tens of villages. And then to a nation. And then to the world. This is, indeed, what has broadly happened; and here we are. Now, that same Christian could argue that it would be essentially arbitrary to stop at any point of demarcation on this spectrum so that even *this* world is not a natural stopping point.

Thus, just as a theodicy (or set of theodicies) works to justify the suffering on Earth, those same theodicies (or even adapted ones) can work, the theist claims, for the whole universe.

As the skeptic claims that creation is either a net evil or a net good that is *sub-optimal* (i.e., it's good, but it really could be better), the theist merely claims it appears either of those ways but, really, it is just an optimally good universe. Indeed, it is just that our knowledge is lacking, and/or we are not privy to the divine calculations.

Theists consistently inoculate themselves from criticism by executing this move to the conceptual possibilities of skeptical theism.

However, it can also be claimed that it is also conceptually possible that this universe is just full of suffering because there is, in fact, no god.

But let us remember that we are trying to move away from a conceptual possibility to an inference to the best explanation. Just as Christianity struggled to deal with Earth not being the center of the solar system, so will it struggle with it not being the theological center of the universe, and we think this latter struggle is justified.

One question to strongly consider is whether our intuitions about suffering are a good basis for rational assent to what they *seem* to tell us. When we look at the world, and this is the same for theists who recognize the problem of evil as a serious problem, we intuitively *feel* that the suffering is at times horrendous and, *prima facie*, seems gratuitous. What can we say about these intuitions?

Phenomenal Conservatism

The position of *phenomenal conservatism* can be brought into play here, espoused by some philosophers, such as Michael Huemer.[1] We referenced him on this subject previously in the chapter on (Christian) Theism but it seems pertinent to revisit the idea as it pertains to the POE. The idea is that we are justified in holding beliefs about the world by the way they "seem" or "appear."

Michael Huemer, in his piece on phenomenal conservatism for the *Internet Encyclopedia of Philosophy*, sets the case out as follows:[2]

> The intuitive idea is that it makes sense to assume that things are the way they seem, unless and until one has reasons for doubting this....
>
> If it seems to S that P, then, in the absence of defeaters, S thereby has at least some justification for believing that P.
>
> The phrase "it seems to S that P" is commonly understood in a broad sense that includes perceptual, intellectual, memory, and introspective appearances. For instance, as I look at the squirrel sitting outside the window now, it seems to me that there is a squirrel there; this is an example of a perceptual appearance (more specifically, a visual appearance). When I think about the proposition that no completely blue object is simultaneously red, it seems to me that this proposition is

---

[1] Originally set out in Huemer (2001), but also variously elsewhere, e.g. more recently in Huemer (2013).
[2] Huemer (n.d.).

true; this is an intellectual appearance (more specifically, an intuition). When I think about my most recent meal, I seem to remember eating a tomatillo cake; this is a mnemonic (memory) appearance. And when I think about my current mental state, it seems to me that I am slightly thirsty; this is an introspective appearance.

Take, for example, the idea that we are living in a real world, and not in *The Matrix* (or that we are brains in vats, or that the world started 20 minutes ago and all of our information and memories were injected into our newly formed brains). Phenomenal conservatives would say, "Well, it doesn't *seem* that this is the case, and I have no good reason to think *it is*, and I have no defeaters against why *it isn't*."

Even though this is a position that many theists favor themselves (well, it *seems* like God exists, so...), it can be applied to the problem of evil and against their position.

Suffering *seems* gratuitous. From stubbing my toe to getting cancer, from every unit of pain and suffering involved in predation to tsunamis killing 230,000 people, from pandemics to genocide, there appears to be *an awful lot* of suffering—far more than one would think necessary.

Remember, on theism (belief in OmniGod), there is no such thing as a tragedy. *All* suffering happens for a reason—for a greater good. Each and every unit of pain and suffering since the beginning of the universe must have been necessary. It couldn't have happened in any better or nicer or less painful way. Otherwise, it would have, given that God had the knowledge and power to have it otherwise.

A theist should not lament any suffering. They should praise God's supreme judgment in every single case, rather like the actions of the infamous Westboro Baptist Church who praise God's judgment even if it means celebrating at airports receiving the dead bodies of fallen US soldiers since this is, to them, clearly part of God's infallible judgment. It's all useful. It's all necessary. It's all the correct design, creation, and judgment by the supreme creator. No whining. This sounds facetious, but under strict classical theistic interpretation of suffering, there can be no such thing as a tragedy: Everything takes place to necessarily bring about a greater good.

Except, it doesn't *seem* that way.

Suffering *seems* gratuitous, and this is *prima facie* evidence against the existence of OmniGod. And phenomenal conservatism says we are justified in this because we are absent any defeaters.

Just saying that "there *could* be a reason" is not a defeater. As mentioned earlier, that's like saying that "unicorns *could* exist" is a defeater for my claim that unicorns *do not* exist. Just saying something "might be the case" does not a justifying argument make.

The only realistic option for the theist is to claim that there are *other* external arguments that lead them to believe that God exists and that, by inference, there must be a reason for such suffering. This is something philosopher Michael Tooley discusses in his *Stanford Encyclopedia of Philosophy* entry on the problem of evil. The issue is that none of those external arguments for God argue for God's omnibenevolence, so they miss the mark. As Tooley states:[1]

> The situation is not essentially different in the case of the argument from order, or in the case of the fine-tuning argument. For while those arguments, if they were sound, would provide grounds for drawing some tentative conclusion concerning the moral character of the designer or creator of the universe, the conclusion in question would not be one that could be used to overthrow the argument from evil. For given the mixture of good and evil that one finds in the world, the argument from order can hardly provide support even for the existence of a designer or creator who is very good, let alone one who is morally perfect. So it is very hard to see how any teleological argument, any more than any cosmological, could overturn the argument from evil.

A similar conclusion can be defended with respect to other arguments, such as those that appeal to purported miracles, or religious experiences. Skeptic Emerson Green, in his piece on skeptical theism, agrees:[2]

> Everyone agrees that there seems to be gratuitous suffering. So until you've offered a defeater for that seeming, then we're rationally justified in affirming that the world is how it seems in this regard. And since the existence of gratuitous suffering is far more likely on naturalism than on theism, the fact that we observe gratuitous suffering is strong evidence against theism.

We have two hypotheses competing here:

(1) **God is love.** OmniGod exists and has designed and created the world. Thus, conforming the world to fit in with his characteristics of being all-knowing, -powerful, and -loving.
(2) **Naturalistic atheism.** The universe is ambivalent to pain and suffering. Evolution works with the tools it has available. "Stuff" happens.

---

[1] Tooley (2015).
[2] Green (2021).

Which thesis best predicts the data, the phenomenal amount of suffering throughout time and space? And, take note that very often this suffering does not involve humanity and is thus unconnected to any reason pinned on humanity. A fawn dying in a forest fire, a baby water buffalo being ripped apart alive by a pride of lions, a tsunami destroying a mangrove ecosystem.

When we look at the sheer volume and type of suffering in this world and extrapolate it over the universe, we are left thinking that if it walks like a gratuitously suffering alien duck and quacks like a gratuitously suffering alien duck, then it probably *is* a gratuitously suffering alien duck. And there is no good reason to think that an all-loving God made it that way.

13 – CHALLENGING THE CHRISTIAN NARRATIVE

Narrative Tension

Rather than a particular doctrine or interpretation of a biblical passage, McIntosh and McNabb consider the source of tension between the possible discovery of ETI and Christianity is the narrative tension it creates. The special story of human life in the universe suggests many things in terms of sin and salvation, but if flying saucers suddenly show up, would anyone be thinking about communing with the divine now that we are just one of a million side-stories in the universe? As they put it:[1]

> The existence of ETI would come as a major shock to Christians, because heretofore ETI simply didn't enter into the story about life, the universe, and everything that they took to be complete. ETI would, at the very least, entail that that story was radically incomplete. More extreme, the existence of ETI might cause some to think we're living in an entirely different story altogether.

When considering the sorts of "plot twists" that ETI could create, McIntosh and McNabb have a useful list:

S1. ETI is so remote or undetectable that any interaction is (nomologically) impossible.
S2. ETI is so remote as to be physically inaccessible, but communication is possible.
S3. Physical interaction with ETI is possible; ETI is peaceable.
S4. Physical interaction with ETI is possible; ETI is hostile but not an existential threat.
S5. Physical interaction with ETI is possible; ETI is hostile and an existential threat.

Each of these scenarios creates all sorts of science fiction possibilities, and so let us engage our imaginations to see what impact they might have on anyone with a Christian background.

---

[1] McIntosh & McNabb (2021), p. 13.

In the first situation (S1), we somehow found out there is an alien civilization so far away we could never talk, see, or even know much about them. Perhaps we develop a telescope powerful enough to look into another galaxy and find evidence of the construction of a super-structure, such as a Dyson swarm. This could be noted by the sort of light spectrum such a gigantic piece of architecture would produce, absorbing the energy of an entire star and then radiating it out as infrared in such a way that the spectrum would have the tell-tale signs of a giant, artificial construction. If those structures are in another galaxy, then our chances of ever visiting them are all but non-existent. Interstellar distances are already mind-boggling, but intergalactic distances are orders of magnitude grander. Similarly, a radio message between the worlds would be nigh-impossible to detect, and the information transfer would require millions of years for just sending a simple "hello" statement, let alone awaiting a reply.

But if ETI exists in another galaxy, and their technology is detectable to us, then they must be a vastly more powerful civilization. To alter the structure of a solar system, or perhaps solar systems (!), these aliens would be numerous as well. Billions, if not trillions, of thinking, feeling creatures.

Is there no tension knowing they exist and the belief that the universe has our species in mind? There seems to be more than what McIntosh and McNabb suggest. There is now a greater power in the universe than us, yet one that is not supernatural. They have become masters of stars, and their mere presence can be seen across the vast oceans of space and time. Compare that to us, who barely cling to one spinning rock. If such an ETI exists and is known to exist, then doesn't that inherently make us small, perhaps even pathetic, in the grand scheme of the universe (some might say)?

For example, if we were to look down the strata of species and pick out a capuchin monkey, would we look at the universe and proclaim that the universe was made for the capuchin, that this was the species God had in mind? This would be highly unlikely. Thus, if we are to the capuchin on Earth what other aliens might be to us in the universe, what does this now say about our position in the narrative?

If our lives and spiritual development are of paramount importance to the God of the universe, then why is some other civilization so very much more developed? In fact, if these ETI visibly affect their stars in a distant galaxy, then they have existed for millions of years, long before our ancestors even began to sharpen rocks.

However, S1 really falls into the issues of divine deception discussed earlier, so there is not a new tension here. Where McIntosh and McNabb and we agree is when ETI is hostile and is coming for us implying there is a very strong tension.

Suppose we discover the first signs of a fleet of craft approaching Earth, zooming towards our planet at an appreciable fraction of the speed of light. Either by means of tracking their motions or by reading their radio messages, we figure out that they have come to conquer or destroy us. This is the *War of the Worlds* scenario, and the religious implications of that conflict appear in one of the memorable scenes of the film adaptation in 1953. A pastor, Matthew Collins, hopes to make peace with the invading Martians, and he steps away from the human army, walking in his vestments and carrying a crucifix. "If they're more advanced than us, they should be nearer the creator for that reason." Or so he thinks. Quoting Psalm 23 as he nears the Martians, speaking of having no fear as he travels through the shadow of the valley of death, he is swiftly vaporized by the uncaring, cruel machines of the invaders.

God and his promise look far away and meaningless if ETI comes and has super-weapons. If God could have issues with iron chariots (Judges 1:19), imagine the futility of the Divine against energy shields and antimatter bombs.

Thus, just like with the problem of evil and how to make sense of suffering in a world where God is in control, there needs to be a hostile ETI-theodicy. There must at least be some greater good for us humans if we are part of God's plan and his plan is justified. And this must take into account the possibility of humanity being entirely wiped out. What theodicy could account for that level of suffering? Then again, with no humans about, there would be no need for harmonizing a Christian narrative...

In the '53 movie of *War of the Worlds*, the hand of God is implied as active in the salvation of humanity, given that the salvation happens when the protagonists are in a cathedral when the Martians are defeated. The source of that defeat was outside human control, suggesting a greater plan for humanity that protected it against an even more powerful civilization. As stated in the original novel (and modified in the film), the Martians were slain "after all man's devices had failed, by the humblest things that God, in his wisdom, has put upon this earth."

In this fictional salvation, the theodicy might be that God protects his subjects, at least from extinction. Perhaps a "greater good" could be stated to exist because the humans are humbled. This may look like an abusive relationship with God—he lets us be hurt (and hurt very badly) so that we learn just how powerful and smart God really is. But even if we suggest that this or a similar theodicy resolve the issue, it rests on one major premise: In the end, humanity survives.

This is also an issue with the hostile ETI-theodicies McIntosh and McNabb consider, largely relying on other stories, namely the film *Signs* (2002) and the *Ender's Game* novel series by Orson Scott Card. By some means, God or a connection with divinity leads to final salvation and growth of faith. Finishing their discussion of hostile ETI, McIntosh and McNabb state,

"It's fascinating to us that there are so many works of fiction which explore the relationship between faith and the existence of ETI. In fact, we aren't aware of a single example where the existence of ETI is presented as incompatible with traditional religious belief."[1] We propose that these authors have read insufficiently many science-fiction books.

The oeuvre of Arthur C. Clarke provides numerous examples of ETI as undermining religion, sometimes even famous Christian philosophers. In *Childhood's End* (1953), the peaceful invaders of Earth cause religion to disappear, and the aliens themselves are revealed to look like Christian devils; the myth of the devil was due to an earlier attempt to evolve humanity that did not work out well. So, aliens not only undermine religious faith by their modern arrival, but our religions are misunderstandings of ET's purpose, and "Satan" was the good guy all along. Later, in his short story *The Star* (1955), a Jesuit astrophysicist has a crisis of faith when he discovers a distant and peaceful civilization was destroyed by its star exploding, but that star was the Star of Bethlehem. So, had God destroyed one innocent world in order to make a fancy light for one night in our sky? In another book, *The Fountains of Paradise* (1979), there is a subplot of an alien robotic probe that passes through the solar system, communicates with humans, and in the process disproves the logical rigor of Thomas Aquinas's *Summa Theologica*, undermining both proofs of theism and the reasonableness of Christianity.

When it comes to specifically hostile ETI and issues with religion, the scenario envisioned in *The Killing Star* (1995) by Charles R. Pellegrino and George Zebrowski is a clear counter-example to what Christian philosophers hope for. In the novel, all but two people on Earth are killed in a matter of hours, as the aliens bombarded the planet with relativistic colliders, vaporizing parts of our world and extinguishing human civilization. The remnants of humanity in a comet, in the minor planet Ceres, and in the Neptunian moon of Titan, are all destroyed, thanks to the invaders. Two humans on Earth are taken as specimens for observation. The only ship that escaped the aliens are led by clones of Buddha and Jesus, the latter who gives the last humans the hope not of ever-lasting salvation but becoming vengeance itself. The Prince of Peace, or at least his facsimile, only promises the sword.

Another example of hostile ETI comes from the *Remembrance of Earth's Past* series by Chinese author, Liu Cixin. The antagonists for most of the books, the Trisolarins from Alpha Centauri, not only outmatch humanity, but their intentions are our extinction. In one space battle, a single probe of theirs destroys the entirety of humanity's space fleet. However, the Earth is not destroyed by *them*, but another, even more powerful ETI that sends a weapon that compresses the entire solar system into two dimensions,

---

[1] McIntosh & McNabb (2021), p. 15.

annihilating everyone on Earth, everyone hiding behind Jupiter, and even those hiding on Pluto. The universe is nothing but hostile, and in the end the universe must also die, if just for the possibility of another universe coming into existence. In this vision of the cosmos, it is not the Living God that rules, but Death.[1]

These science-fiction stories are of varying levels of plausibility, but they capture what is more likely the case than that seen in other films and books. If ETI is able to travel to Earth, they will be vastly more powerful than us, and if they are hostile, then we *will* be destroyed. As estimated earlier, the lowest-level civilization that can travel to our solar system at relativistic speeds is a million times more powerful than us. Since we can destroy ourselves in a matter of hours, greater alien technology by several orders of magnitude all but guarantees we would be gone before we knew we were under attack.

What theodicy makes room for us not just encountering a hostile ETI, but us losing completely? When the First Temple of Solomon was destroyed in 586 BCE, there was still the Jewish people, and a promise of renewal was given. But in the case of existential destruction, there would be no one for God to make promises to. If ETI were to bombard the Earth with relativistic impactors, all of us would be vaporized before our eyes could even register there was something that passed through the sky. There would not be the time to say "oh my God!"

The theodicy that McIntosh and McNabb imagine requires that no ETI could, even in principle, do this to us. Somehow, "the seeds of a successful hostile ETI theodicy would be sown" in the outcome of an S5 event,[2] so it must necessarily be the case that ETI cannot annihilate us, just like how God was supposed to have protected humanity against the Martians in *The War of the Worlds*. The Christian philosophers believe this is so because of independent reasons for believing in the truth of Christianity, but whatever assessment one gives for the evidence of the Resurrection of Jesus, for example, that evidence is inherently weaker than the evidence for the laws of physics. We have far better experimental evidence for conservation of momentum and other laws, and those laws make clear: Relativistic impactors would end us, without fail. No microbes will stop it.

Still, this is a theodicy that we might want to be true, Christian and atheist alike. Let's put it this way: we would rather the theodicy be correct and we are not driven to extinction by ETI than the atheist be proved right by means

---

[1] The title to the last book in the series, *Death's End* (2010, 2016 English publication), is not a literal rendition of the Chinese title, which would more accurately be "The God of Death is Immortal."

[2] ETI is hostile and an existential threat.

of our complete vaporization. Whatever your views on God, we don't want to get to the answers by such a mad (or MAD)[1] scenario.

Let us now consider one more fantastical alien possibility that does not fall into the five scenarios above. Instead of ETI being hostile or helpful, what if it is simply beyond us, even in spiritual terms? Consider the scenario in the famous novel by Olaf Stapledon, *Star Maker* (1937). Not only are there various sorts of ETIs, but there is a greater ETI above it all, the creator known as Star Maker. This entity has more in common with the Demiurge of Plato than with Jehovah, but the incredible feature of Star Maker is that he made many universes, including one that consisted of a universe with a land of the living as well as a heaven and hell, a hell which at some point became so overcrowded that Star Maker entered the universe to bring salvation. But that universe was just one of many trials in universe-creating Star Maker had done. Star Maker would create universes with more than one time dimension, universes where creatures took every action at once, as if they were a quantum superposition of all possibilities. Then came the final, ultimate cosmos, that was beyond the narrator's comprehension.

While this is fiction, we may wonder: If we discovered that our world, either life on this planet or the universe itself, were created by some other intelligence, powerful but naturalistic—an omega-level[2] alien—would that not undermine the Christian narrative? Consider C.S. Lewis's response to Stapledon's book. He wrote to Arthur C. Clarke and told him the ending was "sheer devil worship."[3] So, there is apparently another avenue in which ETI undermines the Christian narrative: if *ETI* instead of us created God.[4] Would it not be the ultimate plot-twist if ETI visited us and told us there is no heavenly father, but declared "I am your father"?

At this time, it looks unlikely that ETI created life on our world. We don't even know if ETs exist, and we have no reason to suggest they started life here 3.8 billion years ago. The Fine-Tuning Argument may be more consistent with aliens than with gods, but it is still a problematic method to prove any sort of universe-creation. While the science may be against so-called directed panspermia and universe-tinkering, the imagined scenario indicates that there are many ways that ETI could be quite unexpected on the Christian worldview, and both philosophers and authors have not yet exhausted the possibilities and conflicts that might arise if we discovered ETI.

And this may indicate the overall issue with aliens and the Christian narrative. If ETI were expected, then there would have been a long literature

---

[1] Mutually assured destruction.
[2] A hypothetical extension of the Kardashev scale, beyond any other numerical value.
[3] Edwards (2007), p. 54.
[4] It would seem then the plot of the film *Prometheus* (2012) should also count as an example of science-fiction that does not comport well with Christian faith.

of philosophical speculations of life in the cosmos. If the story of the life of Jesus and the plan for salvation were not parochial but truly universal, or at least galactic, then ETIs would have been speculated upon by great minds in the Christian tradition. There could have been a magnificent chapter in Augustine's *City of God* by considering a secular world among the stars. There could have been deductions about alien souls and their nature in Aquinas's lengthy theological writings, and Giordano Bruno's speculations about other worlds would not have been part of the inquiry into his heresies. But instead, both Augustine and Aquinas[1] denied the possibility of other worlds, and Bruno was burned alive for his opinion on the matter. Rather, the tradition of Christianity and aliens is very modern and often in reaction to secular thinking. The best way to see that aliens do not fit the Christian narrative is seeing that no one was talking about the Bible and ET until the modern era. Aliens are a symptom of secular thought, not Christian.

Out of Time

There is more still of the Christian narrative that McNabb and McIntosh hadn't considered: Are ETIs awaiting deliverance? If they had their own incarnation of Jesus (see chapter 10), then might they have also been told by their savior "I will return," and that the "end of the age" was upon them (Matt 24:3; 28:20; Rev 3:11; 22:20)? Christians have been waiting for Jesus' return for nearly two millennia, and the excuses for the delay of the End Times is often noted by skeptics of the Christian faith and by biblical scholars to find developments in Christian thinking in the New Testament as later generations explained away all things continuing as they always had (2 Peter 3:3-4).

Let us consider how this will work for another civilization who were also given the message of Jesus' return.[2] The alien disciples would have witnessed Jesus' majesty after his return to life, seeing and holding his hands, giving him food, and watching him return to where he came from. These alien Christians would await the promised day, when the living and the dead would be judged, and a new heaven and a new Earth (or "Earth") would be created. They kept to the word of the Lord, and many generations came and went, just like they have for terrestrial Christians.

However, those ET Christians may be gone. Again, considering the Copernican principle, alien civilizations (if they exist) likely will come into existence in the billions of years to come, and they already existed in the astronomically grand past. That also suggests a terrible conclusion: Some

---

[1] Martinez (2016), p. 355.
[2] This was also considered by Thigpen (2022), p. 348-52.

civilizations have gone extinct. Perhaps a hundred generations of Christians have come and gone, but some ETIs will have been dead before the first letter of Genesis was put to parchment. Astronomical antiquity suggests there are potentially many civilizations who are long-gone.

When we are travelling the stars, we may discover alien civilizations, not by their signals, but by their tombs. Tombs that may have millions of years of dust covering them.

What do we say about the civilizations in the past, if they received the promise of Jesus that they would be saved at the end of the age, but the age of that world had come and gone, and Christ was nowhere to be found? This would be comparable to the last days of humans, and as the final members of our species closed their eyes, and salvation had not come. If life on Earth had ended before Jesus' return, that would seem to be a clear refutation of the prophetic claim that Christ would arrive and be seen by living members of his church. That was the promise (Matt 16:28, 24:34). At least someone holding to the faith would be alive when Jesus returned.

But for some ET civilizations, their existence turned to dust before humans discovered (or invented) monotheism. If Christ visited them a hundred thousand years ago, they may well be long-gone, given the exponentially decreasing probability of a civilization lasting so long (see chapter 1 and the projections from the Drake Equation). If there was any hope that they had received the promise that allegedly Jesus gave to his Disciples, then that promise was broken.

And we have every reason for the promise to be broken for humanity. Again, assuming Jesus must incarnate in every world with intelligent aliens, then Jesus' job of dying for sins will last for millions of years to come. New stars are born every year, and the worlds forming with those stars will, on occasion, bring about intelligent forms of life. This process will carry on in the observable universe for billions of years. Not merely thousands more, nor millions, but billions. The Sun will have died, and new stars will still be forming.

Our planet will be a cinder, gobbled up by our star, which itself will burn out, after several billion years. And after that, somewhere else, life will likely form on another planet. That life has a chance of becoming intelligent and having the *imago Dei*, and they will also need to be a part of the divine plan, requiring Jesus to sacrifice himself again, then promising to return at the end of the age.

That promise will be given to a civilization, when every part of our world no longer exists, having been ripped apart and ionized by the Sun. We may hope that humanity finds a way to exist for billions of years, but probability is against us. And if our existence is unlikely by the time Jesus is promising his return to another star system, then the promise of Jesus' return to us is irredeemably broken.

The Second Coming is incoherent if Jesus has to promise it to millions of worlds over billions of years, and never fulfilling that promise before those civilizations die. Never mind the claim that Jesus was "coming quickly" (Rev 3:11). Such a delayed return, after all are long-gone, is a bad joke, and far less funny that the version of Jesus who literally appears at the last minute of the existence of the universe.[1]

However, one might avoid this conundrum if one instead supposes that there is only one incarnation of Jesus, the one on Earth, so the promise is only made once instead of millions of times to people over vast eons. However, this also seems to be difficult to square with the great periods of time that civilizations come and go. If only the terrestrial incarnation exists, then the most ancient ET civilizations will have come and gone before salvation was even made known. Earlier, the theologian Celia Deana-Drummond stated that Jesus' resurrection from the dead would give hope to all creatures,[2] but this is impossible if the events of Jesus' life took place millions of years after the extinction of some of those creatures.

The Christian narrative of Jesus' death, resurrection, and hoped-for return at the end of time is incoherent when we include civilizations not only far away in space, but also far away in time. Either God's promises are broken on millions of worlds for billions of years (given multiple incarnations), or the promise is simply denied to civilizations that, for no fault of their own, existed before humans did (given one incarnation).

Surprisingly, this possible conflict between Christianity and extraterrestrials was not even considered by McIntosh and McNabb, nor have most other astrotheologians noticed the issue. When Robert John Russell was discussing eschatology, he failed to discuss the actual eschaton—the end times—but he seems to place it still in the future for the whole universe,[3] so he fails to see the problems we highlighted above. More recently, Paul Thigpen also discussed aliens and the end times, but he also believes in a final and universal consummation,[4] which carries the same problems we discussed. Nonetheless, let us consider a few options for ways a theologian might save eschatology from the problem of deep time and aliens. We will consider how aliens who participate in the *imago Dei* can be saved, and what challenges those solutions might have.

First, on the assumption of a unique incarnation of Christ, let us consider the possibility that before humanity existed the entire cosmos was not

---

[1] Douglas Adams, *The Restaurant at the End of the Universe* (1980).
[2] Deane-Drummond (2009), p. 179.
[3] Russell in Peters (2018), p. 312
[4] Thigpen (2022), p. 452.

in a fallen state (i.e., before the sinful act by Adam and Eve). In which case, salvation was not needed, and there was no need for Jesus to incarnate among those old civilizations. While they might have gone extinct, they had never fallen away from God, so they were not cursed to live and die without even having the promise that was given to humanity.

The issue here would be with the general issue of non-fallen aliens: If it is possible that humans did not need to be in a fallen state, then it is even more strange that God created us to fail. If ancient ETs could avoid such a fate, then humanity's failure is inexplicable. So, this first solution will only work if other problems with why God's creatures are imperfect in the first place are resolved. This is tantamount to solving the problem of evil (POE, see chapter 12), so this avenue is unlikely to be successful in the near future, given the POE is still a major problem in theology generally and Christianity in particular after thousands of years of introspection.

Second, on the assumption of multiple incarnations, perhaps Jesus *did return* to those other worlds. Before those species went extinct, Christ came back as he promised to his followers. The dead were raised, the Devil was overthrown, and...

This leads into a new problem. The eschaton, the final event of the Christian narrative, wherein the cosmos is completely transformed, is literally a *final* cosmic event. Satan, for example, is supposed to be defeated, along with all of his minions, in the final battle of Armageddon. The key point is the finality of it all, wherein evil is forever overthrown and death is no more. That has clearly not happened on Earth, so it cannot have happened for the entire universe. A return of Jesus to one planet would not be *the* eschaton, unless every world has its own Satan.

This new scenario has a lot of oddities to swallow. Evil and death are overthrown not across the universe, but on one planet at a time. There are some worlds that have been transformed, and others that will be transformed in billions of years. The heavenly new Jerusalem (Rev 21) will descend to each planet. There must also be millions of Satans, one for each intelligent species that emerges in the cosmos, so each one must be defeated. This all makes the Christian narrative of a *final* eschaton incoherent.

However, it might also be the only observable solution to the theological problem. Suppose that around the star HD 140283, one of the oldest stars known to astronomers (and nicknamed the Methuselah star), there is a planet, and in the past, it had a civilization. Suppose also that this world had a species that encountered Jesus, and the eschaton took place for their world. How would that planet look different than a planet that had not been transformed when God created a new heaven and 'earth'? The planet would have to have fundamentally different physics than a normal planet, since death and decay are prevented in this sanctified domain.

This leaves the possibility of using advanced telescopes to not just find technosignatures of ETIs, but *theosignatures*, where the light of such worlds leads to new physics that break our understanding of fundamental laws of nature. Perhaps the reflected light from such worlds does not have signs of increased entropy after radiating from the surface of the planet—an impossibility for light radiated from our world, but perhaps expected on a world without decay.

If this scenario is not just heresy, it would not only alleviate problems with the Christian narrative, Jesus' incarnation, and ETIs, but it would also provide scientific evidence for God's powers to change the nature of reality. We could have verifiable evidence of a miracle on a planetary scale. Such would be powerful evidence for theism. But again, this assumes that a multiplicity of eschatons is not theological nonsense.

We leave it to the astrotheologians to consider these possibilities. Without taking seriously one the of these solutions above, the Christian narrative of Jesus' final act of salvation is made into nonsense.

Put another way, the existence of ETIs renders the Christian narrative incoherent.

PART IV: FROM ALLAH TO VISHNU – HOW OTHER RELIGIONS MIGHT BE AFFECTED

14 – JUDAISM

According to the Talmud, God visits aliens all the time. Or so it has been suggested by various sources online.[1] This is an odd claim, since the cosmology of the Bible, both the Hebrew Bible and the Greek New Testament, seems to exclude even the possibility of other worlds inhabited by ETIs. So, how could a compendium of debates on interpreting that book come up with the idea that God is a space-traveler? Let's see what all the fuss is about.

The text in question is in *Avodah Zarah* 3b in the Babylonian Talmud. The Talmud is a collection of rabbinic traditions, interpretations, and arguments. The Babylonian Talmud is the version of this collection that was compiled in Mesopotamia by the end of the 6th century. There are also interpretations of interpretations, and we will want to focus on the particular interpretation of this Talmudic text by a late medieval rabbi, Hasdi Crescas. The entire Talmudic tract is about Jews living among the Gentiles; the tract name itself means "foreign worship." So, this already is a strange place to discuss God going on a star trek. Nonetheless, let's look at the part in question. Within the text, there is a discussion of what God does during day and night, and it is said that during the 12 hours of night, God gets on a cherub and flies to 18,000 worlds. The number is derived from a reading of Ps 68:18, and the reasoning is a bit convoluted. But if it says God travels to thousands of worlds every night, then doesn't that mean the Almighty is traveling to the stars?

First, this is a forced reading that ignores the cosmographic context the discussion is found in. Note that the rabbis describing what God does during the day and the night, in particular during each of the 3-hour periods of the day. These scholars are inadvertently assuming something about day and

---

[1] Weintraub (2014), p. 76-82.

211

night. If God is studying the Torah during the first three morning hours as suggested by Abba Arikha (also known as Rav), then whose morning? Sunrise is different in Jerusalem than in Babylon, for example. To say what God does during the day versus the night suggests a universal day and night for everyone on Earth. That only works on the most primitive versions of a flat earth cosmology—yes, even more primitive than the things at modern flat earth conferences. If the Talmudic scholars are assuming the most archaic view of the cosmos, then how are they simultaneously advancing the idea of many Earths, a generally modern view?

When we say the Talmudic scholars assume a flat earth, that isn't quite right. Elsewhere, they are explicit in holding to the ancient cosmology. In *Pesachim* 94b, there is discussion of where the Sun goes during the night. Some believe that the Sun is above the firmament and thus hidden by the thickness of the heavenly canopy. However, the rabbis say that the better interpretation is that the Sun goes below the earth, which is why wells and springs are warmer at night than during the day. Even well into the medieval period, rabbinic sources held to the old view, though there were exceptions such as the influential Maimonides.[1] In other words, looking at *Avodah Zarah* 3b and declaring it has a modern cosmic view in mind is specious.

There seems to be a misreading of this discussion in the Talmud. The term used to mean 'worlds' (*olamim*) can mean a plurality of Earths, but it can also mean periods of time or places on Earth, including places humans can go to, as can be seen in other Talmudic passages. For example, the righteous will each be given 310 "worlds" by God in the future;[2] this is not likely to mean each good Jew will get 310 planets, but rather control of regions on Earth. At least that is a more plausible interpretation of what folks in the 5th century would be discussing.

Also, if God is traveling by chariot, even with 18,000 of them, he's not going to get from planet to planet in one 12-hour period, let alone to all of them. There is again no sense of the scale of the universe in this passage, if we force the text to mean it's about traveling to the scientific planets. Instead, the passage could be about other places in the Hebrew heavens that God resides in. Elsewhere, God's city is said to be 18,000 reeds in circumference,[3] which is an oddly specific (and familiar) number. In fact, the number may come from not places but years, since the Great Year of Greek philosophical thought was also 36,000 years long, so half of it would be 18,000, which was also related to a golden era.[4] The flexibility of the term *olam* meaning places and periods of time could allow for the transformation from this

---

[1] Brown (2013).
[2] *Sanhedrin* 100a.
[3] *Sanhedrin* 97b.
[4] Laurent (2015), p. 84.

philosophical speculation of eras to the less precise Hebrew term, allowing for 18,000 years to become 18,000 periods and then to 18,000 worlds. (The Greek word *aion* has a similar semantic spread.) More likely than not, the number held mystical value to the writers of the Talmud, which is why it was forced into their interpretation of the Hebrew scriptures. At the very least, it seems to have nothing to do with visiting the planets.

While this has little to do with aliens, others allege that another Talmudic passage is about the inhabitants of a star. Discussing the curse against Meroz in Judges 5:23, some commentators indicated uncertainty about what "Meroz" was. The context of the passage suggests a city or location in northern Israel, but other rabbis in the Talmud suggested that Meroz was either a great person or a star.[1] If the inhabitants of a star are cursed, doesn't that mean aliens are in the Talmud?

Let's again consider the passage these scholars are looking at. Judges 5, which contains the Song of Deborah, describes a battle against a Canaanite general, Sisera. It is mentioned that, along with the Kishon River, the stars fought against Sisera, and he was defeated at the battle of Mount Tabor. By stars, most commentaries would suggest this is a reference to angels. At the very least, literal stars do not actually come down to Earth and form a phalanx. So, if there was a star named Meroz being punished, this would be an angel and not an alien civilization. Conversely, if it really were the case that ETIs were a part of a Bronze-Age battle, then one would think it would have a bit more of a presence in the biblical story. Where are the laser guns, for Pete's sake! Deborah composes this whole song about a battle, and she only mentions in half a verse that interstellar forces were on the side of the Jews?!? To say such a silence is unlikely is to be generous.

What we are seeing in these attempts to find ET in the Talmud is not unlike what happens on shows like *Ancient Aliens* and other dubious "history" shows: All of the historical context is removed to make some text or artifact look strange, and the modern imagination forces it into the shape of a bulbous, grey body as we commonly imagine ET to have. But when recontextualized, it looks less alien and more human, or at least within the sorts of thoughts pre-modern people had.

And this is exactly what the people on *Ancient Aliens* do with one passage in particular from the Hebrew scriptures, Genesis 6:2-4. This is the story of angels coming down to Earth and having sex with women, who then give birth to giants, and this was one of God's motivations for sending the Flood to wipe out all these wicked things. The suggestion is that the angels (or sons of God, more literally) were aliens, and the offspring of these angels and women were alien-human hybrids.

---

[1] *Moed Katan* 16a.

There are oh, so many scientific issues with this, but let's just name a few. First, your chances of getting pregnant having sex with an alien are lower than having sex with a tree and getting pregnant. You are genetically closer to any living thing on Earth, since you have a common ancestor with all living things here. But for an ET, you have no genetic commonality, so the chances that your DNA and theirs (even assuming they have DNA) will combine together is all but zero. And why would you want to have sex with an ET, or vice versa? Consider our closest animal cousins, the bonobos. If you don't find them particularly exciting, then why think a species that is even less like you would be an attractive romantic partner? The same is true for the aliens; they aren't likely to find human sexual characteristics to be that enticing in the same way we don't find red butts of baboons to be sexy.

Plus, if we take it as given that these creatures in Gen 6 are aliens, because they are called "sons of God," then God must be a space alien, as he is the father to these aliens. Is that a likely view of the authors of the Torah? We think not.[1]

In summary, there are about as many mentions of ETI in the Old Testament and Talmudic literature as there are non-kosher recipes recommended in the same books. Nonetheless, if aliens landed in the Holy Land tomorrow, what should be the response of the rabbis? How will the ETs be considered? In particular, could an alien become a member of children of Abraham?

The various laws of Judaism may have little meaning to an ET. For example, the practice of male circumcision may be incoherent to a species that does not even have a penis, let alone a foreskin. Numerous purity laws in the Torah concern the physical form and status of the human body (e.g., Lev 21:16-24; Deut 32:1), so these could not be applied to an alien body. At the very least, significant reinterpretation is needed to bring non-humans into the fold, assuming an ET wanted to convert to Judaism.

However, it seems more likely that Judaism is likely to be only for humans. That seems to be the more common opinion given by several modern thinkers on the subject.[2] Others have found some humor in the idea of non-humans wanting to become a part of the religion, as seen in a modern story "On Venus, Have We Got a Rabbi!" (1974). In this, the Bulbas from Rigel IV wish to be seated and accepted at the conference of interstellar Jews, but their alien form is causing trouble. In the end, it is decided "there are Jews—and there are Jews. The Bulbas belong in the second group."[3] That's probably about as definitive as one can get in favor of alien Judaism.

---

[1] We might say let's ask a rabbi to know what is the most-likely interpretation, but as the saying goes: two Jews, three opinions.
[2] Weintraub (2014), p. 80-82.
[3] Tenn (2016).

15 – ISLAM

According to the Qur'an, Allah created many creatures, such as various equines. But, according to Surah 16:8, he also created "what you do not know." Is the Almighty being coy about ETI in this verse? Do we have knowledge of aliens by inuendo?

These and several other verses have been used by some Muslims as evidence of knowledge about aliens, and perhaps such insights are an indication of the prophetic abilities of Muhammad. We will quickly look at several claimed passages,[1] what other interpretations seem to fit, and the general trends we see, especially after what we saw in Christian and Jewish narratives in the prior chapters.

Very early in the Holy Qur'an, Allah is said to be the Lord of "all worlds" (Surah 1:2). Noting the plurality of worlds, does this mean many planets like the Earth? Consider first that there is elsewhere the reference to the multiplicity of heavens, in particular seven heavens (17:44, 42:29, 65:12), but also otherwise there is only ever reference to the Earth in the singular; so the multitude of planets does not seem to be in view. In fact, the "seven heavens" indicates that the cosmology used in the Qur'an is one derived from Greco-Roman astronomical theory, wherein the planets orbiting the Earth are each in their own sphere and their own level of heaven. This was already noted as part of the New Testament, such as with Paul's visit to the third heaven, as well as other Judeo-Christian literature of the first centuries of the Common Era.

Conversely, in modern cosmology, there is not a multitude of heavens, let alone seven. There are no layers, and there are more than seven planets, none of which orbit the Earth. Instead of this verse indicating a plurality of habitable planets, it seems to be consistent with the problematic and premodern cosmology of late antiquity.

But it is still suggested that, in Surah 42:29, there are living creatures in those heavens. These creatures are "spread forth in both" heaven and Earth. Would not spreading creatures to heaven mean aliens? The creatures in the heavens could be a reference to supernatural angels (17:44) and not naturalistic ETIs, but even assuming that the verse is not about angels, there is another interpretation. In the cosmology of late antiquity described by Greek and Latin sources, the first heaven was the sky, sometimes called the firmament—the upper reaches of what we considered the atmosphere where air exists. So, there is a creature that could be spread forth in the firmament:

---

[1] Sardar (2020).

215

birds. This not unlike how birds are described in Genesis 1:20, which fly across the "expanse of the heavens" (notice the plurality).

Then again, in Surah 65:12, it is said Allah created the Earth in the likeness of the heavens. Doesn't that imply there are things in the heavens similar to those on Earth? This again, in the older cosmology, is not what moderns think. In Platonic thinking in particular, the heavens are the world of perfection and the true forms of things, while on Earth are mere copies of those perfect forms. So, to say there are things on Earth like they are in heaven can be just the same as this old cosmology: We have crude copies of the perfect things in heaven. However, most commentaries on this verse suggest it means that Earth has seven levels, just like heaven does. In either case, these interpretations are more likely that interpreting there are earths in the heavens, again considering the cosmology of the Qur'an is derivative of the late antique model which had no planets going around other stars.

And those are all of the Qur'anic verses some look to find ETI in Islam's holiest text. Instead, the contextual clues place them in the same vein as sources found elsewhere in late antiquity, which is also contrary to modern astronomy and cosmology.

Nonetheless, modern Islamic interpreters may find a clue to the specialness of humans compared to any such ETIs. In Surah 95:4-5, humans were created as the best sort of creature, but were also made low. This is interpretable as humans being wonderfully made living beings, given both beauty as well as intelligence,[1] but that we are also destined to suffer and die in this world, and those who do not do good will go to hell (95:6). This relatively high-status for humans would suggest that Earthlings will be superior to any discovered aliens among the stars.

If one takes this interpretation of the Qur'an, then they would be running against the Copernican principle—we expect to be common or average, not exceptional. Moreover, if we do discover ETIs, those aliens would more likely than not be scientifically and technologically more advanced than us. If they are very advanced and can travel among the stars, on what grounds could we claim that we humans are superior?

A homo-centric view is likely to cause at least as much consternation as the *geo*-centric view that Copernicus helped overturn did. In which case, Muslims that do not take humans to be just one more intelligent species among many are more likely to have strong cognitive dissonance if we discover ETIs.

However, there is also a general approach as seen by many interpreters of Islamic texts and traditions concerning angelic creatures and the demonic creatures known as *jinn*,[2] and this approach may avoid the worst of the

---

[1] *Maarif-Ul-Quran* 95:4; *Tazkirul Quran* 95:4-6.
[2] Determann (2021).

homo-centric perspective. In particular, it seems that numerous authors in the predominantly Islamic world have interpreted these supernatural beings as a form of alien. Sometimes it is to make the aliens more theological; sometimes, it is to make the angels and demons more scientific.

For example, some used *jinn* as an explanation for the sightings of UFOs in the 1980s.[1] This would be an example of making UFOs into something theological. The main issue here, though, is that UFOs (or UAPs) are probably not aliens, as was discussed in chapter 3. A theological explanation for a collection of non-alien objects seems unjustified.

On the other hand, books such as Swiss author Erich von Däniken's *Chariots of the Gods?* (and other books that claimed extraterrestrial influences on early human culture) was somewhat popular in countries such as Iran before its revolution in 1979 and remained so for decades.[2] In this interpretation, the angels and demons are actually ETIs that were given supernatural attributes because of their superior technology and the ignorance of ancient humans. While the ancient astronaut hypothesis has little scientific support, this Islamic perspective at least does not mean humans are special and thus conforming to the Copernican principle.

In many ways, the Islamic world has been a mirror image of what was seen in the West. The old religions need to find ways to conform to the new Zeitgeist, and if that includes beings from the stars, then one must discover a way of finding those star men in the old texts.

However, this cursory look at that effort suggests Islamic scholars have been no more successful than their Western counterparts trying to fit ETIs into scripture. In which case, the actual discovery of aliens would have some of the same difficulties found in Christianity, especially in terms of eschatology. We also wonder if multiple Muhammads are required for sharing God's (or Allah's) message, just as multiple incarnations of Jesus seemed to be needed for Christianity. Moreover, since Islam reveres Jesus and Mary, then solutions to how many Jesuses exist must be provided in order for Islam to have a coherent response to the discovery of ETI. Perhaps this will lead to interesting future interfaith dialogs among earthlings—not to mention, dialogs with the aliens themselves, if possible.

---

[1] Determann (2021), p. 110, 136-37.
[2] Ibid., p. 127-28.

16 — HINDUISM AND OTHER EASTERN SPIRITUALITIES

One of the issues with applying the question of "aliens and religion" to many beliefs and practices from the Eastern Hemisphere, especially southeast Asia, is that the concept of "religion" breaks down.[1] The idea of religion as a set of beliefs or doctrines, codified in a holy text, comes out of modern European history, and then explorers and colonists forced non-Europeans into the framework of "true" and "false" religions. In ancient languages, there is no real equivalent to the modern term "religion," and there are serious distortions to get words like *dharma* (literally, "laws") in Sanskrit or *jiao* (literally, "teaching") in Chinese to mean the same thing as the Western term of "religion."

However, if instead of the strictures of the term, which are still debated in religious studies, we simply look at the sorts of beliefs that normally get grouped together as "religious beliefs," then we can meaningfully ask how those beliefs would be impacted by the discovery of extraterrestrial life. The traditional beliefs in reincarnation, in particular, become more interesting when one considers possibilities that one does not have if life *only* exists on Earth.

So, let us consider some of the beliefs that are normally categorized into Hinduism, Buddhism, Sikhism, Zoroastrianism, Taoism, and the like.

Let's first consider Zoroastrian beliefs, since they have strong ties to Christianity and Judaism. According to Matt 2, members of the priesthood of Zoroastrianism visited the infant Jesus, and Old Testament scholars find influence on Judaism from Zoroastrian beliefs. Perhaps the most pronounced of those beliefs are the world-savior figure (the *Saoshyant*), the resurrection and judgment of the dead, and the final battle between the forces of good and evil. However, also like Judaism and Christianity, the events of those final events are quite parochial. Everything happens on Earth. The Saoshyant is born of a virgin who is impregnated by the sperm of the prophet Zoroaster, left behind in a lake in Iran. The signs of things to come are also very Earth-centric in the sacred texts. The stars are holy and even divine (in particular, Sirius called Tishtrya), while the comet (named Mush-Parig) is an evil entity let loose occasionally from the Sun and is part of the final judgment.[2] In modern astronomy, comets are simply balls of water-ice, carbon dioxide, dust, and traces of other materials—not demons.

---

[1] Nongbri (2013).
[2] *Bundahishn* 28, 30.

Under Zoroastrianism, there is no obvious place for aliens. The stars are not locations to visit in the *Avesta* literature, but divinities to praise. Also, there is little to no conversion to Zoroastrianism, since most believe one must come from Zoroastrian parents to be a part of the faith community.[1] If fellow humans have a hard time getting into the club, then there is little hope for an ET to be a part of the group.

If ETI are interested in human spiritual beliefs, they may instead focus on the oldest still-practiced set of beliefs with millions of followers today: Hinduism. Again, we must not make the mistake that Hinduism is a unified set of theological beliefs or doctrines, but there are common features found in the spiritual beliefs and practices located especially in India. Besides the great plurality of divinities in the subcontinent, the belief that may be the most interesting to consider if aliens exist is the concept of life after death in the form of reincarnation.

Traditionally, the world we inhabit is cyclic, in that we are born into it, do our deeds, die, and then we are reborn, and the status of our reincarnated self depends on what happened in that previous life. This cycle of *samsara* is unending, except if one gains enough *karma* in past lives. Buddhism takes this idea but provides a route to escape *samsara* with the eightfold path. Now, assuming one has not raked up enough *karma* points, the next life might be higher or lower than the current one. In fact, even plants and animals participate in *samsara*. So, if you behave poorly, in the next life you might be a poor beggar, a slave, or perhaps even a house fern. Next time, do better!

But now consider this question: In the next life, could you wake up not in the body of a human, fish, or pine, but in the tentacled body of an alien species? And conversely, might an alien find their next life on Earth?

This seems to be plausible to some modern Hindus. For example, C. Bhaktivedanta Swami Prabhupada, the founder of the Hare Krishnas, provided a cosmology that included countless planets with life forms.[2] Then again, the description of that cosmos does not seem to match well the universe of modern science. Prabhupada speaks of the Sun being driven by a chariot, that there are higher planets (there is no "high" in space), and that when the universe is destroyed after about 300,000 billion years, those higher planets will remain and contain the spirits of all the good Hare Krishnas. This setup suggests that there are planets outside of the universe, which makes no cosmological sense, at least not in modern astronomy. So, there appears to be some forced readings of traditional texts to seem more modern yet holding to ancient assumptions of the structure of the universe and reality.

Nonetheless, beliefs about *samsara* do not appear to have a particular issue with extraterrestrials. If a human *atman* (self or soul) can transfer to a

---

[1] *Encyclopedia Iranica* (1993), Conversion vii.
[2] Weintraub (2014), p. 172-174.

rat and vice versa, then there is no theological reason it cannot transfer to or from an alien lifeform. However, the magnitude of the cycle of life and death is far larger than what could have been imagined when the *Rig Vedas* were composed. The mythological texts still held to the pre-Copernican universe, so even if Hinduism can be made consistent with an ET entering *nirvana*, one cannot say it specifically endorsed or predicted the idea.

Making eastern beliefs about the supernatural recycling of the *atman* more modern by including aliens leads to interesting results. For example, not unlike western UFO cults, there are stories of people receiving messages from aliens, but with some attachments to Eastern religious artifacts. For instance, one fairly recent story out of Thailand speaks of an alleged ET hovering over a statue of the Buddha, giving important messages about the approaching world war, and about how salvation can be achieved by typically Buddhist concepts and practices.[1] Sightings of the aliens and the feelings they give to believers seem commonplace compared to Western versions: apocalyptic messages, feelings of energy, doctrinaire. Founding figures in this local UFO movement speak of aliens causing them to spin about. Holy rollers, but with more outer space.

One can only speculate what will happen when ETIs are confirmed to exist. If Eastern and Western UFO cults are any indication (and see the next chapter about them), the old-time religions will be re-packaged. Instead of the colors of traditional objects of veneration, the cults will come with shiny chrome trappings.

---

[1] Ehrlich (2019).

17 – OTHER RELIGIONS OF NOTE

UFO Religions

Depending on one's point of view of what is an alien and what a religion is, there are different answers as to what religion was the first to have aliens at its core.[1] Perhaps one could point to the 19th-century esoteric religion known as Theosophy, with its primary founding figure, Helena Blavatsky. In her most important Theosophical tract, *The Secret Doctrine* (1888), she "channeled" a legendary book of secret wisdom, and she promoted the idea of extraterrestrial life, in part basing those speculations on scientific rather than just esoteric grounds. Later theosophists would expand these ideas, including populating the planets of the solar system with life.

These fantastic beliefs were then parodied and molded into the horrific fictions of HP Lovecraft, whose Old Ones are extraterrestrials who come to Earth in the distant past, creating and evolving life as they saw fit. This is chiefly seen in his *At the Mountains of Madness* (1936), wherein a team of scientists in Antarctica discover the lost remains of the extremely ancient, alien civilization, along with the records of their accomplishments, wars, and decline. Here, aliens were gods, including the creator gods.

While Lovecraft's mythos is more popular now than in his own time, the same cannot be said for Theosophy, so it is not much of a cult of alien gods today. Once the UFO craze came into full swing in the 1950s, various new cults sprung up that centered on contact with ETIs. In the United Kingdom in 1954, George King started to claim he was in communion with such entities, and the Aetherius Society began the next year. Among the beliefs of the Society is that there are Cosmic Masters that reveal great truths, and among these Masters from the heavens who have come to Earth were Buddha, Lao Tzu, and Jesus. That Jesus lived on another planet would be a tenet in another UFO religion, Raëlism. Its leader, Raël (formerly Claude Vorilhorn) began the cult in the 1970s, and it is stated that Jesus, Moses, Buddha, and Muhammad live on the Planet of the Eternals.

Perhaps the largest UFO religion is that of Scientology, created by science fiction writer L. Ron Hubbard in the 1950s. Many of the tenets and scriptures were closely held secrets, but leaks of those mysteries show a mythology right out of mid-twentieth century science fiction. Unsurprisingly. The story includes bringing billions of people to Earth in spacecraft that

---

[1] More analysis on UFO religions can be found in Lewis (2003).

looked like mid-century aircraft (specifically the DC-8), the use of hydrogen bombs, and tricking people with income taxes. Lord Xenu of the Galactic Confederacy is perhaps better-known through the parody on *South Park* than by direct reading of the hand-written tale by Hubbard.

It would take up too much space to get into the epistemic issues with belief in any one of these UFO cults. Instead, we shall consider the following: What will happen to these religious movements if we discover ETIs by scientific means?

This will most likely depend on the kind of detection, represented by the Rio Scale. If there were a weak radio signal that indicated an artificial source, or if we were to find the technosignature of some other kind of ET hardware, then the existence of aliens would be an abstract notion, though they would be real. On just that point, UFO believers could say they were vindicated—science would have *proven* what they believe in is real! However, more careful considerations indicate that such announcements would be going too far. After all, the UFO believer suggests not just that aliens exist anywhere in the universe, but that they have made contact and have visited our world. The mere detection of aliens lightyears away proves none of that. And given the astronomical distances between us and any likely ET civilizations, there is still good reason they haven't come here.

These UFO groups also make many other claims that are dubious, even if aliens exist *and* can travel to Earth. Scientology gives a history that goes back over a trillion years, which is older than the universe itself.[1] Not even proof of a Galactic Confederation will fix this problem. The claim that Jesus was a space-man and lives on another planet with other religious figures is also unjustified by the mere existence of ETI, and it is contradicted by all of historical Jesus studies.

On the other hand, if there are proven aliens out among the stars, we expect more people to claim that they have had contact with them. The more real aliens become, the more potent claims of alien knowledge will become. While the current UFO religions might die out, we suspect that new ones will replace them.

Battlestar Mormonica

While the earlier chapters discussing Christianity in the light of ETI's discovery found significant issues with doctrine and traditions, this was largely focused on the Catholic Church and mainline Protestant denominations. There is another, non-trinitarian version of the Christian religion that is

---

[1] Kent & Raine (2017), p. 11.

unique in ways that might make it more accepting of ETI. Let's consider the case of the Church of Jesus Christ of Latter-day Saints, better known as the Mormons.

The American religion, founded in the 1820s by Joseph Smith, include numerous tenets that are not considered to be enamoring to other Christians. Infamous are the ideas of polygamy (though not actively practiced by the official LDS church), the direct impregnation of Jesus' mother Mary by God, and the general physicality of God. That physicality is actually a feature that seems to make Mormonism more compatible with ETIs than other religions. In a sense, in Mormonism, God is an alien from the star or planet Kolob, as seen in Mormon scripture.[1]

The plurality of worlds and its inhabitants is found elsewhere in the Mormon canon. Within the next well-known book used in Mormonism, *The Pearl of Great Price*, there is a new rendition of the books of the Bible, and in "The Book of Moses" 1:35, it states

> For behold, there are many worlds that have passed away by the word of my power. And there are many that now stand, and innumerable are they unto man.

The original Genesis story only told about our world, but there are many, many more, according to this new version. This already makes the Mormon cosmology more in-line with modern astronomy than the previous, Hebraic cosmos of Genesis 1 2.

The existence of many worlds with intelligent residents is stated in other canonical texts,[2] so this vision of the universe is not a small part of Mormonism. Additionally, early leaders in the church after Smith's death stated that various bodies in the solar system, including the Moon and Sun, were inhabited.[3]

Perhaps it is no wonder that this favorable view on the existence of aliens has allowed Mormonism to go into (fictional) space. In the sci-fi novel and TV series, *The Expanse*, it is the Mormons that built the first interstellar colony ship to spread their faith to another star system. But perhaps most gloriously, in the original incarnation of the TV series *Battlestar Galactica* (1978-1979) does Mormonism thrive among the stars. In this famous series, the humans from the twelve colonies are escaping destruction by evil robots (guided by Lucifer) to find their new homeland, and along the way they rediscovered the old planet of the gods, Kobol. The show was in effect reenacting

---

[1] Book of Abraham 3:2-3; Facsimile No. 2.
[2] I.e., Doctrines and Covenants 76:24, Book of Moses 7:30.
[3] *Journal of Discourses* 13:271.

the overarching story of the Book of Mormon, but with a *Star Wars* flare—not so many Babylonians soldiers, but plenty of Cylon Centurions.

It seems then, if any denomination of Christianity is going to be on-board with the discovery of aliens, it will be the Mormons.

However, if one were to use the existence of aliens as proof of Mormonism, they would have numerous historical problems to deal with. The Latter-day Saints accept the stories of the Bible as well as the Book of Mormon, and there are plenty of things in the Bible that are historically dodgy. And Mormons must also posit an escaping group of Jews that came to the Americas and populated the land, had large wars with metals that were not used in manufacturing at that time, and so many other things that make being a Mormon archaeologist one of the least-happy jobs out there.[1] We must also note that aliens are not more expected on Mormonism than on atheism, so ET couldn't be evidence for the followers of Joseph Smith anyways.

Nonetheless, we suspect the Mormons will have the easiest time of the religions we have looked at incorporating intelligent beings from the stars into their theology. They might also be some of the first to proselytize to them.

---

[1] For many good observations about the issues with Mormonism, see Fitzgerald (2013).

18 – ATHEISM: WHAT IF ALIENS ARE CHRISTIANS?

Given the previous chapters, one would likely come to the conclusion that aliens are trouble for the traditional religions. So, shouldn't the fall-out from all that be that ETI would increase the probability of atheism? As noted when discussing the problem of evil, aliens exacerbated the problem, so aliens did add evidence for atheism.

However, it would be a mistake to assume ETI proves atheism or that there aren't issues for nonbelievers. There are numerous science fiction scenarios that we might consider that could actually reinforce the old-time religions, even the ones with apparently deep theological problems when trying to fit aliens into the narrative. So, if ET lands tomorrow, her first words may not be what an anti-theist might expect.

As noted in chapter 4, our expectations were that if aliens had religious beliefs, they would be very different from any of the commonly held faiths in Earth history. For all we know, the aliens could have no sense of religion at all. However, the converse may also be true. One can imagine a strong faith tradition, perhaps one that allowed an ETI civilization to combine its powers and resources to explore and conquer the galaxy. The scenario exists in numerous examples of science fiction, from the Covenant in the *Halo* video game franchise to the Krill from *The Orville*. There is nothing far-fetched with the notion that aliens have and use religious convictions to organize. After all, religions seem to have been an important organizing force in early human civilization.

If theism is as common among aliens as it is among humans, perhaps that is because there *are* supernatural entities in the universe, and all sentient lifeforms have some awareness of them. Initially, a universal belief in gods is more likely on theism than atheism.

But let us imagine an even more stark example of alien religion: What if our first encounter with an ETI is with a missionary from a Jesuit order? That is, what if there are already Christians among the aliens? Not only are they technologically advanced, but they are also likely to be advanced in their theology as well, having had their Thomas Aquinas centuries before ours learned to read.

The chances of two distant planets, with completely different ecosystems, night skies, and cultural histories, having had no prior contact, developing the same basic theology is far too remarkable to be a chance event. If the aliens tell us that Jesus came to their world, was executed, rose on the third day on their calendar, and preached his return, there might well only be one explanation: Jesus is real, and he does travel the universe bringing salvation to all worlds.

When we were discussing issues with multiple incarnations of the Messiah, we seemed to run into several issues. However, if we have direct evidence that Jesus actually did live on multiple planets, then it is a fact, and we just have to do the theological and philosophical work to make sense of the strange new fact of the universe. Just because quantum mechanics seems absurd when described doesn't mean it's wrong, given that QM's predictions have turned out fantastically correct. The same would be true if we found Jesus on planet Kolob. We don't have an explanation for it, and it seems bizarre. But if that's the reality, we will have to change our minds and figure out a better worldview.

However, this scenario is extremely unlikely, if for no other reason than Earth history gives no such signs of universally-revealed religious truths, especially those specific to the Catholic catechism. There would have to be a lot of explaining to be done, as we figure out why Jesus' message wasn't globally revealed and yet it was intergalactically delivered.

Of course, all of these issues for atheists are also issues for religionists. Likewise, there would also be issues for Christians if aliens were found to be Muslim or Hindu or adherents of any other belief system.

One final point to consider: Let us entertain the idea that the ETIs we are likely to discover are more advanced than us in terms of science and technology. This might also indicate that they are more advanced in their philosophy. Not only might they have resolved metaethical issues more robustly than any human philosopher has done, but they may also have an active philosophy of religion. In which case, they might have new and better solutions to the problem of evil. Perhaps their version of William Lane Craig has *actually* conclusively *proven* the impossibility of an actual infinite and demonstrated the necessity for a Creator.

If we accept the ETIs are more knowledgeable than we are in terms of science, we might also have to initially accept their conclusions in philosophical matters. So, if an alien is a philosopher and a theist, we might again need to add weight to a generally theistic worldview. This would be similar to finding out that human philosophers were mostly theists; shouldn't that influence our convictions in our philosophical deductions if we know they proved unconvincing after several centuries? Likewise with aliens who are centuries ahead of us.

Again, we have no initial reason to think an ETI will have these religious views. Nonetheless, we ought to be open to having our minds changed if we were to receive such awesome new evidence. Perhaps we would not be instantly convinced by ET's arguments, and we cannot accept those beliefs based on mere authority, but we would listen.

Really, ETI. If you're out there, we're listening!

19 — ALIEN APOTHEOSIS

On April 9th, 2021, a god gave up the ghost. He was a prince who lived in a far-away land, according to those who believed in the deity. Members of the religious movement possess several icons depicting the deity. In one image, he wields a *nal-nal*, a wooden club traditionally used on pigs. With great sadness did these people learn of the death of the son of a mountain spirit, a son named Prince Philip, the Duke of Edenborough and husband to Queen Elizabeth II.

The Prince Philip Movement is one of several modern examples of living people in recent times receiving divine honors.[1] The late prince isn't even the only Englishman in modern history to have been so elevated. The explorer Captain James Cook was conflated with the Hawaiian god Lono, and this appears to be serendipitously due to a set of coincidences. Cook's initial circling of the island and then landing during an important religious festival corresponded to rituals and myths. Cook's impressive ship perhaps added to his mystique, though his poor manners and brutal actions against the native Hawaiians had more to do with his demise than a sacrificial ritual, as earlier historians had suggested. The humanity of Cook was apparent when he died on the coast of Kealakekua Bay.

But one is led to speculate: If instead of a man who sailed to Hawaii in a vessel of timber, what if the ship were of shimmering metal, flying instead of floating, and aboard was no human, but something far more alien to the Hawaiians than some chap from Yorkshire? Perhaps instead of a niche cult or short-lived conflation, a wide-spread religion could have emerged.

Throughout this book, we have largely focused on the mere existence of extraterrestrials, with the occasional note to consider how people would react if we discovered the presence of ET close by, or even in our solar system. On the Rio Scale, the factors that went into its calculation were the nearness of an alien signal, the prospect for communication, how likely ET would be aware of these communication possibilities, and lastly how likely the signal is really from ET and not something else.

If the aliens were far away with limited communication prospects, then there would not be as much of an impact. However, if the aliens were here on Earth, then the impact would be extraordinary. Let's consider both scenarios for the effects of religious beliefs on the general population, investigating this question, "Could ETs become deities?"

---

[1] Subin (2021).

In one sense, aliens cannot, even in principle, be gods. The general definition of a god is some sort of supernatural entity, and ETs are going to be made up of atoms and following natural laws. ET also cannot be the cause of existence or the ground of being, as God is imagined in Thomistic theology, for example. However, definitions only capture how people use a word in a given era and are subject to change. After all, plenty of humans were said to have become gods upon their deaths, such as Egyptian Pharaohs or various Roman emperors. Even within Judaism, a living person could be given divine adulations, such as in the case of Herod Agrippa I in 44 CE.[1] The boundaries between divine and non-divine things have been blurry. Conversely, the ideas that gods are powerful, or they have done something to make them worthy of worship, do not cleanly limit the scenario to the supernatural. Indeed, many natural forces were considered to be gods themselves. Our changing conception of the world shifted the boundaries as to what is and is not god-like. A visitation by ET would almost certainly cause those boundaries to be renegotiated.

As was determined earlier, the most likely scenario is that alien civilizations are far away, potentially even in another galaxy. This would suggest that the impact of discovering alien life and how it would change religious thinking is much more likely to come by means of remote sensing. The less likely scenario is ET arriving here on Earth. While the latter would be more impactful, its lower probability means we should only consider its effects secondarily. If the aliens do manage to make it here, then how those creatures interact with us will have a considerable impact on how they shape our society. If ET were to treat us like Captain Cook did the Hawaiians in the 1770s, or instead like Mick Leahy treated the Papuan people in the 1930s, the outcomes could well be wildly different.

The Remote Religion

If we do prove the existence of intelligent alien life in the universe, and it is far away, groups of people would likely think highly of these creatures even without meeting them. If Prince Philip could be considered a god without even having been seen by his worshippers, why not the same for someone believing in the power of the Tralfamadorians living in the Small Magellanic Cloud 200,000 lightyears away? The example of Prince Philip may be the most relevant historical case study to explore discovering a distant, technological power.

---

[1] Josephus, *Antiquities of the Jews* 19.343-50; cf. Acts 12:20-23.

The religious movement started on the island of Tanna in the South Pacific, and this movement is in fact a splintering of a previous cargo cult religious movement that started in the late 1930s.[1] Before Philip, there was another alleged European figure who was revered—a man named John Frum. His name is perhaps a shortening of "John from America," with his other epithet, *Rusefel*, sounding similar to the name Franklin D. Roosevelt.[2] John Frum was encountered not because the US President was making trips to the small island, but because of visionary experiences on the part of some of the islanders. The hope was that John Frum would remove all of the white colonizers and bring the wealth and prosperity of America to their island. This clearly messianic cult would go on to splinter into factions, and by the 1970s, one such group on the island of Tanna would say that Prince Philip was a brother of John Frum and now their center of worship.

While the British prince had not visited the island or made overtures to them, the founders of the movement first attempted to contact the monarch. Over the years, gifts were exchanged, including a signed portrait of Prince Philip and a traditional pig-killing club. There have been significant other forms of contact between Western sources and the Tanna people, so the myths about Philip have been fast evolving. This begins to break down the analogy between the Prince Philip Movement and what is likely to be the case with discovering an alien civilization.

The most likely discovering-ETI scenario is finding some sign of advanced technology, be that radio signals, exoplanetary atmospheric pollution, or waste heat from megastructures; two-way communication is highly unlikely, and near-impossible if ET is thousands of lightyears away. What would be concluded and reported in the media might look like this: ET exists, they are far away, and depending on the technosignature, perhaps we can say something about their technology level. The back-and-forth contact between the islanders in the Prince Philip Movement and the English is quite unlike what will happen with the remote detection of ET.

However, the earliest formation of the Tanna religious movement remains a useful analog. There was some pre-established belief in a sort of messianic figure, which itself has a long history in those South Pacific islands. The incorporation of Prince Philip was serendipitous, in that he could be seen as a higher power and worked into prior beliefs about mountain spirits and John Frum, and the connection between Frum and Prince Philip was done by the chance discovery of the monarch by indirect means.

Following the example from the cargo cults, what might happen next could well depend on the existing theologies and peculiarities about what is discovered about the aliens, and when. The timing of the detectable

---

[1] See Worsley (1968), Trompf (1990), and Steinbauer (1979).
[2] Steinbauer (1979), p. 86.

transmission might be seen as significant, much like the coincidental timing of Captain Cook's arrival on Hawaii. If the transmission were found on an important holy day, that could be taken as significant. For example, if a signal were detected on Easter Sunday, that could be interpreted by some Christians that the signal were related to Jesus and his return to life. Similarly, if it were detected on Ascension Thursday, then that could be interpreted as Jesus returning in some form. Almost certainly someone would find a way of construing the timeline prophecy from the Book of Daniel in light of this discovery of an alien signal.

The same could be possible is Islam. Laylat al-Qadr is the highest holy day, as it is said to be when the Qur'an was first sent from heaven. Laylat al-Qadr, also known as the Night of Destiny or the Night of Power, could be an opportune moment for the discovery of a heavenly message. As for Hindus, perhaps an ET signal arriving to Earth on Rama Navami, the birthday of Rama, the seventh avatar of Vishnu, could be seen as the coming of another avatar. Which major religion might capitalize the most on detecting an alien message might come down to a lucky hit on the calendar.

However, there are other religious believers who could even more easily absorb and grow after discovering the existence of ET. Even without scientific evidence that aliens exist, let alone having already come to Earth, there are and have been numerous UFO religions. The founders of these movements often claim to have some sort of revelatory event, allowing them to bring a message from extraterrestrial powers to us mere terrestrials. George King of the Aetherius Society, Claude Vorilhon of the Raëlians, the authors of the *Urantia Book*—all are examples of people claiming either telepathic messages or direct visitations by extraterrestrial entities, and their movements still exist today. Even if there is not a direct revelation, the interpretation of old scriptures through a science fiction lens can be enough. The Bible, *Star Trek*, and novels by Robert Heinlein were a major part of Heaven's Gate; the mere sighting of a comet and one visual artifact was taken as enough of a sign to the followers of this cult to commit mass suicide in 1997. Clearly, some people are ready to accept ET as a higher power, even forfeiting their lives to such desires.

All of these UFO religions have had trouble becoming mainstream religions for a variety of reasons. Perhaps the most obvious one is not everyone thinks aliens are real. After all, it's hard to believe in something if you initially doubt the object of worship *even exists*. However, if science is able to show that ETI is a reality, this barrier to UFO movements would come tumbling down. If science proves there are aliens, but it fails to find Yahweh, then one of the most potent socio-intellectual forces in the world will decidedly be in favor of one kind of belief over the other. Either these UFO believers would gain new followers, or the old religions would need to adopt ET into their beliefs very quickly.

As noted earlier, the Church of Jesus Christ of Latter-day Saints will probably more easily absorb news of ETI than other Christian churches. Early Mormon documents show an easy acceptance of the possibility of aliens, and the generally corporeal understanding of God makes the barrier between divinity and advanced civilizations much thinner than in other denominations. This isn't to say that Mormons will begin to worship ET, but they could point to those discovered aliens as examples to members of the church of what God has created and what Mormons can themselves become.

While highly speculative in the particulars, the outcome of the discovery of the mere existence of ETI would cause changes to popular religion. Decades after the discovery, there would plausibly be a surge in UFO-religion adherents, and the most alien-friendly versions of older religions would grow, likely taking away from the churches, synagogues, mosques, and temples that can't accept higher powers in the heavens that aren't their God or gods. The timing of that discovery might also be more promising for some groups than others, so there is a significant amount of contingency in what would be the outcome of detecting ETI.

This scenario outlined is all premised on the aliens remaining extremely distant. It would be like the cargo cults, which might have praised John Frum, even if he never visited (or even existed).

But what if John came to the galactic island that is Earth?

Gods Among Us

The image of the flying saucer has become a common staple of "classic" sci-fi, and vessels of silver or chrome are quintessential to what we perceive as The Future. This is even captured by the current work of Starship, the large rocket developed by SpaceX. At this point, we might feel let down if ET arrives in a ship without at least a bit of shimmer. Then again, when the aliens leave their vessel, there will be no doubt about their strangeness, and their travels across the galaxy will instantly make their powers known to us.

As we previously noted, if an alien spaceship can travel to Earth at significant fractions of the speed of light, the civilization that sent the ship would be at least a million times more powerful than us. Their technology would be so beyond ours that it would have no precedent in the history of encounters between peoples. It is worth referencing sci-fi writer Arthur C. Clarke's Third Law here: "Any sufficiently advanced technology is indistinguishable from magic."[1] Such technology would almost certainly look like magic to us. Perhaps even divine.

When the Conquistadors came to the Americas with matchlock muskets, they had a notable weapons advantage. But to have a million-to-one factor in power, the Spanish soldiers would have needed more than horses, plate armor, and gun powder. Only nuclear weapons versus the spear would be an apt comparison. And this million-to-one factor that the arriving aliens have is most likely an underestimate. ET would be the equivalent of a force of nature.

At one time, lightning was attributed to gods like Zeus and Thor. A creature with more power than lightning should at least be in the same pantheon as these old gods, if not their chief. An ETI here on Earth would have been godlike to our ancestors, so what about now?

While we have been using the term "god" here for Earth-bound aliens, there are other terms that might also be applicable, as some religious traditions may not be willing to apply such divine terminology. Also, if the aliens were not benevolent, then the moniker "god" may be hard to accept, leading to preference for other terms. The aliens could be conflated with either angels or demons in the Abrahamic tradition.

This has already happened, to a degree. When trying to explain the supernatural events of the Bible, some Christians have suggested that it was all aliens, and the angels of the Bible were aliens and their technology.[2] This is largely an attempt to rationalize the miraculous with the seemingly scientific. However, some other Christians have concluded that UFOs are in fact demons.[3] This is in part because of the negative abduction stories, probes and all. This pessimistic view might be more common of these two opinions about what UFOs are in a modern Christian context.

Here is a taster of this line of thinking, from Brother Michael Dimond's *UFOs: Demonic Activity & Elaborate Hoaxes Meant to Deceive Mankind.*[4]

> These kinds of false signs and wonders were predicted to take place on the earth in the last days of the world. The Devil is powerless against God, and can only do what God allows him to do. However, to test the world in these last days, God has allowed the Devil to have significant power to attempt to deceive mankind. These false signs, such as UFOs, are done by demons "...according to the working of Satan with all power, signs and lying wonders" (II Thessalonians 2:9).

---

[1] Clarke (2013), ePub reference 33.7.
[2] I.e., Downing (1968).
[3] I.e., Dimond (2008).
[4] Ibid., p. 49.

This approach is fraught with theological issue, and makes God look like a trickster, deceiving humanity, though (one assumes) for some greater good.

Similarly, as noted in the chapter on Islam, some Muslims believe UFOs are piloted by *jinn* or other demonic powers. If Christians and Muslims can claim that non-alien UFOs can be demons, then there is every reason to believe they could think the same thing about *actual* aliens doing terrible things to people (assuming the aliens are malevolent).

Conversely, if ET comes, and they bring us the supplies and philosophies that can bring flourishing, then would we not call them angelic, if not divine? After all, anything that happens in light of OmniGod happens by dint of God's will, active or omissive.[1] The hope of Christianity has been that Jesus returns and brings about the Kingdom of God. If the aliens come and bring world peace, then forget about John Lennon's *Imagine*. The aliens are bigger than Jesus!

The nature of the arriving ET will have the biggest impact in how they would be interpreted, but the details of when and how they arrive would also be interpreted within those prior religious frameworks. As noted earlier, if the detection of ET were to take place on a high holy day, that could be significant to that religious group; this interpretation would be even stronger if ET *arrived* at such a symbolic time.

If these "gods" were to be here, among us, then UFO religions would have explosive growth, and their growth would take members from other religious groups. What might also be interesting is the reaction of non-believing atheists. Most non-religious people might well balk at theological claims because they seem so unevidenced and contrary to the laws of the universe (i.e., bread becomes flesh in the Catholic mass, Jesus returned from the dead). But if the aliens are demonstrably real and perform the seemingly miraculous (see Clarke's Third Law), even if all such things were consistent with the laws of physics, would atheists begin to clamor for communion with the revealed ETI?

To put it another way: if you were a non-believer, and God clearly revealed himself to you, would you believe in him? If God also performed amazing, good deeds, again clearly so that there is no need for faith that God has done the wonderous, would you worship him? We think that your chances of answering in the affirmative are non-negligible. Some fraction of atheists would "convert" to ET worship.

The rate of conversion may depend on the technologies available to ET. While certainly more advanced than ours, there would still be limitations. The aliens would not have infinite energy devices, though they would

---

[1] And this has all the usual implications: Everything, form genocide to a stubbed toe, is God's will.

have access to tremendous amounts of a nonetheless finite quantity of power generation. We also doubt the aliens would have all the technologies imagined in *Star Trek*, such as teleportation and warp drive, because of significant physics problems with these sci-fi concepts. However, if the aliens were to have access to machines that could repair biological tissues with high fidelity to the point they could resurrect people, then the most potent force in traditional religions might disintegrate.

If the aliens could save you and your loved ones from death, and you could see that they were saved in the here and now, then why continue with a religion of an uncertain afterlife? You can follow the aliens with that (tangible) resurrection technology available *now*. Why fear Hell if you could be indefinitely resuscitated, your body de-aged, and the Grim Reaper always pushed away? Similarly, why hope for Heaven with your last breath if ET could give you Heaven on Earth?

However, this all assumes that ETI were to have these technologies and would be willing to share them with us. These are assumptions on top of assumptions, all based on the least likely scenario for how we might discover ETI. We cannot expect this to be the future of human religions.

Nevertheless, it does pose a future issue for religion. With advances in human technology, perhaps *we* can become these imagined aliens. If we developed a cure for death, for example, and it could be delivered widely, and especially in an equitable fashion, that could have a similar effect as if aliens provided this technology themselves. How humanity might react differently between these scenarios is an interesting speculation—what would be the sociological differences between humanity inventing these great technologies versus humanity receiving them from another power? That question cannot be explored here in this book, but an investigation answering it might provide insight into the nature of religion and humanity.

Ad Astra, Inter Pares

We don't know if or when we will discover ETI. That uncertainty is a function of both the frequency that ETIs emerge and how long they last. As we note in Appendix I, the biggest limitation to how many ETIs might exist in the Milky Way now is the lifespan of such civilizations. That limitation is dependent on our ignorance; our only source of how long an advanced civilization lasts is by looking at our own. Perhaps we will exist for millennia; perhaps we will ruin ourselves within a few generations. We hope for greatness, and we would like to imagine what would happen if humanity were to exist well into the distant future.

Suppose human civilization were still to exist in a million years. This is a remarkably long time for our technology to improve, and the likelihood that we would become an interstellar species would increase. With a million years of travel time, even with vessels moving a small fraction the speed of light, we could have humans living on opposite sides of the galaxy. If we are alone in the Milky Way, humanity could become a galactic civilization all to ourselves.

However, if humanity can exist for a million years, then the likely average lifespan of other advanced civilizations also grows. If we are an average civilization (following the Copernican principle), then us lasting for a million years means the average advanced civilization is similarly enduring. If anyone is out there, then probability would favor their civilization as being long-lived. The longer they live, the longer they could make their presence known, and the longer we could develop methods of discovering them. As time goes on, assuming we do not destroy ourselves, not only is it more likely that we find our cosmic neighbors, but the more we become like the aliens we are currently imagining. If we are able to galivant about the galaxy, then we are so very much like the creatures we imagine as existing out there now.

If in that future we discover ETI, we might be technologically superior to them. Or the opposite. Or there is so little room for technological advancement at a certain point that, even with an extra millennium of development, there is little in the way of progress,[1] meaning we are more likely to be like ETI in terms of technical prowess. If we meet ET while we are among the stars, perhaps we will not see them as so strange as we might find them now. And if ET is not greatly more powerful than us, then they are demystified. When meeting them, we would not see them as gods, angels, or demons. But as people. They would be a group of people in the same way we might look at another country's population.

This opens up new ethical questions, as most systems of morality consider human nature and rights. The addition of Cybermen, Andorians, and Jawas will make secular *human*ism a racist ethical structure. A term for a broader moral system suggested by Stephen Dick, *astroethics*, seems fitting.[2] The underlying premise is that all sentient life will have moral worth, though how far that extends is uncertain. The moral worth of non-human animals is still debated, after all. This ethical discussion will be particularly tricky if any of the ETIs are not biological but digital.

There will be immense diversity, and we may need to convince and be convinced by fellow ETIs what is right and wrong. Perhaps on some worlds, there is a seemingly defensible form of chattel slavery; on that world, there

---

[1] As noted earlier, there are limits to the growth of civilizations, which would suggest that there would be an upper limit to the abilities of a future human or ETI civilization. See McInnes (2002); Haqq-Misra & Baum (2009).

[2] Dick (2020), p. 191-206. Dick also coined the term "astrotheology."

are two intelligent species, but one sentient species is inferior to the other. One group has a slave nature, in the way Aristotle defended the practice.[1] Could we humans tell this culture that it is acting unethically? That is one challenge that future astroethicists would need to handle.

Within the story of human history, the ideas of rights and equality have been one of the common refrains. We might then extend this to the strange but sentient creatures we discover, and we might find the differences between ourselves and those ETIs to lack a moral difference. Ethical systems on other worlds may be more similar than we might first guess. After all, a society is unlikely to thrive if it based on anti-social behavior, so deception and murder are likely to be universally denounced practices. Perhaps, when we do meet ETIs, they would be our equals in technology and moral capacity, then, and wouldn't be so strange.

The future we imagine may be weird beyond our comprehension now. But perhaps once that future becomes the present, we will not be so shocked and confused.

When the angel of the Lord was seen by Moses in the form of a burning bush, as the story goes (Exodus 3:2-3), it was a "great sight." It was such a great or "strange" sight[2] because of its unusualness and seeming impossibility. However, if humans normally communicated by means of burning bushes, then there would be nothing so strange, and God would not seem so spectacular but rather mundane. Similarly, if aliens were come down in their flying saucers to communicate telepathically, then it would be a similarly great sight or wonder. Such is Clarke's Third Law. However, if we meet ET in our own flying saucers and use the same means to commune, then the mystique is gone. Perhaps the following could be a corollary to the Third Law: two sufficiently advanced civilizations will recognize the technology of the other. That is, one advance civilization will see another advanced civilization not as using magic, but machinery; they are not gods, but good engineers.

Our future decedents are not as likely to become worshippers of ETI as people today, so not only must we consider how we discover ETI, but when. There are so many factors that will affect future human religion with respect to aliens, and it is so highly contingent and unpredictable, we can only outline some possibilities. But our best hope is not to so much *worship* ET, but to *become* like them, or what we imagine they would be like. Instead of waiting for yet another divinity to come, we should work towards our own apotheosis.

---

[1] Aristotle, *Politics* 1254b16-21.
[2] NIV version of Ex 3:3.

CONCLUSION

20 – WRAPPING THINGS UP

There are several elements to this project. First, how do we think Christians will react to the discovery of ETI. This is a very different question than the one concerning whether the discovery will cause any doctrinal (or any other) conflict with Christian theology and thought.

Perhaps the most interesting question for theists reading this book is the following: Does the existence of extraterrestrial intelligent life make the existence of God more or less likely?

Let us look at these three questions in order.

Cognitive Dissonance and Christian Coping Mechanisms

We have already explicitly discussed cognitive dissonance reduction earlier in this book, and for very good reason. It is arguably a moot point as to whether Christian doctrine and thought will be affected because, at the end of the day, religious belief has a tenacious stickability to it.

Let us digress for a short while. The psychology and sociology (and perhaps even biology) of religious belief is a fascinating and fertile ground for understanding why we believe. It is no surprise that, given scientists (and eminent scientists far less so[1]) are much less likely to believe than the average person in, say, America, psychologists represent the subset of scientists who believe the very least.[2] Psychologists also represent the least religious of American professors (an already fairly heathen collection).[3] When people understand *how* people believe, the *what* becomes somewhat less important in understanding the causation.

One only needs to look at a demographic map of belief in the world to understand that it is primarily geographical. Religious belief is largely a result

---

[1] Stirrat & Cornwell (2013).
[2] Ragan, Malony & Beit-Hallahmi (1980)
[3] Gross & Simmons (2009).

241

of family and wider community and society. People don't assent to a religious belief by reading the *Companion to Natural Theology* and all of the holy books in the world, and then choosing the one that is most rational and plausible. Rather, it happens the other way round. Religious belief appears to be assented to for psychological and social reasons, and then the believer spends time thereafter searching for the rational arguments to defend that belief. This is called *post hoc rationalization.*

Such psychological examination shows that religious content is not really what leads people to religious belief. The process of "religious transmission" is one that "confirms a strong relationship between religious beliefs and practices of parents and those of their adult children."[1] We might see this in the mantra "You can't reason someone out of a position they never reasoned themselves into." This isn't to say that no one ever converts to religion based on rational thought, or that people's religious beliefs aren't rationally justified. But, in the main, people arrive at religious belief for psychological reasons.

This may include *motivated reasoning.*

What is motivated reasoning? Similar to the cognitive bias of confirmation bias, motivated reasoning can be defined as the tendency to find arguments in favor of conclusions we want to believe to be stronger than arguments for conclusions we do not want to believe. The reward or consequence for believing or not believing something is what drives our evaluation of the evidence and reasons for that belief. The desired outcome filters our processes of evaluation.

The Christian or theist has an awful lot more to lose from the skeptical arguments against the existence of God being on point. They are motivated to believe in the existence of God—there is a lot on the line.

There can be no greater thing in human conception than heaven because heaven is whatever an individual can imagine to be the greatest thing. Forever and ever.

Likewise, there can be no worse thing in human conception than hell. Because hell is whatever an individual can imagine to be the worst thing. Forever and ever.

We don't need to argue about the theology and history of these ideas. Heaven and hell mean whatever they mean to, say, a Christian. And they are there, in the consciousness or subconsciousness of every Christian, bathing every moral action someone makes in their light, or threatening them with eternal darkness and torture. Every single moral action, including the simple act of belief in God, is informed—tainted—by these twin bribes. They promise, at the end of the day, eternal life, and a way of coping with the inescapable

---

[1] Smith (2021).

prospect of death (known as *terror management theory* in the context of *mortality salience*[1]).

If we then add to these incredibly potent concepts ideas of social networking and support, including the reality for many that their families, friends, and jobs are wrapped up in their faith and faith communities, then we can see that there are many reasons that motivate people toward believing.

This is why apologists such as William Lane Craig argue so vehemently for different components of their belief, such as the empty tomb—they have everything to lose.

Does this point about motivated reasoning prove anything by itself? No, it has no bearing, *ceteris paribus*, on truth values or truth claims. But it should give you context and give you an understanding of why a Christian or other theist might weight a particular reason with a completely different evaluation than a skeptic.

Many Christians are fairly desperate for, say, the empty tomb to be true because it is *absolutely necessary* that the Resurrection is true, for which the empty tomb arguably provides a pivotal foundation.

How is this relevant for our present project?

If we discover intelligent alien life in the universe, then this may well cause cognitive dissonance with the theist, in particular with the Christian. That dissonance cannot remain since it is uncomfortable. They have a choice, given the acceptance of ETI: They can either ditch their belief in the Christian God because of the aforementioned issues, or they can retain belief in God.

The Christian has far more reasons to retain the belief than to give it up. They are thoroughly *motivated* to continue believing in God. Indeed, the belief may even give them comfort and support in the light of existential terror at the thought of far more sophisticated and dangerous aliens.

It is far more likely that cognitive dissonance reduction will lead the believer to mold their belief to the existence of ETI than have ETI-belief trump God-belief. This will almost certainly be the case even if the existence of ETI fundamentally challenge the doctrinal and theological tenets of that belief. After all, if it worked for Harold Camping...

Therefore, we would predict, leaning on the psychology of cognitive dissonance reduction and motivated reasoning, that the Christian or theist would, by and large, find ways to continue believing in God. Theologians might actually find their time and work in much greater demand in the advent of ETI discovery as they would feverishly work out ways in which religion could work in a universe populated by beings perhaps far more advanced than us.

---

[1] See Moore (2022) for a good overview.

The Threat to Theology

We believe that we have shown that the discovery of ETI in the universe *is* a challenge to all aspects of belief, including theological doctrine and interpretation of the scriptures.

If we discovered such intelligent alien life out there, and given the Copernican and other principles we have discussed, it appears to us that our privileged place in creation that most Christians adhere to would be challenged. Furthermore, ideas of sin and salvation, atonement and redemption, would be rendered problematic in light of such a discovery. God will have been at best neglectful and at worst deceitful in his revelations about the universe. More accurately, though, the existence of ETI would render the existence of God a precarious belief to hold. More on that in the next section.

There are problems with particularity concerning Christianity and especially Judaism that play merry havoc with the idea of ETI. From a narrative standpoint, ETI simply does not fit, and serious *post hoc* rationalizing needs to be carried out to shoehorn such concepts into the biblical world. *Scripturally*, such ideas are missing in a way that is not easily explicable, and *theologically* and *doctrinally*, Christianity and ETI are lightyears apart from any coherent conceptual union.

The problems Christianity has had with ETI are so bad that apologists have done what they can to paper over or just bury that history. As we demonstrated, the biblical cosmos had no room for naturalistic alien species, as heaven was the realm of angels and not Vulcans. The story of God and his Chosen People was so parochial that it ruled out the possibility of other species gaining from that same relationship. The message of salvation was in fact used as an argument *against* the existence of ETI, and it was considered a heretical view worthy of deadly condemnation, as seen in the case of Giordano Bruno. Literally, for over one and a half thousand years, the Bible was used to *prove* the impossibility of other inhabited worlds, even to the point of executing people to disagreeing with this theological deduction.

Only in modern times, when alien life seems to be likely, have theologians now claimed that there is no incompatibility between their faith and the existence of ETI, and there never was. This story told by such apologists is a false one, and similarly we find the arguments to make Christianity fit into the new image of the cosmos problematic. To steal a phrase from C. S. Lewis, the old, biblical model of the universe is the "discarded image," now replaced by the scientific view of our place in the vastness of space and time.

As it stands already, the revelations of the Bible are deeply troubling in terms of the cosmological claims it makes. This should already be enough to render the reliability of the book dubious, as we summed up in the section on biblical cosmology.

Those wanting to uphold Christian tradition have not always been kind to people who have posited the idea of other planets, systems, and potential life, despite what some apologists will claim. Though tradition is not a prescriptive ideal, it does present a barrier for simple acceptance of the existence of ETI. There will be some particular groups, denominations, and churches who would struggle to accept the idea of such alien life in much the same way they struggle to accept the scientific facts of the age of the Earth and the process of evolution.

All of these components of the jigsaw that we have analyzed paint a picture of the probability associated with the existence not of intelligent alien creatures or life, but of God. It is to that we shall turn.

The Existence of Extraterrestrial Intelligent Life Makes the Existence of God Less Likely

Theological arguments, the sort with which we have been dealing, have ramifications when concerning the existence of God. In the absence of them being neutral on the matter (we claim, on the whole, they are not), these arguments either support the God hypothesis or render its conclusion less probable. We emphatically declare that ETI discovery would make the existence of certainly the Christian god, but likely also most major religions' deities, less probable.

This evaluation is affected to some significant degree by the problem of evil or suffering argument that already presents such an issue for OmniGod theists. As Samuel Ruhmkorff has pointed out, if ETI exists in the universe, then this fact together with the problem of suffering would render the existence of OmniGod less probable. If a world with only stubbed toes or rare deaths is easier to explain than our current world in the light of such a god, then a universe with untold levels of suffering, far greater than what we are aware of, is *even more difficult to explain away.*

We think that, when comparing the two competing hypotheses of (Christian) OmniGod theism and naturalistic atheism, the latter does a far better job of explaining the presently hypothetical data of ETI existence. This is certainly the case when considering the particularities of individual religions, with the holy book revelations that they entail.

On atheism, if ETI exists, then we find that humanity is one more result of natural laws creating everything we see. Blind processes, indifferent to love and pain, peace and destruction, life and death, are the underlying components of the universe. Sometimes, those processes produce a world that is just about capable of microscopic machines that can self-replicate with

245

modification. Evolutionary mechanics take these machines the rest of the way to animals that think about how they got to where they are. If natural processes have created ETIs, then we humans can see ourselves as part of that grand natural process, companion pieces in the jigsaw that is this incomprehensibly large universe.

If there are intelligent aliens, we are not special. That is at the heart of the Copernican revolution and the Copernican principle. We are the middle child of the universe.

And that is okay.

In fact, it's amazing. Of all the ways humans could have been, we have every reason to think we would not be exceptional, but we would be part of a grander population than we would ever have imagined. This would be like when a child discovers there are other cities, and then other countries, with other amazing and diverse people. But this doesn't diminish our own achievements. No, we are not at the peak of what could be, but that also means there is so much to explore and to become.

There is hope and wonder in this godless universe. We can be part of something larger than ourselves, yet still be a collection of atoms that, by billions of coincidences over billions of years, looked at the sky and wondered who else might be looking back.

There may be no gods out there in the cosmos. But there is *us*. Out there is the final frontier, and finding another intelligent species is perhaps one more way to find out who *we* are. Our place in this universe is what we make it, and discovering ETI helps us realize what we could make of ourselves. We understand by comparison and finding another society will help us know what it really means to be human. The religions of the world have tried to do that, and they have failed.

Let's see what we can find out for ourselves.

Perhaps the final frontier is really all about self-actualization... Let's make it so.

AFTERWORD: THE LIFE ∴ RELIGION?

In some ways this book reminds me of a giant of the not-too-distant past: Isaac Asimov, professor of biochemistry, was one of the "big three" science fiction writers of his day (along with Robert A. Heinlein and Arthur C. Clarke). He wrote and had published over 500 books; he wrote on religion (his *Guide to the Bible* is grand), dozens of writings on science, some touching on philosophy, and scads of science fiction novels and hundreds of science fiction stories.

Pearce and Adair may stay prolific enough to rival Asimov for quantity. They already, with this book, remind me of him in breadth of interest and keen, deep insights.

UFO enthusiasts and all manner of alien mystical conspiracy theorists will find interesting points here, though not the gullible acceptance some of them crave. Scientific and mathematical theorists fare better, with effective analysis of major currents of thought, long established dilemmas and paradoxes, and of how real aliens, if they are ever encountered, are likely to impinge on our understanding of such things.

Though many of them would probably not admit it, theologians of every sort would likely envy the sharp perceptions Pearce and Adair bring to these subjects in a wonderfully original way. They leave open questions that deserve to be left open and show where pointless speculation should be avoided.

Where there's life, there's religion? Perhaps, but what does possible (or actual) life elsewhere in the universe tell us about *our* religion(s)? If such was discovered, would human religions (some or all) cope or adjust? Could they without revising or distorting their "truths"? Aliens and Religion not only has the answers, but it offers the right questions to help us think all this through. Whatever religion you may follow or are interested in—Judaism, Islam, Christianity (especially), or Hinduism and more—this book will keep helping you thinking about these issues for years to come. (If you've finished reading it now, put it somewhere on your bookshelf—virtual or literal—where you can easily find it again later.)

Pearce and Adair are deep, creative thinkers with flair who can keep leading us into meaningful contemplation and analysis.

Ed Bucker, former President, American Atheists, and writer at *Letters to a Free Country*

"Somewhere, something incredible is waiting to be known."

– Sharon Begley, "Seeking Other Worlds (Profile of Carl Sagan),"
Newsweek

APPENDIX 1

Estimating the Population of Current ETIs in the Milky Way

In the first chapter, we considered the arguments for the parameters in the Drake Equation from Anders Sandberg at Oxford, along with his collaborators.[1] Here we will repeat their analysis, argue for different probability distributions for the factors in the Drake Equation, and what can be said about the probability that we are alone in the Milky Way.[2]

One of the greatest problems with the Drake Equation is what sorts of numbers one should put into the expression. While some can be understood well with current astronomical knowledge, others are completely unknown. Currently, there is significant uncertainty about the origins of life on Earth and just how abundantly those conditions exist elsewhere in the Milky Way. To get a sense of the results of those uncertainties, along with the current null results from SETI, a sampling of the ranges of possible values for the Drake Equation was performed by a group of researchers at Oxford. The method used to sample is a common one used for simulations based on probability distributions: It is called the Markov Chain Monte Carlo (MCMC) method or algorithm.

The setup for the MCMC process requires probability distributions for each factor in the Drake Equation. The way the Oxford researchers established most of their distributions was to look at the orders of magnitude that the variables in the Equation could take. For example, perhaps the range of values for the probability of one planet in a solar system being habitable is as low as one in a thousand (10^{-3}) or as high as one in one (1, or 10^0). The range of magnitudes then would between -3 and 0. Between these values, samples were drawn in what is called a uniform distribution, meaning it is as likely to pick a number between -3 and -2 as it is between -1 and 0. This sort of process of setting a range of magnitudes and then pulling from a uniform distribution was conducted for most of the variables.

How those ranges were determined entailed looking at the ranges that existed in the scientific literature. Some papers might suggest that most solar systems have a habitable planet, while others might say it is very rare.

---

[1] Sandberg, Drexler, & Ord (2018).
[2] The code for this analysis is found at https://github.com/adairaar/drake_simulator. You can run the code yourself in your web browser for free here: https://colab.research.google.com/drive/1g2ejjH0oZL94dj5Tt6M9qm8gzlvlpXvm?usp=sharing.

249

However, if put into a uniform distribution, then the sampling equally weighs every suggested value, not the ones with the best scientific support, either in terms of evidence or consensus. So, one issue that we see is that the ranges of values will be too wide, especially for values in the Drake Equation that are well-supported by empirical observations.

There is an additional complication in the analysis when it comes to the probability of life emerging on a habitable world. Instead of following the same order of magnitude and uniform distribution method, the Oxford group used a much more complex model, using a normal distribution as well as an exponential function. This complexity allowed for some interesting results. While the average value from this probability distribution would be 50%, the individual probabilities of life emerging could be lower than 10^{-200}. These are odds of 1 against a 1 followed by two hundred zeros! It's the equivalent of the same 25 people winning the lottery ten billion times in a row. However, these super-low values were rarely sampled, and the probability of life could be as high as 100%.

In other words, the analysis does allow either life to be extremely commonplace, or so rare it would take more than a googol (10^{100}) observable universes to see life emerge even once. Given the large error ranges, it means it is quite possible that life on Earth is truly special.

Putting all of the proposed probability distributions together, the estimates from the sampling found that most simulations of the Drake Equation indicate that Earth is the only place in the Milky Way with a civilization. In fact, there may be no ETIs in the entire observable universe, according to their results. We really might be alone and never detect another intelligent kind of life even with the greatest telescopes the future may offer.

However, as noted, this is a result of the gigantic uncertainties in the parameter of life emerging. In fact, in the supplemental articles with the main paper, the authors note that the uncertainty in just one factor, namely the probability of life emerging, drives the simulations to suggesting we are alone in the Milky Way. There are reasons to suspect that perhaps the ranges on the probability of life emerging are too wide, which first instigated our efforts to repeat the analysis.

First, we found that the extreme improbability of life is too low, given some recent work in abiogenesis. Looking at one paper on a minimally assembled, random string of functional RNA,[1] the pessimistic estimate is more

---

[1] Totani (2020). Totani assumed an RNA sequence of minimum viable length of 40 nucleotides, which he estimates would require 10^{39} stars for one occurrence to be expected; a strand of 55 nucleotides needs about 10^{70} stars. However, these results may also be too stringent because he only considers the RNA world hypothesis, while other pre-biotics may also achieve the initial steps for life, such as peptide nucleic acids (PNA). Therefore, this paper should only give a lower limit for one hypothetical approach to abiogenesis.

on the order of 10^{-70}, which is still a small number, but nowhere near as small as the previous bound. Rerunning the analysis of the Oxford researchers,[1] the prior probability of us being alone in the Milky Way drops from 52% (slightly likely) to closer to 43% (a bit unlikely). The median value of the simulations' number of civilizations also grew from less than 1 to about 8. The effects of this better-justified lower bound on abiogenesis can be seen graphically as well, when we plot out the frequency of simulations that came up with extremely small values of the number of civilizations, N. In the original paper by Sandberg *et al.*, the nearly-flat left tail of the distribution continues on beyond to very low values, while in our reanalysis those low values of N are much less frequent.

So, even with significant uncertainty in how life might have emerged and even allowing for it to be very improbable, we might still expect ETI to have emerged more than once and still exist in the Milky Way.

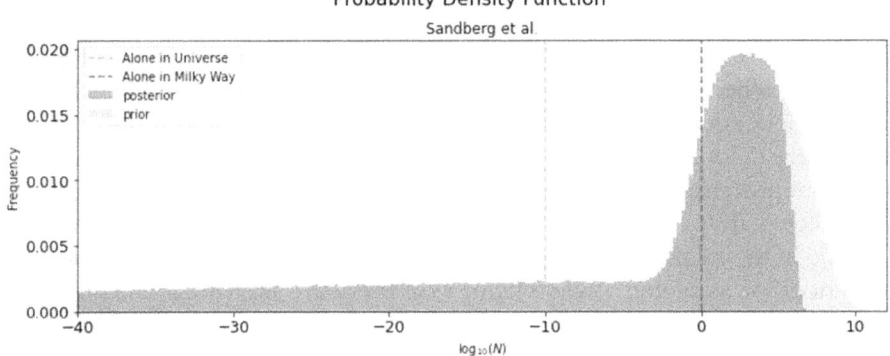

Figure 1. Probability density of N civilizations currently existing in the Milky Way. The lighter gray represents the prior simulations before considering no detection of ET signals. The darker gray represents the density after accounting for no ET signal detections. For values of log(N) less than 0, we are alone in the Milky Way. For values of log(N) less than -10, we are alone in the observable universe.

---

[1] All parameters were kept the same, with the only difference coming from the probability distribution that was applied to the probability of life. The original researchers used a log-normal distribution with a mean of 1.0 and standard deviation of 50, which is very wide; this was then plugged into an exponential decay function to get the probability of life. This allowed for probability less than 10^{-200}. By using a standard deviation of 15 instead put the lower bound of the probability closer to 10^{-70}.

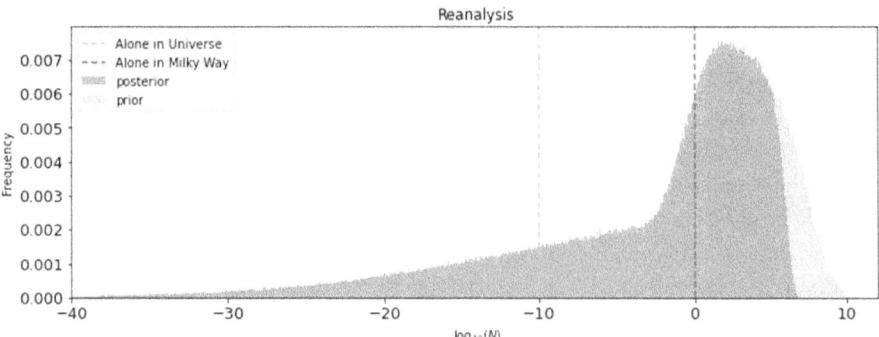

Figure 2. Probability density of N civilizations currently existing in the Milky Way, with better lower bounds on the probability of life emerging. Otherwise, all parameters in Sandberg et al. (2018) are maintained.

However, the next step in the analysis is accounting for the lack of success when searching for ETI (SETI is discussed in more detail in an earlier chapter). When the original researchers included this evidence, it was very model-dependent, such as how the aliens were spread about the galaxy, if they were colonizers, and so on. With different models, the researchers got different posterior probabilities for us being alone in the galaxy. The least assumption-laden approach is that other civilizations are randomly spaced around in the galaxy, in what is called a spatial Poisson distribution. Given the likelihood of us seeing a signal from ET if there are N civilizations randomly distributed across the Milky Way, higher values of N were less likely, and those simulations were removed proportionally to their unlikeliness.

This process provided some evidence for us being alone, but not much. When we repeated the analysis, we similarly found that the probability of us being alone increased, but more strongly since our model suggests ETI signals are more probable, and thus the missing signals are more unexpected. So, our posterior results, the probability that anyone else is in the Milky Way, is a toss-up, with a probability of just about 50% for us being alone, while the median number of civilizations was barely more than 1. These ambivalent results are still largely because of our uncertainties about life starting on other worlds.

In fact, there are several technical issues with the modeling done in the first place. For the probability of life emerging and the probability of life eventually becoming intelligent, the distribution of these values are not the sorts of things one should use when being extremely uncertain about the values. Instead, the distribution should be something that indicates we don't really know what the values should be (an *uninformative prior probability*), which is often done in Bayesian statistical analyses.[1] This again calls upon the Jeffreys prior, which was discussed in "Ruhmkorff and the Copernican Principle" in chapter 12. For values that fall between 0 and 1, that Jeffreys prior is given as the uninformative Beta function, written as B(0.5, 0.5). Because we have no viable, statistical observations for the probability of abiogenesis, intelligence emerging, and intelligence developing radio communications or other means of being detected at astronomical distances, we argue that the Beta function is the appropriate distribution for these parts of the Drake Equation.

However, for the longevity of a civilization, this is also an unknown value, and the uniform distribution employed by Sandberg et al. is not justified; it only shows the opinions (and perhaps merely the dreams) of some scientists. Instead, we must realize that the probability of a civilization existing longer and longer periods of time decreases with time, because every passing year allows for some event that ends a civilization. In other words, the probability of a civilization living to Y number of years will follow an exponential decay function.[2] Here, there is not a range of values to plug in, as was done with the uniform distribution. The exponential decay only incorporates one parameter, which relates to the average lifetime of a civilization.

To figure out that average lifetime, given we do not have a sampling of ETIs to pull from, we use a version of what is called the Doomsday argument. This argument considers our own observation of our existence, how long we have been around, and how likely we would have found ourselves in the early, middle, or end times. This is a similar thought process to the Copernican principle. Such reasoning, however, suggests that the lifetime of a communicative civilization is about 300 years.[3] This value will change with more data—in particular, if our civilization lasts for many centuries, as we hope it will.

There is a considerable amount of literature discussing the issues with the Doomsday argument, but here we will attempt a modified version of the argument, and we will use the mathematical approach similar to what is done in discussing the problem of evil with aliens, as described in Appendix 2. We note that, given the expectancy of a civilization continuing to survive into the next year goes as an exponential decay function, we can use this as a likelihood of our survival past a certain date. Our prior probability of us coming to this year as a communicative civilization will follow the Jeffreys prior of 1/x, where x is the number of years. Combining these points together, as demonstrated in Appendix 2, leads to this equation for the probability of a civilization surviving to a year after so many years have gone by:

---

[1] Kipping (2020).
[2] Kipping, Frank, & Scharf (2020).
[3] Kipping (2020). The author notes that this result is likely conservative.

$$P(X^* > x|X) = X\lambda \cdot Ei(-\lambda X) + e^{-\lambda X},$$

where X^* is the number of years of us surviving, X is how many years we have survived so far, λ is the decay constant (the inverse of the mean lifetime), and $Ei(z)$ is the exponential integral function. If we take ourselves as being in a typical place, following the Copernican principle, then the probability of us lasting as long as we have is 50%, so P is 0.5.

We now note that X, the number of years we have existed as a communicative civilization with radio signals into space, is about 90 years; the first of the signals that could go into space would be the transmissions for the 1936 Olympics. If we use a computer to numerically solve for λ, we get a value of about 0.00297. The inverse of λ is the mean lifetime of a communicative civilization, which is about 336 years. This is in close agreement with the previous analyses that also suggest a mean lifetime for such advanced civilizations at around 300 years.[1]

Note that the above does not suggest that after 300 years, no civilizations will exist, only that the probability of existing as a communicative civilization continues to decrease, either because of a terrible pandemic, war, climatic change, asteroid impacts, or some other mechanism. Perhaps after 300 years, it's just not worth sending our radio signals, but we continue to live on just fine. None of these are physical laws of nature, so it is certainly possible to keep a civilization lasting well beyond a few centuries. In fact, the probability of surviving past this average are not terrible, though they are below 50% (and closer to a 1 in 3 chance). Here we can only describe the odds, given current knowledge and statistical reasoning. We leave it as an exercise to the reader to keep humanity alive well beyond this time.

So, once again, we incorporate both the uninformative prior probability for life, intelligence, and becoming communicative, and we use an exponential distribution for the longevity of a communicative civilization. Additionally, we can step beyond the ignorance encoded into our uninformative prior distributions. We also incorporate the fact that life emerged on Earth very quickly, while intelligence took billions of years to emerge, requiring most of the time the Earth can have life on its surface. The former is suggestive of life coming into existence is easier than one might first guess since life started about as soon as it could. Conversely, intelligence in the form

---

[1] Also note, as time goes on and we continue as an advanced, communicative civilization, the estimated value for the average civilization lifetime also increases. So, approximately 3 centuries is what we can estimate based on current evidence of our own existence, and that estimate is continuously updated as we continue to live. Another way to think of this is, 300 years for the average lifetime of a communicative civilization is the longest average we can currently justify without additional evidence; it is not a clock that is ticking down, because as time goes on and we survive, the clock ticks *up*.

of humans took billions of years, and most of the available time that life can exist on Earth before the Sun makes our planet uninhabitable has been exhausted, suggesting that intelligence is more difficult to get started than not. Both of these facts provided weak evidence in either direction,[1] and the weight of these lines of evidence are included in our analysis, updating the simulation results based on this relatively weak evidence.

Lastly, the distributions used for the rate of star formation and the other parameters in the Drake Equation in the original paper are faulty as they do not follow statistical distributions that are justified by observations or first principles (they were samplings from guesses by scientists). Instead, we have changed those into distributions that follow generator functions whose parameters are grounded in empirical data. In particular, we have relatively good data concerning the rate of star formation, and the number of stars forming in a given year will follow a gamma distribution.[2] The fraction of stars with planets and the number of habitable planets will also follow a gamma distribution, and the parameters in the distributions of those values are governed by the observations of the *Kepler* space telescope. Currently, astronomers have confirmed about 3000 planets, of which about 10% seem to exist in the habitable zone.[3]

The fraction of stars with planets is described by the beta function, which gives a probability that a star will have a planet, and it is shaped by the number of planets found by *Kepler* as a fraction of the stars observed (noting that only a fraction of stars will even have solar systems detectable by the telescope).

---

[1] The Bayes factor for life's early emergence was figured to be about 3, while the lateness of intelligence was given a factor of 2/3, as determined by Kipping (2020). Values less than 5 are usually labelled as barely worth mentioning, but they are included in the simulations for completeness.

[2] A gamma distribution is continuous valued probability distribution, which describes the count of some event happening. It is related to the Poisson and exponential distributions.

[3] What is considered the habitable zone is contentious, with no truly agreed-upon standard, but we have taken the suggested counts of planets from the NASA Exoplanet Archive and their catalog of confirmed planets in the *Kepler* dataset.

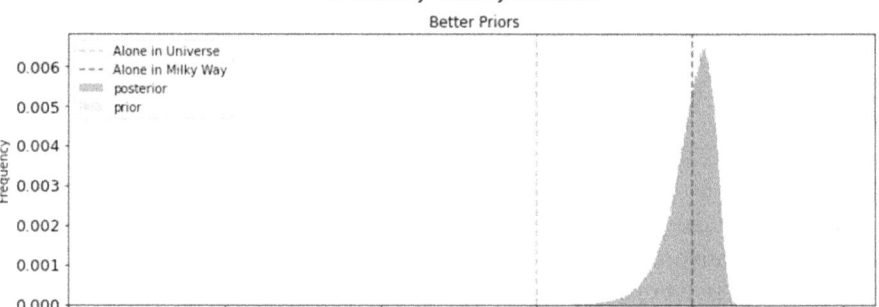

Figure 3. Probability density of N civilizations currently existing in the Milky Way with better justified prior probability distributions for factors in Drake Equation.

The results again suggest basically a 50-50 chance that we are alone in the Milky Way, but the simulations were not so widely spread. This would suggest that if we are not alone in this galaxy, there are still unlikely to be even 100 ETIs. This low number means that the lack of success of SETI has almost no evidentiary value—the aliens are too far away for us to have noticed them.

While this 50-50 chance does not greatly differ from the original analysis by the Oxford team, the overall distribution of simulations does better represent both our current knowledge and our current ignorance about the parameters in the Drake Equation. This can also be seen in how wide-spread the simulation results are, which is a feature not well-considered in the original paper.

However, we are not yet done. The original analysis has an important missing step, and that is determining the credible interval of the number of civilizations, N. Basically, how big are the error bars? Any science paper that fails to express the error (or confidence) range is missing important information. This can be fixed by calculating the credible intervals. In order to complete a Bayesian analysis, one looks at the distributions of posterior results, and then look at what range contains most of the results. The credible interval is then where, say, 95% of the results lay, and so say we have 95% confidence that the real result is in that range.

When we performed the calculations to get the 95% credible interval (calculating highest density interval or HDI) for the original simulations, the range was gigantic, incorporating values of N to less than one civilization in the universe to many alien civilizations in the galaxy right now. This is a result of the long, low probability tail in the distribution, and even the "hump" in the higher end of the probability range is not all that large. However, in the simulation with what we argue has the most-justified ranges for the parameters in the Drake Equation, given both our ignorance and statistical reasoning, the

95% HDI is much tighter. It still allows for us having neighbors in the Milky Way and for us being galactically alone, but it excludes the possibility that we are alone in the observable universe.

Taking this altogether, the best we can say is we don't know if we are alone in the Milky Way or not. We are likely not alone in the universe, but it is basically 50-50 if there is anyone we are likely to meet while gallivanting around the galaxy.

However, let us also consider the best-case scenario that might be allowable given our current ignorance of the development of life and intelligence. Let us suppose that every solar system has a habitable planet, and every habitable planet develops life, and every life-bearing world develops intelligent lifeforms, and every such planet's intelligent beings build radio transmitters. In this absolutely best-case scenario, the rate that just civilizations emerge would be governed by the number of new stars forming each year in the Milky Way. This is a small number, between 1 and 10. So, using almost all of the factors in the Drake Equation, the number of new civilizations coming about is small. The only way there would be many civilizations currently in existence is if they are, on average, long-lived. However, given the average lifespan we found using the Doomsday argument, there would be, at best, a few hundred civilizations. With this small number, and their Poisson distribution around the Milky Way, it is unlikely their radio signals would be detectable.

Conversely, this means if we did detect such a civilization, they would need to be close-by, which suggests there are many such civilizations. If here are many such civilizations, then the longevity of such communicative aliens would need to be much more than 300 years, perhaps even hundreds of thousands of years. This would also be favorable to human civilization, since a larger average civilization lifetime means we have a greater civilization lifetime than the Doomsday argument would support. We thus believe that detecting an ET's signal in the Milky Way would be good news for humanity.

Until we make such a detection, our conclusions are as follows. The best estimate is that there is perhaps one civilization currently alive and transmitting radio signals in the Milky Way galaxy. While it is plausible that we are alone in our part of the universe, the simulations we conducted make it very unlikely that we are alone in the entire universe. The average result from our calculations suggests there is a civilization in every major galaxy at the moment, which could mean there are billions of inhabited worlds. Perhaps one of them is close enough for us to detect, but it is currently unlikely one is close enough to see their technological signatures. However, we welcome more time searching, because even the faintest evidence that aliens are out there would be auspicious for humanity's future.

APPENDIX 2

Updating Probabilities in the Problem of Evil and ETIs

In chapter 12, on the problem of evil, we discussed how to use a Jeffreys prior to see the probability of an even greater evil existing for aliens, and then we could update that prior with the observation of the suffering we have seen on Earth. However, we did not come to a final calculation of how expected evil becomes when God exists. Here we take the previous analysis from chapter 12, and then see how to calculate the likelihood of evil in the universe.

To reach our conclusions, we will first look at how likely an evil event takes place given God's desire to reduce suffering or evil. Next, we consider how probable evil of an even greater magnitude found on Earth is. Combining the results, we have an equation that describes how probable the worst thing in the universe is, given theism and the history of terrible things on Earth.

First, we will consider the probability of evil on theism. Suppose evil happens naturally at a given level described by a function $f(x)$, where x is the magnitude of evil. The likelihood of seeing evil $\mathcal{L}(x)$ is simply based on the function, $f(x)$, thus $\mathcal{L}(x) = f(x)$. As x increases, there is a higher expectation for God to prevent such an evil. This is the basic meaning of being all-good: the omnibenevolent being would want to reduce evil. Suppose the function that describes God suppressing evils of level x is $g(x)$. How much God would want to prevent evil only depends on his level of desire against evil, so it is independent of how much evil happens without divine input. The likelihood of evil, given theism (T), is the combination of evil expected without God, multiplied by the factor God would prevent that evil. Thus, the likelihood $\mathcal{L}(x|T)$ goes as $g(x)f(x)$.

In future steps, we will use Bayes' theorem, which can be written as

$$P(A|X) = \frac{P(X|A)}{P(X)} P(A) = B(X|A) \cdot P(A),$$

where A is the initial hypothesis, X is the evidence (or background assumption), then $P(A)$ is the prior probability of A, $P(X)$ is the probability of seeing evidence X, $P(X|A)$ is the probability of seeing X given A, and finally the posterior probability $P(A|X)$, the probability of A given evidence X. The ratios of

Aliens and Religion: Where Two Worlds Collide

probabilities involving X will act as the Bayes factor, B.[1] This will be the ratio of the likelihoods of seeing X given A and seeing X independent of A being true.

The Bayes factor, comparing the likelihood of God preventing evil and a God that doesn't stop evil, is the ratio $B = L(x|T)/L(x)$, so

$$B = \frac{g(x)f(x)}{f(x)} = g(x)$$

Now, this factor depends on how much God would prevent evil, and greater evils are expected to be prevented more strongly than lesser evils. The simplest model of this would be a constant rate that evils are undesirable by God, so for the rate that evils are allowed should be negative and proportional to the magnitude of such evil given God—the more terrible the evil, the more God will not desire such evil to persist and will do something about it. In other words, the greater the evil, the more God would want to change the amount of evil, reducing it to some lower level. This is written as

$$\frac{dB}{dx} = -\lambda B,$$

where λ is a constant that describes the relation between the rate of likelihood change and the likelihood. The greater λ, the more strongly God would prevent evils. Rearranging the equation, we get

$$\frac{dB}{B} = -\lambda dx,$$

which, by applying the fundamental theorem of calculus, can be easily integrated on both sides, giving us (with integration constant c)

$$\ln(B) = -\lambda x + c,$$

where ln is the natural logarithm. Solving for B by taking the inverse of the natural logarithm to both sides (the exponential function),

---

[1] Usually in the literature, the Bayes factor is the ratio of likelihoods of the evidence given competing models. There is no term generally used for the ratio between the conditional probability $P(X|A)$ and the marginal probability $P(X)$, so our choice of calling this the Bayes factor is not the standard; some have used the term "normalized likelihood" for the ratio, but no clear convention exists. However, we are nonetheless using the Bayes factor in a fundamentally similar way; the main difference is the Bayes factor is used in the odds form of the theorem, while we are using it in the traditional probabilistic form.

$$B = e^{-\lambda x + c} = e^{-\lambda x} e^c = C e^{-\lambda x}$$

B is thus a function of x, along with two constants, C and λ. Now, we expect that there is no suppression of evil if the level of evil is zero (x=0), so the Bayes factor for x=0 is 1 (meaning, events are just as expected on theism as on no good God). Then, we solve for C:

$$B(0) = Ce^0 = 1$$

Thus C=1, and

$$B(x) = e^{-\lambda x}$$

We do not propose a value for the suppression factor, λ, at this point. However, we find it can be fixed in the following way. Suppose one considers the evilest event that ever happened in Earth history. How likely that evil event of badness value X was on theism would be p. We can then solve the equation

$$B(X) = e^{-\lambda X} = p$$

$$\ln(p) = \ln(e^{-\lambda X}) = -\lambda X$$

$$\lambda = -\frac{\ln(p)}{X}$$

In this expression for evil suppression λ, the probably p of such evil on theism is between 0 and 1, while X is the level of evil in the worst event in Earth history. For high values of X, the suppression factor is small; for high probabilities that God would allow such evils (p is close to 1), then the suppression factor is also small. Conversely, if one believes that such an evil is highly unexpected on theism (p close to 0), then the value for λ will be high. As an example, if the evilest event was 100 SBUs (standard badness units), and this has only a 1% chance of happening on theism, then λ is 0.046; the units of λ would be inverse standard badness units or SBU^{-1}. The inverse of λ indicates the average level of evil that God would prevent. In this example, the average evil event God is expected to prevent has a value of just less than 22 SBUs.

Another way to look at λ is to consider how evil something must be for God to act half the time such an evil persisted (p=1/2). This would be the evil half-event η, given as

$$\eta = \frac{\ln (2)}{\lambda}$$

For a lower λ, the greater the evil must be for God to intervene in half of such occasions. We will consider half-events later.

Given the equation relating λ to a set probability for the worst evil on Earth, we can then look at the probability of the evidence of evil in the entire universe, given there exists a population of alien civilizations.

For our next stage of analysis, we determine how probable there is an event X* that is worse than anything we have ever seen on Earth. The analysis to get to that step was done in chapter 12, but we repeat it here while generalizing the results.

Suppose that the worst thing that ever happened on Earth (in terms of pain or suffering) is something we can measure, or at least put on a scale and give it a value X in standard badness units (SBUs). While X is for the worst thing in *Earth* history, we must consider other worlds. Could there be an event X* that has a value of more than X? The prior probability distribution that there exists X* greater than X can be given by the Jeffreys prior.

For something that follows a continuous normal distribution, like badness, Jeffreys tells us the prior probability distribution is proportional to $1/x$, where x is the value of badness in this case. We can write the prior probability as $P(X)dx \propto dx/x$, where dx is a small interval of x which we would look at to see the chances that something of badness x is in that interval. However, this is the probability density given no background knowledge. We know that some bad things have happened, and the worst thing ever had a badness value of X SBUs, so we update our prior probability distribution (of how likely there is an event worse than X), to a new posterior distribution. This is where we employ Bayes's Theorem. For our purposes, we are using a simplified version of Bayes's Theorem wherein we multiply the prior probability of there being greater suffering outside of Earth by the likelihood of observing the evidence of what evil we see on Earth.

That new, posterior probability is the prior probability (proportional to dx/x) multiplied by the likelihood of seeing such evidence, which is also going to follow a probability distribution of $1/x$. This posterior probability distribution is proportional to dx/x^2.

So, the probability distribution of the worst thing in the universe being that terrible thing we have experienced (X) is proportional to $1/x^2$, but where x must be at least X. This looks like

$$P(X^* > x|X)dx = \frac{c}{x^2}dx$$

where c is a proportionality constant. We next provide a value for c using a lower limit on X^* and the basic axioms of probability theory.

The second axiom of probability theory is that the sum of all probabilities must sum up to 1. The summation of all probabilities is done using an integral from 0 to infinity (all possible values for badness, x). However, the lower limit of the integral will not be 0 but X, because X is the smallest amount that X^* could be. Remember that X is the worst event in Earth history, while X^* is the worst thing in the history of the universe. Therefore, $X^* \geq X$, and the lower limit of the integral must be X, because we know X^* cannot be lower than that. In other words, $P(X^* < X) = 0$, so the sum of all probabilities of X^* less than X is 0. With these considerations our integral becomes:

$$\int P(X^* > x|X)dx = \int_X^\infty \frac{c}{x^2}dx = 1$$

Completing the steps for integration:

$$\int_X^\infty \frac{c}{x^2}dx = -\frac{c}{x}\Big|_X^\infty = \frac{c}{X} = 1$$

Now we can see that $c/X = 1$. With one step of algebra, we realize c is equal to X.

We now return to our probability density of X^*, given X.

$$P(X^* > x|X)dx = \frac{X}{x^2}dx$$

This will act as the prior probability density of such an evil existing if there are aliens, and using Bayes theorem, we multiply this by the Bayes factor found above to get the posterior probability of seeing such evil given theism.

$$P(X^* > x|X,T)dx = B(T|X^*) \cdot P(X^* > x|X)dx$$

$$P(X^* > x|X,T) = \int_X^\infty \frac{X}{x^2}e^{-\lambda x}dx$$

This is a more difficult to perform integral, but looking at integral tables,[1] we note that the solution comes out as

---

[1] Jeffrey & Dai (2008), p. 176.

$$-\left(X\lambda \cdot Ei(-\lambda x) + \frac{Xe^{-\lambda x}}{x}\right)\Big|_X^\infty,$$

where $Ei(z)$ is the exponential integral function, $\int_{-\infty}^{z} \frac{e^{-t}}{t}\, dt$.

Inputting the limits of the integral, we find

$$X\lambda \cdot Ei(-\lambda X) + e^{-\lambda X}$$

Inputting our value of λ, the expression simplifies to the final form of the probability of X^* on theism:

$$P(X^* > x | X, T) = p - \ln(p)\, Ei(\ln(p))$$

Again, p is the probability that God would allow the worst evil seen in Earth history. The equation for P then determines what is the probability of such evil, given aliens exist, and then God exists and allowed event X to happen, which has probability p of happening on theism.

Unfortunately, the exponential integral function of a logarithm $(Ei(\ln(p)))$ is neither intuitive, nor can it be derived from elementary functions. However, we can make a few notes about how this overall function works and see its implications.

Because p is between 0 and 1, the value of $\ln(p)$ is always 0 or negative. $Ei(x)$ is also negative for all values less than 0, but $Ei(x)$ approaches 0 as x goes to negative infinity. This means for all values of p between 0 and 1, the left half of the equation is a negative multiplied by a negative, and thus positive. That means more generally, P is equal to $p - y$, for some positive y. Therefore, P is always less than p for all p within 0 and 1 and equal to p only when p is 0 or 1. In other words:

$$P < p, \text{ for } p \in (0,1)$$
$$P = p, \text{ if } p = 1, 0$$

This means that the probably of seeing evils on theism, given aliens, is always less than the probability of seeing evils on theism, but aliens don't exist. P is not smaller only in the case where p is 100% or 0%. This makes intuitive sense again: If aliens experience evils, and more likely than not their evils are worse than ours, then the problem of evil is even more severe; but, if evil is always expected on theism, then alien suffering is also expected. And if suffering is impossible on theism (p=0), then more suffering among the aliens brings about the same conclusion (P=0).

Nonetheless, how much P is less than p is given by a few complicated functions, and the trend is non-obvious. However, the shape of the expression above can be roughly approximated by x^2, as seen when graphing the two results. Graphing also shows that all values of P are less than p (the diagonal line) except at $p = 0, 1$.

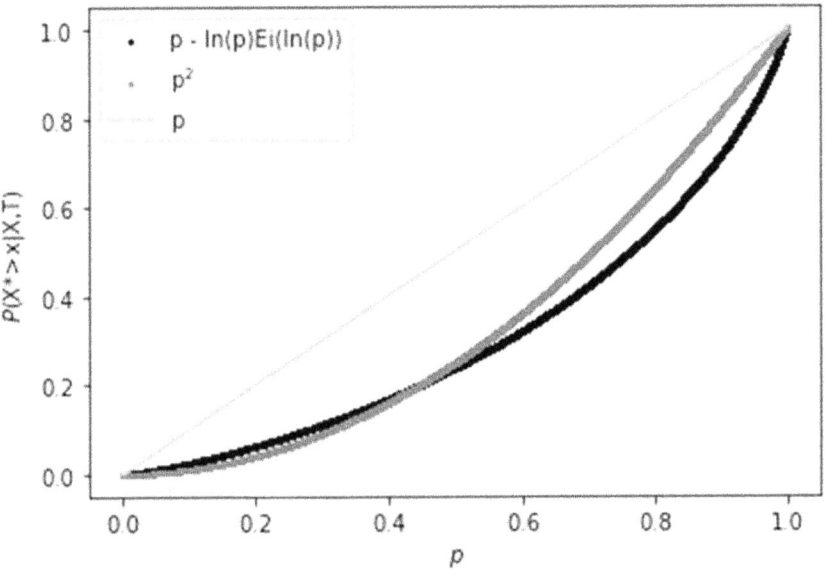

Figure 4. Probability of evil given theism vs probability of greater evil existing for aliens, given evils on Earth and theism. The gray curve is a crude approximation for the actual function representing greater evil existing, given theism and known earthly evils (dark black curve).

For a more precise calculation, it is best to use the full expression, but for faster calculations and understanding the general relationship between the variable x and the result, it is best to use the parabolic function, x^2. However, this simplification works best for the middle range of probabilities, so it may be best to use a machine to compute values generally.

Thus, the easier expression suggests that, for whatever likelihood one gives to the worst event in Earth history given theism, the probability of the evidence for evil in the universe goes as this probability squared. If one said that the likelihood of such earthly evil was 50% ($p=0.5$), then the likely greater evils existing in the universe gives an expected probability of about 25% (with more strict calculations, $P \approx 0.2375$). If one were more likely to believe the worst evil in the world was very unexpected on good-God theism, such as one in a million ($p=10^{-6}$), then the expected probability of evil in the world is extremely low, nearly one in a billion ($P \approx 10^{-9}$).

However, if one believes that any evil, no matter how great, is not un-expected on theism, then the math also indicates the same thing; for $p=1$ (meaning $\lambda=0$), the expected probability of seeing such universal evils is also 100% ($P=1$). However, this is perhaps impossible to believe, because it suggests that any evil, no matter how terrible, is perfectly expected on good-God theism. That would include infinite evils; that would include the Devil winning. To suggest that any horrendous suffering is actually expected on theism means one cannot say that evil will not control the universe. Literal Hell-on-Earth is compatible with omnibenevolence.

Consider again the concept of the half-event η: an event of such evil that an all-good God would prevent it half of the time. Perhaps there are reasons he doesn't always stop such evils, but at this level he would have a 50% chance of doing so. But if there is always a reason (a.k.a., an excuse) for why God would not prevent such an evil, then the value of η goes to infinity—God never intervenes to stop evil. Evil can literally grow unbounded, and God would do nothing.

To put it another way, would God prevent a Holocaust half of the times such events were transpiring? If no, then that is how little God cares to prevent evil. If yes, then why do such evils transpire?

If infinite evil is consistent with omnibenevolence, then we don't know what words mean. And Christians do not actually believe this is true. The main claim of Christian eschatology is that God, through Jesus, will come to our world in power and glory, defeat the Devil once and for all, and create heaven on Earth. But such an event is God preventing evil—in fact, all evil for all time. Letting the Devil rule without end is the only situation consistent with $\eta \to \infty$. If God *does* stop Satan, then η *must* be finite. Therefore, $\lambda > 0$, $p<1$, and $P<1$. And so, the probability of an all-good God existing must be affected by evil in the world. To say otherwise is to abandon eschatology.

Are skeptical theists also skeptical of Jesus' promised return? We will see.

BIBLIOGRAPHY

Adair, Aaron (2012), "Reexamining the Outer Limits and Alien Abductions", *Fleeing Nergal, Seeking Stars*, https://gilgamesh42.word-press.com/2012/09/25/reexamining-the-outer-limits-and-alien-abductions/ (Accessed 08/29/2022).

Ashkenazi, Michael (1992), "Not the sons of Adam", *Space Policy*, November 1992, p. 341-349.

Atran, Scott (2004), *In Gods We Trust: The Evolutionary Landscape of Religion*, Oxford: Oxford University Press.

Avalos, Hector (2013), *Slavery, Abolitionism, and the Ethics of Biblical Scholarship*, Sheffield, Sheffield Phoenix Press.

Bains, Williams & Schulze-Makuch, Dirk (2016), "The Cosmic Zoo: The (Near) Inevitability of the Evolution of Complex, Macroscopic Life", *Life (Basel)* (Jun 30), 6 (3), 25.

Bakker, Frederik A. (2016), *Epicurean Meteorology: Sources, Method, Scope and Organization*, Leiden, Brill.

Barnes, Julian E. (2022). "Many Military U.F.O. Reports Are Just Foreign Spying or Airborne Trash", *New York Times* Oct 28, 2022.

Begley, Sharon et al., (1977), "Seeking Other Worlds (Profile of Carl Sagan)", *Newsweek*, August 15, 1977, Volume 90.

Belgau, Ron (2013), "Friendship and the Scandal of Particularity", *First Things*, https://www.firstthings.com/blogs/firstthoughts/2013/04/friendship-and-the-scandal-of-particularity (Accessed 07/11/2022).

Bergner, Jennifer B. & Seligman, Darryl Z. (2023), "Acceleration of 1I/'Oumuamua from radiolytically produced H2 in H2O ice", *Nature* 615, p. 610-613.

Blackmore, Susan (1998), "Abduction by Aliens or Sleep Paralysis?", *Skeptical Inquirer*, Vol. 22, No. 3 (May/Jun3 1998), p. 23-28.

Block, Daniel L. (1998), *The Book of Ezekiel, Chapters 1-24*, Grand Rapids, MI: William B. Eerdmans.

Bobrick, Alexey & Martire, Gianni (2021), "Introducing physical warp drives", *Classical and Quantum Gravity* Vol. 38, No. 10.

Brown, Jeremy (2013), "The Talmudic View of the Universe", *New Heavens and a New Earth: The Jewish Reception of Copernican Thought*, Oxford Academic, p. 27-41.

Browne, Lewis (1931), *Since Calvary: An Interpretation of Christian History*, New York: The Macmillan Company.

Bruns, J. Edgar (1960), "Cosmolatry", *The Catholic World*, 191.1 (August) 286, p. 143–59.

Bryan, C. D. B. (1995), *Close Encounters of the Fourth Kind: Alien Abduction, UFOs, and the Conference at M.I.T.*, New York: Knopf.

Burgess, Andrew J. (1976), "Earth Chauvinism." *Christian Century*, 93, p. 1098.

Carlson, Eric Merle (2004), "The Holy Hush Of Ancient Sacrifice: An Analysis Of The Legitimacy Of The Non-Centralized Cult In Iron Age Israel", University of Georgia, https://getd.libs.uga.edu/pdfs/carlson_eric_merle_200412_ma.pdf (Accessed 10/04/2022).

Carrier, Richard (2006), "Carrier's Opening Statement: Naturalism Is True, Theism is Not", *The Secular Web*, https://infidels.org/library/modern/richard-carrier-carrier-wanchick-carrier1/ (Accessed 20/12/2020).

Carrier, Richard (2012), *Proving History: Bayes's Theorem and the Quest for the Historical Jesus*, Amherst, NY: Prometheus Books.

Chisholm, R.M., & Feehan, T.D. (1977), "The intent to deceive", *Journal of Philosophy*, 74, p. 143–59.

Cixin, Liu (2016), *The Dark Forest (The Three-Body Problem Series, 2)*, New York: Tor Books.

Clarke, Arthur C. (2013), "Hazards of Prophecy: The Failure of Imagination" in the collection *Profiles of the Future: An Enquiry into the Limits of the Possible*, London: SF Gateway.

Cline, Eric H. (2014), *1177 B.C.: The Year Civilization Collapsed*, Princeton, NJ: Princeton University Press.

Collins, R. (2015), "Extraterrestrial Intelligence and the Incarnation", in Kraay (ed.), *God and the Multiverse: Scientific, Philosophical, and Theological Perspectives* (p. 211-26), Abingdon: Routledge.

Colavito, Jason (2005), *The Cult of Alien Gods: H.P. Lovecraft and Extraterrestrial Pop Culture*, Amherst, NY: Prometheus Books.

Colavito, Jason (2012), *A Critical Companion to Ancient Aliens Seasons 3 and 4: Unauthorized*, Jason Colavito.com Books.

Colavito, Jason (2021), "How Washington Got Hooked on Flying Saucers", *The New Republic*, May 21.

Consolmagno, Brother Guy (2000), "Would You Baptize an Extraterrestrial?", *BeliefNet*, https://www.beliefnet.com/news/science-religion/2000/08/would-you-baptize-an-extraterrestrial.aspx (Accessed 07/26/2022).

Coyne, George V. (2002), "The Evolution of Intelligent Life on Earth and Possibly Elsewhere: Reflections from a Religious Tradition", In *Many Worlds*, p. 177–88.

Crowe, Michael J. (2011), *The Extraterrestrial Life Debate, 1750-1900, The Idea of a Plurality of Worlds from Kant to Lowell*, Cambridge: Cambridge University Press.

Dawkins, Richard (1998), *Unweaving the Rainbow: Science, Delusion, and the Appetite for Wonder*, Boston: Houghton Mifflin.

Deane-Drummond, Celia (2009), *Christ and Evolution: Wonder and Wisdom*, Minneapolis: Fortress.

"Declaration on the Relationship of the Church to the Non-Christian Religions", *Nostra Aetate* 4. In *The Documents of Vatican II*, edited by Walter M. Abbott, SJ., New York: America Press.

Deliop, Ilia (2012), *Christ in Evolution*, Maryknoll, NY: Orbis.

Determann, Jörg Matthais (2021), *Islam, Science Fiction, and Extraterrestrial Life: The Culture of Astrobiology in the Muslim World*, London: I.B. Tauris.

d'Huy, Julien (2013), "A Cosmic Hunt in the Berber sky: a phylogenetic reconstruction of a Palaeolithic mythology", *Les Cahiers de l'AARS, Saint-Lizier: Association des amis de l'art rupestre saharien*, p. 93-106.

d'Huy, Julien (2015), "Polyphemus: a Palaeolithic Tale?" *The Retrospective Methods Network Newsletter*, Department of Philosophy, History, Culture and Art Studies, University of Helsinki, p. 43-64.

Diaconus, Rusticus (1528), *Contra Acephols*, in *Antidotum Contra Diversas Omnium Fere Seculorum Haereses*.

Dick, Stephen J. (2020), *Many Worlds: New Universe Extraterrestrial Life*, Philadelphia: Templeton Foundation Press.

Dick, Stephen J. (2020), *Space, Time, and Aliens: Collected Works on Cosmos and Culture*, Cham: Springer.

Dill, K.A. & Agozzino, L. (2021), "Driving forces in the origins of life", *Open Biology*, Vol. 11, No. 2.

Dimond, Michael (2008), *UFOs: Demonic Activity & Elaborate Hoaxes Meant to Deceive Mankind*, Fillmore, NY: Most Holy Family Monastery.

Dodd, M.S. et al. (2017), "Evidence for early life in Earth's oldest hydrothermal vent precipitates", *Nature*, Vol. 543, p. 60-64.

Dominik, Martin & Zarnecki, John C. (2011), "The detection of extra-terrestrial life and the consequences for science and society", *Philosophical Transactions of the Royal Society A*, 369, p. 499-507.

Downing, Barry H. (1968), *The Bible and Flying Saucers*, Philadelphia: Lippincott.

Drake, Nadia (2018), "How Would We React to Finding Aliens?", *The National Geographic*, https://www.nationalgeographic.com/science/article/how-would-people-react-alien-life-discovery-aaas-space-science (Accessed 06/28/2022).

Edwards, Bruce L (2007), *C.S. Lewis: Life, Works, and Legacy*, Westport, CN: Praeger.

England, Jeremy L. (2015), "Dissipative adaptation in driven self-assembly", *Nature Nanotechnology*, Vol. 10, p. 919–23.

Ehrlich Richard S. (2019), "The UFO seekers flocking to a remote Thai hilltop in search of Buddhist aliens", *CNN*, (5th October 2019), https://www.cnn.com/travel/article/thailand-ufo-buddhist-aliens/index.html (Accessed 09/14/2022).

Firpo, Luigi (1993), *Il Processo di Giordano Bruno*, Rome: Salerno Editrice.

Fitzgerald, David (2013), *The Mormons*, CreateSpace.

Forgan, Duncan H. (2011), "Spatio-temporal constraints on the zoo hypothesis, and the breakdown of total hegemony", *International Journal of Astrobiology*, Vol. 10, No. 4, p. 341-47.

Forgan, Duncan H. (2017), "The Galactic Club or Galactic Cliques? Exploring the limits of interstellar hegemony and the Zoo Hypothesis", *International Journal of Astrobiology*, Vol. 16, No. 4, p. 349-54.

Frazier, Kendrick, Karr, Barry & Nickell, Joe, eds. (1997), *The UFO Invasion: The Roswell Incident, Alien Abductions, and Government Coverups*. Amherst, NY: Prometheus Books.

Garcia-Escartin, Juan Carlos & Chamorro-Posada, Pedro (2013), "Scouting the spectrum for interstellar travellers", *Acta Astronautica*, Vol. 85 (April-May), p. 12-18.

George, Marie I. (1996), "Aquinas on Reincarnation", *The Thomist: A Speculative Quarterly Review*, Vol. 60, No. 1 (January), p. 33-52.

"Giordano Bruno", *The Catholic Encyclopedia*, https://www.newadvent.org/cathen/03016a.htm (Accessed 09/06/2022).

Gott, Richard J. (1993), "Implications of the Copernican principle for our future prospects", *Nature*, 363 (May 27), p. 315-19.

Green, Emerson (2021), "Skeptical Theism", *Emerson Green: Walden Pod / Counter Apologetics*, https://emersongreenblog.wordpress.com/2021/03/01/skeptical-theism/ (Accessed 07/07/2022).

Gross, Neil & Simmons, Solon (2009), "The Religiosity of American College and University Professors", *Sociology of Religion*, Vol. 70:2, p. 101-29.

Haarsma, Deborah (2019), "What would life beyond Earth mean for Christians?", *BioLogos*, https://biologos.org/articles/what-would-life-beyond-earth-mean-for-christians) (Accessed 07-/08/2022).

Haqq-Misra, Jacob D. & Baum, Seth D. (2009), "The Sustainability Solution to the Fermi Paradox", *Journal of the British Interplanetary Society*, Vol. 62, No. 2 (February 2009), p. 47-51.

Hanson, Robin et al. (2021), "If Loud Aliens Explain Human Earliness, Quiet Aliens Are Also Rare", *The Astrophysical Journal*, Vol. 922, No. 2, Art. 182.

Heron, Germain (1954), *On Learned Ignorance by Nichola Cusanus*, Yale University Press.

Howell, Elizabeth (2018), "How Would Humanity React If We Really Found Aliens?", *Space.com*, https://www.space.com/40435-finding-aliens-humanity-reaction.html (Accessed 06/26/2022).

Huemer, Michael (n.d.), "Phenomenal Conservatism", *The Internet Encyclopedia of Philosophy*, https://iep.utm.edu/phen-con/ (Accessed 07/14/2022).

Huemer, Michael (2001), *Skepticism and the Veil of Perception*, Lanham: Rowman & Littlefield.

Huemer, Michael (2013), "Phenomenal Conservatism Uber Alles." In Chris Tucker (ed.), *Seemings and Justification: New Essays on Dogmatism and*

Phenomenal Conservatism (p. 328-350), Oxford: Oxford University Press.

Huemer, Michael (2007), "Compassionate Phenomenal Conservatism." *Philosophy and Phenomenological Research*, 74, p. 30-55.

Isbouts, Jean-Pierre (2019), "Cain and Abel's clash may reflect ancient Bronze Age rivalries", *The National Geographic*, https://www.nationalgeographic.com/culture/article/cain-abel-reflects-bronze-age-rivalry (Accessed 10/04/2022).

Jackson, Alan P. & Desch, Steven J. (2021), "1I/'Oumuamua as an N2 Ice Fragment of an exo-Pluto Surface: I. Size and Compositional Constraints", *Journal of Geophysical Research: Planets*, Vol. 126, Iss. 5.

Jeffrey, Alan & Dai, Hui-Hui (2008), *Handbook of Mathematical Formulas and integrals, 4th Edition*, Amsterdam: Elsevier.

Katz, J.L. (2021), "Oumuamua is not Artificial", arXiv preprint arXiv:2102.07871.

Keel, Othmar (1977), *Jahwe-Visionen und Siegelkunst: Eine neue Deutung der Majestätsschilderungen in Jes, Ez 1 und 10 und Sach 4*, Stuttgart: Katholisches Bibelwerk.

Kelly, Terence J. & Regan, Hilary D. (2002), *God, Life, Intelligence and the Universe*, Adelaide: Australian Theological Forum.

Kent, Stephen A. & Raine, Susan (2017), *Scientology in Popular Culture: Influences and Struggles for Legitimacy*, Santa Barbara, CA: Praeger.

Kipping, David (2020), "An objective Bayesian analysis of life's early start and our late arrival", *PNAS*, Vol. 117, No. 22 (May 18, 2020), p. 11995-12003.

Kipping, David, Frank, Adam & Scharf, Caleb (2020), "Contact inequality: First contact will likely be with an older civilization", *International Journal of Astrobiology*, Vol. 19, No. 6, p. 430-37.

Klass, Philip J. (1983), *UFOs: The Public Deceived*, Buffalo, NY: Prometheus Books.

Kwon, Jung Yul, et al. (2018), "How Will We React to the Discovery of Extraterrestrial Life?", *Frontiers in Psychology*, Vol. 8, https://www.frontiersin.org/article/10.3389/fpsyg.2017.02308 (Accessed 06/28/2022).

Landsman, Klaas (2016), "The Fine-Tuning Argument: Exploring the Improbability of Our Existence", in *The Challenge of Chance*, ed. Landsman & Wolde, New York: Springer, p. 111-129, https://link.springer.com/chapter/10.1007/978-3-319-26300-7_6 (Accessed 06/30/2022).

Langston, Mark C. (2013), "The Accidental Altruist: Inferring Altruism from an Extraterrestrial Signal", in Vakoch (ed.), *Extraterrestrial Altruism: Evolution and Ethics in the Cosmos* (p. 131-140), Berlin: Springer.

Laurent, Régis (2015), *An Introduction to Aristotle's Metaphysics of Time: Historical Research into the Mythological and Astronomical Conceptions that Preceded Aristotle's Philosophy*, Paris: Villegagnons-Plaisance éditions, D.L.

Lesnick, James G. (1990), *Contemporary Authors: a bio-bibliographical guide to current writers in fiction, general nonfiction, poetry, journalism, drama, motion pictures, television and other fields*, Farmington Hills, MI: Gale Research Company.

Levine, W. Garrett et al. (2021), "Constraints on the Occurrence of 'Oumuamua-Like Objects", *The Astrophysical Journal*, Vol. 922, No. 39.

Lewis, James R. (2003), *Encyclopedic Sourcebook of UFO Religions*, Amherst, NY: Prometheus Books.

Loftus, John. W. (2016), *Christianity in the Light of Science: Critically Examining the World's Largest Religion*, Amherst, NY: Prometheus Books.

Losch, Anreas & Krebs, Andreas (2015), "Implications for the Discovery of Extraterrestrial Life: A Theological Approach", *Theology and Science*, 3:2, p. 230-44.

Lushnikova, A. V. (2003), "Ursa Major: ot losya do medvedya (Ursa Major: from elk to bear)", *Istoriko-Astronomicheskie Issledovaniya*, Vol. 28, p. 189-216, 342-43.

MacDonald, Dennis R. (2000), *The Homeric Epics and the Gospel of Mark*, New Haven, CT: Yale University Press, p. 63-76.

Mahon, James Edwin (2015), "The Definition of Lying and Deception", *The Stanford Encyclopedia of Philosophy*, https://plato.stanford.edu/entries/lying-definition/ (Accessed 07/02/2022).

Marcy, Geoffrey W., et al. (2022), "A Search for Monochromatic Light Toward the Galactic Centre", arXiv preprint arXiv: 2208.13561.

Martinez, Alberto A. (2016), "Giordano Bruno and the heresy of many worlds", *Annals of Science*, 73:4, p. 345-74.

Martinez, Alberto A. (2018), "Was Giordano Bruno Burned at the Stake for Believing in Exoplanets?", *Scientific American*, https://blogs.scientificamerican.com/observations/was-giordano-bruno-burned-at-the-stake-for-believing-in-exoplanets/ (Accessed 09/06/2022).

Marty, Peter W. (2014), "Who Gets Saved?", *The Lutheran*, Vol. 27, No. 3.

Mayer, Adrienne (2001), *The First Fossil Hunters: Dinosaurs, Mammoths, and Myth in Greek and Roman Times*, Princeton, NJ: Princeton University Press.

McColley, Grant (1936), "The Seventeenth-century Doctrine of a Plurality of Worlds", *Annals of Science*, 1, 4, p. 385-430.

McHugh, L.C. (1961), "Life in Outer Space? An Interview with Rev. L.C. McHugh, SJ", *Sign* 41.5, p. 29.

McInnes, C.R. (2002), "The Light Cage Limit to Interstellar Expansion", *Journal of the British Interplanetary Society*, Vol. 55, p. 279-284.

McIntosh, C.A. & McNabb, T.D (2021), "Houston, Do We Have A Problem? Extraterrestrial Intelligent Life and Christian Belief", *Philosophia Christi*, 23 (1), p. 101-124. Our pagination refers to the standalone copy found in the

PhilPapers archive: https://philarchive.org/rec/MCIHDW (Accessed 06/29/2022).

Moore, Marissa (2022), "Understanding Terror Management Theory", *PsychCentral,* https://psychcentral.com/health/terror-management-theory (Accessed 10/28/2022).

Nesteruk, Alexei V. (2023), "From Deep Incarnation to Deep Anthropology: Hypostatic Union and the Universe in the Image of Imago Deï", *Theology and Science,* 21 (1), p. 81-95.

Nickell, Joe (1996), "A Study of Fantasy Proneness in the Thirteen Cases of Alleged Encounters in John Mack's Abduction", *Skeptical Inquirer,* Vol. 20, No. 3 (May/June 1996), p. 18-20, 54.

Nongbri, Brent (2013), *Before Religion: A History of a Modern Concept,* Yale University Press.

Norenzayan, Ara et al. (2012), "Mentalizing deficits constrain belief in a personal God." *PloS One,* Vol. 7, No. 5.

Office of the Director of National Intelligence (2023), *2022 Annual Report on Unidentified Aerial Phenomena.*

Ohlheiser, Abby (2014), "Pope Francis Says He Would Definitely Baptize Aliens If They Asked Him To", *The Atlantic,* https://www.theatlantic.com/international/archive/2014/05/pope-francis-says-he-would-definitely-baptize-aliens-if-they-wanted-it/362106/ (Accessed 09/06/2022).

Okwuosa, L.N., Nwaoga, C.T, & Uroko F.C. (2017), "A Critique of John Hick's Multiple Incarnation: Theology and Christian Approach to Religious Dialogue", *Mediterranean Journal of Social Sciences,* Vol 8, No. 5 September 2017.

Peacocke, Arthur (2000), "The Challenge and Stimulus of the Epic of Evolution to Theology", *MW,* p. 89–118.

Pearce, Jonathan M.S. (2021), *The Nativity: A Critical Examination,* Fareham: Onus Books.

Pearce, Jonathan M.S. (2021), *The Resurrection: A Critical Examination of the Easter Story,* Fareham: Onus Books.

Pearce, Jonathan M.S. (2021b), *Why I Am Atheist and Not a Theist: How to Do Knowledge, Meaning, and Morality in a Godless World,* Fareham: Onus Books.

Pearce, Jonathan M.S. (2022), *30 Arguments Against the Existence of "God", Heaven, Hell, Satan, and Divine Design,* Fareham: Onus Books.

Pearce, Jonathan M.S. (2022b), "Can you even calculate a best possible world?" *Only Sky,* https://onlysky.media/jpearce/can-you-even-calculate-a-best-possible-world/ (Accessed 10/17/2022).

Peters, Ted (2011), "The implications of the discovery of extra-terrestrial life for religion", *Philosophical Transactions of the Royal Society A* (2011) 369, p. 644–655.

Peters, Ted (2018), *Astrotheology: Science and Theology Meet Extraterrestrial Life,* Eugene, OR: Cascade Books.

Pittenger, Norman (1959), *The Word Incarnate*, London: Nisbet.

Plackett, Benjamin (2021), "How many early human species existed on Earth?", *Live Science*, https://www.livescience.com/how-many-human-species.html (Accessed 06/30/2022).

Plantinga, A. (2011), *Where the Conflict Really Lies*, Oxford: Oxford University Press.

Polkinghorne, John (2004), *Science and the Trinity: The Christian Encounter with Reality*, New Haven: Yale University Press.

Pritchard, Andrea et al., eds., (1994), *Alien Discussions: Proceedings of the Abduction Study Conference Held at MIT*, Cambridge, MA: North Cambridge Press.

Prothero, Donald R., & Callahan, Timothy D. (2017), *UFOs, Chemtrails, and Aliens: What Science Says*, Bloomington, IN: Indiana University Press.

Puccetti, Roland (1969), *Persons: A Study of Possible Moral Agents in the Universe*, New York: Herder & Herder.

Ragan, Claude, Malony, H. Newton & Beit-Hallahmi, Benjamin (1980), "Psychologists and Religion: Professional Factors and Personal Belief", *Review of Religious Research*, Vol. 21, No. 2 (Spring, 1980), p. 208-17.

Ruhmkorff, Samuel (2019), "The Copernican principle, intelligent extraterrestrials, and arguments from evil", *Religious Studies*, Vol. 55, Special Issue 3: Religious Experience and Desire, September 2019, p. 297–317.

Russell, Stuart and Norvig, Peter (2016), *Artificial Intelligence: A Modern Approach*, Upper Saddle River, NJ: Prentice Hall.

Sagan, Carl (2000 [1973]), *Carl Sagan's Cosmic Connection: An Extraterrestrial Perspective*, Cambridge: Cambridge University Press.

Sagan, Carl (1997), *The Demon-Haunted World: Science as a Candle in the Dark*, New York: Random House.

Sandberg, Anders Drexler, Eric & Ord, Toby (2018), "Dissolving the Fermi paradox", arXiv:1806.02404.

Santiago, Jessica, Schuster, Sebastian, & Visser, Matt (2022), "Generic warp drives violate the null energy condition", *Physical Review D*, Vol. 105, No. 6.

Sardar, Nouri (2020), "What does Islam say about Aliens? A Look at Quranic Verses and Hadiths", *TMV*, https://themuslimvibe.com/faith-islam/science/what-does-islam-say-about-aliens-a-look-at-quranic-verses-and-hadith (Accessed 09/04/2022).

Shackelford, Jole (2009), "That Giordano Bruno Was the First Martyr of Modern Science", in Numbers (ed.), *Galileo Goes to Jail and Other Myths about Science and Religion* (p. 59-67), Harvard.

Schetsche, Michael (2005), "SETI (Search for Extraterrestrial Intelligence) and the Consequences: Futurological Reflections on the Confrontation of Mankind with an Extraterrestrial Civilization", translated by *Astrosociology.com*, http://www.astrosociology.com/Library/PDF/Contributions/SE-TIandConsequences_ENG.pdf (Accessed 06/28/2022).

Sheaffer, Robert (1998), *UFO Sightings: The Evidence*, Amherst, NY: Prometheus Books.

Sheaffer, Robert (2015), *Bad UFOs: Critical Thinking about UFO Claims*, North Charleston, SC: CreateSpace.

Shermer, Michael (2012), *The Believing Brain: From Spiritual Faiths to Political Convictions – How We Construct Beliefs and Reinforce Them as Truths*, London: Robinson.

Shklovskii, Iosif S. & Sagan, Carl (1966), *Intelligent Life in the Universe*, San Francisco: Holden-Day.

Siraj, Amir & Loeb, Abraham (2022), "The mass budget necessary to explain 'Oumuamua as a nitrogen iceberg", *New Astronomy*, Vol. 92 (2022).

Smith, Craig B. (2004), *How the Great Pyramid Was Built*, Washington: Smithsonian Books.

Smith, Jesse (2021), "Transmission of Faith in Families: The Influence of Religious Ideology", *Sociology of Religion*, 2021 Autumn; 82(3), p. 332–356, https://www.ncbi.nlm.nih.gov/pmc/articles/PMC8204683/ (Accessed 10/28/2022).

"Somewhere, Something Incredible Is Waiting To Be Known", *Quote Investigator*, https://quoteinvestigator.com/2013/03/18/incredible/ (Accessed 11/14/2022).

Steinbauer, Friedrich (1979), *Melanesian Cargo Cults: New Salvation Movements in the South Pacific*, St. Lucia, Q'ld: University of Queensland Press.

Stirrat, Michael & Cornwell, Elisabeth (2013), "Eminent scientists reject the supernatural: a survey of the Fellows of the Royal Society", *Evolution: Education and Outreach*, Vol. 6, Article 33.

Subin, Anna Della (2021), *Accidental Gods: On Race, Empire, and Men Unwittingly Turned Divine*, New York: Henry Holt and Company.

Tallet, Pierre & Lehner, Mark (2022), *The Red Sea Scrolls: How Ancient Papyri Reveal the Secrets of the Pyramids*, London: Thames & Hudson.

Tenn, William (2016), "On Venus, Have We Got a Rabbi!", *Tablet*, https://www.tabletmag.com/sections/arts-letters/articles/on-venus-have-we-got-a-rabbi (Accessed 10/26/2022).

Tillich, Paul (1976 [1963]), *Systematic Theology, Volume 3: Life and the Spirit: History and the Kingdom of God*, Chicago: University of Chicago Press.

Thigpen, Paul (2022), *Extraterrestrial Intelligence and the Catholic Faith: Are We Alone in the Universe with God and the Angels?*, Gastonia, NC: TAN Books.

Tooley, Michael (2015), "The Problem of Evil", *The Stanford Encyclopedia of Philosophy*, https://plato.stanford.edu/entries/evil/ (Accessed 09/04/2022).

Totani, Tomonori (2020), "Emergence of life in an inflationary universe", *Science Reports*, Vol. 10, Art. 1671.

Tough, Allen (2000), *When SETI Succeeds: The Impact of High-Information Contact*, Washington: Foundation for the Future.

Trompf, G.W. (1990), *Cargo Cults and Millenarian Movements: Transoceanic Comparisons of New Religious Movements*, New York: Mouton de Gruyter.

Vakoch, D.A & Lee, Y.-S. (2000), "Reactions to Receipt of a Message from Extraterrestrial Intelligence: A Cross-Cultural Empirical Study", *Acta Astronautica*, Vol. 46, No. 10-12, 2000, p. 737-744.

Vakoch, D.A. & Harrison, A.A. (2011), *Civilizations Beyond Earth: Extraterrestrial Life and Society*, New York: Berghahn Books.

Vidal, Clément (2015), "A Multidimensional Impact Model for the Discovery of Extraterrestrial Life" in *The Impact of Discovering Life Beyond Earth*, edited by Steven J. Dick, Cambridge: Cambridge University Press.

Ward, Peter D. & Brownlee, Donald (2000), *Rare Earth: Why Complex Life is Uncommon in the Universe*, New York: Springer.

Wathey, John (2019), *The Illusion of God's Presence: The Biological Origins of Spiritual Longing*, Amherst, NY: Prometheus Books.

Watson, Philip J. (1987), *Egyptian Pyramids and Mastaba Tombs of the Old and Middle Kingdoms*, Princes Risborough: Shire.

Webb, Stephen (2015), *If the Universe Is Teeming with Aliens ... WHERE IS EVERYBODY?: Seventy-Five Solutions to the Fermi Paradox and the Problem of Extraterrestrial Life*, Cham, Switzerland: Springer.

Weidemann, Christian (2016), "Did Jesus die for Klingons, too? Christian faith and extraterrestrial salvation", Ruhr-Universität Bochum, https://www.researchgate.net/publication/330161960_Did_Jesus_die_for_/Klingons_too_Christian_faith_and_extraterrestrial_salvation, (Accessed 08/26/2022).

Weintraub, David A. (2014), *Religions and Extraterrestrial Life: How Will We Deal With It?*, Cham, Switzerland: Springer International Publishing.

West, Mick (2022), "Gimbal UFO - A New Analysis", *Metabunk.org*, https://www.metabunk.org/threads/gimbal-ufo-a-new-analysis.12333/ (Accessed 08/29/2022).

Wilkinson, David (2013), *Science, Religion, and the Search for Extraterrestrial Intelligence*, Oxford: OUP, p. 103.

Worsley, Peter (1968), *The Trumpet Shall Sound: A Study of 'Cargo' Cults in Melanesia*, London: MacGibbon & Kee.

Wright, J. T. et al. (2014), *The Astrophysical* Journal, Vol. 792, No. 26.

Yates, Frances A. (2001 [1964]), *Selected Works, Volume II: Giordano Bruno and the Hermetic Tradition*, NY: Routledge.

www.ingramcontent.com/pod-product-compliance
Lightning Source LLC
Chambersburg PA
CBHW070800170426
43200CB00007B/848

Greco-Roman and Jewish Tributaries to the New Testament

Festschrift in Honor of Gregory J. Riley

Greco-Roman and Jewish Tributaries to the New Testament

Festschrift in Honor of Gregory J. Riley

Christopher S. Crawford, *editor*

CLAREMONT STUDIES IN NEW TESTAMENT
AND CHRISTIAN ORIGINS 4

Greco-Roman and Jewish Tributaries to the New Testament
Festschrift in Honor of Gregory J. Riley

©2018 Claremont Press
1325 N. College Ave
Claremont, CA 91711

ISBN 978-1-946230-35-5

Library of Congress Cataloging-in-Publication Data

Greco-Roman and Jewish Tributaries to the New Testament
    Festschrift in Honor of Gregory J. Riley /Christopher S.
    Crawford
    xii + 224 pp. 22 x 15 cm. –(Claremont Studies in New
    Testament and Christian Origins 4)
    Includes bibliographical references and index.
    ISBN 978-1-946230-35-5

    Bible. New Testament Criticism, interpretation, etc
    BS 2361.2 C739 2019

Cover: *An image of the dome of the Pantheon in Rome, a structure commissioned by Marcus Agrippa and completed by the Emperor Hadrian to honor all of the gods and later converted into a Christian church.*

Table of Contents

Contributors   ix

Abbreviations   xi

Introduction   1
    Christopher S. Crawford

The "Journey of the Soul" in Two Lucan Parables   5
    Nicholas J. Frederick

Kings of the Jews   27
Herodian Collaboration with Rome through a Marcan Lens
    Margaret Froelich

Mark's Transformation of the Love Commandment   37
Controversy in the Lost Gospel
Logoi 6:18-21 and Mark 12:28-34 and 10:17-22
    Dennis R. MacDonald

Evidence for a Relationship Between Mark and Q   45
    James Van Dore

A Woman Caught in Adultery? Or a Wandering   71
Teacher Trapped Between Roman and Jewish Law?
John 7:53-8:11 in Light of Quintilian and Seneca
    Thomas E. Phillips

Romans 1:26-27 in its Rhetorical Tradition   83
    Brett Provance

New Directions in the Gospel of Thomas   117
Oxyrhynchus as Test Case
    Thomas A. Wayment

A Separate Son of Man L. Arik Greenberg	135
Eschatological Perspective in the **Heikhalot Rabbati** Marvin A. Sweeney	181
Bibliography	193
Indices	211

Contributors

Christopher S. Crawford, PhD candidate (Claremont School of Theology), is currently working on his dissertation on the Book of Revelation.

Nicholas J. Frederick, PhD (Claremont School of Theology), is Assistant Professor of Ancient Scripture at Brigham Young University and the author of *Joseph Smith's Seer Stones* (Deseret Book, 2016) and editor of *His Majesty and Mission* (Deseret Book, 2017) and *Give Ear to My Words* (Deseret Book, 2019).

Margaret Froelich, PhD (Claremont School of Theology), is reference librarian at Claremont School of Theology and co-editor of *Christian Origins and the New Testament in the Greco-Roman Context* (Claremont Press, 2017).

L. Arik Greenberg, PhD (Claremont Graduate University), is a Senior Lecturer in Theological Studies at Loyola Marymount University. An ardent advocate of religious tolerance and interfaith dialogue, he is the founder of the Institute for Religious Tolerance, Peace and Justice and author of *"My Share of God's Reward" Exploring the Roles and Formulations of the Afterlife in Early Christian Martyrdom* (Peter Lang, 2009).

Dennis R. MacDonald, PhD (Harvard University), is Research Professor of New Testament at Claremont School of Theology and author of several groundbreaking books on the New Testament and mimesis criticism, including *Two Shipwrecked Gospels: The Logoi of Jesus and Papias's Exposition of Logia about the Lord* (SBL, 2012) and *The Dionysian Gospel: The Fourth Gospel and Euripides* (Fortress, 2017).

Thomas E. Phillips, PhD (Southern Methodist University), is Professor of New Testament and Theological Bibliography at Claremont School of Theology and the author of several volumes on Luke and Acts, including *Paul, His Letters and Acts* (Hendrickson, 2009) and *Reading Acts in Diverse Frames of Reference* (Mercer University Press, 2009).

Brett Provance, PhD (Claremont Graduate University), is Professor of Humanities at California Baptist University and author of *Pocket Dictionary of Liturgy and Worship* (IVP, 2009).

Marvin A. Sweeney, PhD (Claremont Graduate University), is Professor of Hebrew Bible at Claremont School of Theology and the author of nearly 20 scholarly books, *including Jewish Mysticism: From Ancient Times Through Today* (Eerdmans, 2019), *The Pentateuch* (Abingdon, 2017) and *Isaiah 40-66* (Eerdmans, 2016).

James Van Dore, PhD (Claremont Graduate University), is senior lecturer at Old Dominion University and a scholar of the Gospels.

Thomas A. Wayment, PhD (Claremont Graduate University), is Professor of Classics at Brigham Young University and author of several books, including *Go Ye Into All the World* (Deseret, 2002) and *"Behold the Lamb of God"* (Deseret, 2008).

Abbreviations

AB	Anchor Bible
AGAJU	Arbeiten zur Geschichte Antiken Judentums und des Urchristentums
ANF	Ante-Nicene Fathers
ANRW	Aufstieg und Niedergang der römischen Welt
BADG	Bauer, W., W.F. Arndt. F.W. Gingrich, and F.W. Danker. *Greek-English Lexicon of the New Testament and Other Christian Literature*. 2nd ed. Chicago, 1979.
BTB	*Biblical Theology Bulletin*
CBR	*Catholic Biblical Review*
DDD	*Dictionary of Deities and Demons in the Bible*
HTR	*Harvard Theological Review*
ICC	International Critical Commentary
JBL	*Journal of Biblical Literature*
JRE	*Journal of Religious Education*
JSOT	*Journal for the Study of the Old Testament*
JW	Josephus, *Jewish War*
LNTS	Library of New Testament Studies
NHC	Nag Hammadi Codex
NHLE	Nag Hammadi Library in English
NovT	*Novum Testamentum*
NovTSup	Novum Testamentum Supplements
NTS	*New Testament Studies*
OTK	Ökumenischer Taschenbuchkommentar zum Neuen Testament
RVV	Religionsgeschichtliche Versuche und Vorarbeiten
SC	*The Second Century*
SNTSMS	Society for New Testament Studies Monograph Series

SBLECL	Society of Biblical Literature: Early Christianity and Its Literature
SVTP	Studia in Veteris Testamenti pseudepigraphica
TSAJ	Texts and Studies in Ancient Judaism
VC	*Vigilae Christianae*
WBC	Word Biblical Commentary
WMANT	Wissenschaftliche Monographien zum Alten und Neuen Testament
WUNT	Wissenschaftliche Untersuchungen zum Neuen Testament

Introduction

Christopher S. Crawford

Dr. Gregory J. Riley has been a staple in the Claremont community of religious studies for over two decades and has made essential contributions to biblical studies, not only through his academic interests and publications, but also with his countless hours in the classroom implanting the ideals of hard work and enthusiasm for the field into the minds of future religious leaders and academics. Throughout his tenure at the Claremont School of Theology, Claremont Graduate University, and The Episcopal Theological School at Claremont (Bloy House), Dr. Riley has taught courses and led seminars discussing nearly every aspect of the New Testament and the world in which it was produced.

Throughout his academic career, Riley has made a concerted effort to look beyond the New Testament and better understand the world in which its authors lived, in order to comprehend what they were writing and why they wrote it. This is clearly seen in his works such as *Resurrection Reconsidered*, *The River of God*, and *One Jesus, Many Christs*. The common thread throughout is that in order to fully comprehend Christianity, one must look beyond Palestine and Judaism to the Greeks, Romans, Egyptians, Babylonians, and Persians. It is just like Dr. Riley's masterful metaphor of the river of God: as we follow it downstream, we see the impact each of these cultures have had as tributaries that lead into the river of God, which in turn has influenced the development of religion in general, and Christianity in particular.

Year after year, as new students enter Riley's classroom, they not only increase their knowledge of the gospel of Jesus and the New Testament. More than that, he guides them through the world of the mighty heroes of the Homeric epics, the classical works of the Greek tragedians, the philosophical works of Plato, and the often-strange imaginations of the Gnostic writers. He helps his students see the important contributions that all of these made to what would later become Christianity. Riley is able to make these texts and their often-foreign ideas understandable to students who are encountering them for the first time and is able to spark the curiosity of many to continue their studies in these topics. This expansive catalog of courses and topics that Dr. Riley has taught and studied can clearly be seen in the diverse topics that are taken up in this *Festschrift*.

Nicholas J. Frederick opens this work perfectly by discussing a phrase every student of Dr Riley has heard countless times in his lectures: the journey of the soul. In his essay, "The 'Journey of the Soul in Two Lucan Parables," he examines this important concept through the lens of the parables of the good Samaritan and the prodigal son in the Gospel of Luke. He compares these parables with the Nag Hammadi writings the *Hymn of the Pearl* and *Exegesis on the Soul* and argues that the archetype of the journey of the soul is found in both these gospel parables.

Margaret Froelich continues our examination of the Synoptic Gospels by looking at the Gospel of Mark in her essay, "Kings of the Jews: Herodian Collaboration with Rome through a Markan Lens." Froelich argues that if Mark is read as a political document with a specific emphasis on its use of the term βασιλεύς, it can be understood to compare Jesus and the Herodian dynasty. Mark does not use the personal name Antipas, but instead uses Herod as a dynastic name and portrays him as weak, cowardly, and violent, thus implanting these characteristics in the mind of his audience and making the entire dynasty a foil for the Jesus the true βασιλεύς.

Continuing on the theme of the Markan gospel, we are pleased to present the contribution of Dr. Riley's colleague of over a decade in the New Testament department, Dennis R. MacDonald. In his essay, "Mark's Transformation of the Love Commandment: Controversy in the Lost Gospel," MacDonald provides detailed analysis and utilizes his Q+/Papias Hypothesis to determine that the exchange between Jesus and another teacher of the Mosaic law concerning the greatest commandment to conclude that this pericope was adopted by all three evangelists from Q/Q+.

James Van Dore concludes our Markan section by looking at Mark's relationship with Q. Taking Harry T. Fleddermann's reconstruction of Q as his starting point, Van Dore argues that it is nearly impossible to separate the Q from the Gospel of Mark unless one examines the longer passages, where multiple elements were collected into compositions; it is in these sections where parallel constructions and overlaps are found.

Thomas E. Phillips, another esteemed colleague of Dr. Riley's in the New Testament department, agrees with Riley's insistence on reading the New Testament as a product of the Greco-Roman world and uses this as a framework to better understand the Johannine pericope of the woman caught in adultery. In his essay, "A Woman Caught in Adultery? Or a Wandering Teacher Trapped Between Roman and Jewish Law?" Phillips argues that trying to understand this often-debated scene solely through a Jewish lens is not adequate. A full grasp of this scene requires Greco-Roman context. Phillips does this by examining the Roman laws concerning adultery as enunciated by Quintilian and Seneca and comparing them to the Mosaic Law, concluding that Jesus's actions are consistent with the former.

Brett Provance's essay, "Romans 1:26–27 in Its Rhetorical Tradition," moves us away from the Gospels and towards Paul's epistle to the Romans. In this essay, Provance examines one of the more controversial aspects that modern interpreters have used as an indictment against same-sex sexual activity. Provance argues that this interpretation is incorrect. After guiding the reader through similar passages in the Hebrew Bible pseudepigraphal writings, Provance concludes that in order to arrive at the optimal interpretation, we need to look at general rhetorical tradition of the dual-judgment topos.

Thomas A. Wayment's essay, "New Directions in the Gospel of Thomas: Oxyrhynchus as Test Case," moves us out of the canonical gospel and into the realm of Gnosticism. In particular, Wayment looks at the Gospel of Thomas, the topic of one of Riley's earliest publications, *Resurrection Reconsidered*. Wayment considers the use of the Gospel of Thomas in the Oxyrhynchus Christian community by examining the physical evidence through a papyrological lens. Based on the evidence, Wayment concludes that the Gospel of Thomas was not read in public but instead was used in private settings.

Staying within the Gnostic realm of religious studies, L. Arik Greenberg leads the reader through an extensive study of the term "Son of Man" in his essay, "A Separate Son of Man." By closely examining the use of the term in Q, Babylonian literature, the Enoch traditions, and the Gnostic/Thomas traditions, Greenberg comes to the conclusion that the Son of Man as a title existed separate from the Christian tradition. The early Jesus traditions were also familiar with this theological/apocalyptic figure and reimagined Jesus, incorporating it into his teachings. As time passed, the figure of Jesus and the Son of Man were conflated into the current traditions of Jesus referring to himself as the Son of Man.

We conclude this *Festschrift* with a contribution by Marvin E. Sweeney, another long-time colleague of Dr. Riley's in the Hebrew Bible department. In his essay, "Eschatological Perspective in the *Heikhalot Rabbiti,*" Sweeney moves us further in time to the Tannaitic period and into the genre of *hekhalot* literature (i.e., the ascents to the heavenly palaces). Sweeney examines *Heikhalot Rabbiti,* one of the many *hekhalot* works, noting that scholars have debated the fundamental concern addressed in this text. Sweeney argues that this work is concerned with *both* the mystical experience of the sage *and* the interpretation of the Torah.

It is with great pleasure and pride that we present this collection of essays submitted by former students and current colleagues. It is our hope that these will add to the legacy that Dr. Riley has established of diligent and committed research in religious studies. Like the many influences that have added to the river of God through the centuries we, as dedicated scholars, hope to add to the scholarly conversations and make meaningful contributions to our various fields of studies. Dr. Riley has surely paved the way for us.

The "Journey of the Soul" in Two Lucan Parables

Nicholas J. Frederick

One of the revolutionary concepts theorized and developed by the ancient Greek philosophers was the immortality of the soul. Because the soul was immortal, it stood to reason that the soul was in the midst of a journey, with life on this world merely a temporary stop. The Greeks of Homer's time had conceived of Hades as largely a place where the shades of men and women mingle, but without any teleological purpose.[1] By the fourth century BCE however, Plato would challenge these beliefs and introduce a radical revision of the notion of both a pre-mortality and a post-mortality. For Plato, the soul was a resident of the realm of the absolute, the constant and the invariable. It was immortal, pure, and changeless.[2] The soul existed before it came to earth, and thus when humanity "learns" something they are only recollecting or remembering something already learned in a prior state. According to Plato, once the soul is placed into a body of flesh it becomes tainted to a variable degree, and the fate of the soul upon death is a direct correlation to how much it is weighted down. So those who have cultivated gluttony while in the flesh assume the forms of donkeys upon their return to a carnal state. Those who lived good lives but were unable to gain any significant philosophical understanding return as wasps or even as people. Those who have been fortunate enough to encounter philosophy, however, may "attain to the divine nature."[3] In the *Phaedrus,* Plato beautifully illustrates this concept through the example of a charioteer, who guides the immortal soul into a cavalcade of gods and purified souls.[4]

With an understanding that the soul has a pre-mortal and a

---

[1] See, for example, the famous exchange between Odysseus and Achilles, upon the former meeting the latter in Hades: "No winning words about death to me, shining Odysseus! By god, I'd rather slave on earth for another man–Some dirt-poor tenant farmer who scrapes to keep alive–Than rule down here over all the breathless dead" (Homer, *Od.* 11: 554–558).

[2] Plato, *Phaed.* 79d

[3] Plato, *Phaed.* 82a–c

[4] Plato, *Phaedr.* 246a–254e

post-mortal existence, some thinkers began to look at existence as part of a larger picture. Questions quickly arose: Where did we come from? Why are we here? Where are we going? What emerged was the idea that the soul was on a journey of sorts, one that originated prior to this world and would continue beyond it. A useful description of this archetype of the "journey of the soul" is provided by Bentley Layton:

> Both popular belief and certain kinds of academic philosophy (especially Platonism and Pythagoreanism) accepted that the soul had its "origin" in a nonphysical "realm" from which it "had come"; that its incarnation in a material body hindered it from contemplating the good or god, and was generally harmful; that it might be saved from this unfortunate fate, e.g. by acquiring the self-knowledge taught by wisdom or philosophy; and that the result might be an existence free of the body's influence. The problem of why in the first place the soul had ever "fallen" into existence in a body was a topic of philosophical discussion.[5]

For Plato and some of his contemporaries, it was only reasonable to assume that "the body was heavy and belonged to the earth; the soul was heavenly and its proper destiny was to return to its original home."[6]

The idea that the soul was in the midst of a journey became a "commonplace myth"[7] during the Hellenistic period and became

---

[5] Bentley Layton, *The Gnostic Scriptures*. The Anchor Bible Reference Library (New York: Doubleday, 1995), 367. Layton also provides a useful diagram of how the soul's journey factors into the various documents of the Thomas Literature beyond the *Hymn of the Pearl*, such as the *The Gospel according to Thomas* and *The Book of Thomas the Contender*, 368.

[6] Gregory J. Riley, *The River of God: A New History of Christian Origins* (New York: HarperOne, 2001), 153. The concept of the "soul" can become a little tricky, especially when Gnostic texts are brought into play. As Birger Pearson notes, Gnostics generally believed that "Human beings...are split personalities. The true human self is as alien to the world as is the transcendent God. The inner human self is regarded as an immaterial divine spark imprisoned in a material body. The human body and the lower emotive soul belong to this world, whereas the higher self (the mind or spirit) is consubstantial with the transcendent God from which it originated," Birger Pearson, *Ancient Gnosticism: Traditions and Literature* (Minneapolis, MN: Fortress Press, 2007), 13. For the purpose of this paper, when "soul" is used in the context of Gnostic texts I have in mind what Pearson terms the consubstantial "mind or spirit."

[7] Layton, *The Gnostic Scriptures*, 367.

more widespread in the era leading up to the emergence of Christianity, particularly among the middle Platonists.[8] The Roman orator Cicero, writing in the first century BCE, believed that although the soul and the body could work together here on earth, the body was nonetheless still a "tomb," and by living a life in which they devoted "their brilliant intellects to divine pursuits during their earthly lives," men could escape from the body and return to the divine spheres of the universe.[9] The Jewish exegete Philo, writing at roughly the same time, even went so far as to claim that the sacred nature of the soul dictated that it must be 're-virginized' by God, freed from its mortal contaminants: "But when God begins to consort with the soul, He makes what before was a woman into a virgin again, for He takes away the degenerate and emasculate passions which unmanned it and plants instead the native growth of unpolluted virtues."[10] Finally, one of the most explicit statements regarding this concept of the soul's journey is found in the fragments of Theodotus, as preserved by Clement of Alexandria: "But it is not only the washing that is liberating, but the knowledge of who we were, and what we have become, where we were or where we were placed, whither we haste, from what we are redeemed, what birth is and what rebirth."[11]

Due to the prominence of the idea of the "journey of the soul" in Classical literature, it would not be surprising to see these same ideas present in the New Testament and other early Christian literature, a corpus that drew heavily upon the classical tradition.[12]

---

[8] For a fine study of how the different phases of the soul's "journey" evolved in the centuries leading up to and including the early centuries of Christianity, see Benjamin P. Blosser, *Become Like the Angels: Origen's Doctrine of the Soul* (Washington, D. C.: The Catholic University of America Press, 2012), esp. 145–264.

[9] Cicero, *Rep.* VI. 18

[10] Philo, *Cher.* 50

[11] Clement of Alexandria. *Theo.* 78.2

[12] Of all the gospels, the Gospel of Luke would be a likely candidate for a Greek idea such as the 'journey of the soul' to find roots, due to Luke's apparent familiarity with the Greek literary tradition. His concise Greek, his use of a prologue, and his extensive employment of speeches all suggest a familiarity with the Greek historiographical tradition, emulating classical writers such as Herodotus and Thucydides. As Robert Tannehill observes, "Luke differs from the other Gospels in that it begins with a formal preface similar to other Greek writings of its time . . . In the preface the author suggests that the following writings is not the product of a reclusive sect but a work deserving the attention of a broad audience, including those with some claim to Greek culture." Robert C. Tannehill, *Luke: Abingdon New Testament Commentaries* (Nashville: Abingdon Press, 1996), 33.

While the idea of the soul's journey has been recognized clearly as an important element of Gnostic literature,[13] less space has been devoted to exploring how the archetype of the soul's journey may have been integrated into the New Testament. With that in mind, this paper will argue that the archetype of the "soul's journey" is present in two specific Lucan parables, the good Samaritan (Luke 10:30–35) and the prodigal son (Luke 15:11–32)[14]. This is not to say that the author necessarily *intended* to incorporate this specific archetype into these two parables, only that the parables reflect the archetype. In order to more clearly illustrate the "journey of the soul" archetype within these two parables, they will be examined against the backdrop of two Gnostic[15] texts, the *Hymn of the Pearl*

---

[13] "Gnostic knowledge, then, relies on lived mystical experience, on knowledge of the whole timeline of the world, past, present, and future, and on knowledge of the self—where we have come from, who we are, where we are going—and of the soul's journey," *The Gnostic Bible: Gnostic Texts of Mystical Wisdom from the Ancient and Medieval World*, ed. Willis Barnstone and Marvin Meyer (New Seeds: Boston & London, 2006), 8. For a monograph-length study of how the soul's journey is present in one specific text, see Ulla Tervahaufa, *A Story of the Soul's Journey in the Nag Hammadi Library: A Study of Authenikos Logos (NHC VI,3)* (Gottingen: Vandenhoeck & Ruprect, 2015). Her study examines how the soul's journey is present primarily in the *Authentikos Logos* but also indirectly deals with the *Exegesis on the Soul*.

[14] While there are no explicit links made between the "journey of the soul" and the parables of the good Samaritan or the prodigal son among early Christian interpreters, it appears that this type of particular allegorical interpretation of both parables was present, interestingly enough, among certain early gnostic movements. Of the parable of the good Samaritan, Francois Bovon writes, "The oldest extant—albeit indirect—interpretation to which we have access would have to be that of the second century Gnostics, perhaps also Marcion's interpretation, and it is allegorical. *Adam fell like a traveler into the hands of demons who were downright bandits*. Blinded from that point on, he was no longer able to know God, in whom he had his origin," *Luke 2: A Commentary on the Gospel of Luke 9:51–19:27*, ed. H. Koester, trans. D. S. Deer (Minneapolis, MN: Fortress Press, 2013), 62 (Italics added). Of the second parable, that of the prodigal son, Bovon observes, "The oldest exegesis, which can be reconstituted indirectly, seems to be that of the Valentinians (second century CE). Its originality was in seeing in the older son angels who were jealous of the redemption of the human race, represented by the ultimate fate of the younger son. In the view of the Valentinians, the perdition of the young man corresponded to the fall of humanity or of *the soul in the world of matter*," *Luke 2: A Commentary on the Gospel of Luke 9:51–19:27*, 430. (Italics added). The connection between the parable of the prodigal son and Valentinian Gnostics is a useful one, seeing as how the *Exegesis on the Soul* appears to specifically reflect Valentinian tendencies such as the "bridal chamber."

[15] The problems of using "Gnostic" or "Gnosticism" as umbrella terms for non-orthodox literature written during the 2nd–4th centuries CE has been well documented. See discussions in Karen L. King, *What is Gnosticism?* (Cambridge:

and the *Exegesis on the Soul*, documents that demonstrate a more fleshed-out and realized view of the "journey of the soul."[16] While composed (at least in a final, redacted form) after the Gospel of Luke, both the *Hymn of the Pearl* and the *Exegesis on the Soul* cogently demonstrate how the concept of the soul's journey had been successfully appropriated within Gnosticism in the early stages of Christianity and thus provide a suitable lens for reconstructing how this same archetype may be present in Luke's gospel.[17]

The *Hymn of the Pearl* is a poignant story preserved within the text of the *Acts of Thomas* and depicts "a Hellenistic myth of the human soul's entry into bodily incarnation and its eventual disengagement from the body."[18] In the *Hymn*, a young child is sent away by his parents to Egypt in order to procure a pearl. The boy's parents remove his "robe" (symbolic for his divine identity) and send him on his journey. After a series of events in which the young boy falls "asleep," he finally obtains the pearl and upon his return also receives a new robe, measured to fit his stature. This short vignette serves to illustrate how the soul, in the form of the "boy," left his divine heritage, entered into "Egypt" or the world, temporarily lost his way, but fortunately obtained enough wisdom, in the form of the "pearl," to be able to reclaim his divine place once more (the second "robe"). Hellenistic influence is apparent, both in the immortal/divine nature of the soul and the concept of the "robe," or the "form" of the young man which exists in the di-

---

Belknap Press, 2003), 191–236; and Michael A. Williams, *Rethinking "Gnosticism": An Argument for the Dismantling of a Dubious Category* (Princeton: Princeton University Press, 1996), esp. 263–266. For the current state of the debate, see David Brakke, *The Gnostics: Myth, Ritual, and Diversity in Early Christianity* (Cambridge: Harvard University Press, 2010), 19–28.

[16] Although this paper will make primary use of the *Exegesis on the Soul* and the *Hymn of the Pearl*, they are by no means the sole representatives of the "journey of the soul" in Gnostic literature. The *Authentikos Logos*, *The Book of Thomas the Contender*, *The Second Treatise of the Great Seth*, and *The Teachings of Silvanus* also narrate different dimensions of the soul's journey.

[17] Establishing dates on both texts has proven tricky. Birger Pearson suggests that the *Exegesis on the Soul* "was probably composed in Alexandria, Egypt, sometime in the late second century," and lengthy quotations from First Corinthians and Ephesians support the later date, even if obvious parallels to the Sophia myth hint at a possible pre-Christian Jewish influence. As for the *Hymn of the Pearl*, Pearson suggests that "the Parthian details found in the hymn indicate that it was probably composed sometime before 165 in Edessa" and was "secondarily embedded" in the text of the *Acts of Thomas*" (Pearson, *Ancient Gnosticism*, 228, 261).

[18] Layton, *The Gnostic Scriptures*, 366.

vine realm. By obtaining the "robe," the young man has, to borrow the imagery of Philo, been "re-virginized."

A second text, the *Exegesis on the Soul*, presents a similar Gnostic myth, although this time the soul is feminine and falls to earth from a divine realm on account of her own actions. After undergoing sexual abuse and being taught false religion, the soul finally repents and partakes of the Father's grace. After she is "re-virginized" by having her womb turned inward and cleansed of all external pollution,[19] she is joined by the bridegroom, and the two become "a single flesh" (*Exeg. Soul* 131.20–133.2).[20] The story is an obvious, although touching, allegory for "the existential itinerary of every Gnostic seeking for his or her true origins."[21]

Using these two texts as a template, four characteristics of the soul's journey can be drawn out which can then be applied to our two Lucan parables: 1) The soul originates in a divine realm, but at some point, willingly or unwillingly, the soul descends down to earth, where 2) it experiences a great deal of hardship, until 3) the soul remembers or recollects its divine origin, often due to divine intervention of some sort, and 4) returns back home to its divine home, reborn and wiser then when it left.[22]

Part 1: The Departure and Descent of the Soul

Both parables begin, as most journeys do, with a departure. For example, in the parable of the good Samaritan,[23] that departure

---

[19] Or, in other words, "the soul has a womb, but prior to redemption this womb is turned inside out, so that it resembles male genitalia because of its externality. The repentance of the soul is in turning inward once again, a return to 'natural' femaleness," Michael A. Williams, "Variety in Gnostic Perspectives on Gender," in *Images of the Feminine in Gnosticism*, ed. Karen L. King (Harrisburg, PA: Trinity Press International, 1998), 15.

[20] All references from the *Hymn of the Pearl* (*Hymn Pearl*) are from Layton, *The Gnostic Scriptures*, 371–375. All references from the *Exegesis on the Soul* (*Exeg. Soul*) are from Madeleine Scopollo and Marvin Meyer, "The Exegesis on the Soul," in *The Nag Hammadi Scriptures*, ed. Marvin Meyer (New York: HarperCollins, 2007), 227–234.

[21] *The Nag Hammadi Scriptures*, 225.

[22] A. F. J. Klijn, although speaking specifically of the *Hymn of the Pearl*, provides a nice summary of the soul's journey in general: the soul travels "from its pre-existence with God till its' coming back again to God. Between these two periods the soul has to fulfill a heavenly charge. His is able to fulfill it, but not without a call from God." A. F. J. Klijn, "The So-Called Hymn of the Pearl," *VC* 14 (1960): 164.

[23] Bentley Layton in his discussion on the *Hymn of the Pearl* gives a very nice illustration of the power of these allegorical stories: "For the most part, the myth

takes the traveler "from Jerusalem to Jericho" (Luke 10:30). The key to understanding this departure is the direction: the man "went down."[24] This man's story will be, then, a *katabasis*, or descent, from a higher realm to a lower one. A common mythological motif, the aim of a *katabasis* was usually the acquisition of knowledge, or at least the retrieval of something lost. Odysseus, following his time with Kirkē, journeyed to Hades to speak with his mother Anticleia.[25] Aeneas also descended to Hades to visit with a parent, in this case his father Anchises.[26] In both stories the sons gain knowledge which will assist them in their coming trials. Although the journey is treacherous, it is necessary, and the men gain heroic status because of it. Heracles, Orpheus, and even Dante's Pilgrim will make similar journeys. Whether this journey takes the traveler from the above-world to the underworld or from Heaven to Earth, the pattern is a simple one: the soul originates in a heavenly realm, often under some form of supervision, and then departs to a lower, more profane, world that is unfamiliar and alien and, crucially, possibly alluring or deceptive enough that the soul risks becoming permanently lost.[27]

---

of salvation is not expressed literally in HPrl but, rather, is hidden behind a figurative fairy tale or folktale. To perceive the myth, an ancient reader would have needed to reinterpret the tale allegorically . . . Starting from this clue, an ancient reader could work back through the story at another level, retelling it as an account or model of the quest for self-knowledge and salvation. It must be emphasized that, except for the one explicit clue, the text itself provided ancient readers no more than a figurative representation of this hidden message. Readers had to supply or construct the rest of the deeper interpretation." Bentley Layton, *The Gnostic Scriptures: Ancient Wisdom for a New Age* (New York: Doubleday, 1987), 366. Layton's illustration of the power of allegory works very well for the parables of Jesus as well. Jesus mentions a small detail such as the direction of the journey or the loss of raiment, and allows his listeners to unpack the story themselves.

[24] Luke's use of "Jerusalem" and "Jericho" serves to give a "real-world" sense to this story and successfully illustrates the dangers such a "descent" entails: "Over a distance of approximately eighteen miles there was a descent of about 3,300 feet. The road passes through desert and rocky country (Josephus *War* 4.474), and Strabo (16.2.41) reports Pompey's dealings with robbers there. The location is suitable for robbers and for traveling priests and Levites, quite a number of whom lived in Jericho and traveled up to Jerusalem for their periodic responsibilities at the Temple. For the dynamics of the story we need to know nothing of the man except that he has been reduced to a state of desperate need." John Nolland, *Luke 9:21–18:34* (Dallas: Word Books, 1998), 593.

[25] *Homer Od.* 10.

[26] Virgil, *Aen.* 6.

[27] This departure motif relates directly to a fundamental element of many

In the *Exegesis on the Soul*, Sophia begins in a complete, androgynous state "alone with the Father" (*Exeg. Soul* 127:24). At some point, however, she "fell down into a body and entered this life" (*Exeg. Soul* 127:26-27). Sophia has fallen from the heavenly *pleroma* to earth. Likewise, in the *Hymn of the Pearl*, the young man dwelt "in my father's palace" (*Hymn Pearl* 108:1), until he finally went "down to Egypt" (*Hymn Pearl* 108:12). Both of Luke's protagonists will follow similar paths. The traveler in the parable of the good Samaritan departs from "Jerusalem," the holy city, to travel to "Jericho," symbolic for the "World."[28] In the parable of the prodigal son, Luke tells us that the younger of two sons took his inheritance and "traveled to a distant country" (Luke 15:13). Sometimes the destination is named. In the case of the parable of the good Samaritan, the destination is Jericho. For Sophia, it is earth and the opportunity to live "this life." The young man in the *Hymn of the Pearl* travels east, to Egypt. For the prodigal son, it is enough to state that he went "far" away.[29]

The crucial thread throughout these stories is that the traveler is alone; he or she has forsaken the comforts of home and parents and now stands ready to acquire knowledge and experi-

---

Gnostic texts, namely the idea of alienation. As Hans Jonas famously described it, "The alien is that which stems from elsewhere and does not belong here. To those who do belong here it is thus the strange, the unfamiliar and incomprehensible; but their world on its part is just as incomprehensible to the alien that comes to dwell here, and like a foreign land where it is far from home. Then it suffers the lot of the stranger who is lonely, unprotected, uncomprehended, and uncomprehending in a situation full of danger. Anguish and homesickness are a part of the stranger's lot. The stranger who does not know the ways of the foreign land wanders around lost; if he learns its way too well, he forgets that he is a stranger and gets lost in a different sense by succumbing to the lure of the alien world and becoming estranged from his own origins." Hans Jonas, *The Gnostic Religion* (Boston: Beacon Press, 2001), 49.

[28] The imagery of this descent from "Jerusalem" to "Jericho" is enriched by the fact that in traveling from Jerusalem to Jericho one goes from being 2500 feet above sea level to 770 feet below it. In all, the distance covered would be about 18 miles through rather stark terrain; Joseph A Fitzmyer, *The Gospel according to Luke X–XXIV* (Garden City: Doubleday & Company, Inc., 1985), 886.

[29] Cf. Luke 19:12, where εἰς χώραν μακρὰν is used in a similar fashion. Joel Green notes that the "far country" "already suggests the non-Jewish world, and this identification is helped along by the prominence of pigs, abhorrent to Jewish sensibilities, in the story" (Joel B. Green, *The Gospel of Luke* (Grand Rapids: William B. Eerdmans Publishing Company, 1997), 580). Thus in a similar parallel to the parable of the good Samaritan, the journey is undertaken from a Jewish world into a non-Jewish or Gentile world. Read symbolically, this would entail traveling from a holy or divine world to one that is impure and mortal.

ence, unobtainable by any other means. This knowledge, whatever it is, cannot be found in the heavens or the realm of the divine, so the soul must descend into a body and dwell in profane space; the soul may be immortal and, on some level, immutable, but the physical body in which it dwells is not, and is therefore susceptible to the trials of mortality.

Part 2: The Trials of the Soul

Once the traveler has departed from the higher realm to the lower realm, obstacles and trials must inevitably follow, as part of the purpose of the journey is to obtain the knowledge and experience that will be pivotal to spiritual growth.[30] In the *Exegesis on the Soul*, Sophia "fell into the hands of many robbers" who proceed to "pass her from one to the other" (*Exeg. Soul* 127:27-28), the result being that they "defiled her, and she [lost her] virginity" (*Exeg. Soul* 127:31-32).[31] In the *Hymn of the Pearl*, the youth encountered "a mixture of cunning and treachery" by his Egyptian hosts until "I did not (any longer) recognize that I was a child of the (Great King)" (*Hymn Pearl* 109:32-33).

In a similar fashion, the traveler in the parable of the good Samaritan "fell into the hands of robbers, who stripped him, beat him, and went away, leaving him half dead." (Luke 10:30). Several key points emerge here. First, no motivations or causes are given by Luke for the assault upon the traveler. There is no mention made of money or jewels being taken. Clearly, Luke's emphasis is

---

[30] This trait of depicting "Heaven/Spiritual" and "Earth/Physical" in antithetical fashion is typical of many Gnostic texts and is often termed "radical anticosmic dualism." However, see King, *What is Gnosticism?* 192-200, and Williams, *Rethinking Gnosticism*, 96-115.

[31] Madeleine Scopollo argues that the details of the soul's sexual abuse stem not from a suspicion of carnal, physical activity but from its roots as a Hellenistic romance tale: "As to the *Exegesis on the Soul*, the most detailed parts of the treatise concern the earthly adventures of Soul. These can be summarized by one word: prostitution. Soul's deceptions are fully related by the author. Her life in the world gives the gnostic writer the possibility of displaying his Romanesque taste: thieves, brigands, and bandits are inserted in the novel and intensify its effect. The scenes consist of places of ill repute, of brothels and bedrooms where Soul is deceived by her lovers. More than that, she is painted as a slave subject to her master's desires. Filthy gifts, tricks, and a final storm are used to grasp the attention of the reader," (Scopello, "Jewish and Greek Heroines," in King, *Images of the Feminine*, 77-78). However, see also Pheme Perkins: The *Exegesis on the Soul* "understand(s) redemption as a return to a state of undefiled virginity." Perkins, *Gnosticism and the New Testament* (Minneapolis: Fortress Press, 1993), 172.

upon the condition of the traveler and specifically the loss of raiment; other details, though interesting, would obscure the true purpose of the story.³² Second, the fact that the man "fell into the hands of robbers" brings to mind the fall of Adam. Just as evil entered the world on account of Adam's fall, so the fall of the soul into the body allows both the soul to enter into a world of evil.³³ The presence of the Greek verb περιπίπτω in verse 30 suggests more than just falling, but falling into very "discomfiting circumstances," even to the extent of having to "suffer tortures."³⁴ For the soul, which had been used to an existence *sine corpore*, the presence of a physical body alone would certainly feel "discomfiting." The experience of being beaten, robbed, and especially raped (as Sophia had been) would only have multiplied the pain. Third, the loss of his "raiment" at the hands of the robbers is reminiscent of the loss of the "robe" in the *Hymn of the Pearl*, although the youth in the *Hymn* loses his robe prior to his fall. But the robe remains a crucial part of the young man's search through the *Hymn*. His parents encourage him to "Call to mind your garment shot with gold." (*Hymn Pearl* 110.46). At the conclusion of the journey, the young man received a new robe, this one fitted to match his current stature. In the *Hymn*, the young man's raiment indicates his identity, and thus the loss of raiment coincides with being lost in the world.

This insight is important for understanding the state of Luke's traveler, who has been violently "unclothed" (ἐκδύσαντες) by the thieves upon his leaving "Jerusalem."³⁵ As a result of losing his raiment, the traveler is now "half-dead." This adjective "haldead" is a rendering of the Greek ἡμιθανής: the prefix ἡμι meaning "half," and θανής, from the verb θνήσκω, meaning to die in the sense of "passing from physical life," but also implying "losing

---

³² "What was stolen from the man apart from his clothing attracts no comment; it does not contribute to the man's immediately pressing need." Nolland, *Luke*, 593.

³³ "While the particular actions of the robbers can be given credible explanations, their only role in the narrative is to <u>render the man entirely needy</u>" Nolland, *Luke*, 593. Italics added.

³⁴ Walter Bauer, *A Greek-English Lexicon of the New Testament and Other Early Christian Writings*, translated and adapted by William E. Arndt and F. Wilbur Gingrich (Chicago: University of Chicago Press, 1957), 804. Hereafter *BDAG*. The reference Bauer gives for "suffering tortures" is *1 Clem*. 51:2.

³⁵ The verb ἐκδύω primarily means "to remove clothing from the body," but can also be understood as figuratively referring to "the body as a garment" for the soul (*BDAG*, 301).

one's relationship to God," or *spiritual death*.³⁶ The combination of falling from "Jerusalem," "falling" amongst thieves, and losing his "raiment" have severed the relationship the traveler has with his former home; he is now lost in "Jericho" and forced to deal with the hardships and trials that come with that situation.

Following a similar vein, the young man in the parable of the prodigal son, having left his house and father to travel into a "far country," also endures difficult times. Luke relates that the younger son διεσκόρπισεν τὴν οὐσίαν (Luke 15:13). The Greek verb διασκορπίζω literally means "to waste, or to squander," but the curious word in this phrase is οὐσίαν, a noun stemming from εἰμί and thus meaning "that which exists or has substance."³⁷ While οὐσίαν could be a reference to the young man's inheritance, there is perhaps an additional meaning. As the soul travels through different stages of existence, the one thing that constantly "exists" is the soul itself; the body comes and goes but the soul has been and will continue to be. By traveling to the "far country" the younger son has now begun to squander or waste his own soul.³⁸ Luke illustrates this squandering of the οὐσίαν by relating that "a severe famine took place throughout that country," causing the young man "to be in need" (Luke 15:14). These tragic circumstances highlight the theme of loss discussed above. Sophia had her virginity snatched away in the *Exegesis on the Soul*. The young man in the *Hymn of the Pearl* lost his robe, and the traveler in the parable of the good Samaritan lost his "raiment."

In the next verses of the parable of the prodigal son, Luke informs his audience that the loss inflicted upon the younger son will be no less traumatic or serious. Although Luke never specifically states whether or not the younger son was from a Jewish family, it is likely that he was, and for the Jews, who believed that pigs were unclean (Lev 11:7), the feeding of pigs "was thus about as low as Jews could go."³⁹ Yet, bereft of money and with nowhere

---

³⁶ *BDAG*, 457

³⁷ *BDAG*, 740.

³⁸ "The son throws away his wealth through an undisciplined, wild life. The next verse pictures a young man on a spending spree for things of no value. His approach to life will lead to his downfall. He will quickly come into dire straits," (Darrell L. Bock, *Luke*, 1311).

³⁹ I. Howard Marshall, *The Gospel of Luke* (Grand Rapids: William B. Eerdmans Publishing Company, 1978), 608. Bock adds, "In effect, the son has taken the lowest job possible—one that no Jew would even want. He is clearly taking whatever he can get," (Bock, *Luke*, 1311).

else to go, that is the path the younger son is forced to take. Luke tells us that the prodigal son "he went and hired himself out to one of the citizens of that country, who sent him to his fields to feed the pigs" (Luke 15:15). Now unclean himself, the younger son has breached his relationship with God. He has been stripped of his identity as a member of the Jewish nation, God's covenant people.[40] Like Sophia and the traveler journeying to Jericho, the younger son stands naked and exposed. Like the child in the *Hymn of the Pearl* he too has lost something invaluable and precious. His soul has truly lost its way.

Predictably, events continue to deteriorate for the young man. Luke states in the next verse that "He would gladly have filled himself with the pods that the pigs were eating; and no one gave him anything" (Luke 15:16). If feeding swine was "as low as Jews could go," wanting to share their food was "the nadir of degradation."[41] At this point, the prodigal son finds himself absolutely alone, as the οὐδεὶς in verse 16 rather emphatically asserts. Once again, the state of Sophia following her sexual assault provides a useful parallel, as Sophia "became a whore and gave herself to everyone" (*Exeg. Soul* 128.1–2). Both Sophia and the younger son in Luke's parable quite literally cannot reach any lower state of existence than that which they currently enjoy. Their souls are truly lost, and unless circumstances drastically change, they will remain that way.

Part 3: The Awakening of the Soul

One of the crucial points in the journey of the soul is the "a-ha" moment, the point in time when the soul realizes that it has a divine origin and attempts to extricate itself from the miserable world in which it lives.[42] For Sophia, this moment arrives when, at

---

[40] "The link with the foreigner may already be seen as compromising his Jewish loyalties (cf. Acts 10:28) and as interfering with his practice of the faith (e.g., sabbath)." Nolland, *The Gospel according to Luke*, 783.

[41] Marshall, *The Gospel of Luke*, 608. Darrell L. Bock suggests that the swine's food may have been "a sweet bean from a carob or locust tree or a bitter, thorny berry." Darrell L. Bock, *Luke Volume 2: 9:51–24:53* (Grand Rapids, MI: Baker Academic, 1996), 1311.

[42] Kurt Rudolph, speaking specifically of the *Exegesis on the Soul*, describes this "a-ha" moment as a two-fold step: "for one thing as a 'resuscitation' of the spark of light from 'forgetfulness' and 'ignorance' (both of which are understood as a figure of death) through the call of the redeemer and through self-knowledge, and secondly as an 'ascent' of the spark of light to the pleroma . . . Both aspects often merge with one another because the liberating 'knowledge' can already signify an

some point following her prostitution, she "sighed deeply and repented" (*Exeg. Soul* 128.7). But even then, she can't escape subjugation from men "who fooled her into thinking they respected her like faithful, true husbands" (*Exeg. Soul* 128.14). More is required. Fortunately, in a moment of clarity she remembers her "Father on high" and "began to call on him for help, and [she signed] with all her heart" (*Exeg. Soul* 128.27-33). Mindful of all that she has lost over the course of her journey, her prayer rings with poignancy: "My Father, save me. Look, I shall tell [you how I] left home and fled from my maiden's quarters" (*Exeg. Soul* 128.35-129.1). With her final words, she expresses her desire to return back to her former home: "Restore me to yourself" (*Exeg. Soul* 129.2). Significantly, her Father, having viewed Sophia's fallen state, now will "consider her worthy of his mercy" (*Exeg. Soul* 129.3-4). Clearly, Sophia can't return home under her own power or by her own means; she requires the assistance of her "Father." The "a-ha" moment for the king's son in the *Hymn of the Pearl* plays out in slightly different fashion, in this case the son's reception of a letter from his father. Upon receiving the letter, "I gave a start when I perceived its voice. And I took it up and kissed it, and I read. But what was written there concerned that which was engraved in my heart. And on that spot I remembered that I was a child of kings and that my people demanded my freedom" (*Hymn Pearl* 111.53-56). Although the son has not been sullied by the world to the extent that Sophia has, he still requires his "father" to set in motion his return to his former home.

In the parable of the good Samaritan, the "a-ha" moment is downplayed as the traveler assumes a passive role in favor of the parable's other protagonist, the "good Samaritan." When last we encountered the traveler, he was "half-dead," in need of someone to come along and restore his "raiment." As Luke tells the story, "Now by chance a priest was going down that road; and when he saw him, he passed by on the other side. So likewise a Levite, when he came to the place and saw him, passed by on the other side." (Luke 10:31-32). Like Sophia prior to her repentance, the traveler has also encountered 'false husbands' in the course of his journey.

---

anticipation of the end and its realization is already achieved in time." Kurt Rudolph, *Gnosis: The Nature and History of Gnosticism* (San Francisco: HarperCollins, 1987), 190.

In the *Exegesis on the Soul* the "false husbands" likely represented the popular Hellenistic philosophies such as Epicureanism or Stoicism. They are closer to the truth than those whom Sophia had encountered earlier in the story, but they still are lacking the ability to bestow mercy and elevate Sophia back to her former state. For the traveler in Luke's parable, the identity of those who pass by as "priest" and "Levite" suggest that, in a similar fashion, the "priest" and the "Levite" (and whatever they represent) lack the capability to fully restore what was lost to the traveler during his assault.[43] The priest and Levite pass by because any effort to help would be a futile one. Perhaps that is why Luke omits a motivation for the priest and Levite not assisting the traveler; none is needed. Any attempt to rationalize would deprive from the overall impotence of Luke's characters.[44] Even worse, the fact that they move "to the other side" suggests a stubbornness or reluctance to evaluate their role in the larger picture. Additionally, the priest, like the traveler, κατέβαινεν from Jerusalem. Thus the priest and Levite are taking the same journey but, unlike the traveler, whose descent downward has been halted, the priest and Levite continue descending downward, unaware of where their own journey is, or is not, leading them.

It is not until the arrival of the "Samaritan" that the traveler finally receives assistance: "But a Samaritan while traveling came near him; and when he saw him, he was moved with pity" (Luke 10.33). Significantly, the verb καταβαίνω is not used here, but in-

---

[43] One solution is offered by Darrell Bock: "A second refusal by a supposedly exemplary person is a literary way to speak of a generalized condemnation of official Judaism. The lawyer, as a part of this group, would recognize this. At a minimum, it shows that the priest's response in Luke 10:31 was not unique. Official, pious Judaism had two tries to respond and did not. The drama remains, 'Who will love this dying man?'" (Bock, *Luke*, 1031).

[44] Joel Green observes that "The stark reality is simply that they do nothing for this wounded man," after noting that "it is remarkable and probably significant that no inside information regarding the incentives of the priest and Levite is provided" (Green, *The Gospel of Luke,* 430). Nolland suggests that "the priest's failure to help is more likely to be motivated by fear of the robbers, though it is just possible that there may be a hint of a distinction between formal religious practice and covenant obligation to one's neighbor. We are not to tar every priest with this brush (priestly worship is very positively represented in the Infancy Gospel); his priesthood should have made this man a good candidate for coming to the aid of the needy man, but in this case such an expectation was not borne out in practice. In the story the role of the priest is to raise hopes and then to dash them" (Nolland, *Luke*, 593–594).

stead the Samaritan is merely journeying, ὁδεύων. The Samaritan does not travel on the same downward descent as the priest, the Levite, or the traveler. This suggests that the Samaritan is enlightened enough to know the proper course, that his presence upon the road is not due to a fallen state, as is the traveler's. This realization then begs the question of the true identity of the "Samaritan:" is he a literary trope employed by Luke to represent the necessity of mercy among peoples who hate each other, as the Jews and Samaritans did, or is there a deeper level of meaning? The Samaritans, at least in the minds of some Jews, were "half-breeds," containing the blood of both Israel and gentiles.[45] Likewise, so was Jesus a "half-breed," the son of a mortal mother and a divine father. Like Sophia, who needs her divine father to show her "mercy," the traveler finds salvation through the "compassion" of his "father," Jesus Christ.[46]

Further insight is gained from the following verse, in which the Samaritan "went to him and bandaged his wounds, having poured oil and wine on them. Then he put him on his own animal, brought him to an inn, and took care of him." (Luke 10:34).[47] The

---

[45] "The origin of the split between the Samaritans and the Jews is shrouded in mystery. Traditionally they have been taken to be the descendants of the mixed population settled in Israel after the Assyrian conquest of the Northern Kingdom," (Nolland, *Luke*, 535). See R. J. Coggins, *Samaritans and Jews: The Origins of the Samaritans Reconsidered* (Oxford: Blackwell, 1975).

[46] The verb found in Luke 10:33, σπλαγχνίζομαι, means to "have pity or feel sympathy," and is used on two other occasions by Luke: when Jesus raises the son of the widow of Nain (7:13) and in the parable of the prodigal son (15:20) (*BDAG*, 938). The application of this verb to Jesus in an earlier setting strengthens the argument for the Samaritan's identity to be that of Jesus as well.

[47] It is clear that some early Christians viewed the "good Samaritan" as a veiled reference to Jesus: "From Marcion and Irenaeus, through the Middle Ages and the Reformation period, until the nineteenth century, it has often been given a christological explanation (Christ is the good Samaritan), an ecclesiological explanation (the inn is the church), a sacramental explanation, or an extrinsic soteriological explanation." Joseph A Fitzmyer, *The Gospel according to Luke X–XXIV: introduction, translation, and notes* (New Haven: Yale University Press, 2008), 885). For example, Origen relays that "One of the elders wanted to interpret the parable as follows. The man who was going down is Adam. Jerusalem is paradise, and Jericho is the world. The robbers are hostile powers. The priest is the law, the Levite is the prophets, and the Samaritan is Christ." Ambrose would later suggest something similar in his *Exposition of the Gospel of Luke*: "Here the Samaritan is going down. Who is he except he who descended from heaven, who also ascended to heaven, the Son of man who is in heaven?" See A.A. Just, *Luke: Ancient Christian Commentary on Scripture* (Downer's Grove, IL: InterVarsity Press, 2005), 179–180). However, see also Bock, who states, "The text gives no basis for reading the para-

acts of binding up the wounds and the pouring in of oil and wine are clearly salvific images.⁴⁸ Finally, the placing of the traveler upon the "beast" of the Samaritan indicates that, like Sophia and the young boy, the traveler requires divine assistance, an active intervention, to halt his *katabasis* and begin his *anabasis*. Although the traveler does not actively seek out the Samaritan, as Sophia had done in the course of her repentance, he does humbly acquiesce to the actions of the Samaritan and allow himself to be carried for part of his journey; unlike the priest and the Levite, the traveler understands the futility of making the journey alone.

Not surprisingly, the "a-ha" moment for the younger son in the parable of the prodigal son follows a similar course. Last seen having been stripped of his identity to the extent that he desired to eat with the pigs, Luke relates that eventually the son "came to himself" and proclaimed "How many of my father's hired hands have bread enough and to spare, but here I am dying of hunger!" (Luke 15:17). In the Greek, the phrase "came to himself," εἰς ἑαυτὸν δὲ ἐλθὼν, literally means "to come to one's senses" and may be representative of a Semitic phrase meaning 'to repent.'⁴⁹ Thus, like Sophia, the younger son reaches a point in his travels where he rejects the world and repents, although Luke does not specify to whom. But even more important than the repentance is the idea that he came "to his senses," that, like the young child in the *Hymn of the Pearl*, something has sparked in his mind a sense of what his true identity and ancestry actually are.⁵⁰ He has rediscovered ἑαυτὸν. It is in this moment of self-discovery, this epiphany, that the younger son realizes just how nice the conditions were back in his father's house; even the servants had their fill of bread, while his soul, lost in the world, continually aches. The contrast between

---

ble symbolically. Jesus' exhortation to go and do likewise shows that the point is not christological. Jesus is not telling the lawyer to look for him in the parable, but to be this kind of caring person." *Luke*, 1033.

⁴⁸ Bock makes the intriguing suggestion that in the process of binding the traveler's wounds "This might have involved the Samaritan's ripping up some of his own clothes for bandages (his head cloth and linen undergarment would be likely candidates." *Luke*, 1032. In the context of the soul's journey, where ones "robe" represents a distinct identity, to use one's own "robe" to bind the wounds of someone else adds an additional layer of pathos to this account.

⁴⁹ Marshall, *The Gospel of Luke*, 609. Green agrees, stating that "shades of repentance are clearly evident" in that phrase (Green, *The Gospel of Luke*, 581).

⁵⁰ "The idiom εἰς ἑαυτὸν ἔρχεσθαι, "to come to himself," is known also in extrabiblical Greek as an idiom for coming to one's senses," (Nolland, *The Gospel according to Luke*, 783). Cf. BDAG, 311).

the world he now lives in and the remembrance of what his life used to be like prior to his journey cause the son to contemplate a return home, even if it requires him to submit to a life of servitude. The reader can easily hear the echoes of the prodigal's lament and that of Sophia: "I will get up and go to my father, and will say unto him, Father, I have sinned against heaven and before you; I am no longer worthy to be called thy son; treat me like one of thy hired servants" (Luke 15:18-19).

The first line of v. 20 is equally intriguing: And he arose, and came to his father." The construction used by Luke for "he arose" is the aorist participle ἀναστὰς. While the verb ἀνίστημι generally means "to rise up" or "be erect," in the aorist tense the verb also carries the possible interpretation of returning to a living state from a dead one.[51] This interpretation is most fitting for the younger son, who was essentially "dead" insofar as he had lost knowledge of his true identity, but after "returning to himself" has emerged re-born and equipped with the knowledge of his former home now prepares for his *anabasis*[52]

Part 4: The Ascent and Return of the Soul

The fourth element of the soul's journey relates to the return home. Now that the soul has endured its time below and has acquired the requisite knowledge and experience, the time has come for the return journey. In the *Exegesis on the Soul*, this return comes in the form of a restoration to the form Sophia held before her descent, when "she was virgin and androgynous in form" (*Exeg. Soul* 127.24). Thus, when the father bestows his mercy on Sophia, he "will make her womb turn from the outside back to the inside, so that the soul will recover her proper character," causing Sophia to be "at once...free of the external pollution forced upon her" (*Exeg. Soul* 131.20-21, 30-31). In language that parallels the plight of the

---

[51] *BDAG*, 83. In Luke's gospel ἀναστὰς was used in 9:8 and 9:19 to refer to Herod Agrippa's fear that John the Baptist had returned from the dead. The infinitive ἀναστῆναι is later used by Luke to describe Jesus' resurrection in 24:7. Green calls this "the central verbal form of this chapter," and believes that "these words begin to signal his return to life from death" (Green, *The Gospel of Luke*, 582).

[52] ἀνίστημι was an especially common verb for denoting the beginning of a journey, and was used by Luke previously in 1:39 when relating the story of Mary's journey to Elisabeth. The use of the verb here strengthens the sense that this is the beginning of the younger son's return or *anabasis*. See J. Reiling and J.L. Swellengrebel, *A Translator's Handbook for the Gospel of Luke* (London: E.J. Brill, 1971), 64.

traveler in the parable of the good Samaritan, this restoration is compared to putting on new clothing or "raiment," just as "dirty [clothes] are soaked in [water and] are moved about until the dirt is removed and they are clean" (*Exeg. Soul* 131.33-34). Finally, Sophia's prior androgynous state is restored through her union with the bridegroom in the bridal chamber: "These partners were originally joined to each other when they were with the Father, before the woman led astray the man, her brother. This marriage has brought them together again, and the soul has joined her true love and real master" (*Exeg. Soul* 133.5-10).[53] Sophia's journey, though hard and at times severely demeaning, has ingrained in her the knowledge of her Father's goodness and the necessity of his mercy. She has learned, in a truly difficult manner, the trials that come from isolation and loneliness, and now that she has regained her place with her father and assumed her former state, she is prepared for the next phase of her journey, wherever that might take her.[54]

In a similar fashion, the young man in the *Hymn of the Pearl* also desires a return to his former state in his father's house. Once again the image of "raiment" is prominent as the young boy, now grown, finds that the robe which he had left behind as a small boy has increased in size yet still fits him perfectly: "I saw my garment reflected as in a mirror, I perceived in it my whole self as well, And through it I recognized and saw myself. For, though we derived from one and the same we were partially divided; and then again we were one, with a single form" (*Hymn Pearl* 112.76-78). Perhaps as a sign of his new status, the robe is decorated with beautiful gemstones, and dressed in this new robe the young boy offers thanks to his father: "Once I had put it on, I arose into the realm of peace belonging to reverential awe. And I bowed my head and prostrated myself before the splendor of the father who had sent it to me" (*Hymn Pearl* 113.98-99). Due to his diligence in completing his journey, the young boy, who had descended down to the world and had temporarily lost his identity before being reminded of his divine status, now "mingled at the doors of his archaic royal build-

---

[53] There exist several different interpretations of what role was played by the enigmatic "bridal chamber" in Valentinian thought. See discussion in April D. DeConick, "The Great Mystery of Marriage, Sex and Conception in Ancient Valentinian Traditions," *Vigilae Christiane* 57.3 (Aug. 2003): 307-342.

[54] The passages in 135.25-30 and 136.17-27 which form the conclusion of the work suggest that this was the primary knowledge meant to be gained by "Sophia," or by any other soul currently desiring to return to the "Father."

ing" (*Hymn Pearl* 113.101b). His father, equally pleased, "took delight in me, and received me with him in the palace" (*Hymn Pearl* 113.102). Significantly, the young man's journey is not over; even though he has completed one step, another journey awaits, further knowledge must be obtained. The task of finding the pearl was preparatory for even greater experiences: "He suffered me also to be ushered in to the Kings' Court in his company: So that with my gifts and the pearl I might make an appearance before the king himself" (*Hymn Pearl* 113.105).

In the parable of the good Samaritan this return home is realized when the Samaritan carries the wounded traveler to an inn, at which point "he took out two denarii, gaven them to the innkeeper, and said, 'Take care of him'" (Luke 10:35). Furthermore, the Samaritan insures that the traveler will find the required help by promising the innkeeper that "when I come back, I will repay you whatever more you spend." (Luke 10.35). At first glance, the arrival at the inn does not seem like much of a return. The traveler began in Jerusalem, but the location of the inn could be anywhere. However, a clue to finding the location of the inn, and thus the present location of the traveler, may be detected from the amount of money left by the good Samaritan. The annual temple tax that Jewish men were required to pay happened to be two *denarii*, precisely the amount left by the "Samaritan" with the inn-keeper.[55] If Luke did intend for this payment to coincide with the temple-tax, then the "inn" would be symbolically representing the temple, which would place the traveler's location back in "Jerusalem." The traveler has, in fact, returned home, and the Samaritan did not merely provide the traveler with temporary relief, but a permanent restoration in the presence of his "Father."

The story of the return home in the parable of the prodigal son is likewise a successful one. Upon seeing his younger son "still far off," his father "was filled with compassion; he ran and put his arms around him, and kissed him" (Luke 15:20).[56] After reciting

---

[55] Ekkehard W. and Wolfgang Stegemann, *The Jesus Movement: A Social History of Its First Century*, trans. O.C. Dean, Jr. (Minneapolis: Augsburg Fortress Press, 1999), 121. At the very least, "The two denarii would have provided for very basic board and lodging for about two weeks." Nolland, *Luke*, 596. Bock adds the additional detail "That the Samaritan plans to foot the entire bill is made clear by ἐγώ (*egō*, I) plus με (*me*, me), which has an emphatic force. The sense is, 'I will repay, not the man.'" Bock, *Luke*, 1033.

[56] As noted earlier, the verb σπλαγχνίζομαι suggests to 'have pity or feel sympathy,' and is used on two other occasions by Luke: when Jesus raises the son of the

his prayer of repentance, the father "said to his slaves, 'Quickly bring out a robe—the best one—and put it on him; put a ring on his finger, and sandals on his feet" (Luke 15:22). Just as the Father had brought Sophia back and restored her following her humble prayer, the father in Luke's parable acts in a similar, merciful fashion. Once again, restoration takes the symbolic form of putting on clothes; in addition to a new pair of shoes, the younger son also receives his new "raiment" in the form of a robe. However, this robe is not just any old robe, but στολὴν τὴν πρώτην. The adjective πρώτην could be interpreted as "best," but it could also be rendered as "first."[57] Furthermore, the placement of the adjective πρώτην in the attributive position with a definite article while πρώτην lacks an article places even greater emphasis upon the πρώτην.[58] While the robe plays an important part, especially in establishing identity, the primary point Luke may be making is that the robe was the *first* robe worn by the younger son.[59] Thus the father is not merely giving his recently-returned son a beautiful robe to wear, but in fact restoring to him the robe that he had worn prior to his departure.[60] The parallel with the *Hymn of the Pearl* is striking: both sons shed their robes at the beginning of their journey, and both sons receive the same robes upon their return. Significantly, both robes somehow fit, although a great deal of time has presumably passed. The other two items bestowed upon the younger son also bear significance. Rings were signs of power, often acting as a seal for important men.[61] Contrary to guests, who

---

widow of Nain (Luke 7:13) and in the parable of the good Samaritan (Luke 10:33). The fact that this specific verb is used in both of these parables and on only one other occasion by Luke suggests a parallel thread running through both parables, as well as the idea that the 'father' in this parable shares the same identity as the Samaritan, namely Christ. The fact that both characters who display mercy are divine fits the pattern of the divine parentage of both Sophia in the *Exegesis on the Soul* and the young boy in the *Hymn of the Pearl*.

[57] BDAG, 893.

[58] According to Daniel B. Wallace, this is the least frequent of the attributive positions and occurs only a few times with adjectives in the New Testament (Daniel B. Wallace, *Greek Grammar Beyond the Basics: An Exegetical Syntax of the New Testament* (Grand Rapids: Zondervan, 1996), 307).

[59] "στολὴν τὴν πρώτην may be 'the best robe,' and thus perhaps the best of the father's own wardrobe (cf. Esth 6:8), or just possibly 'the former robe,' and thus the clothing that marked the son's place in the family before his departure, (Nolland, *The Gospel according to Luke*, 785).

[60] However, as Marshall notes, this point would be strengthened with the inclusion of αὐτοῦ (Marshall, *The Gospel of Luke*, 610).

[61] Cf. Gen. 41:42: "Then Pharaoh took his signet ring from his hand and put it on

customarily remove their shoes upon entering a house, the younger son puts shoes on, suggesting that he has reclaimed his former status, perhaps even more.[62] The father's joyous words at the return of his younger son summarize well the journey of the soul and could just as easily apply to any of the four texts examined in this paper: "For this son of mine was dead, and is alive again (ἀνέζησεν); he was lost and is found" (Luke 15:24).[63]

By examining how the parables of the good Samaritan and the prodigal son fit within the framework of the "journey of the soul" tradition, readers can gain a deeper appreciation for rich cultural background that lies behind Luke's text. Luke 10 functions as more than just an injunction to love my neighbor; rather it reminds me why I must. The traveler represents each one of us who has left a divine home and currently exists in a fallen, "wounded" state, praying that the "good Samaritan," Jesus himself, will find us and provide the necessary assistance to return us to our home. Likewise, the prodigal son represents a humanity that has lost its inheritance, but can at any "come to themselves" and realize not only what they have lost, but also carry with them the knowledge that all, and perhaps more, can be restored. Reading these two Lucan parables within the framework of two other texts with a similar message allows us to widen the scope of the good Samaritan and

---

Joseph's hand, and arrayed him in garments of fine linen, and put a gold chain about his neck." Bock writes, "The ring may contain a seal and thus represent the son's membership in the family, but it stops short of being a transfer of authority," (Bock, *Luke*, 1315).

[62] Francois Bovon also favors the idea that the bestowal of these three items upon the prodigal Son signify his restoration back to his former status, leading him to conclude that the robe "had been carefully put in order and kept ever since his departure" (Francois Bovon and Gregoire Rouiller, *Exegesis: Problems of Method and Exercises in Reading (Genesis 22 and Luke 15)* Trans. Donald G. Miller (Pittsburgh: The Pickwick Press, 1978), 57-58). Green writes: "The embrace, the kiss, and the gifts of the robe, ring, and sandals-these are all emblematic of the son's honorable restoration to the family he had snubbed and abandoned" (Green, *The Gospel of Luke*, 583).

[63] The verb ἀναζάω, like ἀνίστημι, suggests the restoration to life of one who was either physically dead, and who was thus resurrected, or one who was spiritually or morally dead. Either interpretation works well with the journey of the soul archetype (*BDAG*, 62). Of the nature of the language in 15:24, Nolland writes, "The language here is indeed striking, and while it does not actually break the bounds of the story, it comes closer to being immediately symbolic than at other points of the parable" (Nolland, *The Gospel according to Luke*, 786).

the prodigal son beyond the present, to understand Jesus's words as applying not only to our sojourn through mortality, but also to our souls as they continue their journey home.

Kings of the Jews
Herodian Collaboration with Rome through a Markan Lens

Margaret Froelich

There is a steady and growing interest in the relationship between the Gospel of Mark and the First Roman/Judean War of the 60s and 70s CE. Two recent works frame the war as a central concern in the Gospel,[1] and many others propose fundamental connections between Mark and the war or Roman imperialist domination in general.[2] Further, a majority of scholars date the Gospel to sometime closely following the defining event of the war: the destruction of Jerusalem and its temple in 70 CE.[3]

Focus on Mark as, at least in part, a political document—that is, a document in some way responding to and shaped by the political and historical events that surrounded its composition—opens up a number of possibilities beyond traditional ones for under-

---

[1] Gabriella Gelardini, *Christus Militans: Studien zur politisch-militarischen Semantik im Markusevangelium vor dem Hintergrund des ersten judisch-romischen Krieges*, Supplements to Novum Testamentum 165 (Boston: Brill, 2016); Christopher Zeichman, "Military-Civilian Interactions in Early Roman Palestine and the Gospel of Mark" (PhD diss., University of St. Michael's College, 2017).

[2] E.g., Richard A. Horsley, *Hearing the Whole Story: The Politics of Plot in Mark's Gospel* (Louisville, KY: Westminster John Knox Press, 2001); John S. Kloppenborg, "Evocatio Deorum and the Date of Mark," *JBL* 124, no. 3 (2005): 419–50; Tat-siong Benny Liew, *Politics of Parousia: Reading Mark Inter(Con)Textually*, Biblical Interpretation Series 42 (Leiden, Boston: Brill, 1999); Dieter Lührmann, "Markus 14.55-64: Christologie und Zerstörung des Tempels im Markusevangelium," *NTS* 27, no. 4 (1981): 457–74; Joel Marcus, "The Jewish War and The Sitz Im Leben of Mark," *Journal of Biblical Literature* 111 (1992): 441–62; Adam Winn, *The Purpose of Mark's Gospel: An Early Christian Response to Roman Imperial Propaganda*, WUNT 245 (Tübingen, Germany: Mohr Siebeck, 2008).

[3] Besides many of those already cited, see Walter Schmithals, *Das Evangelium nach Markus*, 2 vols., OTK (Gütersloh: Gütersloher Verlagshaus Mohn; Würzburg: Echter-Verlag, 1979), 2:558; William Telford, *The Theology of the Gospel of Mark*, New Testament Theology (Cambridge, UK; New York: Cambridge University Press, 1999), 13; Burton L. Mack, *A Myth of Innocence: Mark and Christian Origins*, (Philadelphia: Fortress Press, 1991), 315; Craig A. Evans, *Mark 8:27–16:20*, WBC 34B (Nashville, TN: Thomas Nelson, 2001), 509.

standing its plot, rhetoric, and characters. With the theology of the text understood as functionally related to its historical mooring, Mark can be read as an attempt to cope with the dissolution of the status quo in Judea and to maintain a sense of divine control in spite of the failure of rebellion and messianic hopes. Details previously thought small or unimportant can take on new meaning. Thus, this paper will examine the word βασιλεύς in Mark and the Gospel's striking comparison, by means of this vocabulary, between Jesus and the Herodian dynasty. In this context, the meanings of words like βασιλεύς are crucial for understanding Mark's response to Roman domination—such terms are not used lightly or casually.

Βασιλεύς in Mark

The word βασιλεύς appears twelve times in the Gospel of Mark. In 13:9 it is a generic reference to the "kings" and other rulers who will persecute Jesus's followers; five times in chapter 6 it refers to Herod; and in chapter 15 Jesus is six times named either "King of the Judeans" or "King of Israel."[4] Regarding chapter 6, scholarship most frequently addresses issues surrounding John the Baptist, as well as the foreshadowing of Jesus's death.[5] Discussions of Herod himself as relate to the Gospel focus on Mark's account of John's execution versus Josephus's account in the *Antiquities* (18.116–19), and on the historical circumstances surrounding Herod's relationship to Herodias, as well as the legality of their marriage.[6] I hope to buck the trend somewhat by addressing my remarks to the relatively unexamined comparison between Jesus and Herod embodied in the term βασιλεύς.[7]

---

[4] Mark 6:14, 22, 25, 26, 27; 15:2, 9, 12, 18, 26, 32.

[5] E.g., Mark McVann, "The 'Passion' of John the Baptist and Jesus before Pilate: Mark's Warnings about Kings and Governors," *BTB* 38, no. 4 (2008): 152–57; Adela Yarbro Collins, *Mark: A Commentary*, ed. Harold W. Attridge, Hermeneia (Minneapolis, MN: Fortress Press, 2007), 303–15; S. Nortje, "John the Baptist and the Resurrection Traditions in the Gospels," *Neotestamentica* 23 (1989): 349–58.

[6] Harold W. Hoehner, *Herod Antipas*, SNTSMS 17 (Cambridge: Cambridge University Press, 1972), 110–71; Peter Richardson, *Herod: King of the Jews and Friend of the Romans*, Studies on Personalities of the New Testament (Columbia, SC: University of South Carolina Press, 1996), 307–08. In addition, there is some work on the historical relationship between Jesus and Herod. For example, Joseph B. Tyson, "Jesus and Herod Antipas," *JBL* 79 (1960): 239–46.

[7] Gabriella Gelardini has addressed this connection in a paper similar to my own ("The Contest for a Royal Title: Herod versus Jesus in the *Gospel According to*

Three facts, two historical and one literary, form the basis of this discussion. First, historically, the Herod of chapter 6 is not Herod per se but Antipas, son of Herod the Great. Mark has substituted the dynastic name for the personal one. Second, Antipas never officially carried the title βασιλεύς, but was rather the τετραάρχης of Galilee and Perea, the region east of the Jordan spanning roughly from just south of Pella to the Dead Sea.[8] Josephus tells us that Antipas eventually sought the title βασιλεύς from Caligula but was denied and, due to his brother's political machinations, exiled to Gaul (Jos. *Ant.* 18.240-55). Third, our literary consideration, in chapter 15 of Mark the title βασιλεύς is always applied to Jesus by characters who mean it in mockery; it is never sincere. The later Synoptics highlight Mark's use of βασιλεύς by diluting it considerably, both by adding further generic usages (e.g., Matt 11:8; Luke 10:24) and by applying the title to Jesus in sincerity (e.g., Matt 2:2; Luke 19:38). Further, Matthew and Luke completely eliminate references to Herod Antipas as βασιλεύς, preferring the ruler's official title (Matt 14:1; Luke 9:7).

In chapter 15, characters call Jesus both "king of the Judeans" and "king of Israel." R. T. France understands this to represent an ethnic divide: Roman characters such as Pilate and the soldiers use "king of the Judeans" (e.g., 15:2, 18), and the Judean crowd uses "king of Israel," here apparently synonymous with "Messiah" (15:32; see also the high priest's accusation at 14:61).[9] In all cases characters use the title to emphasize what Jesus is not: a victorious and politically powerful ruler capable of reigning over and protecting the ancestral homeland of Israel.

The irony of this usage, of course, is that Mark's Jesus is powerful and will conquer and reign. The eschatological Son of

---

Mark [6,14-29; 15,6-15]," *Annali di storia dell'esegesi* 28, no. 2 [2011]: 93-106). Her argument focuses on Herod's failure to recognize Jesus as a competitor and the subsequent implications for the plot, whereas mine turns attention to the juxtaposing characterizations of the two "kings" and how they reflect the political situation of the 60s and 70s. For a discussion of Mark's Herod as a tyrant, without reference to the comparison to Jesus through kingship vocabulary, see Abraham Smith, "Tyranny Exposed: Mark's Typological Characterization of Herod Antipas (Mark 6:14-29)," *Biblical Interpretation* 14, (2006): 259-93.

[8] See Hoehner, *Herod Antipas*, 106n3, for a set of inscriptions that specifically name Antipas as tetrarch rather than king.

[9] R. T. France, *The Gospel of Mark: A Commentary on the Greek Text*, The New International Greek Testament Commentary (Grand Rapids, MI: W. B. Eerdmans Publishing Company, 2002), 628.

Man passages, especially 13:26-27 and 14:62, portray Jesus not as the absolute sovereign but as that figure's right hand and trusted general, leading the angels on a search and rescue mission to "gather the elect." The word δύναμις in verse 9:1 (which reads, "Some standing here will not taste death until they see the kingdom of God has come ἐν δυνάμει") might even plausibly be read not as the "power" that is either innate in Jesus or at least at his command, but to the "force" or army that he leads. This military usage of δύναμις is not primary, of course, but it is well attested in ancient literature, including Josephus.[10]

Meanwhile, Mark's Herod is weak. He arrests John to squelch criticism of himself, but secretly admires and fears him, although he does not understand what John teaches. The entire scene in the middle of Mark chapter 6 is a showcase of Herod's inability to command his own narrative. His marriage faces public scrutiny represented by John. There is the strong implication that he is sexually attracted to his step-daughter, a situation that Paul, at least, would have found unconscionable (1 Cor 5:1-2). Attraction or not, Herod allows himself to be tricked into promising the death of John, which we already know he is loath to perform. But since he has promised this thing before a number of high-ranking guests, he cannot back out. Essentially and somewhat crudely, John the Baptist dies to preserve Herod's ego.[11]

Long after the deaths of Jesus, John, and Antipas, what does this contrast between mighty and weak kings add to Mark's narrative?

Herod the Great

Let us go back to the name "Herod," and Mark's choice to use the dynastic name rather than Antipas's personal name. The Gospels bear witness to the name's use as a title: Luke and Matthew follow Mark in referring to Antipas by the dynastic name alone, and Acts refers to Agrippa I simply as Herod (Acts 12:19), and Agrippa II by the personal name (Acts 25). Josephus often re-

---

[10] E.g., Dem. Ex. 21.3; Dem. 4.28; Hyp. 5.3; Jos. JW 1.66; Xen. Anab. 1.1.6.
[11] M. H. Jensen reaches a similar conclusion, that in Mark, Herod's weakness functions to make him dangerous. Morten Hørning Jensen, *Herod Antipas in Galilee: The Literary and Archaeological Sources on the Reign of Herod Antipas and Its Socio-Economic Impact on Galilee*, WUNT 2. Reihe 215 (Tübingen: Mohr Siebeck, 2006), 112-14.

fers to Antipas as Herod and calls him "Herod who was called Antipas" (*JW* 2.167) but does not use the dynastic name for either Agrippa. It is plausible, therefore, that "Herod" had become a colloquialism or semi-official title, much like "Caesar" in the late first century. The example of Acts demonstrates that to some extent the dynastic name was interchangeable with personal names.[12] Further, while there is the possibility that Mark is simply ignorant of that particular Herod's personal name, I find that angle of inquiry unhelpful. A growing body of scholars situates Mark not only as an eastern text but a Galilean one,[13] suggesting that the author would be familiar with some of the finer details of local political history or have access to people who were—his Galilean audience certainly would have been. If we are reading Mark through the lens of local politics, moreover, importance rests not on whether the author is ignorant of the name Antipas, but on what associations he draws with the name Herod.

From the beginning of the dynasty, the Herods were associated with Rome. Antipater, allied with one branch of the declining Hasmoneans, originally gained approval from Julius Caesar in 47, being named procurator of Judea compared to Hyrcanus's more limited high priesthood (Jos. *Ant.* 14.143). After Caesar's death, Antipater's son Herod—who would give the dynasty its name—supported Cassius against Antonius, gaining favor by being the first to collect his share of the high tribute that Cassius demanded from Judea and the surrounding areas (*Ant.* 14.272-75). Thus, the early days of the Herodian dynasty were marked by support of Rome at the expense of the local population. After the fall of Cassius and Brutus, Herod was able by much the same means to gain favor with Antony, who named him a tetrarch and later supported his elevation to king (*Ant.* 14.327).

Herod the Great's relations with the Hasmonean family were, according to Josephus, a series of intrigues and political stratagems by which he expanded his power and gained greater

---

[12] Hoehner, *Herod Antipas*, 105-108, details the use of Herod as a dynastic name, and suggests that it, like the official titles βασιλεύς and τετραάρχης, was dependent on permission from the emperor.

[13] Dean W. Chapman, "Locating the Gospel of Mark: A Model of Agrarian Biography," *BTB* 25 (1995): 24-36; Yarbro Collins, *Mark*, 8-9; Zeichman, "Military-Civilian Interactions," 30-51; Christopher B. Zeichmann, "Capernaum: A 'Hub' for the Historical Jesus or the Markan Evangelist?" *Journal for the Study of the Historical Jesus* 15 (2017): 147-65.

influence with Roman personalities. After Actium, Herod had Hyrcanus executed and then, having done away with one of his last Hasmonean rivals, went to Octavian. In Josephus's account Herod cleverly managed this audience, spinning his strong and material support for Antony as an example of how he could be expected to deal with allies (*Ant.* 15.161–93). His close relationship with Caesar and other high-ranking Romans was marked by extravagant gifts and receptions, military support, and his sons' residence in aristocratic homes in the city of Rome (*Ant* 15.342f). He was appointed king by the senate in 37 CE (*Ant.* 14.384–85), a title that was later confirmed by Augustus (*Ant.* 15.195).

Herod the Great's associations with Rome extended into the enforcement of Roman, or Roman-style, law, often in conflict with Torah.[14] This, combined with his non-Judean heritage, conflicts with the high priesthood, and extravagant support of Hellenism both inside and outside of his realm, gave him a reputation as an outsider, if not an outright foreigner. Peter Schäfer calls it an "irrefutable fact that, in the eyes of his Jewish subjects, he was a usurper who had destroyed the legitimate dynasty of the Hasmoneans in order to grab power for himself," although he notes that the Hasmoneans displayed similarly Hellenizing tendencies.[15]

The Herodian Dynasty

Herod's successors, although for the most part lacking the grandiosity of his own reign, continued to rule at the pleasure of Rome. In the immediate aftermath of Herod the Great's death, the emperor settled the successional dispute by dividing what had been a single kingdom under Herod into three regions, each ruled by one of Herod's sons. Each one of them—Philip, our Antipas, and Archelaus—were successful to the extent that they managed to satisfy both Roman and local interests. Antipas, as tetrarch over Galilee, found himself dealing with the revolutionary movements there, represented in Josephus and the New Testament by John the Baptist. Perhaps his more pressing problem, however, was the conflict he brought about with the neighboring Nabateans when he

---

[14] Peter Schäfer, *The History of the Jews in the Greco–Roman World: The Jews of Palestine from Alexander the Great to the Arab Conquest*, Rev. ed. (London: Routledge, 2003), 88–89.

[15] Schäfer, *History of the Jews*, 97.

divorced their king's daughter in favor of Herodias (Jos. *Ant.* 18.109-115). In the Gospels, as we have seen, Antipas is portrayed as weak-willed and ambitious; in Josephus's *Antiquities* he is similarly manipulated by his wife, here to plead for the title βασιλεύς (*Ant.* 18.240-256), and inseparably intertwined with Rome. Every mention of his career—the building projects (*Ant.* 18.26-28, 36-38), the war with Aretas (*Ant.* 18.111-125), and his exile after the failed petition for kingship (*Ant.* 18.240-252)—involves some plea to or flattery of the imperium.

Agrippa I, who lived extensively in Rome, likewise owed his title (truly βασιλεύς; *Ant.* 18.237) and authority to Roman favor and was a close acquaintance of the emperor himself. Josephus credits him with allaying the Caligula crisis: in *Antiquities* 18.297 he is able to leverage his favored position into a request to halt the planned dedication of Caligula's statue in the Jerusalem temple. Not much later he is portrayed as instrumental in the stand-off between Claudius and the senate that followed the death of Caligula (*Ant.* 18.236-273), for which he was rewarded with kingship over Judea and Samaria in addition to parts of Lebanon and Cilicia (*Ant.* 18.274-276).

Agrippa II, also βασιλεύς in Josephus and Acts, did not receive his father's kingdom upon the latter's death, but was eventually rewarded the kingship over his uncle's dominions in Chalcis, including control over the Jerusalem priesthood (*Ant.* 20.104, 179). Around 52 CE Nero removed him from Chalcis but granted him Philip's tetrarchy—that is, the regions northwest of the sea of Galilee—and a few other holdings including Tiberias (*JW* 2.247, 252).

As we can see, the fortunes of the Herods are tied closely to Rome for the life of their dynasty.

Agrippa II and the Judean War

Agrippa II is of particular interest to us as the ruling scion of the Herodian dynasty during the Judean War, and as one of Josephus's apparent sources. He had grown up in Rome at court with Claudius (*Ant.* 19.360) and over his career would be close with Vespasian and Titus. He is perhaps most famous for the lengthy speech that Josephus attributes to him in *War* 2.345-401. The logic of this speech is realist: Judea is too weak to successfully rebel against Rome, so it would be foolish to try. It acknowledges some of the abuses of Rome, and admits that freedom would be better, but argues that ultimately even Athens has submitted to servitude;

Judea cannot hope to fare better.

There is a hint at Agrippa's unpopularity in *Antiquities* 20 (211-12), when his building projects in Caesarea Philippi draw popular ire; and, perhaps, at his lack of political prowess, when his appointment of Ismael to the high priesthood apparently results in conflict between the priests and the Jerusalem aristocracy (20.179-80). Tessa Rajak has suggested that Josephus's understated accounts of the second Agrippa's scandals mask the historian's knowledge of his friend's incapacity for rule.[16] If this is the case, the next few lines of the account come more into focus. Josephus reports that they listen to him at first and begin to collect the overdue tributes and rebuild some destroyed structures in Jerusalem. But Agrippa seems to overstep, and when he urges the people to come into obedience to the procurator Florus, they turn against him and drive him out of the city. This ambivalent and even fickle reception is in line with the less than glorious portrayal of Agrippa in the *Antiquities*, and perhaps exposes the speech, and its single-handed if momentary persuasion of the crowd against rebellion, as primarily the post-facto apologia of Josephus himself.[17]

More damningly, if one takes the side of the rebels, are the second Agrippa's contributions of troops, first to the "moderate," or anti-revolution, Judeans (*War* 2.421), and then to the Roman general Cestius (2.500). At other points he is again seen unsuccessfully urging the rebels to surrender; he sends ambassadors, whom the rebels kill (2.523-526), and at Gamala goes so far as to approach the wall of the city, where he is injured by a stone (4.14-15). Throughout Josephus, then, he is seen as functionally, though not always full-throatedly, pro-Roman and an ineffective advocate for the status quo.

How accurate this picture of Agrippa might be dependent on Josephus's reliability overall. Disagreements between his works and questions regarding his agendas and memory are well known to the student of Josephus studies. The so-called classic view—that Josephus is only as trustworthy as his sources, to which he owes the bulk of his content, and that the man himself is a "tendentious

---

[16] Tessa Rajak, "Friends, Romans, Subjects: Agrippa II's Speech in Josephus's *Jewish War*," in *The Jewish Dialogue with Greece and Rome: Studies in Cultural and Social Interaction*, Arbeiten zur Geschichte des antiken Judentums und des Urchristentums 48 (Leiden: Brill, 2001), 151-54.

[17] Rajak, "Friends, Romans, Subjects," 158.

hireling" and apologist for Rome[18]—has been variously tempered and revised in the twentieth and twenty-first centuries, but unquestioned acceptance of his accounts is beyond possibility. The most helpful approach, in my view, is the one exposited in detail by Steve Mason in 2016: that Josephus should be understood in the context of ancient historians, and thus more concerned with "truth" than with accuracy.[19] Mason concedes that Josephus is not reliable per se, but insists that reliability is the wrong expectation for ancient historiography. "We can only try to understand his work as a product of its time," states Mason.[20] Similarly, Michael Tuval claims that we can use Josephus not to understand what actually happened in any given period that he narrates, but how that history was important for Josephus in the period he was writing it.[21]

How might this affect our assessment of the Herods? Rajak points out that, at the very least, Josephus's account must be *believable* to his audience.[22] Given the circumstances of the second Agrippa's upbringing and career, Josephus could reasonably assume that his audiences comprised some people who were quite familiar with the king. Thus, it is reasonable for us to regard his portrayal of Agrippa, which is neither especially flattering nor especially disparaging, as more or less accurately capturing the perception, if not the reality, of his character and his behavior during the war.

Conclusion

Let us finally return to Mark. The evangelist's motives are different from Josephus's, but to some extent he may be held to the same standard: the Gospel must be credible to its audience, and its narration of earlier periods can shed light on the contemporary concerns of its author. If Mark is concerned with the war, the figure of Agrippa II is important. This is the king who gave troops to Ces-

---

[18] Per Bilde, *Flavius Josephus between Jerusalem and Rome: His Life, His Works and Their Importance*, Journal for the Study of the Pseudepigrapha Supplement Series 2 (Sheffield, UK: JSOT, 1988), 127.

[19] Steve Mason, *A History of the Jewish War: A.D. 66–74* (Cambridge: Cambridge University Press, 2016), 60–137.

[20] Mason, *History*, 136–37.

[21] Michael Tuval, *From Jerusalem Priest to Roman Jew: On Josephus and the Paradigms of Ancient Judaism*, WUNT 2. Reihe 357 (Tübingen: Mohr Siebeck, 2013), 19.

[22] Rajak, "Friends, Romans, Subjects," 150.

tius and sided with Vespasian, and who still reigned at the time of Mark's composition. Whatever the author may have thought of the rebellion before the war, in its aftermath he predicts the downfall of Roman hegemony in Judea, and is, at best, ambivalent about the local and regional leaderships who facilitated it.

Josephus's portrayal of the scandalous Herodian dynasty, and especially Agrippa II, then, becomes illuminating for Mark's portrayal of Antipas. Mark's Herod is weak. He is an apologist for the status quo (although not directly for Rome in Mark's account), and his initial support for John is no match for his self-serving desire to look good in front of the local elite class. He is Josephus's Agrippa seen through a far less charitable lens.

Mark's use of the dynastic name rather than the personal name for his Herod results in the occasional anachronism: a reader unfamiliar with the historical situation but quite familiar with the Gospels might be forgiven for confusing the murderous King Herod of Matthew's infancy narrative with the murderous King Herod of Mark's Galilean ministry. This study argues that this is not a bug, but a feature. The ambiguity of the name would have allowed Mark's earliest audiences to read the qualities of cowardice, weakness, and violence into whichever Herod came to mind, and into the dynasty as a whole, including and perhaps especially the most contemporary representative, Agrippa II.

Mark's Transformations of the Love Commandment
Controversy in the Lost Gospel
Logoi 6:18–21 and Mark 12:28–34 and 10:17–22

Dennis R. MacDonald

The exchange between Jesus and another teacher of Mosaic Law over the greatest commandment appears in all three canonical Gospels: Mark 12:28–34, Matt 22:34–40, and Luke 10:25–28. Although Mark was the most ancient of these Gospels and often served as a source for the Matthean and Lukan Evangelists, its version of this dispute certainly is secondary to a version known to Luke. Christopher M. Tuckett and Jan Lambrecht have adduced the following observations for ascribing Luke 10:25–28 to the lost Gospel Q, even though it is absent in all but one published reconstruction of it.[1]

- If Luke's version redacted Mark 12:28–34 (or Matt 22:34–40), one would expect to find it after the controversy over marriage and resurrection, that is, after, 20:40 (cf. Mark 12:27 and Matt 22:32), but it occurs ten chapters earlier in ch. 10!
- If one assumes that Luke merely redacted Mark, one must explain why he omitted the *shema* (Mark 12:29) and the devaluation of sacrifices (Mark 12:33). These verses similarly have no equivalents in Matthew.
- In Luke, it is not Jesus but the interlocutor who evokes the love command; Jesus merely affirms it. In Mark and Matthew Jesus himself gets credit for do-

---

[1] Christopher M. Tuckett, *The Revival of the Griesbach Hypothesis: An Analysis and Appraisal* (Cambridge: Cambridge University Press, 1983), 125–39, and Jan Lambrecht, "The Great Commandment Pericope and Q," in *The Gospel Behind the Gospels: Current Studies on Q*, ed. Ronald A. Piper, NovTSup 75 (Leiden: Brill, 1995), 73–96. The only published reconstruction of the lost Gospel to include this controversy is mine in *Two Shipwrecked Gospels: The* Logoi *of Jesus and Papias's Exposition of Logia about the Lord*, SBLECL 8 (Atlanta: Society of Biblical Literature, 2012).

ing so.
- Mark duplicates the biblical citations that comprise the love commandment and by so doing unnecessarily bloats the episode. Matthew cites it only once, as a statement of Jesus; Luke cites it only once, as a statement of the interlocutor, reflecting an earlier stage of tradition.
- Furthermore, Mark's version is internally inconsistent. The scribe askes Jesus "what is the first commandment [singular] of all". The first begins with the *shema* followed by the command to love God. Mark relegates the commandment to love one's neighbor to second place. This inconsistent expansion is missing in Luke (and Matthew).

It is one thing to argue that Luke's version is more primitive and independent of Mark and Matthew, but quite another to attribute it to Q, as do Tuckett and Lambrecht. Three other considerations, however, should seal the deal:

- Controversies with Jewish authorities appear elsewhere in content conventionally attributed to Q (e.g., the Beelzebul Controversy in Q 11:14–20 [= *Logoi* 6:22–29]).
- Luke's version appears immediately after other content conventionally attributed to Q (Q 10:21–24 [=*Logoi* 10:26–29]; cf. Matt 11:25–27 and 13:16–17).
- If one insists that the content of Q must satisfy either the criterion of Matthew-Luke overlaps against Mark or of Lukan *Sondergut* (unique content that Matthew may have omitted), Luke 10:25–28 fails on both counts; its agreements with Matthew are minor and the controversy is not unique to Luke; it obviously appears in both Mark and Matthew. But if one uses the criterion of Inverted Priority, as I do throughout *Two Shipwrecked Gospels*, these verses qualify, for the reasons offered already.
- Even though Matthew's agreements with Luke are minor, they are noteworthy and may suggest the Evangelist's reliance not only on Mark but also on Q. These agreements include the following: πειράζων αὐτόν· διδάσκαλε, (Matt 22:35–36), ἐκπειράζων αὐτὸν λέγων· διδάσκαλε, (Luke 10:25), the triple replacment

of Mark's ἐξ with ἐν, and the avoidance of repeating the commandment.

Lambrecht put it well: "It appears impossible to explain adequately Lk 10:25-28 and Mt 22:34-40 without the postulate of a second source, a text which is different from Mk 12:28-34.... There is no reason why this text should not be called a Q passage."[2]

My reconstruction of the lost Gospel appears in *Two Shipwrecked Gospels* and reads as follows:

> 6:18 And behold a certain exegete of the law, to test him, asked, "Teacher, which is the greatest commandment in the law?"
>
> 19 He said to him, "What is written in the law?"
>
> 20 He answered and said, "You will love the Lord your God with all your heart, and with all your soul, and with all your mind, and your neighbor as yourself."
>
> 21 He said to him, "You have answered rightly. Do this and you will live." (*Logoi* 6:18-21)

Here the inquirer cites Deuteronomy and Leviticus:

Deut 6:5	*Logoi* 6:20
"You will love the Lord your God from all your heart, and from all your soul, and from all your strength."	He answered and said, "You will love the Lord your God with all your heart, and with all your soul, and with all your mind,
Lev 19:18	
"You will love your neighbor as yourself."	and your neighbor as yourself."

For Jesus, this commandment not only is the greatest, its observance makes unnecessary obedience to every commandment. Compare the following:

Deut 11:8 (cf. Lev 18:5)	*Logoi* 6:21
"You will keep all my commandments, whatever I command you today, that you may live."	He said to him, "You have answered rightly. Do this and you will live."

---

[2] Lambrecht, "Great Commandment," 95.

If the omission of "all my commandments, whatever I command you today" is intentional, it also is revolutionary. Moses required the obedience of every commandment to "live"; Jesus requires only the double love commandment. One should not, however, assume that the author opposed Torah obedience in general, as one can see from two other passages.

> 4:10 "The law and the prophets «were in force» until John. From then on the kingdom of God is in force. 11 But it is easier for heaven and earth to pass away than for one iota or one serif of the law to fall. 12 So whoever does not do one of the least of these commandments will be called least in the kingdom of God, and whoever does them, this one will be called great in the kingdom of God."

Even though the Law is eternal, its status among those "in the kingdom of God" has radically changed. Those who observe "the least of these commandments will be called great". However, "the kingdom of God" will include those who do not observe all of them, who "will be called the least". *Logoi*'s Jesus makes this clear in his denunciation of Torah authorities:

> 7:1 "Woe to you, exegetes of the law, for you bind burdens and load on the backs of people, but you yourselves do not want to lift your finger to move them. . . . 11 "Woe to you, Pharisees, for you tithe mint and dill and cumin, and give up justice and love. But these one ought to do, without giving up those."

The tithing of spices apparently was among "the least of these commandments", whereas justice and love are essential.

Mark redacts this controversy twice. The second is more obvious than the first. Compare the following:

Logoi 6:18–20	Mark 12:28–31
A certain exegete of the law, to test him, asked, "Teacher, which is the greatest commandment in the law?"	One of the scribes came, heard them disputing, saw that Jesus answered them well, and asked him, "Which is the first commandment of all?"
19 He said to him, what is writ-	29 Jesus answered, "The first is:

ten in the law?" ²⁰ He answered and said,	'Hear, O Israel, the Lord our God is one Lord,'
"You will love the Lord your God with all your heart and with all your soul, and with all your mind, and	³⁰ and 'love the Lord your God from your whole heart, and from your whole soul, and from your whole mind, and from your whole strength.' ³¹ The second is this: 'Love
your neighbor as yourself."	your neighbor as yourself.' There is no commandment greater than these."

Mark added the *shema* and numbered the commandments—"the first is . . . the second is . . ."—even though his enquirer asked only about one. The awkward expansion surely is secondary.

Secondary too is Mark's unnecessary repetition of the love commandment by the scribe:

> ³² "Teacher, you rightly say in truth that there is one [God] and no other than he, ³³ and to love him from one's whole heart, from one's whole understanding, and from one's whole strength, and to love the neighbor as one's self is greater than whole burnt offerings and sacrifices." (Mark 12:32-33)

The lost Gospel did not contrast love with sacrifices but with requirements that one must observe every commandment, as in Deut 11:8. Mark's Jesus, however, replaces "you will live" with further requirements.

Logoi 6:21	Mark 12:34
He said to him, "You have answered rightly. Do this and you will live."	When Jesus saw that he had answered him thoughtfully, he said to him, "You are not far from the kingdom of God." And no one dared to question him again.

A hostile testing in *Logoi* becomes an admiring question in Mark, even though the tagline implies that the scribe, like *Logoi*'s exegete, intended to trick him but dared not do so, a trace of the story in the lost Gospel (cf. *Logoi* 6:18).

The Markan Evangelist had already imitated this controversy in ch. 10, and here too his transformations emphasize Jesus's fidelity to Jewish Law:

Logoi 6:18	Mark 10:17
A certain exegete of the law, to test him, asked, "Teacher [ἐπηρώτησεν ... αὐτόν· διδάσκαλε], which is the greatest commandment in the law?" ... [cf. 21b, where Jesus answers that one need only observe the love commandment: "Do (ποίει) this and you will live (ζήση)."]	A man ran up to him, kneeled, and asked him, "Good teacher [ἐπηρώτα αὐτόν· διδάσκαλε], what must I do [ποιήσω] to inherit eternal life [ζωήν]?"

In both Gospels, Jesus responds to the question with one of his own.

Logoi 6:21b	Mark 10:18
And he said to him [ὁ δὲ εἶπεν πρὸς αὐτόν], "What is written in the law?"	And Jesus said to him [ὁ δὲ Ἰησοῦς εἶπεν αὐτῷ], "Why do you call me good? God alone is good."

In Mark's other imitation of this dispute, he added the *shema* (12:29: κύριος ὁ θεὸς ἡμῶν κύριος εἷς ἐστιν), which resembles "God alone is good [εἷς ὁ θεός]".

Logoi's exegete answers with the double love command (6:20), which Jesus affirms as the only commandment one need follow "to live". Mark's Jesus, however, requires much more: "You know the commandments"; he then lists six of the Ten Commandments (10:19) and praises the man for having observed them from his youth.

Logoi 6:21b	Mark 10:21a
He said to him [εἶπεν δὲ αὐτῷ], "You have answered rightly."	Jesus looked at him, loved him [for keeping the commandments], and said to him [καὶ εἶπεν αὐτῷ] ..."

For Mark, even the keeping of the commandments was not sufficient, and to make the point he redacts yet another passage from the lost Gospel. Compare the following:

Logoi 10:43-44	Mark 10:20-22
	He said to him, "Teacher, I have observed all these things from my youth." ²¹ Jesus looked at him, loved him, and said, "You lack one thing.
"Do not treasure for yourselves treasures on earth, . . .	Go, sell whatever you have and give to the poor,
but treasure for yourselves treasures in heaven [θησαυροὺς ... ἐν οὐρανῷ]. . . . ⁴⁴ For where your treasure [θησαυρός] is, there will also be your heart."	and you will have treasure in heaven [θησαυρὸν ἐν οὐρανῷ],
[Cf. *Logoi* 3:10] [Cf. *Logoi* 10:48: "Soul, you have many good things (ἔχεις πολλά ἀγαθά).")]	and come, follow me." ²² He was shocked by this saying and left in sorrow, for he had lots of possession [ἔχων κτήματα πολλά].

Here Mark has embedded a saying from *Logoi* within a narrative. A logos has become a chreia.³ The combination of the verb ἔχω with πολλά appears in the Synoptic Gospels only in Luke 12:19, his redaction of *Logoi* 10:48, and in its parallel in Mark 10:22. The Markan Evangelist thus added to *Logoi*'s double love command the observance of the Ten Commandments and the distribution to wealth to the poor.

The Matthean Evangelist apparently conflated elements from both *Logoi* 6:18-21 and Mark 12:28-34 in 22:34-40, but his alterations show somewhat more sympathy for Mark's more Torah observant version. He omits Mark's *shema* but adopts Jesus's reply with two commandments: first is the command to love God, the second is the command to love one's neighbor. Like Mark, he credits Jesus for evoking the biblical love command, but omits the criticism of "whole burnt offerings and sacrifices" (Mark 12:33). Instead he adds that "on these two commandments hang all the law and the prophets" (Matt 22:40). What *Logoi*'s Jesus seemed to exclude, Mark's and Matthew's reasserts.

---

³ Luke, too, transformed the prohibition of laying up treasure into a positive command to sell and give to the poor; compare *Logoi* 10:43-44 and Mark 10:21 to Luke 12:33-34.

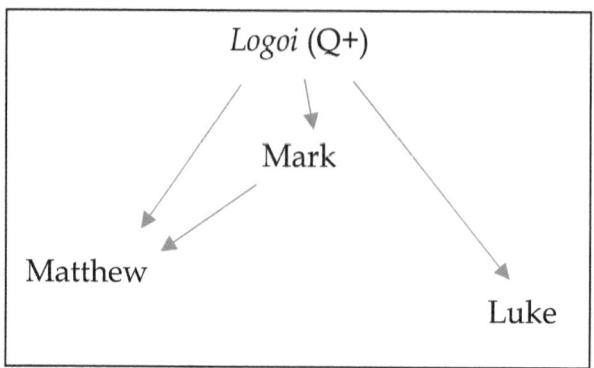

If this assessment is correct, its implications are significant. First, it adds to the lost Gospel a passage that only modestly satisfies Mathew-Luke agreements and does not qualify as *Sondergut* insofar as analogies appear in Mark and Matthew. The controversy does, however, satisfy the criterion of Inverted Priority. Luke's version preserves its earliest meaning, is coherent with Q/Q+, and appears immediately after another logion from *Logoi*.

Second, although the lost Gospel surely was Jewish—it apparently excluded a Gentile mission—the author's attitude to Jewish law was revolutionary. Some in the kingdom of God, who "will be called the least," will not observe all of "the least of these commandments"; all one needs to observe to "live" is the double love commandment.

Third, ironically, when this dispute found a new home into the Gospels enthusiastic about a Gentile mission—Mark and Matthew—its radical attitudes to Jewish law were tamed to make them more Torah observant. Mark added observance of the Ten Commandments and generosity to the poor; Matthew implied the authority of all "the law and the prophets."

Evidence for a Relationship Between Mark and Q[1]

James Van Dore

New Testament scholars have devoted their lives to staking out one position or another on the synoptic problem, the problem of how to account for the similarities and differences of material among the Gospels of Matthew, Mark, and Luke. The majority of these scholars have accepted that the Gospel of Mark was written first and was used as a primary source by Matthew and Luke. These scholars also theorize that along with Mark, Matthew and Luke used a second source (generally called Q), now lost, that primarily consisted of a collection of sayings material attributed to Jesus, which Matthew and Luke inserted into the primarily narrative base of Mark.[2] Considering the great bulk of scholarly effort put to the synoptic question, very few scholars have seriously considered the question of whether there is a relationship between Mark and Q, most often assuming that any material shared by Mark and Q these two authors independently found in earlier Christian tradition.[3]

---

[1] My first paper was a disaster. When I started my studies at Claremont Graduate University, I had no idea how to use sources critically and my first paper reflected that. Dr. Riley tore that first paper apart, filling the last page with note after note, detailing all the many mistakes I had made. And then he literally tore the paper apart, tearing off that last page before returning the paper to me. He knew that such correction could be devastating to a young scholar-in-training. So instead of letting me read such harsh, but merited, criticism, he removed it and met with me face-to-face to explain how real scholarship works. I will be forever grateful to Dr. Riley for that kindness. The bulk of the following paper is from the dissertation that Greg Riley made possible by that act and through his teaching and mentorship.

[2] This "two-source" hypothesis was put into its classic form by B. H. Streeter. See Burnett Hillman Streeter, *The Four Gospels: A Study of Origins* (London: MacMillan, 1924), 149–200.

[3] This position goes back to at least Adolph Harnack who argued that Mark and the "second source" were heirs to some fixed traditions, though in different translations. See, Adolf Harnack, *The Sayings of Jesus: The Second Source of St. Matthew and St. Luke*, tr. J.R. Wilkinson (New York: G.P. Putnam's Sons, 1908), 193–229.

A few scholars have challenged this assumption of the independence of Mark and Q, however. A tiny minority, led by Julius Wellhausen, suggested that Q knew and used Mark. Wellhausen's argument was primarily based on his dating of the two documents, which relies on an extremely unlikely interpretation of Matt 23:35 // Luke 11:51 as referring to the destruction of Jerusalem, though Wellhausen also argued that Q was more Christianized than Mark, that Q's discourses were more developed than Mark's independent sayings, and that because Mark strove for completeness he would have included more of Q if he knew it.[4] This position has been all but forgotten by scholarship.

A slightly larger group of scholars have argued the contrary, that Mark knew and used Q. This line of argumentation was initiated by Bernard Weiss, whose primary argument was that in those places where Mark and Q present the same material, Mark's version is always more developed than Q's.[5] Those scholars who have followed Weiss have argued for a wide variety of forms of Q that Mark might have known—some more limited first editions than traditionally assumed, some much more expansive editions of Q including elements in the triple tradition usually assigned to Mark and not Q—but most concentrating on individually examining places where Mark and Q appear to have contained the same material—the so-called Mark-Q overlaps—to demonstrate that Mark knew and used Q.

More recently, and comprehensively, this argument—that examination of the Mark-Q overlaps reveals Mark's knowledge and use of Q—has been made by Harry T. Fleddermann.[6] Fleddermann first reconstructed an original Q version of each discrete saying also found in Mark then attempted to demonstrate that Mark had developed his version directly from the Q version. He argued that all the differences between the two variants were attributable to Markan redaction of the Q original. Fleddermann also

---

[4] Wellhausen believed that that saying dated Q after the destruction of Jerusalem. See, J. Wellhausen, *Einleitung in die drei ersten Evangelien*, 2d ed. (Berlin: Reimer, 1911), 64–79

[5] Bernhard Weiss, *A Manual of Introduction to the New Testament*, tr. A. J. K. Davidson (New York: Funk & Wagnalls, 1889), 219–39.

[6] Harry T. Fleddermann, *Mark and Q: A Study of the Overlap Texts*, Bibliotheca Ephemeridum Theologicarum Lovaniensium CXXII (Leuven: Leuven University Press; Peeters, 1995) and Harry T. Fleddermann, *Q: A Reconstruction and Commentary* (Dudley, Mass.: Peeters, 2005).

identified several elements of Q redaction which appear in Mark's Gospel, including most importantly for determining the dependence of Mark on Q, sayings that were created by, and therefore could only come from, Q itself.

However, Fleddermann failed to take seriously the possibility of the independence of Mark and Q. He decided from the outset of his study to consider independence only if he could not find another explanation for the evidence; this is hugely prejudicial. Fleddermann also failed to take into account the persistence of oral tradition in the Church, saying "by the time we have come to the overlap texts we have long since left the realm of oral tradition."[7] Furthermore, his reconstructions of Q material often differs from the reconstructions of other scholars, putting into doubt whether his analysis of the relationship between the Markan variant and the Q variant is on solid ground. Finally, the reasons for his determination of Markan development of Q material must also be re-examined carefully. For example, Fleddermann argued that parallelism in a Q saying, versus non-parallelism in its Markan counterpart, indicates that the Q saying is in its more original form. This is not necessarily so. The incidence of strict parallelism may in fact indicate that Q has modified the material into a more literary and artificial version than the original form preserved by Mark. Therefore, Fleddermann's arguments, based on a one-to-one analysis of the overlap passages were unconvincing.

Those scholars who actively refute the notion that there is a dependency relationship between Mark and Q follow in the footsteps of Adolph Harnack.[8] Responding to Wellhausen's argument that Q knew and used Mark, Harnack evaluated the overlap passages and determined that in no instance should the evidence lead one to conclude that Q was dependent on Mark and that in some cases the version in Q is more primitive that that in Mark. This "priority discrepancy" — that in the overlap passages sometimes Mark preserves the more primitive, sometimes Q — became a key argument for those who argued for the independence of Mark and Q. For example, more recently, Rudolf Laufen published a comprehensive study on the Mark-Q overlaps following a similar procedure as Fleddermann.[9] He concluded, however, contrary to

---

[7] Fleddermann, *Mark*, 20.
[8] Harnack, *The Sayings of Jesus*, 193–229.
[9] Rudolf Laufen, *Die Doppelüberlieferungen der Logienquelle und des Markusevange-*

Fleddermann, that of the 23 overlap passages he deemed suitable for consideration in some cases Mark preserved the more primitive version, in others Q did, and in yet other cases the results were mixed, that is, Mark preserved some more primitive elements of a passage and Q preserved other more primitive elements within the same overlap. In other words, neither Mark nor Q held a monopoly on priority in the overlap material, so neither could be dependent on the other.

Christopher M. Tuckett pointed out that one of the major problems with the position that Mark knew and used Q was that under that understanding any of the material in Matthew, Luke, or Mark could potentially have come from Q, since it is a source for each one.[10] Tuckett argued that in order to determine direct dependence of Mark on Q one must find close verbal agreement between the two. However, the only evidence of Q is contained within Matthew and Luke, who also knew and used Mark. Therefore, in passages of close verbal agreement, Matthew and Luke may simply be reflecting Mark and not demonstrating Mark's similarity to Q. It is not enough, Tuckett argued, to point out that the differences between Mark and Q in the overlap passages are due to Markan redaction. One must find Q redactional elements in the Markan versions of the overlap passages to determine that Mark knew and used Q.

Frans Neirynck made a similar argument, stating that in order to prove Mark's dependence on Q a scholar has to prove more than that Mark's version of the saying is secondary.[11] Furthermore, Neirynck responded specifically to Fleddermann's work, arguing that he doubted Fleddermann's reconstruction of the original Q texts saying that the author tended to neglect the influence of the Markan text on Matthew and that his reconstructions seemed to be designed to show Markan dependence on Q. Neirynck also challenged whether Fleddermann had proved that Mark's clustering of individual overlap passages reflects his knowledge of Q, as

---

liums (Bonn: Peter Hanstein, 1980).

[10] Christopher M. Tuckett, "Mark and Q," in *The Synoptic Gospels: Source Criticism and the New Literary Criticism*, ed. by Camille Focant, Bibliotheca Ephemeridum Theologicarum Loveniensium CX (Leuven: Leuven University Press, 1993), 149-175

[11] Frans Neirynck, "Recent Developments in the Study of Q," in *Logia: Les Paroles de Jésus -- The Sayings of Jesus: Mémorial Joseph Coppens*, ed. by Joël Delobel (Leuven: Uitgeverij Peeters, 1982), 29-75.

within those Markan clusters, the Markan order always differs from the Q order. Finally, he questioned Fleddermann's method of deciding Markan dependence on a strict alternative of Mark either knowing Q or Mark knowing oral tradition (that Fleddermann believed had all but died by this time), completely neglecting the possibility of oral clusters of sayings or even written documents arising in the tradition which preceded the writing of Q. Therefore, Neirynck remained unconvinced that Fleddermann had proven a dependence of Mark on Q.

This paper seeks to re-evaluate the relationship of Mark and Q. Because absolute determination of the text of Q is impossible and is liable to reconstructions that bias the final analysis, however, this paper will not attempt another reconstruction of the individual texts. Instead, this paper will examine six places where Mark's use of multiple pieces of the tradition overlaps with Q's use of the same multiple pieces. However, unlike Fleddermann's clusters of tradition, these six groupings of tradition are used by Mark and Q in intentional editorial constructions, not mere collections of random independent sayings. These six editorial compositions shared by Mark and Q are John's baptism, the temptation narrative, the composite reference to John as a messenger, the mission discourse, the Beelzeboul accusation, and a collection of sayings which may be called "the demands of discipleship." These parallel editorial constructions are too similar, extensive, and distinctive for Mark and Q to have independently found them in oral tradition. A textual relationship of some kind is demanded. That Mark and Q had independent access to as many as six otherwise unknown written documents is extremely unlikely. Likewise, a single document containing these editorial constructions, looking very much like a proto-Gospel, otherwise unattested, is equally unlikely. The simplest and most logical conclusion is that these six shared editorial constructions are evidence that Mark knew and used Q. The paper therefore turns to examinations of these overlaps in order to show that their constructions reveal an editorial hand that in turn indicates a direct relationship between Mark and Q.

John's Baptism

Q contained several sayings related to John's baptism which Matt 3:7-12 // Luke 3:7-9, 16-17 reproduce very similarly. Among these saying there is a curse against those who come to be baptized, then a warning against relying on being descended from Abraham, a metaphor concerning an axe and a tree about to be cut, a saying about the "coming one," and finally a metaphor about the coming harvest. Of these, it is only the Q saying on the "coming one" (Matt 3:11 // Luke 3:16) that has a parallel in Mark 1:7-8. Minor differences aside, there is clear agreement between Matthew and Luke on the four basic elements of this saying and their order: John's baptism by water, a more powerful coming one, John's inadequacy in comparison, and the coming one's baptism in holy spirit and fire. In Mark, however, the single Q saying is two discrete sayings with the same general content as the single Q saying. The first saying in Mark contains the two central elements of the Q version; the second saying is composed of the first and last elements of the Q version. There are only some small, though theologically significant, differences between the Mark and Q versions.

In its composition, Q presents two separate thoughts interwoven into a chiastic structure. A comparison of John's baptism with the baptism of the "coming one" is separated by the insertion of a comparison of the great status of the "coming one" with the lowly status of John. Mark, on the other hand, presents the same ideas, but in parallel; i.e., a comparison of the status of the "coming one" with the status of John, followed by a comparison of John's baptism with that of the "coming one." But despite the differences in the presentations of Q and Mark, they do present the two essentially independent sayings in conjunction with one another. It is highly unlikely that Mark and Q independently brought these two essentially independent sayings together. Some sort of editorial hand must have. Furthermore, there is too much similarity between the Q and Markan versions in vocabulary and thought for this editorial hand to have been in the realm of oral tradition. The words "more powerful one," "sufficient," and "sandals" are distinctive of both versions. And though Q presents the sayings in a chiasm and Mark presents them in parallel, they present the main elements in the same relative order: I/water—he/Spirit; more powerful one—one not sufficient. Therefore, because Mark and Q present two essentially independent ideas in conjunction with each other and do so with strong verbal and constructional correlations which are too extensive and specific to be the result of Mark and Q being heirs to a common oral tradition, there must be a documentary relationship.

The Temptation Narrative

Scholars often have not included Jesus' temptation in their consideration of the overlaps between Mark and Q. It is certainly a special case, as the nature of the concordance is different from the other overlaps. Mark's account at Mark 1:12-13 is extremely brief and has no sayings material. The Q account in Q 4:1-13 is fairly long and involved, with several discrete sayings elements. Because of the lack of sayings material in Mark, an enumeration of the narrative elements Matthew and Luke have in common, and therefore in Q, and comparison with those in Mark's account is in order.

Q has, first of all, that Jesus was led into the wilderness by the Spirit to be tempted by the devil. Second, Jesus stayed there for forty days after which, when he was hungry, he was approached by the devil. Third, Jesus was tempted three times and refused each temptation with a Biblical citation. The temptations were orders from the devil to turn stones to bread for food, to throw himself down from a height and let the angels protect him, and to bow down to the devil and claim the kingdoms of the world for himself. Finally, both accounts ended with the departure of the devil. There are some minor differences in wording and order between Matthew and Luke, but the basic elements are the same.

The outline of Mark's account at Mark 1:12-13 is fairly similar though it differs in significant details. First of all, Mark, like Q, reports that the Spirit drove Jesus into the wilderness. Second, Jesus remained in the wilderness for forty days tempted by Satan. Third, Jesus dwelled with wild animals. Finally, the angels administered to him.

It is not clear that the temptation of Jesus is the main point of Mark's pericope. The temptation by Satan is given the same narrative weight as Jesus' communion with wild animals and the ministration of the angels to him. It is also not clear in the Markan version that Jesus was driven into the wilderness for the purpose of being tempted, which is explicit in the Matthean version and implicit in the Lukan. Mark seemed to imply that Jesus was tempted by Satan over the entire course of the forty days, while in Q it is clear that it is only after a period of forty days that the temptations occurred. All of the sayings material is found only in the Q temptation account and all of the sayings are biblical citations. Therefore, all of the sayings material is known to have existed outside the Jesus circle and any oral traditions it may have passed on. The creation of these sayings is not attributable to Q, though their use in the temptation account may be attributable to Q. Furthermore,

Mark's account does not enumerate the three specific temptations which elicit these biblical citations in any way. Therefore, if Mark is dependent on Q for his account of the temptation, he has removed the bulk of the account from his telling of it. Essentially, all Mark hypothetically retained from his Q source was the fact that Jesus went into the wilderness, not entirely of his own accord, spent a period of 40 days there, and was tempted.

Mark's second element, the mention of the wild beasts accompanying Jesus, has no equivalent, and thus no potential source, in the Q account. However, if Mark knew and used Q and identified the Q temptation as being in part a reference to Ps 90:11-12 (LXX), then Mark may have borrowed the wild animals from the following verse, Ps 90:13 (LXX).

Finally, Mark had Jesus ministered to by the angels, while in one of Q's temptations the expectation of the ministration of the angels to Jesus is treated somewhat negatively. This last point, though, is complicated by the fact that Matthew agrees with Mark in ending with the angels ministering to Jesus. However, this reference in Matthew may be drawn from Mark. In fact, there is little correspondence between the endings of the temptation accounts in Matthew and Luke, beyond the mention that the devil departed.

Simply put, the two temptation accounts are vastly different. The Markan account has not one distinctive Q element of vocabulary in it. For example, Mark uses the term "Satan," while Q characteristically uses the term "the devil" in this passage.[12] But there are a number of intriguing commonalities to the two accounts, including Jesus being driven into the wilderness, the location in the wilderness, the 40-day time period, and the temptation by the supreme demonic power. It is these concurrences of narrative detail that indicate a close textual connection between the two accounts. If these parallels were due to Mark and Q sharing a common source, it is unclear what that source would be and what its purpose would have been. A temptation account on its own, particularly one as sketchy as one which could have inspired separately both the Markan and Q accounts, would not likely have been circulated. For the temptation account to have any meaning for its audience, it would need to have appeared in a larger narrative. Therefore, it is unlikely that Mark and Q independently borrowed

---

[12] Q does use the term "Satan" in the Beelzeboul accusation passage, so there is no reason he could not have used it here as well.

an entirely self-contained temptation account from oral tradition. By the same token, it is unlikely that there was an otherwise unknown extensive narrative account that included a temptation story as part of a broader narrative and included the specific details that Mark and Q have in common in their two temptation narratives, such as being driven into the wilderness, the location of the wilderness, the 40-day time period, and the temptation by the supreme demonic power. None of these details is absolutely required of the story. Jesus could have gone off on his own, Jesus could have gone to a mountain, the time period could have been any length, and the temptation could have been by multiple figures. The coincidence of these details, combined with the clearly secondary nature of the Markan version, and the unlikelihood that a temptation account of necessary detail and extent to inspire both Gospel accounts circulated prior to the writing of the Gospels indicates again that there is a close textual relationship between the account in Mark and the one in Q. Add to this the likelihood that Mark identified the source of Q's temptation account, Psalm 90, borrowed elements from the source itself for his account, and the general secondary and abbreviated nature of Mark's account, then it is likely that the Markan passage is a heavily revised version of the Q account.

John as a Messenger

The next Q passage for which there is a Markan overlap appears in an extensive Q section on John the Baptist (Luke 7:18-19, 22-28, 29-35; 16:16 // Matt 11:2-15, 16-19; 21:31b-32) after Jesus' great sermon, though the Markan overlap appears at Mark 1:2 with the account of John's baptism early in his narrative. Mark has no other parallels with this section of Q, so it is conceivable that if Mark knew and used Q he could have taken the one element from this section he wanted and combined it with his other John material, including the passage on the baptism of John for which, as already seen above, there is another Mark-Q overlap.

The specific passage for which Mark has an overlap is Matt 11:10 // Luke 7:27, which announces the sending of a messenger to prepare the way and explicitly indicates it is a scriptural reference. However, it should be noted that this citation is not a strict quotation, being a composite of Exod 23:20 and Mal 3:1. Mark presented the same scriptural citation with only one small difference but used an entirely different introduction. The slight difference is that Mark does not use the final two words of the quotation that Q used, but it is unclear whether this was an intentional omission or because he was following Exod 23:20 over Mal 3:1 here. In addition, Mark continued beyond the overlap passage with an additional citation from Isa 40:3.

There are two elements of these overlap passages that indicate a close connection between Mark and Q. First, there is the fact that Mark and Q presented a composite of two scripture passages essentially verbatim. Second, the citation of the scripture is distinctive in its use of the word κατασκευάσει. No known texts of Mal 3:1 use that word. The combination of these two factors firmly suggests a textual relationship between Mark and Q in this passage.[13] Additionally, Mark's apparent ascription of his entire citation, along with the actual citation from Isaiah may indicate Mark's unfamiliarity with the Hebrew Bible, and therefore his reliance on a source. The Markan and Q versions of the combined citation are too similar, being nearly verbatim, and the use of κατασκευάσει is too remarkable to suggest that that source could be independently found by Mark and Q in preceding oral tradition. There must be a close textual relationship between the two versions. The most reasonable source for Mark, then, would be Q.

The Mission Discourse

Both Q and Mark presented a set of mission instructions given by Jesus to a group of disciples. Luke reproduced two versions of the mission instructions, one which relied on Mark at Luke 9:1-6 and a second at Luke 10:1-16 which apparently relied on Q. Matthew recorded only one set of mission instructions at Matt 10, most likely combining elements from both the Q and Markan versions.

The Q version of the mission instructions, as essentially presented in Luke 10:1-16, consists of approximately ten sayings of Jesus. In Luke, these ten sayings are framed by a second dispatch of disciples—the first having occurred earlier in Luke's narrative

---

[13] This is essentially the argument of David Catchpole, though he goes a step further and argues that Q 7:27 was a creation of the Synoptic Sayings Source, and therefore that Mark must have known Q. See, David R.Catchpole, "The Beginning of Q: A Proposal," *NTS* 38 (1992): 211ff. Ismo Dunderberg objected to Catchpole's argument, countering that removing the verse from its context makes incomprehensible the passage, but he had to posit an unknown version of the Septuagint to justify the presence of κατασκευάσει. See, Ismo Dunderberg, "Q and the Beginning of Mark," *NTS* (1995): 501-11.

where Luke had reproduced Mark's mission instructions—and their return. This Q/Luke section on the mission of the disciples includes both instructions on conduct and more general sayings on the nature of missionary work, and climaxes with condemnations directed at several cities. The Q section on mission most likely concluded with a saying identifying the missionaries with Jesus (Q 10:16), though Matthew presented that Q saying independent of his mission discourse at Matt 10:40. The Markan mission instructions at Mark 9:1-6 are much briefer than the Q mission instructions. Mark's composition does not contain versions of the general sayings on the nature of missionary work or the condemnations on the various cities.

Fleddermann offered three reasons why he thought comparing the mission discourses in total revealed Mark's dependence on Q.[14] First, all of the topics in Mark's version also appear in the same order in Q's version, and that Mark appears to be a "sketchy excerpt of a longer composition."[15] Second, though Q consistently had the mission instructions in direct discourse, Mark switched from indirect discourse to direct discourse within the pericope. This, he argued, showed that Mark adapted a source in direct discourse into his own narrative frame. Third, Q's equipment rule prohibited carrying certain items which Mark's version of the equipment rule expressly allowed. Mark appeared, according to Fleddermann, to be reversing an established ban on these items.[16]

Fleddermann's first argument is based on his definition and consideration of the number of topics in both the Q version of the mission instructions and in the Markan version. The Q version has at least five topics: an equipment rule, a command not to move about, a command to accept what is offered, a command to cure the sick, and an instruction on what to do if rejected. Putting aside the possible larger context of the Q mission instructions, of the five explicitly mission instruction topics, Mark shares with Q four of those topics: the equipment rule, the command to not move about, an instruction on what to do if rejected, and a mention of curing the sick. Fleddermann only counted three overlaps, however, as he excluded the Markan narrative verse about healing the sick from

---

[14] Fleddermann, *Mark*, 118.
[15] Fleddermann, *Mark*, 118.
[16] Fleddermann, *Mark*, 118.

consideration, deeming it part of Mark's narrative frame.[17] However, the topic of healing the sick is clearly part of Mark's literary construction and should be included in consideration. If it is, Mark's order is not the same as Q's, in that Q's fourth and fifth topics are Mark's fourth and third topics. In addition, the mission instructions have a certain amount of inherent logic to their order. The equipment rule should come first, as preparation for mission work logically precedes the mission work. Likewise, the instruction to stay in one place, logically precedes the instruction on what to do when rejected. Therefore, the fact that Mark and Q place these topics in the same relative order is not surprising, and does not indicate dependence of Mark on Q.

Additionally, Mark does not have any of the additional sayings material with which Q framed the instructional material of his mission composition anywhere in his Gospel.[18] So Mark betrayed little if any knowledge of the Q context for the mission instructions. Therefore, if Fleddermann is correct, and Mark abbreviated the Q mission composition, Mark has retained no distinctive trace of Q's frame in his abbreviation.

Fleddermann's second argument—that Mark shows evidence of using Q for his mission instructions because he shifts from indirect discourse to direct discourse within his composition—is also weak. Though the shift is curious and may indicate that Mark's version has been secondarily modified, what is necessary to prove Mark's knowledge and use of Q is Mark's retention of an element that could only have come from Q's redactional efforts. Proving Mark secondarily developed a source does not prove that Mark secondarily developed Q.

It is Fleddermann's third argument that carries the most weight. Mark's allowances for a staff, sandals, and two tunics do appear to reverse earlier prohibitions. But only one of the three allowed items, the sandals, appears in Q. There is no strong reason for Luke to drop prohibitions against the other two items, the staff and two tunics, from his Q version of the mission instructions if they did in fact originally appear in Q. Luke had no problem repeating much of the material in the Markan version of the mission instructions in his recounting of the Q version. It is much more likely that Matthew and Luke, taking their cues from the absolute

---

[17] Fleddermann, *Mark*, 118.
[18] With the possible exception of Q 10:16 which may have a parallel in Mark 9:37.

strictness of the Q version of the mission instructions, reversed Mark's allowances and that Q itself did not mention two tunics. So while Mark's version of the mission instructions does show evidence of being a secondary redaction due to the allowance of staff, sandals, and two tunics, because Mark does not retain any specific element that can only be traced only to Q's redactional efforts, it is not evident based on this alone that Mark knew and used Q.

Finally, Fleddermann argued at some length why Mark would have condensed Q in the manner he suggested Mark must have if Mark were dependent on Q.[19] First, he argued that Mark needed to keep the discourse brief so as not to obscure that Mark's two passages on the mission of the twelve frame the death of John. But, the two framing passages are horribly imbalanced as they stand in Mark, with the return of the missionaries only meriting one verse, Mark 6:30. In any case, it is the length of the intermediary passage that serves to make the reader forget the introductory element of the framing device, not the length of the introductory passage itself. Fleddermann also noted that Mark appeared to be condensed from the Q version because Mark lacked the passages of the Q version which do not fit his own literary and theological purposes. But it should be expected that Mark would not present material contrary to his own point of view. It is rather the inclusion of material that is in Q and yet goes against Mark's purposes that would be indicative of Mark's use of Q, and that sort of material is completely absent from Fleddermann's analysis. Therefore, Fleddermann failed to make a persuasive case that comparison of the separate elements of Mark's version of the mission discourse with the separate elements of Q's version demonstrates that Mark knew and used Q on these bases.[20]

Nonetheless, the fact that Mark's version of the mission instructions contains four of the same essential elements—not three as Fleddermann thought—indicates that there is a close connection between the two accounts. Again, Q's five topics are an equipment rule, a command not to move about, a command to accept what is offered, a command to cure the sick, and an instruction on what to

---

[19] Fleddermann, *Mark*, 119-121.
[20] Joachim Schüling thought that there were too many passages in Q's version of the mission instructions absent from Mark's version to allow for a direct literary dependence of Mark on Q. See Joachim Schüling, *Studien zum Verhältnis von Logienquelle und Markusevangelium* (Würzburg: Echter Verlag, 1991), 19.

do if rejected. Mark's account only lacks the command to accept what is offered. Furthermore, Mark's amelioration of the demands of the equipment rule and Q's retention of a stricter equipment rule indicate that Mark is a secondary revision. So, the connection between the accounts must be from Q to Mark. It is not Mark's retention of isolated elements of Q redaction that indicate the relationship, but rather the way in which Mark has clearly taken over and secondarily developed the entire composition as it exists in Q.

The Beelzeboul Accusation

The next Q-composition for which there are significant Markan overlaps, Q 11:14–23, the Beelzeboul accusation, consists of at least three distinct sayings: a saying about a divided kingdom, a saying about a strong man, and a saying about he who is not with me. Mark, however, also presented two of these sayings in the same order and within a similar narrative frame at Mark 3:22–27, which is strong evidence of a close textual connection between the two compositions.

It is clear that this Q-composition must have been introduced by some sort of healing account. Matthew and Luke show evidence of such an introduction, though they agree on little about this miracle beyond the general outline (Luke 11:14 // Matt 12:22). Luke has simply that Jesus exorcised a demon from a mute, that the former mute spoke, and that the crowds marveled. Matthew has a more elaborate miracle, having a blind and mute person brought to Jesus, a healing, a demonstration of the healing, the amazement of the crowd, and a questioning by the crowd as to whether Jesus is the Son of David. The interest in the healing of the blind is a Matthean characteristic,[21] so the demon-possessed was most likely mute in Q, as he is in Luke's version. Also typical of Matthew is the question of the crowd.[22] Therefore, the introduction to the first saying of the Q composition under consideration was probably closer to the Lukan version, though the exact wording is impossible to reconstruct. Confirmation that the Lukan version of the introduction is from Q may be found in that Matthew has a doublet of this narrative, an exorcism at Matt 9:32–34, which involves a demon-possessed mute, and like Luke's introduction here, has the crowds

---

[21] As Laufen noted with the support of much further scholarship. See Laufen, *Doppelüberlieferungen*, 127.

[22] Laufen, *Doppelüberlieferungen* 126; Fleddermann, *Mark*, 44.

"marveling." Therefore, the outline of the narrative introduction in Q to the following sayings material consists of an exorcism of a demon, or perhaps, demons, from a mute; the mute demonstrating his healing; and the crowd reacting with amazement.

This sayings material itself, which begins with the accusation that Jesus has exorcised demons by the prince of demons, Beelzeboul, constitutes a long, tightly-connected composition. Jesus responds to the accusation with the saying: "Every kingdom against itself divided comes to ruin and house against house falls" (Q 11:17). This saying is certainly self-contained enough to have theoretically circulated independently and been applied to a multitude of different circumstances. For example, it could be a warning against disputes within a family or household. It is only the context in which it is found in Q that gives it its association with exorcism.

That saying is certainly followed in Q, as the close agreement between Matthew and Luke makes clear, by a saying about Satan being divided against himself and the ruination of his kingdom because of it. This saying, however, is not self-contained like the previous saying, and therefore could not previously have been an independently circulating saying, as it derives much of its meaning from the general maxim which precedes it. The same can be said about the following two sayings of this grouping. Neither of these two additional sayings can have been independently circulated, as they each depend on their context, particularly the sentence which precedes them, for much of their meaning. The construction in Q 11:15, 17–20 then reads:

> But some from them said in Beelzeboul the prince of demons he casts out the demons.
> But he knowing their thoughts said to them:
> "Every kingdom against itself divided comes to ruin and house to house falls.
> And if also Satan against himself is divided, how will stand his kingdom?
> And if I in Beelzeboul cast out the demons, your sons in whom cast out?
> And if in the finger of God I cast out the demons came upon you the Kingdom of God."

The presentation of the material in this way clearly shows how the core saying was introduced and then expanded by three similarly-introduced sayings.

For comparison purposes, the Markan passage containing the parallel to the core saying at Mark 3:22–26 is presented below in a similar fashion:

> And the scribes...said that he has Beelzeboul and that in the prince of demons he casts out the demons.
> And calling them near he spoke to them in parables:
> "How can Satan cast out Satan?
> And if a kingdom against itself is divided that kingdom cannot stand.
> And if a house against itself is divided that house cannot stand.
> And if Satan rises against himself he cannot stand but has an end."

At first glance, it might appear that Mark follows the same pattern of introduction, core saying, and three similarly introduced commentary sayings. But Mark has done something very different from Q. Mark's first saying of Jesus is not parallel to the first core saying of the Q version. Instead, Mark's first saying reiterates the narrative contextualization for the core saying, which actually appears as the second saying in Mark's account. This means that Mark, in fact, has an introduction, a core saying, and only two commentary additions.

Also, Mark's three similarly-introduced sayings do not function in the same way that Q's fourfold group of sayings do. Q's first commentary saying specifies the general saying to Satan, but the second saying goes off into a new direction by comparing Jesus' exorcisms to those of the Jewish exorcists and declaring them judges of Jesus' accusers. The third saying goes even farther away from the core saying's meaning by making Jesus' exorcisms a sign of the coming kingdom of God. Therefore, Q has used the core saying as the base on which to collect a handful of diverse themes. Mark, on the other hand, does something quite different. Each of Mark's three similarly-introduced sayings states the same essential concept but each narrows the focus from the general truth to a tighter, and yet, even tighter scope: from a broad collection of people and institutions, to a single institution and limited group of people, to a single actor. These are two different literary compositions gathered around the same core saying. Q uses the concept of a kingdom divided against itself as a springboard for several other largely unrelated topics, while Mark explores the implications of the core saying through analogies.

In this way, Mark is a much tighter narrative. Mark's pericope is about a charge that Jesus exorcises by the power of

Beelzeboul, which Jesus counters with a single concept. Q's pericope begins as a narrative about a similar charge, but goes off into different, tangentially related topics. Mark shows no evidence of knowing any of the concepts contained in those expansions as found in the text of Q (i.e., the sons, judges, or kingdom of God). Nonetheless, Mark and Q share a combination of compositional elements—a narrative accusation of expelling demons by Beelzeboul answered by a saying about the weakness of a divided kingdom—that indicates a close connection between the two accounts. Both accounts are more developed than that bare summary, but in entirely different directions. Nonetheless, Mark could have known Q's version and stripped out Q's tangential expansions and replaced them with his contextualizing question and focus-narrowing analogies. It is not, therefore, that Mark's version retains traces of Q's redaction in word choice or theme that argues for Mark's knowledge and use of Q, but rather that the combination of the demonic accusation with mention of the unusual name Beelzeboul being attached to the divided kingdom saying demands more than a shared oral tradition behind Mark's and Q's accounts. A literary relationship is required.

Complicating the question of the relationship between Mark and Q as determined through examination of the Beelzeboul pericope, however, is the saying which follows Jesus' response to the Beelzeboul controversy in both Mark and Q: the saying about the strong man. Despite the fact that Matthew and Luke continued to follow Q after this pericope, and the entire Beelzeboul complex is well within a quite extensive block of Q material in Luke, several scholars have argued that the strong man saying was not in Q—since the agreements between the Matthean version and the Lukan version are slight—and that Matthew and Luke have independently redacted Mark's saying here. Other scholars have noted that Matthew and Luke are largely following Q in this section, and the likelihood that Matthew and Luke independently decided to insert Markan material at the same point, particularly Markan material which they did not pass on faithfully, is small. These scholars assume that Matthew's version, since it closely parallels Mark's, was not heavily influenced by Q's version, while Luke's was.[23]

---

[23] See John S. Kloppenborg, *Q Parallels* (Sonoma, Calif.: Sonoma Press, 1987), 92 for a summary of the debate. See Schüling, *Verhältnis*, 109 for an argument that Matthew and Luke independently redacted Mark's version of this saying. The

Fleddermann, on the other hand, argued that the saying was in Q, but that Matthew has followed Q more faithfully—for if he had to choose between Luke's version in Q or Mark's "blander" version, he would have chosen Q's—and that Luke's version does not accurately reflect Q's version, but is rather primarily a creation of Luke.[24] In support of Luke's creation of his version of the saying, Fleddermann argued that Luke's Strong Man saying shares the same interests as Luke 12:16-21.

Alternatively, several commentators have assigned Luke 12:16-21 not to Luke, but to Q.[25] Therefore, it is not clear whether these similarities actually indicate Lukan interests or Q interests. Fleddermann also noted that six words in Luke's version of this saying are found only here among the synoptic writers.[26] But this determines not that these words are distinctly Lukan, for Luke uses four of them nowhere else and the other two only once and twice more, and rather suggests only that they are peculiar to this passage, which has not yet been proven to be either a Q composition or a Lukan reworking of a Q composition. [27] Therefore, Fleddermann failed to prove that this is a Lukan reworking of the Q original, which is better represented by Matthew. He had merely assumed it. His proof that Matthew follows Q more closely consisted of a series of arguments that Luke could derive his version from Q's by a great number of substitutions and movements of individual words, as is typical of his analyses, but is unlikely to be the way in which real authors worked.[28] Finally, that Matthew's version of the saying has some general Q characteristics, such as being in the form of a question and using "first…and then," is not persuasive, as such turns of phrase are not particularly distinctive of Q alone, and could just as easily be Matthean alterations to his Markan source based on such similar phrasings in the neighboring Q passages with which he was familiar, instead of reflective of the Q

---

possibility that Matthew and Luke coincidentally moved this Q saying from another location in Q to this position based on Mark is rightly dismissed by Laufen. See Laufen, *Doppelüberlieferungen*, 130-1.

[24] Fleddermann, *Mark*, 52-55 and *Q*, 484-8, 496-7.
[25] Kloppenborg, *Parallels*, 128.
[26] Fleddermann, *Mark*, 52.
[27] Maurice Casey, *An Aramaic Approach to Q: Sources for the Gospels of Matthew and Luke* (New York: Cambridge University Press, 2002), 173-4.
[28] Fleddermann, *Mark*, 53-54.

original.²⁹

Therefore, it is quite likely that Q included a version of the strong man saying at this point, which means that both Mark and Q presented the saying about the strong man in conjunction with the Beelzeboul accusation and the divided kingdom saying. If the strong man saying is considered in isolation, there is no absolutely compelling reason to identify the strong man with Satan. The strong man could be any adversary, natural or supernatural. It is only when combined with the Beelzeboul accusation and divided kingdom saying that the strong man receives this identification with Satan. In other words, there is nothing intrinsic to the strong man saying that demanded it be placed with the Beelzeboul accusation and divided kingdom saying.³⁰ That both Mark and Q have these three elements joined into these similar compositions in this way may be explained in a few ways. The first is that they independently made the connection and joined the previously independent pieces of tradition. This seems unlikely, as not only must they have combined the two pieces, but they must have done so in such a similar fashion that defies coincidence, with the strong man saying being directly attached at the end of the divided kingdom saying without any real bridging or narrative contextualization. The second option is that Mark and Q knew and used the same written source. The combination of the Beelzeboul accusation and divided kingdom saying with the strong man saying demands some sort of textual relationship. The points of contact are too numerous and similar for anything else. The third possibility is that Q knew and used Mark or that Mark knew and used Q. As seen above, the idea that Q knew Mark is nearly universally rejected. Therefore, either Mark and Q ultimately rely on a third document, or Mark knew and used Q.

The last element of this large Q cluster beginning with the Beelzeboul accusation for which there is a parallel in Mark is the saying about those with and those against. In Q, this saying follows immediately upon the saying of the strong man. However, Mark has this saying in a different location and in a different form. Clearly this saying is part of the Beelzeboul-Strong Man complex in Q, as Matthew and Luke have it in precisely the same place and have it attached in precisely the same way. However, the basic saying it-

---

²⁹ Fleddermann, *Mark*, 54–55.
³⁰ Laufen, *Doppelüberlieferungen*, 137.

self has no specific, necessary connection to the themes of the previous two sayings group in Q. Therefore, it would not be surprising to discover that Mark, if he knew and used Q, had removed it from its position and placed it somewhere else within his work. However, the Markan version, so differently worded and so differently used, shares very little with the Q version of the saying beyond the basic concept. There is no element in Mark's version that can only be traced to Q's distinctive redaction in vocabulary or theme that would prove that Mark knew and used the Q version. Therefore, an isolated consideration of the two versions of the saying does not indicate that Mark knew Q. Rather it is Mark's sure use of the rest of Q's version of the Beelzeboul pericope which indicates that Mark relied on the text of Q for his version of this individual saying.

In sum, examining the parts individually of the Beelzeboul Accusation-Divided Kingdom-Strong Man complex as contained in Q and in Mark separately provides no concrete evidence that Mark knew and used Q, contrary to what Fleddermann argued.[31] Rather it is the combination of these three elements—none of which necessarily had to be combined with any of the others for their essential meanings—into a single composition in such a similar order and similar manner in both Mark and Q, that is strongest evidence that there is a close textual connection between Mark and the Synoptic Sayings Source. These parallel constructions cannot be the result of independent use of oral tradition and are so similar, extensive, and distinctive that there must be a direct connection between the composition in Mark and that in Q.

The Demands of Discipleship

The final Mark-Q overlap compositions are extremely complicated complexes of sayings. At issue are three discrete sayings. The first saying is about the relationship between Jesus' followers and their families. In the second saying, Jesus states that taking up one's own cross is a requirement of discipleship. The third saying is a seemingly paradoxical saying in which Jesus states that whoever

---

[31] Fleddermann attempted to argue that there were plausible reasons why Mark would have omitted Luke 11:19-20, as well as Luke 11:23-26 (which is not considered here as there is no evidence of overlap between Q and Mark for this section), but he never offered any proof that Mark knew these passages in the first place. See Fleddermann, *Mark*, 59-60.

loses his life will not actually lose it. These three sayings are presented by Matthew in this order one right after the other at Matt 10:37-39. Luke presents the first two of these sayings together, and in the same order, in one of his large blocks of Q material at Luke 14:26-27, but he presents the third saying several chapters later in his work, again within a block of Q material at 17:33. Scholars are divided on whether the third saying originally appeared with the first two, though the majority of scholars have argued that it did.[32] In both Matthew and Luke, however, the sayings are only lightly connected to each other. Matthew's trio is somewhat stylistically connected through each beginning with the indefinite pronoun, but in neither his nor Luke's recounting do the three sayings lose their independence. Each is a complete thought unto itself and no real effort has been made to interconnect them.

Mark has close parallels to the second and third sayings, which he presented in the same relative order as the sayings appear in Q. He, however, has more closely linked the two sayings, with the third saying serving as the maxim which justifies the verity of the second saying. Both Matthew and Luke have separate, strong parallels to this Markan combination of the two sayings where they are clearly copying and editing Mark. The Matthean parallel is at Matt 16:24-25 and the Lukan at Luke 9:23-24. Additionally, Mark presents a saying later in his gospel at 10:29-30 that bears some resemblance to the first saying in the Q trio, though the Q and Markan sayings may not be very closely related.

Mark 10:29-30 contains the same essential message as Q 14:26, that is, that Jesus demands of his disciples the alienation from their families. The two versions are quite different, however, and Mark's version shows no traces of the distinctive editorial work of Q on his version of the saying, such as the command to hate one's relatives or the phrase ἔτι τε καὶ τὴν ψυχὴν ἑαυτοῦ.[33] Both of these elements are certainly due to Q's redaction. Luke's version of the saying is clearly the stronger, more difficult to follow, and more likely to offend version of the command. Therefore, in outline, it is more likely closer to the original, as Matthew has a tendency to ameliorate harsh demands.[34] Luke's additional ἔτι τε

---

[32] See Kloppenborg, *Parallels*, 170.

[33] There is also no evidence of Markan interference in the Lukan version of the Q saying.

[34] For example, in the cross saying. See Fleddermann, *Mark*, 137. See also the

καὶ τὴν ψυχὴν ἑαυτοῦ must also be in the Q original and not an addition of Luke's, as it connects this saying with the third of the Q sayings in this group, a connection Luke himself has severed by moving the third saying to another location in his Gospel.[35] Therefore, analysis of this specific overlap does not demonstrate that Mark knew and used Q.

The second saying in this grouping has to do with carrying one's own cross in imitation of Jesus. Fleddermann asserted that the Q version of this cross saying forms the climax of a grouping of three sayings which he identified as Q 14:26-27. In other words, he considered the alienation from family portion of the passage as two sayings and reconstructed that saying in the general form in which it appears in Matthew, but with the Lukan ending. He did not include the soul saying in Luke 17:33 // Matt 10:38 as part of this Q grouping. The primary assertion that the cross saying constitutes the climax of the grouping and the two related assumptions he made without argument.[36] But, clearly the two halves of Matt 10:37 are not independent, unrelated sayings, but rather one saying and its reiteration, and it only appears in this double form in Matthew, not Luke. The Lukan version of this verse more closely reflected the text of Q, not the Matthean version which Fleddermann had put forward as reproducing the Q original without argument. Because there is clearly only one essential saying here, Fleddermann's primary assertion of a Q collection of three sayings, each beginning with ὃς οὐ, is not convincing. The two Q sayings on alienation from the family and on the cross are clearly two separate sayings. In fact, both Matthew and Luke have made no attempts to link the two sayings and merely present them sequentially as sayings sharing the theme of the demands of discipleship, so the connection between the two is much looser than Fleddermann thought.

Fleddermann argued that Mark's "deny" and "take up" reflect Mark's passion narrative, and on this he might have been correct.[37] But identifying the differences in the Markan version as

---

saying on divorce at Mt 5.32.

[35] Jan Lambrecht came to some of the same conclusions, arguing for the Q cluster of three sayings, despite their separation in Luke, based on the same phrase. See Jan Lambrecht, "Q-Influence on Mark 8,34-9,1," in *Logia: Les Paroles de Jésus – The Sayings of Jesus: Mémorial Joseph Coppens*, ed. by Joël Delobel (Leuven: Uitgeverij Peeters, 1982), 277-304.

[36] Fleddermann, *Mark*, 137-9.

[37] Fleddermann, *Mark*, 138.

due to Markan redaction establishes only that and nothing more. In order to prove that Mark knew and used Q, Fleddermann would have had to have established that Mark retained elements of the Q redaction in his version of the overlap saying. Fleddermann claimed that Q has brought two themes together in what he considered the original composition of Q 14:26-27, and that therefore Mark, by knowing the cross saying, knew Q. But Fleddermann determined that Mark showed no evidence of knowing Q 14:26, and therefore showed no evidence that he knew that Q presented 14:26 together with 14:27, without first proving that Mark knew the entirety of Q. Also, even if Q did place this saying at this point as the climax of a group of three sayings, as Fleddermann asserted, Fleddermann in no way proved that Q must have created the saying as he claimed. Therefore, since Mark's version of the cross saying contains no element of vocabulary or theme that can only be traced to Q's redaction of his cross saying, and since Fleddermann did not prove that Q created the cross saying, merely asserted that he did, Fleddermann did not prove that Mark knew and used Q in his comparison of the two versions of the saying.

The last of Q's three sayings on the demands of discipleship appears immediately after the first two in Matthew, though in a later position in Luke. Above, in discussing the saying on alienation from one's family, it was argued that the phrase ἔτι τε καὶ τὴν ψυχὴν ἑαυτοῦ looks forward to the last of the three sayings, the saying on losing one's life. If this has been correctly identified as the purpose of this Q redactional addition, then these three sayings appeared together in Q, and Luke has moved the final saying.

Fleddermann argued that Matthew moved this saying on losing one's soul after his version of the cross saying due to the influence of Mark 8:34-35, where Mark had his versions of the two sayings follow one after the other. It was also the influence of Mark's version, according to Fleddermann, that caused Matthew to add the phrase "for my sake" to his Q version of the saying. Fleddermann pointed out that Luke retained the phrase in his account of the Markan version of the saying, so Luke had no absolute aversion to the phrase and would have retained it if he found it in Q.[38] Though Fleddermann was most likely correct that Matthew added "for my sake" to his Q version of the saying on the model of the Markan version of the saying, the rest of his argumentation is

---

[38] Fleddermann, *Mark*, 143.

weak. As argued above, this saying was grouped in Q with the other two previous sayings on the theme of the demands of discipleship. However, that does not exclude the possibility that Matthew was influenced by Mark in his presentation of the saying, just that it was not to the extent that Fleddermann asserted. Secondly, Luke's acceptance of the phrase "for my sake" in his copying of the version in Mark proves only that he was not completely averse to the phrase. There are all sorts of other reasons why he might have deleted the phrase from his copy of the Q version, some of which may be related to his relocation of the saying to a later part of his Gospel. Regardless of these criticisms, however, Fleddermann was most likely correct that the original Q version of the saying did not include the phrase, as it disturbs the parallelism of the saying and is implied by Q's grouping this saying with his two previous sayings.

However, the analysis of these three sayings taken as a whole does provide concrete evidence for a close textual relationship between Mark and Q. These three sayings appeared in sequence in Q and Q clearly clustered these three sayings based on their shared thematic interest in the demands of discipleship. However, only the first two sayings speak to that topic in their essence. The third saying is a multi-valent paradox which could speak to all sorts of different topics. It is only when it was attached to the cross saying, and the phrase "for my sake" was added, that the saying directly spoke to the demands of discipleship. Therefore, at some point, some redactor added the losing one's life saying to the cross saying as a commentary.

Mark also had the cross saying attached to the saying on losing one's life. In fact, Mark went to greater lengths than Q to interconnect the two sayings. And yet, despite these redactional efforts, Mark presented these sayings in forms strikingly similar to the ones in which Q presented them. It is unlikely that two separate strains of tradition would independently pass on these sayings in such similar forms and have appended the saying on losing one's life to the cross saying as a commentary.[39] There must be a textual relationship between the two compositions. Since Mark's collection

---

[39] Laufen made a similar argument that there is nothing intrinsic to these two sayings that suggests that they originally belonged together, though he more vaguely locates the association in the "early church." See, Laufen, *Doppelüberlieferungen*, 331.

of the sayings shows more development, particularly in the attachment of the sayings to each other, Mark must have secondarily developed Q, establishing that Mark knew and used Q.

Conclusion

The search for traces of Q's redactional vocabulary and themes in the Markan overlaps is ultimately futile for two reasons. First, those themes and vocabulary that Mark took over from Q have become part of the general synoptic tradition—through Mark's use of them, and through Matthew and Luke's use of Mark—and are no longer identifiable as originating in Q. Second, if Mark knew and used Q, he did so in a very different way from Luke or Matthew. He was much more selective in his use of Q and felt free to re-write completely the material to serve his own needs, while Matthew and Luke were usually much more conservative in their use of Q. In using Q in this way, Mark may have effectively obliterated the evidence—the distinctive vocabulary and thematic interests of Q—that individual comparison seeks.

Nevertheless, it is the longer passages where multiple elements were collected into compositions that prove Mark's knowledge and use of Q, for these places demand a close textual connection between the two texts. There are six identifiable instances within the Mark-Q overlaps where Mark and Q share editorial constructions and establish a close textual connection between the Markan and Q documents. The parallel constructions are too similar, extensive, and distinctive for Mark and Q to have independently found them in oral tradition. Because Mark is always secondarily developed in the overlap passages, the direction of the connection between Mark and Q must be from Q to Mark. Because of the great amount of overlaps, the fact that they appear across Mark and Q, and the great variety of the overlaps—being narrative and sayings material, about Jesus and about John, being biographical and hortatory—Mark must have known and used Q itself.

A Woman Caught in Adultery? Or A Wandering Teacher Trapped Between Roman and Jewish Law?
John 7:53–8:11 in Light of Quintilian and Seneca

Thomas E. Phillips

It's been my distinct pleasure to work with Greg Riley over the last 6 years. With his passion for teaching, his clarity of thought and his scholarly acumen, Dr. Riley is an exemplar for our shared students. Perhaps most prominently, his unwavering commitment to reading the New Testament as a product of the Greco-Roman world has modeled the kind of historical-critical rigor that is sadly waning in the current culture of ideologically latent and historically shallow readings. In sympathy with Dr. Riley's insistence upon reading the New Testament as a series of Greco-Roman texts, I want to reexamine the familiar story of the woman caught in adultery (John 7:53–8:11) in light of the Roman legal tradition.

The Familiar Story and Equally Familiar Questions

In this account, Jesus was teaching on the Mount Olives when he was approached by a cadre of "scribes and Pharisees." These men had a woman in tow who had been caught in the very act of adultery. Even though the narrator has already assured the reader of the woman's guilt (John 8:3), her accusers annunciated her crime again when they forced her to stand (presumably naked and humiliated) before Jesus (vv. 3b-4). They (correctly) asserted that the Mosaic law prescribed death for "such women" (v. 5).[1] They were prepared to immediately stone her. Then the woman's

---

[1] Note that the woman's accusers used the feminine (τὰς τοιαύτας, 8:5) when claiming that the Mosaic demanded the death of the guilty party. Since the default gender in Greek for mixed gender groups is masculine, the use of the feminine was undoubtedly specifically intended to exclude the guilty man from the death penalty. This interpretation of the Mosaic law is strictly speaking a misinterpretation of the Mosaic law which—as we shall see—imposed a death penalty upon both guilty parties (Lev 20:10; Deut 22:22).

apparently self-appointed prosecutors asked Jesus, "What then do you say?" (v. 5). The narrator is explicit that this question is designed to "test" (πειράζω) Jesus (v. 6).

In a literary action unique to this account, Jesus wrote something in the sand. Although the content of Jesus' earthy epistle was not recorded, Jesus' subsequent command is among the best known of all gospel sayings, "The sinless one among you must throw the first stone at her" (v. 7). After Jesus wrote another unrecorded comment in the dirt,[2] the crowd dispersed, leaving the woman alone with Jesus (vv. 8-9). Jesus then asked the woman if there was anyone left to condemn (κατέκρινω) her (v. 10). Having previously spoken nothing in her own defense (she was apparently without defense, vv. 3-4), the woman simply responded, "No one, Lord." Jesus then said that he did not condemn her either and told her to go and sin no more (v. 11).

The story has an unparalleled textual history, being the longest interpolation in any New Testament book (in competition only with the so-called longer ending of Mark).[3] Although there has been mild dissent,[4] the overwhelming consensus of scholarship maintains that this account was a relatively late insertion into John's Gospel.[5] The story almost certainly had an independent oral

---

[2] In a recent Ph.D. dissertation, Chris Keith has argued that this passage was interpolated into the fourth gospel because this story demonstrated Jesus' literacy and that everything which transpired between Jesus, the woman and her accusers simply provided the backdrop against which the interpolator could insert a story to demonstrate Jesus' literacy. According to Keith, his location was chosen within the fourth gospel because of the earlier comment about Jesus' literacy in 7:15. This suggestion has the advantage of explaining both why Jesus wrote in the dirt — and did so twice — and also why the narrator felt no need to reveal what Jesus' inscriptions said. According to Keith, the point of the entire account was merely to assert Jesus' literacy and the content of Jesus' written comments was insignificant to the interpolator's intent. See Chris Keith, "Jesus Began to Write: Literacy, the *Pericope Adulterae*, and the Gospel of John" (Ph.D. diss.; University of Edinburgh, 2008).

[3] See most importantly, Bart D. Ehrman, "Jesus and the Adulteress," *Biblical Theology* 40 (1989): 24-44 and G. M. Burge, "A Specific Problem in the New Testament Text and Canon: The Woman Caught in Adultery (John 7:53-8:11)," *Journal of the Evangelical Theological Society* 27 (1984): 141-48.

[4] E.g., John Paul Heil, "The Story of Jesus and the Adulteress (John 7,53-8,11) Reconsidered," *Biblica* 72.2 (1991): 182-91. More characteristic of scholarship, see Daniel B. Wallace, "The Story of Jesus and the Adulteress Reconsidered," *NTS* 39.2 (1993): 290-96.

[5] Francis Moloney, *The Gospel of John* (Minneapolis: Liturgical Press, 1998), 106, insists: "For sound textual reasons it is universally admitted that the account of Jesus and the woman taken in adultery (7:53-8:11) does not belong to the Fourth

and written history which predates its inclusion within the fourth gospel. Our earliest witnesses to John are completely devoid of this account.[6] Some people have suggested that the account is more Lukan than Johannine,[7] others have argued that the story may go back to the historical Jesus.[8]

The story naturally raises important interpretative questions, most obviously, where is the man? Adultery requires at least two people, but the woman is alone, why?[9] As a follow up to this obvious question, readers have wondered if the absence of a guilty man is significant to the story's interpretation. While most readers have interpreted the story as an illustration of Jesus' profoundly merciful nature,[10] others have read the story as the account of Jesus

---

Gospel." Likewise, Ernst Haenchen (*John 2: A Commentary on the Gospel of John Chapters 7-21*, tr. Robert W. Funk, ed. Robert W. Funk and Ulrich Busse (Hermeneia; Philadelphia: Fortress, 1984), 22) passes over this account, commenting only that "The author did not consider the periscope of the woman taken in adultery to be an original part of the Fourth Gospel." (It is unclear if "the author" refers to Haenchen himself or the author of John's Gospel.) Even a scholar as conservative as Craig Keener (*The Gospel of John: A Commentary* [Grand Rapids: Baker Academic, 2003]) admits that this "passage bears all the marks of an interpolation" (735) and that John 7- 8 forms a coherent unit "once 7:53-8:11 is excised" (701). A very good brief discussion of the textual evidence is offered by George R. Beasley-Murray, although he spoke hyperbolically (and perhaps naively) when he claimed: "It is universally agreed by textual critics of the Greek NT that this passage was not part of the Fourth Gospel in its original form." See Beasley-Murray, *John*, 2nd ed., WBC 36 (Nashville: Thomas Nelson, 2000), 143-44. The authorship of the Fourth Gospel is too complex to speak confidentially about any "original form."

[6] Gail R. O'Day, "John 7:53-8:11: A Study in Misreading," *JBL* 111.4 (1992): 631-40, suggests that "patriarchal prejudices" may have contributed to the story's "canonical marginality" (640).

[7] E.g., Michel Gourgues, "'Moi non plus je ne condamne pas:' Le mots et la théologie de Luc en Jean 8,1-11 (la femme adultère)," *Studies in Religion/Sciences Religieuses* 19 (1990): 305-18.

[8] E.g., J. Duncan M. Derrett, "The Woman Taken in Adultery,"in *Law in the New Testament* (London: Darton, Longman & Todd, 1970), 156-88, here 156, describes this account as "a piece of authentic tradition going back to the beginnings of the Church." Original article: "Law in the New Testament: The Woman Taken in Adultery," *NTS* 10 (1963): 1-26.

[9] Derrett, ("The Woman Taken in Adultery," 162ff) acknowledges that the women's sexual partner may have already been executed, but he speculates that it is more likely that the women's guilty partner had made some deal with the witnesses to be released from punishment. Derrett's speculation cannot be proven or disproven. We only know that the adulterer is not mentioned in this account.

[10] See O'Day ("John 7:53-8:11," 633-35) on the influential readings of Augustine and Calvin.

standing up to the unjust victimization of a vulnerable woman.[11]

In this brief essay, I want to interpret the story a test, that is, in the terms of the literary model which the text itself suggests (v. 6). Specifically, I want to suggest that this account, regardless of its origin, confronts Jesus with the challenge of adjudicating between following Jewish law and Roman law. I believe that reading the story in this manner provides plausible answers to many of the central interpretative questions associated with this text, particularly questions about the absence of the offending man and about a likely *Sitz im Leben* for the story's origin.

The Conflict between Jewish and Roman Law on Adultery

Admittedly, the formalities of the legal processes being presumed in this account are vague. Although both the narrator and the people who brought the woman before Jesus clearly presumed the woman's guilt (8:3-4), the account does not clarify whether the woman had been formally convicted in some kind of official legal setting or not. It seems equally plausible to infer either that the woman had already been tried by some set of Jewish legal authorities or that she was simply being subjected to mob "justice." If the woman had not been previously convicted by some established legal body, then Jesus was being called upon to act as a *de facto* judge. However, if the woman had already been convicted by some established legal body, then Jesus was being called upon to act as a *de facto* executor.

Although the ambiguities surrounding the legal processes in this account are real and seemingly impossible to overcome, these ambiguities are, from the perspective of both Jewish and Roman law, insignificant. Both Jewish and Roman law allowed for the execution of an adulterous woman—and this woman was clearly

---

[11] Michael O'Sullivan, "Reading John 7:53–8:11 as a Narrative Against Male Violence Against Women," *Hervormde Teologiese Studies* 71.1 (2015): 1-8; Jayne Scott, "The One that Got Away," in *Ciphers in the Sand: Interpretations of the Woman Taken in Adultery (John 7.53–8.11)*, ed. Larry J. Kreitzer and Deborah W. Rooke, Biblical Seminar 74 (Sheffield: Sheffield Academic Press, 2000), 214-39. Elizabeth E. Green is correct to warn against feminist readings which end up portraying Jesus as essentially a knight in shining armor who saves the damsel in distress. See "Making Her Case and Reading it Too: Feminist Readings of the Story of the Woman Taken in Adultery," in *Ciphers in the Sand*, 240-67.

guilty (if the narrator is to be trusted).¹² Jewish and Roman law agreed that the woman deserved to die. However, these two bodies of law disagreed on another important point. Let me explain.

On the one hand, Jewish law was clear (as the woman's detractors correctly pointed out). The woman was worthy of death (Lev 20:10; Deut 22:22). The method of her execution, whether by stoning, strangulation or some other means, is less clear, but the outcome was clear. An adulteress should be executed. Undoubtedly, everyone with even the vaguest knowledge of Jewish law and ethics would have been aware that the woman's accusers were correct in their assertion that the Mosaic law called for her death. Most interpreters would concur with J. Duncan Derrett's assertion that this account presumes a knowledge of Jewish law and that:

> a reader of the pericope as it stands today would have little difficulty in understanding what it was all about, provided that he had a general knowledge of Jewish law, and a common acquaintance with the Pentateuch and with contemporary notions of the subjects which were necessarily in question.¹³

Clearly the account does presume a knowledge of Jewish law. In fact, Jesus' dialog partners explicitly refer to the Mosaic law (8:5). Of course, Jewish law also called for the death of her partner, whom we can only assume is an unindicted co-conspirator in this story. Although Jewish law was clear that both guilty parties should die, there was no legal principle which required that they be killed at the same time or in the same manner. Importantly for our purposes, Jewish law did not provide any exemptions from the death penalty for either guilty party.¹⁴ However, it is important to note that Jewish law was not the only legal system of significance

---

¹² A few interpreters (e.g., J. Martin C. Scott, "On the Trail of a Good Story: John 7.53–8.11 in the Gospel Tradition," in *Ciphers in the Sand*, 53–82) have suggested that the woman was actually innocent and that Jesus' statement about casting the first stone was really just a device to demonstrate that there were not really any witnesses against her. Unless the account employs the literary device of an unreliable narrator, there is little to commend this reading. There are no known examples of an ancient author employing an unreliable narrator.

¹³ Derrett, "The Woman Taken in Adultery," 158.

¹⁴ For greater detail on the imposition of the death penalty in the Pentateuch, see Anthony Phillips, *Ancient Israel's Criminal Law* (New York: Oxford University Press, 1970) and Henry McKeating, "Sanctions against Adultery," *JSOT* 11 (1979): 57–72.

in the dawn of the Christian era. Roman law was also important. Unfortunately, Derrett's rapid dismissal of the significance of Roman law as a cultural backdrop for this account is typical of interpreters. This failure to consider Roman law more carefully is, as I shall demonstrate, an important interpretative oversight.

So, on the other hand, Roman law also allowed for the execution of an adulterous woman.[15] However, Roman law, unlike Jewish law, legislated an exemption which could spare an adulteress from execution. She could not be executed for her crime unless the offending man was also executed. An adulteress could be executed only if her co-conspirator was also executed. She could not be executed alone.[16] This law was repeatedly enunciated by Quintilian, perhaps Rome's most famous first century lawyer. In his training exercises for rhetoricians, Quintilian was discussing "arguments based upon circumstances" when he offered the following example:

> For example, in the case of the adulterer priest who chose to use his power of saving one life to save his own, it is a special feature of the *controversia* that we can say "You were saving more than one life, because once *you* were spared it was no longer lawful to kill the woman." (This

---

[15] In the Roman world, the association between adultery and the execution of the offending party was so strong that Valerius Maximus (*Val. Max.* 6.1.12–13) even records a story in which a private soldier named Gaius Lusius killed a military tribune for committing adultery and was exonerated by his general, Gaius Marius. Still, to be sure, not all adulterers in the Roman Empire were executed. Valerius tells of other adulterers being beaten, whipped, and castrated. Valerius insisted that all of these actions were justified and that none of the men who inflicted these severe punishments "suffered a penalty for indulging his anger." The cases described by Valerius are different than the account in John's Gospel, because all of the cases reviewed by Valerius involved the punishment of a man who had impinged upon another man's honor (typically a husband, engaged man, or father) through his sexual indiscretions. Still, none of the cases which Valerius presented depict an adulteress being executed without the offending man also being executed.

[16] This legal tradition derives from the *lex Iulia de adulteriis coercendis* (Justinian *Dig.* 4.4.37; 48.5; *Cod.* 9.9.3; *Inst.* 4.18). On the Roman laws and traditions governing marriage and the role of women in marriage in the Roman period, see Suzanne Dixon, "From Ceremonial to Sexualities: A Survey of Scholarship on Roman Marriage," in *A Companion to Families in the Greek and Roman Worlds*, ed. Beryl Rawson (Malden, MA: Wiley–Blackwell, 2011), 245–61 and with a narrower application to women and marriage in the New Testament, also see Lynn Cohick, *Women in the World of the Earliest Christians: Illuminating Ancient Ways of Life* (Grand Rapids: Baker Academic, 2009), 65–132.

argument follows from the law which forbids an adulteress to be killed without the adulterer.)[17]

This example is important not only because Quintilian clearly states Roman law regarding the necessity of executing both guilty parties in the case of adultery, but also because he presumes that this law is so well-known that he can use this established case law to address other legal questions. The rationale is clear: A priest cannot pardon himself and save himself from the death penalty in the case of adultery because by doing so he would also spare his adulterous accomplice from death since she could not be executed unless they were both executed. Therefore, he cannot exempt himself from death because by exempting himself he would really be exempting two people (himself and his partner) from death, and he is limited to exempting only one person. While the case of the adulterous priest is not particularly significant for the Johannine story, the legal principle behind the case is very significant. Roman law "forbids an adulteress to be killed without the adulterer."[18]

One of the first century's other most famous lawyers, Seneca the Elder, likewise reasoned from this apparently widely known Roman law. In his *Controveriae*, Seneca offered up the following statute: "Whoever catches an adulterer with his mistress in the act, *provided that he kills both*, may go free. A son may punish adultery on the part of his mother."[19] Seneca then proceeded to discuss a specific application of this statute:

> A hero lost his hands in war. He caught an adulterer with his wife, by whom he had a youthful son. He told the son to do the killing. The son refused. The adulterer fled. The husband now disinherits his son (*Controv.* 1.4).

The specific issue of whether or not the father was justified in disinheriting his son need not concern us here. However, this controversy is significant because it again illustrates a prominent first century author presuming the Roman legal principle that the adulterer must be killed along with the adulteress and then draw-

---

[17] Quintilian *Inst.* 5.10.104–05 (LCL, Russell). Italics in original translation.

[18] Quintilian notes one exception to this principle of the equal application of justice. An adulteress can be executed alone when there is clear evidence of her guilt, but the identity of her partner is unknown (*Inst.* 7.2.52). This exception would not apply in the case presented in John's Gospel because the woman was reportedly caught in the very act of adultery (John 8:4).

[19] *Controv.* 1.4 (LCL Winterbottom). Emphasis added.

ing that established case law to address a different legal issue.[20]

In light to the difference between Jewish and Roman law regarding the execution of an adulteress without also killing the adulterer, it seems likely that the real legal issue at stake in the Johannine case of the adulterous woman may well have been the divergence between Jewish law and Roman law. That is, would Jesus choose to follow Jewish or Roman law in regard to how the adulterous woman should be handled.[21] According to Jewish law, her execution was entirely justified, even required. According to Roman law, her execution was forbidden (unless the guilty man was also executed).[22]

Rereading the Controversy Story of the Women Caught in Adultery

Given the tension which existed between Jewish and Roman law in the first century, it seems best, in form-critical terms, to

---

[20] There may have an exception when an offended husband was the executioner. Aulus Gellius (*Noct. Att.* 10.23.1–5) claims to be copying a speech from Marcus Cato in which he acknowledged a husband's right to put his adulterous wife to death without a trial. Nothing is mentioned of the adulterous man. For a helpful survey of Roman treatments of marriage and infidelity (with excerpts from ancient texts), see Matthew Dillon and Lynda Garland, *Ancient Rome: Social and Historical Documents from the Early Republic to the Death of Augustus*, 2nd ed. (New York: Routledge, 2015), 298–355.

[21] Although both Seneca and Quintilian presume the validity of the law and extend their legal reasoning from this legal foundation, some scholars have doubted the scope of the law's legal force in the first century (e.g., Derrett, "The Woman Taken in Adultery, 168). However, if the account is simply a controversy story similar to those provided by Seneca and Quintilian, the historical accuracy of the account to time period in question is no more significant in the Johannine account than is the historical accuracy in the accounts provided by the mid- to late first century accounts provided by Seneca and Quintilian.

[22] Some interpreters (e.g., Raymond E. Brown, *The Gospel according to John* I–XII, 2nd ed., AB 29 (Garden City: Doubleday, 1985), 337) have seen a potential conflict between Jewish and Roman legal authority, but they point toward the (unverified Christian) tradition that the Jewish authorities were unable to impose the death penalty after AD 30. This tradition is likely a Christian invention based on the Gospel's reporting of Jewish culpability in Jesus' death. Besides, as a historical matter, it is unlikely that the Romans would have been particularly concerned over the death of an anonymous adulteress. Brown warns that any concern about the Jewish authorities' inability to impose a death penalty "is highly ingenious but must remain an hypothesis." In addition to relying upon unverified traditions, this suggestion overly historicizes the story. The story is probably a strictly theoretical (or at least dehistoricized) account like those in Quintilian and Seneca.

regard this unit of tradition as a controversy story similar to those commonly found in the Synoptics. Form-critically, this story has tremendous affinity to the controversy stories over divorce and the Mosaic law (Mark 10:2-12; Matt 19:3-12) and paying taxes to Caesar (Mark 12:13-17; Matt 22:15-22; Luke 20:20-26).[23] In each case, the narrator in these stories specifically states that Jesus' dialogue partners were trying to test (πειράζω, Mark 10:2; 12:15; Matt 19:3; 22:18) or trap (ἐπιλαμβάνομαι, Luke 20:20) him. Also analogous to their Johannine counterpart, Jesus' antagonists in these Synoptic stories were attempting to confront Jesus with a "no win" situation by exploiting the cultural hybridity of first century Jews living under the hegemony of the Roman Empire. In each case, Jesus' rhetorical sparring partners knew that Jewish law and Roman law existed in tension and they were trying to force Jesus to side with one tradition or the other. Of course, in both of these Synoptic situations, Jesus outsmarted his opponents by choosing a third option (opting for a world without divorce in the first case and for a world which pleased both God and Caesar in the second case).

I would suggest that this controversy story about a woman caught in adultery follows a similar pattern. Jesus is confronted

---

[23] Although many NT scholars presume the authenticity of the controversy about the coin, this story is exceedingly unlikely to be an authentic story of Jesus. In ancient Palestine, as a concession to Jewish sensitivities about making graven images, most locally minted coins did not have human representations on them. (Palm trees, dates, and other agricultural produce were common.) This fact about Jewish coinage has been known for over 100 years. See G. F. Hill, *Catalogue of the Greek Coins of Palestine (Galilee, Samaria, and Judaea)* (London: British Museum Dept. of Coins and Medals, 1914). More recently, see J. Matiel-Gerstenfeld, *260 Years of Ancient Jewish Coinage: A Catalogue* (Tel Aviv: Kol, 1982) and Y. Mosherer, *A Treasury of Jewish Coins from the Persian Period to Bar Kokhba* (Jerusalem: Yad Ben-Zvi, 2001). In light of the patterns of Jewish coinage, it is unlikely that the Jesus' Jewish antagonists would even have had access to a coin with the emperor's image on it—and it is even more unlikely that Jesus' Jewish opponents would have been willing to handle such an idolatrous coin to transfer it to Jesus. During the first Jewish War when Roman coins (with the emperor's image on them) were flooding the area, several competing Jewish groups minted coins to compete with the imperial coinage. See Robert Deutsch, "Coinage of the First Jewish Revolt against Rome: Iconography, Minting Authority, Metallurgy," and Donald T. Ariel, "Identifying the Mints, Minters and Meanings of the First Jewish Revolt Coins," in *The Jewish Revolt against Rome: Interdisciplinary Perspectives*, ed. Mladen Popović, Supplements to the Journal for the Study of Judaism 154 (Boston: Brill, 2011), 361-71; 373-97. This story likely has Christian origins in the wake of the First Jewish War.

with a choice between Jewish law (and stoning the woman) and Roman law (and freeing the woman on a technicality), but Jesus chose a third option—forgiveness and reclamation. This story, admittedly a story on the fringes of New Testament study, well illustrates the importance of reading the New Testament as a document very deeply enmeshed within the Greco-Roman world. Although what eventually became known as Christianity was clearly a Jewish phenomenon, the Judaisms of the first century all experienced some degree of cultural hybridity. All first and second century Jews in ancient Palestine (and the entire Mediterranean) lived in the Greco-Roman world and found themselves forced to locate themselves within that world. Some, like the Essenes strongly rejected the values of the Greco-Roman world; others, like Philo of Alexandria, appear to have accepted deeper enculturation into the Greco-Rome world.

This story, whether it originated in the first, second or even third century, appears to show the Jesus of early Christianity also being enmeshed in the struggles of a culturally hybrid culture. This account, it seems to me, probably reflects the cultural struggles of an early Christian community which was seeking to position itself as a third sort of thing—neither Jewish nor Roman, but as something different from each—yet as more compatible with Roman than with Jewish values. This kind of distinctively Christian identity could have originated at a number of points. Personally, I would not regard a story like this to have origins any deeper than the first Jewish war (65–73 CE), probably later. I suspect that the first Jewish war (and subsequent events, including the Bar Kokhba Revolt [132–35 CE]) would have incentivized Christians to distinguish themselves from the Jewish groups who were actively warring against the Romans.

In such times of war, this story would have served multiple purposes. On the one hand, the story distinguished Jesus (and, by inference, his followers) from the Jews because Jesus did not follow the Mosaic law. In terms of identity formation, Christians were not to be confused with the Jews who were warring against Rome. On the other hand, the story demonstrated that Jesus' actions (and, by inference, those of his followers) were consistent with Roman law. Even if Jesus and his followers acted on the bases of different motives, their ethical decisions did not directly contradict Roman traditions. The effect of such a controversy story would have been to distinguish the followers of Jesus from the Jews, while also placing the Christians' distinctive traditions in a social location which was at least compatible with Roman traditions. It is quite possible that controversy stories like the account of the woman caught in adultery gave the Christian community the ability to carve out a social space for itself as an ethically distinctive, though legally compatible, tradition with the Roman Empire. This Christian social location was clearly not Jewish, but neither was it fully Roman. It was, however, more compatible with Roman values than it was with the violent Jewish traditions (that's why the story had a life and death question at stake).

Having made this historical suggestion, I am the first to admit that this proposed historical setting for this controversy story is merely one plausible historical suggestion among other equally reasonable possibilities. My primary concern has been to identify this story as a controversy story which drew upon an existing conflict between Roman and Jewish law to create a third option for the followers of Jesus, a distinctly non-Jewish space which was not identical with Roman values, but which was still compatible with Roman values.

Romans 1:26-27 in Its Rhetorical Tradition

Brett Provance

²⁶ For this reason God gave them up to degrading passions. Their women exchanged natural intercourse for unnatural, ²⁷ and in the same way also the men, giving up natural intercourse with women, were consumed with passion for one another. Men committed shameless acts with men and received in their own persons the due penalty for their error.[1]

This essay offers an interpretive perspective on Romans 1:26-27 that differs from the common view that the passage is an indictment of both female and male same-sex sexual activity, and suggests a rhetorical tradition for the passage that is overlooked, namely, reference to the stories found in Genesis 6 and 19, which are often linked in wisdom and apocalyptic contexts in the New Testament and other ancient literature. With this alternative perspective, the paper finds that homosexuality is not the main concern in this passage, and female same-sex sexual activity is not to be found in the passage at all. Rather, the two verses fit naturally in a traditional judgment proclamation.[2]

The Status Quo Interpretation: A Reassessment

The prevalent modern interpretation of these two verses is that they involve female and male same-sex sexual activity,[3] both

---

[1] All English Bible quotations are from the NRSV, unless otherwise noted.
[2] This paper was originally developed for a presentation given at the Society of Biblical Literature Pacific Coast Regional Meeting in 2018. The general interpretation of Romans 1:26-27 this paper presents coincides at points with an interpretation previously made, but unknown to the author when originally presented, by Michael Brinkschröder in his large work *Sodom als Symptom*, published in 2006; see Michael Brinkschröder, *Sodom als Symptom: Gleichgeschlechtliche Sexualität im christlichen Imaginären – eine religionsgeschichtliche Anamnese*, RVV 55 (Berlin: Walter de Gruyter, 2006). The author wishes to thank Dr. F. Stanley Jones and Dr. David Tabb Stewart of California State University, Long Beach, for their criticism and remarks on the original version.
[3] Bernadette Brooten, *Love between Women: Early Christian Responses to Female*

special examples of human waywardness. Female same-sex behavior is thought to be the subject of v. 26, and male same-sex sexual activity that of v. 27. The behaviors mentioned in both verses are understood in Romans 1 as the ultimate reasons for bringing judgment upon humanity, in a passage reflecting "a missional endeavor that encompasses the entire human race that refuses to acknowledge God's preeminence."[4] Homoeroticism is singled out, along with, even connected to, idolatry, as reasons for divine wrath to fall on "all" humanity.[5] As Käsemann observed, "One-sidedly emphasizing sex, he speaks of women and men with scorn."[6] And yet, "This section of the epistle [1:18-3:20] deals with the totality of the cosmos and not just with an aggregate of individuals; hence, it deals with humanity as such and not just with representatives of religious groupings."[7]

However, female homoeroticism is not stated in verse 26; the verse only states that "females" participated in unnatural sexual activity.[8] Writes Jamie Banister:

Homoeroticism (Chicago: University of Chicago Press, 1996), 216, 244. A list of scholars affirming this position is found at James E. Miller, "The Practices of Romans 1:26: Homosexual or Heterosexual?" *NovT* 37 (1995), 1-2, and an extensive list is at Jamie A. Banister, "Ὁμοίως and the Use of Parallelism in Romans 1:26-27," *JBL* 128.3 (2009), 569-570 n. 1. Those affirming among the major commentators are C. E. B. Cranfield, *A Critical and Exegetical Commentary on The Epistle to the Romans*, 2 vols., ICC (Edinburgh: T & T Clark, 1975), 125 (based on ὁμοίως in v. 27), Robert Jewett, *Romans: A Commentary*, Hermeneia (Minneapolis, MN: Fortress Press, 2007), 173, and Joseph A. Fitzmyer, *Romans: A New Translation with Introduction and Commentary*, AB 33 (New York, NY: Doubleday, 1993), 285. See also Richard B. Hays, "Relations Natural and Unnatural: A Response to John Boswell's Exegesis of Romans 1," *JRE* 14.1 (1986), 195-196.

[4] Jewett, *Romans*, 157. Jewett describes the whole of Romans 1:18-23 as a "rhetorical tour de force" (148).

[5] "ἐπι contains a hostile element while πᾶσαν rules out any exceptions. To the intensity of the judgment corresponds the totality of the world which stands under it, so that the statement about Gentiles applies to the heathen nature of mankind as such, and hence implies the guilty Jew as well." Ernst Käsemann, *Commentary on Romans*, trans. Geoffrey W. Bromily Grand Rapids: Eerdmans, 1980, 38.

[6] Käsemann, *Romans*, 48.

[7] Käsemann, *Romans*, 33, bracketed reference added.

[8] On the use of the term χρῆσις for sexual activity, see Brooten, *Love between Women*, 245; cf. Joseph A. Marchal, "The Usefulness of Onesimus: The Sexual Use of Slaves and Paul's Letter to Philemon," *JBL* 130.4 (2011), 749-754, on the "use" of slaves in antiquity.

Romans 1:26 has received much attention because it is possibly the only clear reference to, and condemnation of, female homoerotic activity in the biblical texts. However, female homoerotic activity is not explicitly mentioned in the passage; rather, this specific interpretation rests solely on the content of the subsequent verse (v. 27), in which Paul describes male homoerotic activity in the context of condemning it.[9]

The Grammatical Use of ὁμοίως

This problem, namely, the lack of any mention of female homoeroticism in v. 26, is usually met with the evocation of ὁμοίως in v. 27:

> [26] For this reason God gave them up to degrading passions. Their women exchanged natural intercourse for unnatural, [27] And in the same way [ὁμοίως][10] also the men, giving up natural intercourse with women, were consumed with passion for one another. Men committed shameless acts with men and received in their own persons the due penalty for their error.

Under this interpretive strategy, it is understood that the term ὁμοίως allows the interpreter to read back into v. 26 what is so clearly noted in v. 27, namely, male homoeroticism. It is assumed, then, that since females are the subject of v. 26, that female homoeroticism is the issue of v. 26.

James Miller has opposed finding v. 26 referring to female homoeroticism.[11] Ὁμοίως might simply mean that both the women and the men under consideration were sexually deviant, but in differing ways. Bernadette Brooten has countered Miller's position, stating, "I argue that 'unnatural intercourse' refers specifically to sexual relations between women, because (1) the 'likewise' (*homoi-*

---

[9] Jamie A. Banister, "Ὁμοίως and the Use of Parallelism in Romans 1:26-27," *JBL* 128.3 (2009), 569, who further notes that this position is also taken by Cranfield in his commentary on Romans (569 n. 1; see Cranfield, *Romans*, 125), as well as Brooten, whose position has been noted above.

[10] All references to the Greek text of the New Testament are from *Nestle-Aland, Novum Testamentum Graece*, 28th rev. ed., ed. Barbara Aland, Kurt Aland, Johannes Karavidopoulos, Carlo M. Martini, and Bruce M. Metzger in cooperation with the Institute for New Testament Textual Research (Münster/Westphalia: Deutsche Bibelgesellschaft, Stuttgart, 2012).

[11] Miller, "Practices," 1-2, 8.

ōs) of Rom 1:27 serves to specify the meaning of Rom 1: 26; and (2) other ancient sources depict sexual relations between women as unnatural (Plato, Seneca the Elder, Martial, Ovid, Ptolemy, Artemidoros, probably Dorotheos of Sidon)."[12] Point #2, however, assumes the conclusion in a premise, and thus it would appear that this one word, ὁμοίως, is the primary evidence given by Brooten that substantiates her view on the nature of the activity in verse 26.[13] On this point, she is not alone.[14]

Jamie Banister's article has notably challenged the hermeneutical strategy of reading back v. 27 into v. 26 via the term ὁμοίως. Addressing every New Testament passage involving ὁμοίως, and also noting various uses of ὁμοίως in the LXX, Philo, Josephus and the Apostolic Fathers, she has demonstrated that this reading-back approach is *never* demanded by the NT texts, and at times would be detrimental to proper interpretation.[15] While in a number of the cases the antecedent clause dictated the meaning of the subsequent ὁμοίως clause, the reverse was not the case: "The result is that, of the twenty-eight eligible [NT] passages for consideration, in twenty-five (89 percent) it would be impossible or illogical to use the ὁμοίως clause to interpret its antecedent clause."[16] Of the three passages remaining for consideration,[17] Banister finds no instance in each case of the ὁμοίως clause determining meaning in the antecedent clauses.

Thus it stands that the modern interpretive assumption in finding female homoeroticism in v. 26 is simply not a rhetorical/grammatical approach utilized by the ancient authors. It is a weak inference (warrant) in interpreting Romans 1:26-27. Indeed, it is virtually *unwarranted*. Thus, the "unnatural" sexual action of the women remains *unidentified*.[18] However, there is a well-known leg-

---

[12] Brooten, Love between Women, 248–250.

[13] Cf. also Brooten, Love between Women, 253.

[14] Though my position on Romans 1:26-27 differs from Brooten's, her significant work on this passage is invaluable, to the point that any study of this passage must consult her work.

[15] A notable summary is found at Banister, "Ὁμοίως," 588, wherein it is found that, in "89 percent of the twenty-eight other independent GNT passages in which ὁμοίως is used, it would be either impossible or illogical to attempt to use any part of the *homoiōs* clause to interpret the antecedent clause."

[16] Banister, "Ὁμοίως," 582.

[17] Luke 17:28, John 6:11 and Jude 8 (Banister, "Ὁμοίως," 582).

[18] Others have hypothesized certain non-homosexual activities that the women might have been involved with, such as, e.g., Miller, "Practices," 8-11. They are in

end in ancient Jewish literary history that involves unnatural sexual activity on the part of certain women. Indeed, two of Banister's three remaining passages concern interests of this paper, namely, passages involving the Watchers-Flood and Sodom judgments. The place of ὁμοίως will be reconsidered later in the paper.

Male Homoerotic Activity Singled out in Verse 27
Beyond the initial hermeneutical difficulty of finding female homoeroticism in verse 26 is why male same-sex sexual activity should be so singled out for judgment in verse 27, and in such language:

> Here we have the most egregious instance Paul can find to demonstrate his thesis about human distortion, the arena of sexual perversity that created wide revulsion in the Jewish and early Christian communities of his time. . . . The depiction of a particularly unpopular example for the sake of an effective argument leads Paul to highly prejudicial language, particularly to the modern ear.[19]

Yet Hays notes, referencing Furnish, Käsemann, and Cranfield, that "Romans 1 is neither a general discussion of sexual ethics nor an explicitly prescriptive admonition about the sexual behavior appropriate for Christians."[20] Indeed, he asks, "Why does Paul introduce the reference to homosexual behavior in Rom 1:26-27? If he is not giving moral instruction to his readers, why does the topic arise at all?"[21] Indeed, and Hays offers his interpretation: "The 'exchange' of truth for a lie to which Paul refers in Rom 1:18-25 is a *mythico-historical event* in which the whole pagan world is implicated."[22] And yet, Hays identifies no *particular* mythico-historical

---

the interpretive minority, and the issue still stands that the activity of the women is not stated in v. 26.

[19] Jewett, *Romans*, 173.
[20] Hays, "Relations," 187.
[21] Hays, "Relations," 187-188. So Fitzmyer also asks (*Romans*, 275), and he gives a two-part answer (275-276): (1) Corinth's reputation and (2) according to his understanding of Paul's view, homosexual behavior as a result of idolatry. Both explanations press logic. He cites T. Naph. 3:2-4 (which shall be covered in this paper), but mistakenly interprets the reference to Sodom as meaning that, logically, Sodomite behavior (not described in T. Naph.) is the result of idolatry. Fitzmyer does not include v. 5, which refers to the Watchers, which is part of the argument of T. Naph., and has nothing to do with homosexual behavior.
[22] Hays, "Relations," 200, italics added.

event, no punctiliar occurrence. He continues:

> In the same way, the charge that these fallen humans have "exchanged natural relations for unnatural" means nothing more nor less than that human beings, created for heterosexual companionship as the Genesis story bears witness, have distorted even so basic a truth as their sexual identity by rejecting the male and female roles which are "naturally" theirs in God's created order.[23]

And yet, this is not the experience of most humans. How does a minority account for a universal charge? There is something askew in modern interpretations of Paul's message. Rather, a particular "mythico-historical event" must be *identified*. And there is a mythico-historical event that involves judgment of the whole world and will soon be covered in this paper.

The Shift to Past Tense: Looking to Past Events

The shift from the present to the aorist tense at Romans 1:21 takes the audience into the past: "The shift from present tense verbs in vv. 18-20 to aorist verbs in vv. 21-23 signals a turn to the representatives of an archaic past who turned away from the truth and imposed a grim future on their descendants."[24] Indeed, a stated third person "they" and "them" become the subjects in vv. 19-28.[25] This shift in tense and referencing what appears to be a specific group is critical to understanding the overall passage. It will be shown that the past tense takes the audience to ancient times, to the stories of old.

The Apocalyptic Background of the General Context (Romans 1:18-32)

As this paper seeks to offer a better general explanation of Romans 1:26-27 than the status quo, the general setting of the text should be noted. Romans 1:26-27 is part of a well-defined passage influenced by apocalyptic rhetoric (Romans 1:18-32), beginning with a double revelation given in vv. 17-18:

---

[23] Hays, "Relations", 200.
[24] Jewett, *Romans*, 156, though the shift can be detected as early as v. 19 with ἐφανέρωσεν ("For God *manifested* [it] to them").
[25] Forms of ἀυτός ("them" and "their"), together, occur 14 times in vv. 19-28, as well as third person plural verbs (on the latter, cf. Calvin Porter, "Romans 1.18-32: Its Role in the Developing Argument," *NTS* 40 [1994], 219). The tone is not an abstract discourse on people in general, but the interaction (or lack thereof) between God and "them" points to something more specific.

16 For I am not ashamed of the gospel; it is the power of God for salvation to everyone who has faith, to the Jew first and also to the Greek. 17 For in it the righteousness of God is revealed [ἀποκαλύπτεται] through faith for faith; as it is written, "The one who is righteous will live by faith." 18 For the wrath of God is revealed [ἀποκαλύπτεται] from heaven against all ungodliness and wickedness of those who by their wickedness suppress the truth.

Concerning 1:17, Käsemann writes, "Even if ἀποκαλύπτειν does not necessarily have an 'apocalyptic' sense, in this context that seems most natural." Indeed, "the antithesis between the righteousness which reveals itself in the gospel and the wrath which discloses itself ἀπ' οὐρανοῦ in the stereotyped phraseology of Judaism and primitive Christianity . . . show that the apocalyptic view is also held by Paul."[26] However, apocalyptic in Romans 1:18-32 is carefully utilized, as, in all of Romans 1:18-32, there is no mention of "sin" (ἁμαρτία), "law" (νόμος), or "Gentiles."[27] It is not an "us vs. them" statement. It is a statement about humanity, and from v. 21 on, a statement about humanity past. It is a rather grand statement regarding human depravity and God's action against such.

But Paul's apocalyptic indictment is multifaceted. Paul is not only critical of sexual deviance in Romans 1:18-32 (vv. 24, 26, 27). He also finds fault in humanity for deterioration of cognition (vv. 21, 22, 28) and for idol worship (vv. 23, 25), shunning knowledge of God manifested via the creation (vv. 19-21, 25), all of which are connected with sexual deviance; cf. Wisdom 14:12: "For the idea of making idols was the beginning of fornication, and the invention of them was the corruption of life." The cognitive decline in Romans 1 is stated as an indictment, in that God is revealed, is made manifest, through a consideration of the creation (vv.19-20). These associated areas of human depravity are important for locating Romans 1:26-27 within its rhetorical tradition, which will be discussed below.

An Alternative Understanding of Romans 1:26-27

Returning to 1:18, the section (vv. 18-32) leads with an

---

[26] Käsemann, *Romans*, 38. For this passage, Käsemann (37) references 1 Enoch 91:7.
[27] No "Gentiles"; cf. Brooten, *Love between Women*, 205; contra Käsemann, *Romans*, 33.

apocalyptic pronouncement: "For the wrath of God is revealed from heaven against all ungodliness and wickedness of those who by their wickedness suppress the truth." Brooten asks an important question concerning the language of Romans 1:18:

> How is God's wrath revealed from heaven? There has been no *earthly catastrophe*, such as a devastating earthquake or *the destruction of a city* or enslavement of a people that might be called a sign of God's wrath. Paul may be working within the framework of the Jewish apocalyptic movement, which speculated about the coming wrath of God. A more plausible explanation, however, is that God's wrath is revealed through the human behavior described in vv. 24-32. This would mean that God's handing idolaters over to wretched deeds is the revelation of God's wrath against those who suppress the truth.[28]

Brooten here well notes what we should expect from Paul's apocalyptic language. This paper argues that Paul, in fact, *is* referencing a (super)natural catastrophe and the destruction of a city, for Romans 1:26-27 is an apocalyptic recollection motif involving the antediluvian world and the Flood on the one hand (v. 26), and Sodom and its destruction on the other (v. 27). Both the antediluvian world and the Cities of the Plain were judged from heaven, both spiritually and literally. These stories are recounted in Genesis 6 and Genesis 19, respectively. The primary actors in Genesis 6 are angelic beings and women, and the primary actors in Genesis 19 are angelic beings and men, and as will be shown presently, when the *universal* destruction of wicked humanity is the issue, these two particular biblical events are often provided as examples.

Beyond recognizing the events referenced in vv. 26-27, a rhetorical pattern does exist to locate the concepts in these verses, and that context involves apocalyptic and/or sapiential conceptualization and indictment that highlights human sexual deviance, cognitive deterioration/misdirection, and idolatry. At times, one might also find a catalogue of vices (vv. 29-31). Indeed, this pattern is known to interpreters of Romans 1, as comparisons with Wisdom 12-14 are often presented (e.g., Sanday and Headlam's well-

---

[28] Brooten, *Love between Women*, 221, italics added. Käsemann also suggests, "Our starting point should be that the manifestation of this wrath is described in vv. 24ff." (*Romans*, 37).

known chart comparing particular similarities, esp. with Wis 13-14).[29]

The Antediluvian World Judgment and Sodom Judgment as a Joint Rhetorical Topos

A strong tradition exists in ancient Jewish and Christian literature wherein the judgment stories of Genesis 6 and 19 are often joined as dual rhetorical examples of God's judgment. This tradition is well presented by, *inter alia*, Lührmann in his notable work on Q,[30] who writes,

> [T]here is a fixed tradition of the connection of the Flood—sometimes as the judgment of the fallen Angels—and the destruction of Sodom; both are primarily understood as examples of the punishment of the ungodly, and only secondarily, but then never without the first meaning present, as examples of the salvation of the godly.[31]

This paper will note examples of this tradition and will compare elements of Romans 1:18-32 in order to demonstrate how well Romans 1:26-27 can be understood as participating in this tradition.[32]

Foundational Texts

The foundation of the double tradition is two passages from

---

[29] William Sanday and Arthur Headlam, *A Critical and Exegetical Commentary on the Epistle to the Romans*, ICC (New York: Scribner's, 1902, 1968), 51-52.

[30] Dieter Lührmann, "Exkurs: Noah und Lot (Lk 17,26-29)," in *Die Redaktion der Logienquelle*, WMANT 33 (Neukirchener Verlag, 1969), 75-83. This "common tradition" is briefly noted by Neyrey, who writes, "In varying ways, each document cites a list of examples from the Pentateuch, always in the same order in which they occur in the Scriptures, but differing in function." Jerome H. Neyrey, *2 Peter, Jude: A New Translation with Introduction and Commentary*, AB 37 (New York: Doubleday, 1993), 59.

[31] Lührmann, *Redaktion*, 82 (my translation).

[32] A heuristic approach that helps in determining *intentional similarity* when comparing texts can be found in Dennis R. MacDonald's work on mimesis in the New Testament. This approach involves assessing seven important criteria, namely, "accessibility" (including popularity of the text under consideration), "analogy" (the text's use by others), "density" (number of parallels), "sequence," "distinctive traits," and "interpretability" (*Does the New Testament Imitate Homer? Four Cases from the Acts of the Apostles* [New Haven: Yale University Press, 2003], 2-6).

Genesis and the expansion of Genesis' antediluvian story in 1 Enoch, and these will now be recounted, with particular LXX/Greek expressions given in brackets.

Genesis 6:1–4: The Judgment of the Antediluvian World

> When people began to multiply on the face of the ground, and daughters [θυγατέρες] were born to them, the sons of God [οἱ υἱοὶ τοῦ θεοῦ] saw that they [τὰς θυγατέρας τῶν ἀνθρώπων] were fair; and they took wives [γυναῖκας] for themselves of all that they chose. Then the LORD said, "My spirit shall not abide in mortals forever, for they are flesh; their days shall be one hundred twenty years." The Nephilim [γίγαντες] were on the earth in those days—and also afterward[33]—when the sons of God went in to the daughters of humans, who bore children to them. These were the heroes [γίγαντες] that were of old, warriors of renown.

Where the critical edition of the Septuagint[34] (LXX) reads "sons of God," some other ancient variations of the LXX,[35] as well as other ancient sources utilizing a Greek text of Genesis 6, read "angels,"[36] and the consensus of interpreters is that angelic beings are involved. The LXX notes that these angelic "sons of God"[37] married the daughters of humans, resulting in (implied) offspring labeled "giants" in Greek (the *nephilim* in Hebrew), and this is the tradition known in the first century CE. The mixing of women and angelic beings is usually understood as contributing to the chaos that eventually resulted in judgment of the Flood.[38]

---

[33] See Ronald S. Hendel, "When the Sons of God Cavorted with the Daughters of Men," in *Understanding the Dead Sea Scrolls: A Reader from the Biblical Archaeology Review*, ed. Hershel Shanks (New York: Vintage, 1993), 173.

[34] All citations of the Septuagint (LXX) are from *Septuaginta*, Alfred Rahlfs, ed., 2nd rev. ed., Robert Hanhart, ed. (Stuttgart: Deutsche Bibelgesellschaft, 2006).

[35] E.g., Codex Alexandrinus.

[36] E.g., Philo, *De gigantibus*, 6.

[37] Nephilim (נפילים), which in Hebrew means "the fallen ones"; see Peter W. Coxon, "Nephilim," *DDD*, 618–620.

[38] Hendel notes the mythic-literary connection between the story of the union of angels and women, and the ensuing flood: "Where, then, did Gen 6:1–4 come from? I submit that the story of the mingling of gods and mortals and the procreation of the demigods was originally connected to the flood narrative and functioned as its motivation. The Yahwist detached the story of the demigods from the myth of the deluge in order to preface the flood with a more purely

1 Enoch: Expansion of the Genesis Story
For first century CE studies, the antediluvian world as communicated in 1 Enoch is essential. [39] This paper focuses on an early component of that work, namely, The Book of the Watchers (= 1 Enoch 1-36), especially chapters 1-19,[40] wherein "the Flood [is] a precursor to the judgment of human sinners" (esp. 1 Enoch 10-11)."[41] 1 Enoch was a popular work during the Second Temple period, based on seven significant fragments recovered at Qumran,[42] and the Greek fragments among Christians.[43] In 1 Enoch, the angelic beings who took human wives are often referred to as "Watchers" (Greek ἐγρήγοροι), and sometimes "the sons of heaven," even "holy watchers" (1 Enoch 15:9), and sometimes merely

---

ethical motive: Yahweh's anger at the evil behavior of humanity. This would explain why Gen 6:1-4 directly precedes the flood narrative, and, simultaneously, why it is unconnected from its context." Ronald S. Hendel, "Of Demigods and the Deluge: Toward an Interpretation of Genesis 6:1-4," *JBL* 106 (1987):16-17.

[39] Parts dating to as early as fourth century BCE (Frank Moore Cross, "Light on the Bible from the Dead Sea Caves," in *Understanding the Dead Sea Scrolls: A Reader from the Biblical Archaeology Review*, ed. Hershel Shanks [New York: Vintage, 1993], 164).

[40] 1 Enoch 6-11 is thought to be the most ancient level, based on its third person narrative (Annette Yoshiko Reed, *Fallen Angels and the History of Judaism and Christianity: The Reception of Enochic Literature* [Cambridge: Cambridge University Press, 2005], 25), but by New Testament times, all but perhaps the Similitudes (chapters 37-71) were available in redacted form. Seven copies of the Book of the Watchers were found at Qumran (George W. E. Nickelsburg, *1 Enoch 1: A Commentary on the Book of 1 Enoch, Chapters 1-36; 81-108* [Hermeneia. Minneapolis: Fortress Press, 2001], 9-10). Aside from the passages quoted in full in this paper, the incident is also mentioned in Jub. 4:15 and Bar 3:26-28. Black would limit the Book of the Watchers to just chaps. 1-16, and there is little doubt that a real shift occurs with chapter 17 (Matthew Black, *The Book of Enoch or I Enoch: A New English Edition*, SVTP [Leiden: Brill, 1985], 10).

[41] Reed, *Fallen Angels*, 72.

[42] Florentino García Martínez, *The Dead Sea Scrolls Translated: The Qumran Texts in English*, 2nd ed. (Leiden: Brill; Grand Rapids: Eerdmans, 1996), 246-259. Also found were 5 fragments of the Book of Giants (ibid., 260-262) and four fragments of the Book of Noah (ibid., 263-264). Nickelsburg puts the number as eleven copies of 1 Enoch represented at Qumran. The Dead Sea Scrolls discovery indicates that the original language of the Book of the Watchers is Aramaic.

[43] Reed, *Fallen Angels*, 7. There are two important Greek texts that exist for the Book of the Watchers: (1) the Akhmim Manuscript (Codex Panopolitanus), covering chaps. 1:1-32:6a, dated to the 5th-6th centuries CE; and (2) excerpts contained in George Syncellus' *Chronographia*, written in the 9th century CE (Nickelsburg, *1 Enoch 1*, 12).

"angels."[44] The following is a representative account given in 1 Enoch (Greek text in brackets)[45]:

> 6:1In those days, when the children of man had multiplied, it happened that there were born unto them handsome and beautiful daughters [θυγατέρες]. 6:2And the angels, the children of heaven [οἱ ἄγγελοι υἱοί (sons) οὐρανοῦ], saw them and desired [ἐπεθύμησαν] them; and they said to one another, "Come, let us choose wives for ourselves from among the daughters of man and beget us children. . . . 7:1And they took wives unto themselves, and everyone (respectively) chose one woman for himself, and they began to go unto them [and became defiled by them (μιαίνεσθαι)].[46] And they taught them magical medicine, incantations, the cutting of roots, and taught them (about) plants. 7:2And the women became pregnant and gave birth to great giants [γίγαντας μεγάλους] whose heights were three hundred cubits. 7:3These (giants) consumed the produce of all the people until the people detested feeding them. 7:4So the giants turned against (the people) in order to eat them. . . . 8:1And Azaz'el taught the people (the art of) making swords and knives, and shields, and breastplates; and he showed to their chosen ones bracelets, decorations, . . . beautifying of the eyelids, all kinds of precious stones, . . . 8:2And there were many wicked [ἀσέβεια] ones and they committed adultery [ἐπόρνευσαν] and erred [ἀπεπλανήθησαν], and all their conduct became corrupt. . . . 8:4And (the people) cried and their voice reached unto heaven. (1 Enoch 6:1-2; 7:1-6; 8:1-2, 4; trans. Issac, *OT Pseudepigrapha* 1:15-16, bracketed material added)

The following is also indicative (God speaking, via Enoch, to the Watchers):

> For what reason have you abandoned the high, holy, and eternal heaven; and slept with women [γυναικῶν] and defiled [ἐμιάνθητε] yourselves with the daughters of the

---

[44] Cf. Nickelsburg's excursus on the Watchers in his commentary (*1 Enoch 1*, 140-141). A notable variant description is in the Akhmim MS, wherein the Watchers are called "Titans" (1 Enoch 9:9).

[45] The Greek text of 1 Enoch utilzed, unless otherwise indicated, is from Matthew Black, ed., *Apocalypsis Henochi Graece*, PVTG 3 (Leiden: Brill, 1970).

[46] The bracketed material is demanded by the Greek, and thus is here inserted and translated.

people, taking wives, acting like the children of the earth, and begetting giant sons [υἱοὺς γίγαντας]? (1 Enoch 15:3, ibid., bracketed material added)

Reed has well noted that 1 Enoch adds an important element to the antediluvian story, namely, the communication of secret knowledge to humans by the angelic beings.[47] She writes, "The motif of illicit angelic instruction is central to the Book of the Watchers, shaping its unique approach to issues such as the origins of evil and the limits of human knowledge"[48] She finds a schema of "sex, sin and instruction," or even "sexual impurity, knowledge and violence."[49]

Genesis 19: Lot at Sodom[50]

> [1] The two angels [ἄγγελοι] came to Sodom in the evening, and Lot was sitting in the gateway of Sodom. When Lot saw them, he rose to meet them, and bowed down with his face to the ground. [2] He said, "Please, my lords, turn aside to your servant's house and spend the night, and wash your feet; then you can rise early and go on your way." They said, "No; we will spend the night in the square." [3] But he urged them strongly; so they turned aside to him and entered his house; and he made them a feast, and baked unleavened bread, and they ate. [4] But before they lay down [κοιμηθῆναι], the men of the city, the men of Sodom, both young and old, all the people to the last man, surrounded the house; [5] and they called to Lot, "Where are the men [οἱ ἄνδρες] who came to you tonight? Bring them out to us, so that we may know them [συγγενώμεθα]." [6] Lot went out of the door to the men, shut the door after him, [7] and said, "I beg you, my brothers, do not act so wickedly [μὴ πονηρεύσησθε]. [8] Look, I have two daughters [θυγατέρες] who have not known a man; let me bring them out to you, and do to them as you please [καὶ χρήσασθε[51] αὐταῖς]; only do nothing [wicked] to these men [μὴ ποιήσητε μηδὲν ἄδικον], for they have come under the shelter of my roof." [9] But they replied, "Stand

---

[47] Reed titles one of her sections "Sex, sin and instruction in *1 En.* 6–11" (p. 27).
[48] Reed, *Fallen Angels*, 6.
[49] See her headings at pp. 27 and 30.
[50] Sodom is mentioned 36 times in the Bible, 16 times alone. It was one of the Five Cities of the Plain.
[51] Cf. Romans 1:26 (χρῆσιν) and v. 27 (χρῆσιν).

back!" And they said, "This fellow came here as an alien, and he would play the judge [μὴ καὶ κρίσιν κρίνειν?]! Now we will deal worse [κακώσομεν] with you than with them." Then they pressed hard against the man Lot, and came near the door to break it down. ¹⁰ But the men inside reached out their hands and brought Lot into the house with them, and shut the door. ¹¹ And they struck with blindness [ἐπάταξαν ἀορασίᾳ] the men who were at the door of the house, both small and great [ἀπὸ μικροῦ ἕως μεγάλου], so that they were unable to find the door.
¹² Then the men said to Lot, "Have you anyone else here? Sons-in-law, sons, daughters, or anyone you have in the city—bring them out of the place. ¹³ For we are about to destroy [ἀπόλλυμεν] this place, because the outcry [ἡ κραυγὴ] against its people has become great before the LORD, and the LORD has sent us to destroy [ἀπέστειλεν] it." ²⁴ Then the LORD rained [ἔβρεξεν] on Sodom and Gomorrah sulfur and fire from the LORD out of heaven [ἐκ τοῦ οὐρανοῦ]; ²⁵ and he overthrew those cities, and all [πᾶσαν] the Plain, and all [πάντας] the inhabitants of the cities, and [all πάντα] what grew on the ground. ²⁶ But Lot's wife, behind him, looked back, and she became a pillar of salt.⁵²

As with the antediluvian story, angelic beings are involved. 2 Peter 2 and Jude emphasize that the great offense was not honoring the angelic beings. As to judgment, Philo considered the activity of Sodom to be something with universal repercussions, that they were "corrupting in this way *the whole race of man*, as far as depended on them" (*Abr*. 136 [Yonge], italics added).

Comparative Passages Involving the Dual-judgment Topos

This paper will now begin to review certain biblical and extrabiblical passages that incorporate the rhetorical use of the Antediluvian-Flood and Sodom stories. Comparisons with the text of Romans 1:18-32 will be made along the way, forming a cumulative argument for interpreting Romans 1:26-27 along the lines of the two exemplar universal judgment stories. For initial comparative purposes, an extended portion of Romans 1 is now given.

Romans 1:24-27

---

⁵² More on Sodom: Ezek 16:49-50; Zeph 2:9.

²⁴ Therefore God gave them up [παρέδωκεν] in the lusts [ἐπιθυμίαις] of their hearts to impurity, to the degrading of their bodies among themselves, ²⁵because they exchanged [μετήλλαξαν] the truth about God for a lie and worshiped and served the creature rather than the Creator, who is blessed forever! Amen.
²⁶ For this reason God gave them up [παρέδωκεν] to degrading passions [πάθη ἀτιμίας]. Their women [θήλειαι αὐτῶν, "their females"] exchanged [μετήλλαξαν] natural intercourse [φυσικὴν χρῆσιν] for unnatural [παρὰ φύσιν, "against nature"].[53] ²⁷ And in the same way [ὁμοίως] also the men [ἄρσενες, "males"], giving up [ἀφέντες] natural intercourse with women [τὴν φυσικὴν χρῆσιν τῆς θηλείας], were consumed with passion for one another. Men committed shameless acts with men and received [ἀπολαμβάνοντες] in their own persons the due penalty [ἀντιμισθίαν] for their error [πλάνης[54]].

Jude 5–7

⁵ Now I desire to remind you, though you are fully informed, that the Lord, who once for all saved a people out of the land of Egypt, afterward destroyed [ἀπώλεσεν] those who did not believe. ⁶ And the angels [ἀγγέλους] who did not keep their own position, but left their proper dwelling, he has kept in eternal chains in deepest darkness for the judgment of the great day. ⁷ Likewise [ὡς], Sodom and Gomorrah and the surrounding cities, which, in the same manner as [τὸν ὅμοιον τρόπον][55] they, indulged in sexual immorality [ἐκπορνεύσασαι] and pursued unnatural lust [σαρκὸς ἑτέρας, "other flesh"], serve as an example by undergoing a punishment of eternal fire.
⁸ Yet in the same way [ὁμοίως] these dreamers also defile [μιαίνουσιν] the flesh, reject authority, and slander the glorious ones. . . . ¹⁰ But these people slander whatever they do not understand, and they are destroyed by those things that, like irrational animals, they know by in-

---

[53] On the expression παρὰ φύσιν and its complexity in interpretation, see Hays, "Relations Natural and Unnatural," 197–199.

[54] Πλάνη is found in the judgment passages in Jude and 2 Peter (2 Peter 2:18; Jude 11).

[55] The expression is an adverbial accusative (and solecism). See BDF § 160; BADG, 827: "In the same way."

stinct. ¹¹ Woe to them! For they go the way of Cain, and abandon themselves to Balaam's error [πλάνῃ] for the sake of gain [μισθοῦ], and perish in Korah's rebellion.

Here in Jude the two Genesis judgment events are linked. Important terms to note in this passage are ὁμοίως/ὅμοιος, the former found in Romans 1:27. Also noteworthy is the term μιαίνουσιν ("they defile"), a rare verb in the NT, found also in 1 Enoch 15:3,4.[56] Here it is used of humans; in 1 Enoch it is used of the Watchers. The nominal form of the word is found in 2 Peter's account of the antediluvian world, which will be covered below.[57]

Reflecting on the rhetoric of the passage, Reed writes, "Jude assumes that his audience knows the story of the Watchers' sins and punishment, and he cites aspects of the angelic descent myth drawn from *1 En.* 6–16 (BW), while alluding in no way to Gen 6:1–4."[58] Indeed, 1 Enoch is invoked in verse 14, and these two ancient stories, along with the account of the Exodus generation, constitute "exemplars of the punished wicked."[59] Neyrey understands Jude's use of these two judgment stories as participating in a "common tradition." In his commentary on Jude's use of the "tradition," he writes:

> In varying ways, each document [exhibiting the tradition] cites a list of examples from the Pentateuch, always in the same order in which they occur in the Scriptures, but differing in function (Bauckham, *Jude, 1 Peter*, 46). The Damascus Document and *T. Naphtali* both warn against straying from the tradition, whereas 3 Macc. cites these examples against those who profane the temple; the Mishnah simply lists those "who will have no share in the world to come." Jude is not dependent on any one of these documents, but reflects the common tradition which he redacts for his own purposes.[60]

Further reflecting on ὁμοίως in the Jude passage (v. 8), Banister notes that ὁμοίως here is not essential in understanding the ante-

---

[56] Though the term μιαίνουσιν receives its own section in his commentary (*2 Peter, Jude*, 67–68), Neyrey makes no reference to 1 Enoch concerning this term.

[57] Cf. μιαίνεσθαι (1 En. 7:1, cited above), μιασμοῦ (2 Pet 2:10), μιάσματα (2 Pet 2:20), and μιαίνουσιν (Jude 8).

[58] Reed, *Fallen Angels*, 106.

[59] Reed, *Fallen Angels*, 104.

[60] Neyrey, *2 Peter, Jude*, 59–60. The only correction to be made here is that T. Naph. has the order of the two stories reversed.

cedent clause, and "the passage could also be understood as containing three separate, albeit similar, examples of various groups of beings who indulged in sexually immoral activity."[61] Yes, yet beyond this, what should be further noted about the Jude passage is the use of τὸν ὅμοιον τρόπον in Jude 7, which is functionally equivalent to ὁμοίως in linking the two judgments, especially in conjunction with the preceding ὡς ("Likewise").

2 Peter 2:4–10

> [4] For if God did not spare [οὐκ ἐφείσατο] the angels [ἀγγέλων] when they sinned, but cast them into hell [ζόφου ταρταρώσας] and committed [παρέδωκεν] them to chains of deepest darkness to be kept until the judgment; [5] and if he did not spare the ancient world, even though he saved Noah, a herald of righteousness, with seven others, when he brought a flood on a world of the ungodly [ἀσεβῶν]; [6] and if by turning the cities of Sodom and Gomorrah to ashes he condemned them to extinction and made them an example of what is coming to the ungodly [ἀσεβεῖν]; [7] and if he rescued Lot, a righteous man greatly distressed by the licentiousness of the lawless [8] (for that righteous man, living among them day after day, was tormented in his righteous soul by their lawless deeds that he saw and heard), [9] then the Lord knows how to rescue the godly from trial, and to keep the unrighteous under punishment until the day of judgment [10] —especially those who indulge their flesh in depraved lust [ἐπιθυμίᾳ μιασμοῦ], and who despise authority. Bold and willful, they are not afraid to slander the glorious ones.

Though 2 Peter is often thought dependent on Jude, 2 Peter expands the complexity of the two-judgment rhetorical motif. Interestingly, Reed observes that "2 Peter thus draws an even closer connection between the Watchers and the Sodomites, presenting them as twin paradigms of the sexually impure. . . . In this case, what proves significant is that 2 Peter assumes this knowledge on the part of his audience, implying but not explaining what these angels share with the inhabitants of Sodom."[62] Lexically of interest are the two terms comprising the expression ἐπιθυμίᾳ μιασμοῦ in v.

---

[61] Banister, "Ὁμοίως," 583–584.
[62] Reed, "Fallen Angels," 107.

10, which are found as verbs in 1 Enoch 15:3, 4, and the former draws Romans 1 into the interpretive network.[63]

While this passage is often thought to be derived from Jude, significant differences must be noted, as the author of 2 Peter has adapted the tradition for his own purposes. Ὁμοίως is absent, and the author has jettisoned the prophet Enoch and the disobedient Children of Israel example. The passage incorporates the punitive language found verbatim in Sirach 16:7-9 (to be examined below).[64] The language concerning improper sexual activity remains for the Sodom example, and also appears in the restatement of the antediluvian situation in chapter 3 of 2 Peter, playing on a "water and fire" judgment theme:

> ³First of all you must understand this, that in the last days scoffers will come, scoffing and indulging their own lusts [ἐπιθυμίας] ⁴ and saying, "Where is the promise of his coming? For ever since our ancestors died, all things continue as they were from the beginning of creation!" ⁵ They deliberately ignore this fact, that by the word of God heavens existed long ago and an earth was formed out of water and by means of water, ⁶ through which the world of that time was deluged with water and perished. ⁷ But by the same word the present heavens and earth have been reserved for fire, being kept until the day of judgment and destruction of the godless.

It should be noted that both passages, Jude and 2 Peter, involve angelic beings. Indeed, part of the condemnation of humans in these two passages is that the humans do not "honor" or "respect" the glorious ones, that is, the angels (Jude 8; 2 Peter 2:10). This is a peculiarity of these two passages, which is not out of place with the Genesis accounts.

---

[63] Cf. ἐπιθυμίαις (Rom 1:24), ἐπιθυμίᾳ (2 Pet 2:10), ἐπεθυμήσατε (1 En. 15:4, quoted above); ἐπεθύμησαν (1 En. 6:2, quoted above), from noun ἐπιθυμία and verb ἐπιθυμέω. For μιασμός; cf. μιασμοῦ (2 Pet 2:10), μιάσματα (2 Pet 2:20), μιαίνουσιν (Jude 1:8) ἐμιάνθητε (1 Enoch 15:3, 4, partially quoted above), and μιαίνεσθαι (1 En. 7:1, quoted above), from noun μίασμα and verb μιαίνω (LSJ: aorist ἐμιάνθην).

[64] Neyrey, *2 Peter, Jude*, 196-197, noting "the author [of 2 Pet] taps into a tradition which appears to be a commonplace in antiquity" (196).

Luke [Q] 17:26-29 // Matthew 24:37-39[65]

> Just as it was in the days of Noah, so too it will be in the days of the Son of Man. They were eating and drinking, and marrying and being given in marriage, until the day Noah entered the ark, and the flood came and destroyed all of them [ἀπώλεσεν πάντας]. Likewise [ὁμοίως],[66] just as it was in the days of Lot: they were eating and drinking, buying and selling, planting and building, but on the day that Lot left Sodom, it rained fire and sulfur from heaven [ἀπ' οὐρανοῦ] and destroyed all of them [ἀπώλεσεν πάντας] — it will be like that on the day that the Son of Man is revealed. (Luke 17:26-29)

> For as the days of Noah were, so will be the coming [παρουσία] of the Son of Man. For as in those days before the flood they were eating and drinking, marrying and giving in marriage, until the day Noah entered the ark, and they knew nothing until the flood came and swept them all away, so too will be the coming [παρουσία] of the Son of Man. (Matthew 24:37-39)

According to Lührmann, Matthew has the original Q reading, which lacked a reference to Lot.[67] Kloppenborg, on the other hand, argues that Matthew eliminated the Lot references for his own purposes,[68] and that the Lot reference was part of the Q tradition to some extent.[69] Regardless of Q's original wording, the Noah-Lot connection exhibited in Luke clearly represents the dual-judgment motif, and one notes the utilization of ὁμοίως, once again.

---

[65] Especially analyzed at Lührmann, *Redaktion*, 71-75.
[66] BAGD on Luke 17:28: "in the same way as," taking into consideration καθώς.
[67] Lührmann, *Redaktion*, 82-83.
[68] John Kloppenborg, *The Formation of Q: Trajectories in Ancient Wisdom Tradition*, SAC (Philadelphia: Fortress Press, 1987), 158.
[69] Lührmann was one of the first to elucidate the antediluvian world-Sodom tradition: "Lührmann (*Redaktion*, 75-83) has shown that there existed an established tradition which associated the flood generation with the contemporaries of Lot, and used both as *examples of divine punishment of the wicked*" (Kloppenborg, *Formation of Q*, 158 n. 246, italics added). As to the place of the Noah-Lot tradition in the Q tradition, Kloppenborg writes, "Lührmann himself argues that this tradition influenced a pre-Lucan expansion of Q, but this tradition could just as easily have influenced Q itself" (158 n. 246). Indeed, the destruction-of-all repetition in Luke is in line with the dual-judgment motif, and thus perhaps is a further indication that the dual-judgment motif influenced the Q statement.

Sirach 16:7-8 (part of larger context 16:5-10)[70]

> He did not forgive [οὐκ ἐξιλάσατο] the ancient giants [ἀρχαίων γιγάντων] who revolted in their might. He did not spare [οὐκ ἐφείσατο] the neighbors of Lot, whom he loathed on account of their arrogance.

Here the emphasis is on the offspring of the angels and women, as well as on Sodom, and "wrath" is mentioned in verse 11. The note concerning "He did not spare" is the same expression found in 2 Peter 2:4, 5, which demonstrates that 2 Peter is not drawing on Jude in that document's exhortation.[71] Along this line, in this passage from Sirach the neighbors of Lot are arrogant [ὑπερηφανίαν], while in Wisdom 14:6 it is the giants who are arrogant [ὑπερηφάνων γιγάντων]. Subtle variations can be at work.

Wisdom of Solomon 10:3-6, 8, 9[72]

> But when an unrighteous man departed from her in his anger, he perished because in rage he killed his brother. When the earth was flooded because of him, wisdom again saved it, steering the righteous man by a paltry piece of wood. Wisdom also, when the nations in wicked agreement had been put to confusion, recognized the righteous man and preserved him blameless before God, and kept him strong in the face of his compassion for his child. . . . Wisdom rescued a righteous man when the ungodly were perishing; he escaped the fire that descended on the Five Cities. . . . . [T]hey . . . were hindered from recognizing [γνῶναι] the good, . . . Wisdom rescued from troubles those who served her.

3 Maccabees 2:3-5[73]

> For you, the creator of all things and the governor of all, are a just Ruler, and you judge those who have done anything in insolence and arrogance. You destroyed those who in the past committed injustice [ἀδικίαν ποιήσαντας], among whom were even giants [γίγαντες] who trusted in

---

[70] Lührmann, *Redaktion*, 76.

[71] On the non-forgiveness of the Watchers or giants (they are confused at times), see Reed, *Fallen Angels*, 69-70 and n. 52, who references 1 Enoch 13:4; 14:6-7; and 15:3.

[72] Lührmann, *Redaktion*, 78.

[73] Lührmann, *Redaktion*, 78.

their strength and boldness, whom you destroyed by bringing on them a boundless flood [ὕδωρ]. You consumed with fire and sulfur the people of Sodom who acted arrogantly, who were notorious for their vices; and you made them an example to those who should come afterward.

Jubilees 20:5[74]

And he [Abraham] told them the judgment of the giants and the judgments of the Sodomites just as they had been judged on account of their evil. And on account of their fornication and impurity and the corruption among themselves with fornication they died. (trans. Wintermute, *OT Pseudipigrapha* 2:94, brackets added)

In the Wisdom passage, the two events are included within a list of judgments. The Jubilees passage unites the two great Genesis judgments as a unit. 3 Maccabees cites the two judgments together, and then adds the Exodus judgments on the Egyptians.

Brief excursus: The Women before the Men in Order
All of the preceding manifestations of the dual-judgment motif help to understand a particular issue with Romans 1:26-27. Commentators on Romans 1:26-27 are puzzled why the women are mentioned first.[75] Noting that Paul's term "female" references gender differences in creation (Gen 1:27; Gal 3:28, with female mentioned after male), Jewett offers that "[i]n contrast to these and other examples, however, relations between females are mentioned first by Paul, probably for rhetorical reasons."[76] Of course, it is then anyone's guess as to the rhetorical point being scored by mentioning the females first in Romans 1:26-27. However, the above examples of the tradition suggest an answer – the order of women first makes sense in the antediluvian world–Sodom rhetorical tradition, as this is the order of events. Genesis 6 material chronologically and literarily precedes Genesis 19 material, and

---

[74] Lührmann, *Redaktion*, 77. Jubilees also covers the Watchers–Flood story at 4:22-25; 5:1-11, 20-32; 7:21-23; 10:1-6, and the destruction of Sodom and Gomorrah is also covered at 16:5-6. No Greek fragments of either story survive.

[75] Jewett, *Romans*, 174; Cranfield, *Romans*, 1:125; Brooten, *Love between Women*, 240 n. 73.

[76] Jewett, *Romans*, 174, who appears to favor a view that finds "shock" value in placing the women first (n. 107).

thus most passages utilizing the dual tradition retain this sequence.

Testament of Naphtali 3:4–5: A Striking Parallel[77]

> In the firmament, in the earth, and in the sea, in all the products of his workmanship discern the Lord who made all things, so that you do not become like Sodom, which departed from the order of nature. Likewise [Ὁμοίως] the Watchers departed from nature's order; the Lord pronounced a curse on them at the Flood. (trans. Kee, *OT Pseudepigrapha* 1:812, brackets added)

This passage from the Testament of Naphtali is one of the more striking parallels to Romans 1:26–27, structurally and linguistically. Indeed, the fuller context of the passage is invoked by Jewett to elucidate Romans 1:26,[78] especially as a conceptual parallel of humans "exchanging" the truth of God for idolatry in Romans 1:23. This "exchange" motif can be illustrated with Nickelsburg's translation of T. Naph. 3 in his commentary on 1 Enoch, which is quite thought-provoking, as Nickelsburg approaches the text from a completely different direction than Romans 1. In this passage, Nickelsburg's own Greek term highlights are in parentheses, while bracketed material is added to draw further connections with Romans 1 (his introduction to the text is included):

> This text [i.e., T. Naph. 3] —Christian in its present form — even cites an Enochic source; its very close parallels to 1 Enoch 2–5 may reflect knowledge of that text.
>
> Sun and moon and stars do not change their order (οὐκ ἀλλοιοῦσιν τάξιν αὐτῶν). Thus also you do not change (ἀλλοιώσητε) the law of God in the disorder of your deed (ἐν πράξεων ὑμῶν). The Gentiles who went astray and forsook (πλανηθέντα καὶ ἀφέντα) the Lord changed their order (ἠλλοίωσαν τάξιν αὐτῶν) and followed after stones and rocks [ξύλοις[79] καὶ λίθοις; cf. Wis 14:21], led astray by spirits of error [πλάνης]. But you (do) not (be) so, my children, since you recognize [γινώσκοντες] in the firmament, on earth, and in the sea and in all created things the Lord who made all these things, lest you be like Sodom, which

---

[77] Lührmann, *Redaktion*, 77.
[78] Jewett, *Romans*, 174, who also points to γενεσέως ἐναλλάγη at Wis 14:26 (174–175).
[79] ξύλον usually means "wood."

changed the order (ἐνήλλαξε τάξιν) of its nature [φύσεως]. Likewise ['Ομοίως] also, the watchers ['Εγρήγοροι] changed the order (ἐνήλλαξαν τάξιν) of their nature [φύσεως]. . . . I have read in the sacred writing of Enoch that you too will turn aside (ἀποστήσεσθε) from the Lord [4:1].[80]

Significant linguistic parallels with Romans 1:23-28 are apparent.[81] Πλάνη (Rom 1:28) is found here, as well as ἀφίημι (Rom 1:27), ὁμοίως (Rom 1:27), φύσις (Rom 1:26; φυσικός is used in vv. 26, 27),[82] the important verb ἐναλλάσσω which brings to mind ἀλλάσσω (Rom 1:23) and μεταλλάσσω (Rom 1:25, 26) and γιγνώσκω for ἐπίγνωσις (Rom 1:28). Käsemann comments on the similarity of language between Romans 1:26-27 and Testament of Naphtali 3:2-5: "*T. Naph.* 3:2-5 in its condemnation even uses the Pauline idea of exchanging the divine for the earthly."[83] Testament of Naphtali 3:2-5 and Romans 1:26-27 are, linguistically and interpretively, close. Nickelsburg comes at the Testament of Naphtali passage from an interest in 1 Enoch, for another term for changing, ἀλλάσσω, is found in 1 Enoch 2:1-2, and it is used in a context of understanding the unchanging order (τάξις) of the cosmos.

Miller also compares Romans 1:23-27 with Testament of Naphtali, noting a parallel between the use of the "rare term" μεταλλάσσω in v. 26 and ἀλλάσσω in v.23, and ἐναλλάσσω in Wisdom 14:26 and Testament of Naphtali 3:4-5,[84] the latter especially significant:

---

[80] Nickelsburg, *1 Enoch 1*, 153-154, parenthetical text in the original, and bracketed text added. Bracketed Greek text from R.H. Charles, *The Greek Versions of the Testaments of the Twelve Patriarchs. Edited from Nine MSS., together with the Varients of the Armenian and Slavonic Versions and Some Hebrew Fragments* (Oxford: Clarendon Press, 1908).

[81] Brinkschröder also notes a number of lexical similarities (*Sodom als Symptom*, 519-521).

[82] On Paul's use of φύσις here, Fitzmyer writes, "The noun *physis*, 'nature,' which occurs in the OT only in the Greek deuterocanonical books (Wis 7:20; 13:1; 19:20; cf. 3 Macc 3:29; 4 Macc [various places] . . . ; T. Naph. 3:4-5), occurs here for the first time. Paul will use it again in 2:14, 27; 11:21, 24" (*Romans*, 286, brackets added). He also uses it at 1 Cor 11:14.

[83] Käsemann, *Romans*, 40.

[84] Cranfield references the usual comparisons with Ps 106:20 (LXX) and Jer 2:11, and also Deut 4:16-18, apparently noting the "exchange" (ἤλλαξαν) at v. 23 (*Romans*, 119). He also finds that "μετήλλαξαν echoes v. 25: the thought would seem to be present that there is a correspondence between this exchange and that described in v. 25" (1:125).

Μεταλλάσσω is a rare term, and is used for sexual perversion only in Romans 1. In v. 26 it describes the perversion as a result of the μεταλλάσσω of worship mentioned in v. 25, itself an intensification of ἀλλάσσω in v. 23. As the gentiles perverted their worship with idolatry, so also was their sexual practice perverted. Outside Romans the related term ἐναλλάσσω is more common for sexual perversion. *The best parallels* are in Wisdom 14:26 and Testament of Naphtali 3:4-5. Though Wisdom 14 parallels Romans 1, ἐναλλάσσω in Wisdom 14:26 *is not specific* as to the perversion intended. The most significant usage is Testament of Naphtali 3:4-5 where ἐναλλάσσω is used twice with τάξιν φύσεως ("natural order"), *connected by ὁμοίως* with specific contexts. In verse 4 the perversion of the natural order was done by Sodom, but in verse 5 the similar (ὁμοίως) perversion was done by the Watchers who transgressed the boundary between human and super-human beings by having (heterosexual) intercourse with human females. . . . As with Romans 1, the perversion of Test. Naphtali is linked to gentiles exchanging *idolatry* for true worship (3:3).[85]

What also stands out noticeably in Testament of Naphtali 3 is the use of the term ὁμοίως. This use of ὁμοίως is precisely the use of the synonymous τὸν ὅμοιον τρόπον found in Jude 7, namely, a comparison of Sodom with the antediluvian angels along the lines of unnatural sexual relations.[86] Thus, a rather noticeable hermeneutical network emerges involving Jude, 2 Peter 2, Luke 17 and T. Naph. 3. Romans 1:26-27 is similar to these passages in that all describe a two-event past example of God's judgment, all but one involving sexual misconduct, with the two events connected by the term ὁμοίως (2 Peter excepted).[87] Whereas Romans 1:26-27 remains rather vague as to the nature of the event in v. 26,[88] the ready appli-

---

[85] Miller, "Practices," 3-4, italics added.

[86] Ὁμοίως appears in Jude 8, bringing in the second comparison with the third, "contemporary" group.

[87] Indeed, ὁμοίως, as it applies to the dual-event judgments, appears to have become formulaic; it is the hinge on the diptych, the two primordial stories featured on either panel. Evidence for this is perhaps in the Q-saying (Luke 17:26-29), as Lührmann finds the ὁμοίως clause involving Lot a secondary addition; perhaps ὁμοίως functions as a connector for a rhetorically traditional element. As the times of Noah and Lot were combined into a known rhetorical *form*, ὁμοίως appears to be moving into the realm of being *formulaic*.

[88] Banister, e.g., notes the obvious absence of content as to the nature of the

cation of v. 27 to Sodom (to be examined below) in light of these other ὁμοίως passages points in the direction of v. 26 involving the antediluvian story. As Brinkschröder concludes,

> "Diese Übereinstimmungen mit TNaph begründen die These, dass Paulus in Röm 1,18-32 die Wächter- und Sodom-Eschatologie ebenfalls als imagenäres Substrat voraussetzt."[89]

Wisdom of Solomon 12-14 and a Widening Network of Meaning

Returning to Miller's examination of Romans 1:26-27 in light of Testament of Naphtali 3, not only does he find linguistic similarities and conceptual parallels in the two passages, but he notes that both passages involve the issue of idolatry. Indeed, a pattern of sexual deviation, cognitive misdirection and idolatry is the point of Romans 1:18-32, and, moving to another text, Wisdom of Solomon is frequently noted for parallels with Romans 1 along these lines, particularly Wisdom 12-14[90] (e.g., Wis 12:23-26; 14:12),[91] famously charted in Sanday and Headlam's commentary on Romans.[92] However, what is not noted by the commentaries, likely because the context of Romans 1:26 remains misinterpreted, is that this significant portion of Wisdom includes in its rhetorical statement concerning idolatrous human folly, notably, the example

---

"females" unnatural sexual (παρὰ φύσιν) participation, while the "male" activity is described ("Ὁμοίως," 571). "Apparently the female's sexual partner was obvious but the male specifically abandons the natural partner for an unnatural one" (Miller, "Practices," 2).

[89] Brinkschröder, *Sodom als Symptom*, 521.

[90] F. F. Bruce gives Wis 12-14 (*Romans: An Introduction and Commentary*, TNTC [Downers Grove, Ill.: IVP, 1963], 79), while NA[27] gives Wis 13-15 in the margin. For specific examples, cf. Sanday and Headlam, *Romans*, 51-52, and Stuhlmacher's parallels to what he calls "the apostle's indictment": Rom 1:19-20: Wis 13:1-9; Rom 1:22-23: Wis 11:15; Rom 1:24-32: Wis 14:12-14, 22-31 (Peter Stuhlmacher, *Paul's Letter to the Romans: A Commentary*, trans. Scott J. Hafemann (Louisville, KY: Westminster/John Knox, 1994), 34. Cf., e.g., Fitzmyer, *Romans*, 272, 274, 280, 281, references Wis 13:1-19 (esp. 13:8) and 14:22-31 (esp. 14:22), and also Wis 11:15-16; Brooten, *Love between Women*, 238 n. 67; Käsemann, *Romans*, 35. Also Anthony J. Guerra, *Romans and the Apologetic Tradition: The Purpose, Audience and Genre of Paul's Letter*, SNTSMS 81 (Cambridge: Cambridge University Press, 1995), p. 54, concerning Wis 14:12-31, and p. 56 for the catalogue of vices at Wis 14:23-30. Keck highlights Wis 13:1, 6-9 (Leander E. Keck, Romans, ANTC [Nashville: Abingdon, 2005], 61). Brinkschröder elucidates comparisons at *Sodom als Symptom*, 516-519.

[91] Fitzmyer brings in T. Naph. 3:4, as well (*Romans*, 272).

[92] Sanday and Headlam, *Romans*, 51-52.

of Noah, his wooden ark, and the Giants (14:1-6):

> For even in the beginning, when arrogant giants were perishing, the hope of the world took refuge on a raft, and guided by your hand left to the world the seed of a new generation. For blessed is the wood by which righteousness comes. (Wis 14:6)[93]

Thus Wisdom 12-14, usually invoked to provide interpretive assistance for Rom 1:18-32, also provides precedent for finding the antediluvian story in Romans 1.[94] It appears that Testament of Naphtali 3:2-5, Wisdom 12-14 and Romans 1:18-32 participate in a hermeneutical network of meaning with 1 Enoch 2-5.

Returning to Testament of Naphtali 3, Nickelsburg finds this passage to have "very close parallels to 1 Enoch 2-5" and "may reflect knowledge of that text."[95] Indeed, the appeal to the creation can be found in 1 Enoch 2-5, wherein the reason for appealing to the creation is to demonstrate the unwavering order of the created cosmos that God has made, being made manifest to humanity (2:2;

---

[93] This passage is also noted by Brinkschröder, *Sodom als Symptom*, 522.

[94] It should be noted that four times in Wis 12-13 the Creator is referred to as δεσπότης (Wis 12:16, 18; 13:3, 9), a term referring to the Creator in the Jude and 2 Peter passages under consideration, as well (Jude 4; 2 Peter 2:1). The title is also found at Barn. 4:3, also in connection with the Enoch story, as well as 1 Clem. 9.3-4; 11.1 when noting both Enoch and Lot.

[95] Nickelsburg, *1 Enoch 1*, 153. Nickelsburg finds the T. Naph. text to be "Christian in its present form," but it is not clear if he is referring to the whole book, or only to the passage quoted above. No fragments of this part of T. Naph. were found at Qumran for comparison's sake, though fragments of T. Naph. were found. Dating of the original Jewish document is c. early first century BCE (Charlesworth, *The Old Testament Pseudepigrapha and the New Testament: Prolegomena for the Study of Christian Origins*, SNTSMS 54 [Cambridge: Cambridge University Press, 1985], 153 n. 29). Kee concluded, in reference to the entire Testaments of the Twelve Patriarchs, that "The Christian interpolations, which number not more than twelve, and which occur in the latter part of those testaments that contain them, are conceptually peripheral to the main thrust of the document and are literarily incongruous, so that they may be readily differentiated from the original Greek text" (*Old Testament Pseudepigrapha*, 1:777). T. Naph. 3:2-5 would not appear to fit this criteria. Writes Charlesworth, "These observations lead to the conclusion that New Testament scholars should—indeed must—use 1 Enoch and the Testaments of the Twelve Patriarchs in examining the literary evidence of Early Judaism and its importance for reconstructing Christian origins.... It is now relatively certain that both 1 Enoch and the Testaments of the Twelve Patriarchs (in its original form) were documents used not only by Christian communities, but also by Palestinian pre-Christian Jewish groups" (*Old Testament Pseudepigrapha and the New Testament*, 40).

5:1). The appeal has the same result as Wisdom's and Romans' appeal, namely, the culpability and indictment of humanity, which is brought out in 1 Enoch 5:4-7 (aside from the "elect" in v. 7).

While both Nickelsburg and Matthew Black do not reference Wisdom's creation appeal in discussing the "nature-homily" of 1 Enoch 2-5,[96] two of the closest parallels they cite are both associated with the Flood-Sodom motif, as the appeal-to-creation of Testament of Naphtali 3:2-4 is integrated with the antediluvian and Sodom judgments of Testament of Naphtali 3:3-5, and the appeal-to-creation of Sirach 16:16-28 has the antediluvian and Sodom judgments at Sirach 16:7-8, the two passages separated by a section that deals with God's dual predisposition to mercy and to wrath (vv. 11b-14), which is similar to Romans 1:16-17. So, while Black and Nickelsburg find a creation-appeal in 1 Enoch 2-5 that is paralleled in Testament of Naphtali 3, those commenting on Romans find the sexual misbehavior-idolatry reported in Testament of Naphtali 3 a parallel to Wisdom and Romans passages. What is so well known (connections between Wisdom 12-14 and Roman 1 and T. Naph. and Sirach) is now connected to 1 Enoch 1-5. The greater hermeneutical network of a shared topos emerges. [97]

What has been considered in detail up to this point is that Romans 1:26-27 fits well, lexically, structurally and contextually, with the dual judgment rhetorical tradition.[98] This alignment points toward the identification of the activity performed by the females that was considered "against nature." As to sexual activity that is unnatural, Banister lists those given by Artimedorus (*Onir.* 1.80),

---

[96] See Matthew Black, *The Book of Enoch or I Enoch: A New English Edition*, SVTP (Leiden: Brill, 1985), 109; Nickelsburg, *1 Enoch 1*, 153-154.

[97] Brinkschröder, in his work, has also made certain connections between Romans 1:18-32, 1 Enoch 2-5, Wisdom 12-14 and T. Naph. 3, including the appeal to creation (Brinkschröder, *Sodom als Symptom*, 518-521).

[98] Other examples of the same dual judgment rhetorical topos include Philo, *Abr.* 135 (cited below), certain Sethian literature (*Apocalypse of Adam* V.69,2 - 71,8 [Noah] and 74,26 - 76,7 [Sodom], *Gospel of the Egyptians* III.60, 61, and *Paraphrase of Shem* VII.25 [the Flood] and VII.28-29 [Sodom and Gomorrah], the first two documents noted by Lührmann, *Redaktion*, 79 n. 1), and 4Q252 (CommGen A), which Eisenman and Wise note involves "two Genesis 'salvation of the Righteous' stories," noting similarity to CD 2.18-20 with its "allusion to the "'Heavenly Watchers' and the Noah story just as here" (Robert Eisenman and Michael Wise, *The Dead Sea Scrolls Uncovered: The First Complete Translation and Interpretation of 50 Key Documents Withheld for over 35 Years* [Shaftesbury, UK: Element, 1992], 81).

which includes "a human having sex with a god or goddess."⁹⁹ Indeed, Brooten has elaborated on this:

> Intercourse between the Watchers, who were sons of God, and human women transgressed the order of nature by crossing the boundary between the human and the divine. The Testament of Naphtali 3:4f constitutes a closer parallel to Artemidoros, *Oneirokritika* 1.80, than to Rom 1:26f. Artemidoros classifies both sexual relations between women and sexual relations between a human and a deity as unnatural.¹⁰⁰

This unnatural sexual activity would likely include intercourse with angels, and 1 Enoch clearly states that the angelic beings "defiled" themselves by having sexual intercourse with women (1 Enoch 7:1; 15:3; 9:8 [Syncellus]), and this tradition is maintained at Jubilees 7:21 (cf. Jude 8), activity that is not according to their nature, the Watchers (along with the city of Sodom) having left the natural order of their nature (T. Naph 3:5; ἐνήλλαξαν τάξιν φύσεως αὐτῶν). It follows that the women thus participated in forbidden relations, and this is the perspective that Romans 1:26-27 takes, in language quite reflective of Testament of Naphtali 3. This study will now focus on the gender language of Romans 1:26-27 to strengthen the identification of the two judgments referenced.

The Possible Origin of the Peculiar Gender Language of Romans 1:26-27

"Females"

It has been shown thus far that there is reason to identify the women in Romans 1:26 with the antediluvian women who married angels.¹⁰¹ It should now be noted that Romans 1:26-27 does not refer to women and men by the usual terms (γυνή and ἀνήρ), but by terms meaning "female" and "male" (θῆλυς and ἄρσην, respectively). This section will focus only on θῆλυς:

---

⁹⁹ Banister, "Ὁμοίως," 572 n. 4. This reference is also noted by Brooten, *Love between Women*, 249 n. 99.

¹⁰⁰ Brooten, *Love between Women*, 249 n. 99.

¹⁰¹ Focusing on the parallels in T. Naph. 3, Brinkschröder has well declared, "Wenn man den religionsgeschichtlichen Vergleich mit TNaph 3 als Schlüssel für die Interpretation von Röm 1,18–32 ernst nimmt, eröffnet dies ein neues Verständnis des widernatürlichen Verkehrs der Rauen aus Röm 1,26" (*Sodom als Symptom*, 522).

"For this reason God gave them up to degrading passions. Their women (θήλειαι) exchanged natural intercourse for unnatural" (Rom 1:26)

The word θῆλυς ("female") is found in the New Testament only at Matthew 19:4//Mark 10:6, Galatians 3:28 and Romans 1:26, 27. In the first three references, the term is used, along with ἄρσην ("male"), to designate gender differences in creation (cf. Gen 1:27; 5:2 LXX). Indeed, θῆλυς is not common in the LXX, but is used, e.g., when designating female animals (e.g., Gen 6:19; 7:16), sacrificial or otherwise, or, e.g., regulations concerning menses (Lev 12:5; 15:33). Thus, Paul's use of the term is significant due to its rarity in the NT. In v. 26, the term is in a passage that begins with the causal expression Διὰ τοῦτο ("Because of this"). This is Paul's only use of διὰ τοῦτο in the first three chapters of Romans. With this construction, the passage bears a notable lexical similarity to a rhetorical refrain in 1 Enoch:

> Διὰ τοῦτο παρέδωκεν αὐτοὺς ὁ θεὸς εἰς πάθη ἀτιμίας, αἵ τε γὰρ θήλειαι αὐτῶν μετήλλαξαν τὴν φυσικὴν χρῆσιν εἰς τὴν παρὰ φύσιν (Rom 1:26)
>
> διὰ τοῦτο ἔδωκα αὐτοῖς θηλείας, ἵνα σπερματίζουσιν εἰς αὐτὰς καὶ τεκνώσουσιν ἐν αὐταῖς τέκνα οὕτως, ἵνα μὴ ἐκλείπῃ αὐτοῖς πᾶν ἔργον ἐπὶ τῆς γῆς. (1 Enoch 15:5)
>
> "Therefore I gave them women ["females"], that they might cast seed into them, and thus beget children by them, that nothing fail them upon the earth." (1 Enoch 15:5; trans. Nickelsburg, 1 Enoch 1, 267, brackets added)

In this passage, God informs the Watchers that females were given to men so that humanity would not die out. Remarkably, the language of Romans 1:26 is found as a single unit in 1 Enoch. The only other use of διὰ τοῦτο in 1 Enoch is two verses later:

> καὶ διὰ τοῦτο οὐκ ἐποίησα ἐν ὑμῖν θηλείας· (1 Enoch 15:7).
>
> "Therefore I did not make women among you" (1 Enoch 15:7; trans. Nickelsburg, 1 Enoch 1, 267)

Here, as a refrain, God informs the Watchers that because they are eternal beings, females were not made for them. Thus, here are three refrains, one in Romans and two in 1 Enoch, that begin with the causal διὰ τοῦτο and concern "females" participating in unnatural sexual relations.¹⁰² Διὰ τοῦτο only appears here in all extant Greek 1 Enoch fragments, and θῆλυς appears only one more time in the Akhmim version of Greek 1 Enoch,¹⁰³ and that appearance is in the following chapter, in the penultimate statement of the tale of the Watchers, and highlights the culpability the "females" have in the decline of the times, due to their participation in the secrets taught to them by the Watchers:

> "And through this mystery the women [αἱ θήλειαι; "females"] and men [ἄνθρωποι] are multiplying [πληθύνουσιν] evils [τὰ κακὰ] upon the earth." (1 Enoch 16:3c; trans. Nickelsburg, *1 Enoch 1*, 267, bracketed material added)¹⁰⁴

With these rare linguistic and conceptual connections (one even notes that the "females" are *mentioned first* in 1 Enoch 16:3), it is probable that Paul's language in Romans 1:26 is influenced by the Greek version of 1 Enoch 15-16,¹⁰⁵ and further highlights that Romans 1:26 concerns the condition of the antediluvian world, which involves female culpability. This female culpability can also be found in a passage of Syncellus' version of 1 Enoch: ¹⁰⁶

> And the sons of men made them for themselves and for their daughters, and they transgressed and led astray [ἐπλάνησαν] the holy ones. (2) And there was much godlessness [ἀσέβεια] upon the earth, and they made the ways

---

¹⁰² Cf. 1 Corinthians 11:10: διὰ τοῦτο ὀφείλει ἡ γυνὴ ἐξουσίαν ἔχειν ἐπὶ τῆς κεφαλῆς διὰ τοὺς ἀγγέλους.

¹⁰³ θῆλυς also appears once more in the Syncellus version:
καὶ ἐπορεύθησαν πρὸς τὰς θυγατέρας τῶν ἀνθρώπων τῆς γῆς καὶ συνεκοιμήθησαν μετ' αὐτῶν καὶ ἐν ταῖς θηλείαις ἐμιάνθησαν, καὶ ἐδήλωσαν αὐταῖς πάσας τὰς ἁμαρτίας, καὶ ἐδίδαξαν αὐτὰς μίσητρα ποιεῖν. (1 Enoch. 9:8 Syncellus)
"They have gone in to the daughters of the men of earth, and they have lain with them, and have defiled themselves with the women. And they have revealed to them all sins, and have taught them to make hate-producing charms." (Nickelsburg, *1 Enoch 1*, 202)
Here Paul's θῆλυς is closely associated with Jude's μιαίνω (Jude 1:8).

¹⁰⁴ Cf. Rom 1:29-30: πεπληρωμένους πάσῃ ἀδικίᾳ . . . κακίᾳ, . . . φόνου . . . κακοηθείας, . . . ἐφευρετὰς κακῶν; also 1 Enoch 9:9: καὶ αἱ γυναῖκες ἐγέννησαν τιτᾶνας, ὑφ' ὧν ὅλη ἡ γῆ ἐπλήσθη αἵματος καὶ ἀδικίας (Akhmim). It should be noted that in this Greek version, the offspring of the women and Watchers are called "Titans" (cf. the citation from Judith toward the end of this article).

¹⁰⁵ If so, then Romans 1 is one of the earliest witnesses to the Greek text of 1 Enoch.

¹⁰⁶ Cf. Reed, *Fallen Angels*, 35.

desolate.¹⁰⁷ (1 En. 8:1c-2; trans. Nickelsburg, *1 Enoch 1*, 188, bracketed material added)

Nickelsburg notes, "According to the second clause, these women then led the holy watchers astray.... [T]he angels were seduced by the women."¹⁰⁸ This statement of 1 Enoch is echoed in the Testament of Reuben:

> Accordingly, my children, flee from sexual promiscuity, and order your wives and your daughters not to adorn their heads and their appearances so as to deceive men's sound minds.... For it was thus that they charmed the Watchers, who were before the Flood." (T. Reu. 5.5-6 [Trans. Kee, *OT Pseudepigrapha* 1:784])

This paper has argued that Romans 1:26-27 involves the two great judgments on the antediluvian word and on Sodom. The cumulative evidence thus far suggests that Romans 1:26 references the part certain women legendarily played in the great chaos before the Flood. Now the men of Romans 1:27 will be considered.

"Males"

> "And in the same way [ὁμοίως] also the men [ἄρσενες], giving up natural intercourse [ἀφέντες τὴν φυσικὴν χρῆσιν] with women [θηλείας], were consumed with passion for one another. Men [ἄρσενες] committed shameless acts with men [ἐν ἄρσεσιν] and received in their own persons the due penalty for their error [πλάνης]" (Rom 1:27)

In Romans 1:27, the men are referred to as "males." The term for "male" (ἄρσην, ἄρρην) appears much more frequently in the LXX than θῆλυς. The term can designate gender in humans and male animals (Gen 1:27; 5:2; 6:19, 20; 7:2, 3, 9, 15, 16), including sacrificial animals (Lev 1:3, 10; 3:1, 6; 4:23), but also is utilized in

---

¹⁰⁷ Based on a reconstruction by Nickelsburg involving the Greek version of Syncellus. Cf. Reed, *Fallen Angels*, 36, and also Kelley Coblentz Bautch, "Decoration, Destruction and Debauchery: Reflection on 1 Enoch 8 in Light of 4QEnᵇ," *DSD* 15 (2008): 79-95.

¹⁰⁸ Nickelsburg, *1 Enoch 1*, 195, who references along this line Jub. 4:15. See also William Loader, *Enoch, Levi, and Jubilees on Sexuality: Attitudes toward Sexuality in the Early Enoch Literature, the Aramaic Levi Document, and the Book of Jubilees* (Grand Rapids: Eerdmans, 2007), 10. Loader adds, concerning 1 En. 8:1, "As it stands, the effect is to include women's sexuality, enhanced by cosmetics and jewelry, as a cause of the 'great sin'" (18).

passages concerning circumcision (Gen 17:14, 23; 34:24). In Leviticus, the term appears in certain prohibitions concerning sexual activity (Lev 12:2; 15:33), including male homosexual activity (18:22; 20:13). The strongest parallel to Romans 1:27, though, is found in Philo (*Abr.* 135-137):

> As men, being unable to bear discreetly a satiety of these things, get restive like cattle, and become stiff-necked, and discard the laws of nature [τὸν τῆς φύσεως νόμον], pursuing a great and intemperate indulgence of gluttony, and drinking, and unlawful connections; for not only did they go mad after women [θηλυμανοῦντες], and defile the marriage bed of others, but also those who were men lusted after one another [ἀλλὰ καὶ ἄνδρες ὄντες ἄρρεσιν ἐπιβαίνοντες], doing unseemly things [πρὸς τοὺς πάσχοντας],[109] and not regarding or respecting their common nature [τὴν κοινὴν . . . φύσιν], and though eager for children, they were convicted by having only an abortive offspring; but the conviction produced no advantage, since they were overcome by violent desire [ἐπιθυμίας[110]]. (Philo, *Abr.* 135 [Yonge], bracked material added)

Jewett comments on the similarity between Philo's description of Sodom (from various works) and Romans 1:27:

> In a similar manner, Philo makes prominent use of the terminology Paul employs in this verse, describing the perversion of the Sodomites as a violation of 'nature' (*Abr.* 135-137) and stigmatizing pederasty as ἡ παρὰ φύσιν ἡδονή ('an unnatural pleasure'; *Spec.* 3.39.2) while affirming the 'natural use' (τῆς κατὰ φύσιν χρήσεως) of heterosexual sexuality (*Mut.* 111-12)."[111]

Fitzmyer also is compelled to reference Sodom,[112] as is Neyrey.[113] Though recognizing the similarities between Romans

---

[109] Philo's "unseemly things" (τοὺς πάσχοντας) corresponds to the "dishonorable passions" of Romans 1:26.
[110] A term used in Romans 1:24.
[111] Jewett, *Romans*, 176.
[112] Fitzmyer, *Romans*, 288: "especially the Sodom story," and noting Isa 1:9-10; 3:9; Jer 23:14 and Lam 4:6, as well as the peculiar adaptation of the story at Judg 19:22-26.
[113] Covering Jude 5-7: "Hence, Sodom and Gomorrah cause pollution [μιαίνω] by crossing the lines of acceptable sexual partners. Paul reflects the same pollution code in Rom 1:26-27, where he labels this pollution as 'shameful' (*atimia*)"

1:27 and Philo's passage, an equivalence is never recognized by these scholars. However, this paper argues that Sodom *is* the subject of Romans 1:27, the rhetorical complement to v. 26's account of the antediluvian women, and thus a manifestation of the dual judgment rhetorical topos. Justin Martyr's reference to the antediluvian world (and Sodom?) in the second century CE utilizes language and concepts found in Romans 1:

> God, when He had made the whole world, and subjected things earthly to man, and arranged the heavenly elements for the increase of fruits and rotation of the seasons, and appointed this divine law—for these things also He evidently made [φαίνεται πεποιηκώς] for man—committed [παρέδωκεν] the care of men and of all things under heaven to angels [ἄγγελοι] whom He appointed over them. But the angels transgressed this appointment, and were captivated by love of women [γυναικῶν], and begat children who are those that are called demons; and besides, they afterwards subdued the human race to themselves, partly by magical writings, and partly by fears and the punishments they occasioned, and partly by teaching them to offer sacrifices, and incense, and libations, of which things they stood in need after they were enslaved by lustful passions [πάθεσιν ἐπιθυμιῶν]; and among men they sowed murders, wars, adulteries, intemperate deeds, and all wickedness [πᾶσαν κακίαν[114]]. Whence also the poets and mythologists, not knowing that it was the angels and those demons who had been begotten by them that did these things to men [ἄρρενας], and women [θηλείας], and cities [πόλεις], and nations, which they related, ascribed them to god himself, and to those who were accounted to be his very offspring, and to the offspring of those who were called his brothers, Neptune and Pluto, and to the children again of these their offspring. For whatever name each of the angels had given to himself and his children, by that name they called them. (*2 Apol.* 5.2-6; *ANF* 1:190).[115]

(Neyrey, *2 Peter, Jude*, 61, bracketed word added).

[114] Cf. Rom 1:29: πάσῃ ἀδικίᾳ πονηρίᾳ πλεονεξίᾳ κακίᾳ.

[115] The Greek text utilized is found in Edgar J. Goodspeed, *Die ältesten Apologeten: Texte mit kurzen Einleitungen* (Göttingen: Vandenhoeck & Ruprecht, 1914).

Conclusion

This paper seeks to offer a better general explanation of Romans 1:26-27 than the status quo. This paper has put forth evidence that Romans 1:26-27 is to be understood as participating in the rhetorical tradition of citing the two great divine judgments in Genesis as part of an indictment of, or warning to, humanity. The tradition, a dual-event rhetorical topos that couples the antediluvian world chaos involving women, angels and giants with the story of the city of Sodom and its destruction, was a popular motif that appears in a number of ancient works. The Sodom tradition easily corresponds to the particulars of v. 27 (Philo's account quite similar in certain language). The legend of the sexual interaction of women with angels in the antediluvian world fits the language of v. 26. Paul's order of the women's indiscretions mentioned first in Romans 1:26-27 parallels the biblical and chronological presentation of the stories in Genesis, and Paul's use of the language and rhetoric of 1 Enoch 15-16 solidifies the connection, perhaps one of the earliest, if not earliest, witnesses to the Greek translation of 1 Enoch. Similar language is shared by Romans 1:18-32 with various manifestations of the dual-event rhetorical topos, with ὁμοίως at times the link for the two great judgments.

On a final note, one ought to consider the one named θῆλυς (θήλεια) in the biblical literature, namely, Judith:

> But the Lord Almighty has foiled them
> by the hand of a woman [θηλείας].
> For their mighty one did not fall by the hands of the young men [νεανίσκων],
> nor did the sons of the Titans [υἱοὶ τιτάνων] strike him down,[116]
> nor did tall giants [ὑψηλοὶ γίγαντες] set upon him;
> but Judith daughter [θυγάτηρ] of Merari
> with the beauty of her countenance [κάλλει προσώπου αὐτῆς] undid him.
> Her sandal ravished his eyes, her beauty captivated his mind,
> and the sword severed his neck! (Judith 16:5-9)

---

[116] See notes 44 and 106 above.

New Directions in the Gospel of Thomas
Oxyrhynchus as Test Case[1]

Thomas A. Wayment

The sayings of the Gospel of Thomas, it seems, cannot be interpreted without appealing to the similarities and differences of its sayings with respect to the four canonical gospels.[2] Such contextualized discussions have often envisioned a history where the Gospel of Thomas, either at the stage of its composition or during publication and transmission, maintained goals and interests that were fairly consistent over time.[3] That is not intended as a critical assessment of such approaches, in part, because situating the text in a nuanced historical trajectory must remain part of careful exegetical work of its sayings. But what if usage of the Gospel of Thomas were to be discovered or described in a discrete setting where the original interests of the author, publisher, and tradents were no longer a significant point of influence on how the text was adapted and used by a new, distinct Christian community? This

---

[1] I would like to thank the editors of this volume for bringing the project to fruition. Gregory Riley's work has continued to shape my academic interests and pursuits. I will never forget his classroom discussions, their depth, clarity, and richness. *Resurrection Reconsidered: Thomas and John in Controversy* (Minneapolis: Fortress, 1995) is the work that I would like to engage most directly, although this contribution to the volume also interacts at several points with his legacy for the Gospel of Thomas.

[2] Although not directly related to the scholarship on finding the so-called Thomas community, Sarah Rollens has offered a valuable critique of the efforts to recover the Q community. See Sarah Rollens, "Does 'Q' Have Any Representative Potential?" *Method and Theory in the Study of Religion* 23 (2011): 73–77.

[3] This is particularly true in Robinson and Koester, *Trajectories Through Early Christianity* (Minneapolis: Fortress Press, 1971), 158–204 and Helmut Koester, *Introduction to the New Testament: History and Literature of Early Christianity*, 2 vols., 2d ed. (Berlin: Walter de Gruyter, 2000), 154–58. For a careful study of the various community theories and how they interacted with canonical texts, see Philip Sellew, "Thomas Christianity: Scholars in Quest of a Community," in *The Apocryphal Acts of Thomas*, ed. Jan Bremer (Leuven: Peeters, 2001), 11–34.

occurs, of course, at Oxyrhynchus where three fragmentary pages of papyrus containing the Gospel of Thomas have been recovered, and where usage of that text carried it into new circles of engagement, into a new usage-community, and where the original goals and interests of the text were shaped and adapted.

At the time of Riley's Harvard Dissertation, papyrological interest in Christian texts as community artifacts had not fully taken on a life of its own.[4] It was not until Larry Hurtado's work on Christian text as artifact that discussing the physical text itself became more commonplace and almost an expected desideratum of the discipline.[5] More recently, AnneMarie Luijendijk's exploration of Christian scripture as trash brings us closer to the question at hand, namely whether the Gospel of Thomas was put to new use within a distinct Christian community that was at times connected to the larger Christian world directly through Alexandria and indirectly through its participants.[6] Her study forces the reader to consider the ultimate outcome of the Oxyrhynchus papyri, namely as objects that were discarded, as part of their transmission history. A discussion of the artifacts themselves—the three Oxyrhynchus Thomas papyri—can generate new data points that will push the discussion of Gospel of Thomas usage in new directions. This contribution to the Festschrift will attempt to separate the interests of the author of the Gospel of Thomas from one specific community of end users. It will then move into a contextualized treatment of the Oxyrhynchus artifacts (P.Oxy. 1, 654–655), and finally it will offer a suggestion regarding how the text was used and adapted to fit the needs of a community seeking increased independent identity.

Author, Tradents, and New Users

In the early 1990s, the study of Christian origins at

---

[4] Gregory J. Riley, "Doubting Thomas: Controversy Between the Communities of Thomas and John" (PhD diss., Harvard University, 1990).

[5] See Larry W. Hurtado, "The Earliest Evidence of an Emerging Christian Material and Visual Culture: The Codex, the *Nomina Sacra* and the Staurogram," in *Text and Artifact in the Religions of Mediterranean Antiquity: Essays in Honour of Peter Richardson*, eds. Stephen G. Wilson and Michel Desjardins (Waterloo, Ont.: Wilfrid Laurier University Press, 2000), 271–88; idem, *The Earliest Christian Artifacts: Manuscripts and Christian Origins* (Grand Rapids: Eerdmans, 2006).

[6] AnneMarie Luijendijk, "Sacred Scriptures as Trash: Biblical Papyri from Oxyrhynchus," *VC* 64 (2010): 217–54.

Claremont was still deeply engaged in pursuing the model of Christian "trajectories," a concept that was founded upon Walter Bauer and later given more nuanced articulation by Robinson and Koester.[7] We were in one sense neo-Bauer-ites and his model of Christian origins when transformed in the hands of Robinson and Koester and their trajectories model was at that time particularly productive.[8] But it was also not without its own critics even though it was effective in its clarity and ability to explain diverse manifestations of Christian belief and practice deriving simultaneously from the historical Jesus.[9] This model, however, as effective as it was on a broad level was ineffective in describing Christian origins for individual rural villages and towns, a concept that was not engaged beyond several Syrian cities like Edessa.[10] A smaller town or village, like Oxyrhynchus, shows very few signs of participating in the wider conversations and controversies of the broad Christian world, or the world of the Christian elites like Irenaeus, Justin, and Origen, or even of the communities that theoretically authored Christian texts.[11] Instead, the rural villages of Egypt were largely the recipients of texts that were already authored and published, and fixed in ways that made them largely off-limits to adaptation and adjustment, although smaller changes to manuscripts were made. In other words, while some communities of Christians were the originators of text, other communities, like Oxyrhynchus, were the recipients.

---

[7] James M. Robinson and Helmut Koester, *Trajectories through Early Christianity* (Philadelphia: Fortress Press, 1971) and Walter Bauer, *Rechtgläubigkeit und Ketzerei im ältesten Christentum* (Tübingen: Mohr/Siebeck, 1934); ET: *Orthodoxy and Heresy in Earliest Christianity* (Philadelphia: Fortress Press, 1971).

[8] Sellew, "Thomas Christianity," 13.

[9] Thomas A. Robinson, *The Bauer Thesis Examined: The Geography of Heresy in the Early Christian Church* (Lewiston, NY: Edwin Mellen Press, 1988); Daniel J. Harrington, "The Reception of Walter Bauer's Orthodoxy and Heresy in Earliest Christianity During the Last Decade," *HTR* 73 (1980): 289–98; Michel Desjardins, "Bauer and Beyond: On Recent Scholarly Discussions of *Hairesis* in the Early Christian Era," *SC* 8 (1991): 65–82.

[10] Some of the beginning moments of criticism in this trajectory can be seen in the work of Ron Cameron, a Harvard alumnus of Riley, who criticized some of the early points of emphasis that were made with respect to the Gospels of Thomas and John being in conversation. See Ronald D. Cameron and Merrill P. Miller, *Redescribing Christian Origins*, Society of Biblical Literature Symposium Series 28 (Atlanta: Society of Biblical Literature; Leiden and Boston: Brill, 2004), 93–98.

[11] Although, see Lincoln H. Blumell, "PSI IV 311: Early Evidence for Arianism at Oxyrhynchus?" *Bulletin of the American Society of Papyrologists* 49 (2012): 279–99.

Mapping more recent conversations about the origins of the Synoptic Gospels onto the trajectories model raises a number of important criticisms while also raising new points of conversation. Importantly, discussions about the origins and dissemination of the Christian gospel note the preponderance of letter carrying in the first centuries of the common era with an interest to suggest that Christians communicated extensively with one another.[12] The evidence for Christian interconnectedness is certainly significant, but by placing the discussion of Christianity at Oxyrhynchus within that context it becomes simply another part of a larger interest in Christian origins and it subordinates discussions of a discreet community to the larger interest in describing an emerging global Christianity.[13] Instead of pursuing a broad scope, this paper will focus its attention on the texts discovered in a single city and how they were used by a community of Christians that likely grew out of a few small networks of believers and that potentially had limited contact with Christian missionaries and the larger Christian world in its first few centuries. It will also assume a network theory of Christian growth similar to the way Roman voluntary associations functioned, which grew from shared interactions, interests, and collaborative efforts.[14]

In addition, Matthew Larsen's recent work on authorship,

---

[12] Lincoln H. Blumell, *Lettered Christians: Christians, Letters, and Late Antique Oxyrhynchus* (Leiden and Boston: Brill, 2012), 89-61; Abraham Malherbe, *Social Aspects of Early Christianity* (Baton Rouge: Louisiana State University Press, 1977), 62-70; Harry Y. Gamble, *Books and Readers in the Early Church: A History of Early Christian Texts* (New Haven: Yale University Press, 1995), 82-143; Eldon J. Epp, 'New Testament Papyrus Manuscripts and Letter Carrying in Greco-Roman Times,' in *The Future of Early Christianity: Essays in Honor of Helmut Koester*, ed. Birger A. Pearson (Minneapolis: Fortress Press, 1991), 35-56; Lincoln H. Blumell, "Christians on the Move in Late Antique Oxyrhynchus," in *Travel and Religion in Antiquity*, ed. Philip Harland, Studies in Christianity and Judaism/Études sur le christianisme et le judaïsme 21 (Waterloo: Wilfred Laurier University Press, 2011), 235-54.

[13] Sellew, "Thomas Christianity," 18-27 uses textual production as a means of reconstructing the communities who composed the canonical gospels.

[14] The scholarly literature on voluntary associations is immense, see Richard S. Ascough, Philip A. Harland, and John S. Kloppenborg, *Associations in the Greco-Roman World: A Sourcebook* (Waco / Berlin: Baylor University Press / de Gruyter, 2012); Markus Öhler, "Greco-Roman Associations, Judean Synagogues and Early Christianity in Bithynia-Pontus," in *Authority and Identity in Emerging Christianities in Asia Minor and Greece*, eds. C. Breytenbach and J. M. Ogereau (Leiden and Boston: Brill, 2018), 62-88.

publication, and text-dissemination convincingly demonstrates a need to adopt more nuanced terminology when speaking of books, publication, and authorship.[15] With respect to Oxyrhynchus, it is important to note that it received texts that were already authored, published, and distributed, and that were in many ways largely presented to them as completed products. On the other hand, Christians at Oxyrhynchus also engaged in the practice of authoring new texts that were potentially never intended for distribution (e.g. P.Oxy. 2070 "Anti-Jewish Dialogue," etc.), or, if they were later published and distributed, then only a draft stage of the text exists. There also appear to be Christian texts from the city that were circulated in pre-publication form, and it remains an open question what freedom copyists felt to intentionally adapt or modify published texts. From the surviving evidence, it is clear that published texts that were part of the larger world of Mediterranean Christianity functioned as a normalizing force to bring local communities into dialogue with the wider community, whereas locally authored texts represent the concerns of the community that authored them. Additionally, unique local adaptation, quality of the text produced, willingness to adapt already published texts all become signifiers of local community interests.

It remains true that the modern scholar cannot remove from consideration those texts that may have been carried to Oxyrhynchus from places like Alexandria, but it can distinguish between texts that have been intentionally published and circulated drawing upon the financial resources Oxyrhynchus Christians had to offer and those that were of a lesser quality, that were still undergoing correction and adaptation and were not intentionally circulated for public use. These two types of texts can be categorized as public (published, circulated texts) and private (non-published and non-circulated to the entire community) texts. These designations are not qualitative evaluations of the impact of any given text, but for this discussion they represent signifiers of intended usage. A private text may be highly influential as a document used to teach the Christian gospel, train catechumens, or offer exegesis of Christian teachings, whereas a public text may have been created to help meet the liturgical needs of a community, and therefore important in a different way. These two divisions

---

[15] Matthew Larsen, *Gospels before the Book* (Oxford: Oxford University Press, 2018), 1–78.

also cannot be taken as a signal of the number of individuals who used the text. A private text could have been circulated among a broad network of Christians (e.g. P.Oxy. 4365 where Christian books were shared) whereas a public text may have been the public copy owned by a church and only read by a few but heard by many.[16] The interest of this exercise is to recover circles of usage within a community. Although the comparison is imperfect, a modern reader can quickly distinguish between a published and non-published product by the presence of a binding, book covers and a copyright page, whereas a single loose-leaf page or one that has printed and handwritten text is clearly a draft composition. Distinguishing between the two types is important and making them all equal clearly overlooks an opportunity for further consideration.

Looking only at the extra-canonical texts during the period when the three fragments from the Gospel of Thomas appear in Oxyrhynchus, a developing picture emerges of usage and intent.[17] Using the criteria of type, size, quality of handwriting, size of handwriting, preparation of the writing surface, attention to physical layout, reuse of writing substrate, and the presence of corrections, it quickly becomes apparent that the Oxyrhynchus texts from the first two centuries fall into two distinct categories in which further divisions may be possible. The following papyri can be categorized as private texts, which are defined as being copied in a non-literary hand, mixing literary and documentary hands, disregarding bilinearity in writing or other attempts at elegance, having few or no corrections, and showing signs of being produced in haste.[18] The texts that fall into the category of private texts are the Gospel of Peter (P.Oxy. 4009), an unknown gospel with Synop-

---

[16] An important example of this practice is the early fourth century letter that reads, "To my dearest lady sister, greetings in the Lord. Lend the Ezra, since I lent you the little Genesis. Farewell in God from us." Translation from Lincoln H. Blumell and Thomas A. Wayment eds., *Christian Oxyrhynchus: Texts, Documents, and Sources* (Waco: Baylor University Press, 2015), 511. See also AnnMarie Luijendijk, *Greetings in the Lord: Early Christians and the Oxyrhynchus Papyri* (Cambridge, MA: Harvard University Press, 2008), 71-74.

[17] Many of the canonical texts assigned to approximately the first three centuries were clearly professional productions (e.g. P.Oxy. 4404, 3523, 4403 (part of a multi book codex + 4405 and 2683), 4401 (but not elegant and lacking bilinearity, and preserving some rapidly formed letters).

[18] Roger Bagnall, *Early Christian Books in Egypt* (Princeton: Princeton University Press, 2009), 50-69.

tic parallels (P.Oxy. 5072), the Gospel of Mary (P.Oxy. 3525 and P.Rylands Gr. III 463), and a copy of the Shepherd of Hermas (P.Oxy. 4705). In the second category, the category of professional texts, fall almost all of the early fragments of the Shepherd of Hermas (P.Oxy. 4706, 3527-3528, 4707), a gospel text that has affinities with the Gospel of Peter (P.Oxy. 2949), and an unknown gospel text (P.Oxy. 210). The three fragments of the Gospel of Thomas fall into both the category of private and professional productions with some interesting unique variations (more below).

The Gospel of Thomas as Private Text

P.Oxy. 1, a single sheet from a papyrus codex, shows careful attention to formatting and layout with consistent margins, spaces between words, a moderately well executed literary hand that is not elegant with letters that often appear to hang from the top of the line rather than being placed on top of a line below the letters. The codex itself measured approximately 27+ cm x 13+ cm, thus making it a rather tall and narrow codex that can be classified as Turner's group 8.[19] Hurtado notes that the handwriting is "workaday and certainly not calligraphic," with no attempt at bilinearity and a hand that is often inconsistent.[20] Line endings are often noted by alternate letter forms such as an elongated ο or an enlarged letter like υ or ι. Overall, the handwriting is rather small (approximately 2 mm in height) while some letters are approximately 3 mm in height.[21]

---

[19] Eric G. Turner, *The Typology of the Early Codex* (Philadelphia: University of Pennsylvania Press, 1977), 20. See also Larry Hurtado, "The Greek Fragments of 'The Gospel of Thomas' as Artefacts: Papyrological Observations on P.Oxy 1, P.Oxy 654, and P.Oxy 655," in *Das Thomasevangelium: Entstehung – Rezeption – Theologie*, eds. J. Frey, E. E. Popkes, and J. Schröter (Berlin and New York: Walter de Gruyter, 2008), 19-24. Other codices from Oxyrhynchus of this size were P.Oxy. 64 4404 (Gospel of Matthew); P.Oxy. 15 1780 (Gospel of John); P.Oxy. 65 4445 (Gospel of John); P.Oxy. 65 4448 (Gospel of John); P.Oxy. 71 4803 (Gospel of John); P.Oxy. 71 4805 (Gospel of John); P.Oxy. 11 1355 (Romans); P.Oxy. 10 1229 (James), etc. The size was apparently quite popular at Oxyrhynchus. Hurtado, "The Greek Fragments of 'The Gospel of Thomas' as Artefacts," 21-22 notes that $\mathfrak{P}^{46}$ was also similar in size.

[20] Hurtado, "The Greek Fragments of 'The Gospel of Thomas' as Artefacts," 22; Harold W. Attridge, "The Gospel of Thomas. Appendix: The Greek Fragments," in *Nag Hammadi Codex II,2-7 Together With XIII,2*, BRIT. LIB. OR. 4926(1), and P.OXY. 1, 654, 655*, ed. B. Layton, 2 vols. (New York: Brill, 1989), 96-97 refers to it as an informal literary type hand.

[21] Hurtado, "The Greek Fragments of 'The Gospel of Thomas' as Artefacts," 22.

With respect to the preparation of the manuscript for publication or circulation, the lines of writing are straight and evenly spaced even though the handwriting can devolve into cursive (i.e. documentary) forms. The scribe appears to have sought to achieve a straight right-hand margin, and when lines did not quite reach the right-hand margin a carrot shaped sign ">" was used to fill up the remaining spaces (see, lines 3, 9, 17-18). Diaeresis is used sparingly and not as a signal of a rough breathing. The single surviving page is numbered (11), which may signal that 11 pages of text preceded the surviving page, or perhaps the number of bifolios that preceded the surviving page.[22]

As an artifact of the Oxyrhynchus Christian community, P.Oxy. 1 was prepared by a person or community of modest means. The scribal hand was middle of the road, and while some attention was paid to preparation, the small size of the handwriting and quality of the hand strongly hint at a private text. Whether an individual produced this text or whether the community produced it for study and personal reading is difficult to determine, but the text does fall into the category of private texts that were not intended for public presentation or to represent the finest quality the community had to offer.

P.Oxy. 654 is more obviously a private text that was written on the reverse of a land survey. Despite its impoverished beginnings, the scribe intentionally adorned the text with wedge or carrot shaped signs ">" to create a more consistent right-hand margin. The land register on the reverse side dates to the second or third century and may connect directly to the owner, although it is equally possible that the used papyrus was simply purchased for use by a later client. The individual logia are separated by enlarged spaces, and a horizontal line at the beginning of some lines indicate the beginning of a new saying (lines 6, 10, 22, 28, 32). The height of the individual letters varies, although it can be described as smallish.[23] There is no attempt at bilinearity, and the overall preparation

---

[22] Hurtado, "The Greek Fragments of 'The Gospel of Thomas' as Artefacts," 24 urges caution in using this page number to suggest the length of the original codex, going so far as suggesting that it could be a quire number.

[23] See Bernard P. Grenfell and Arthur S. Hunt eds., *New Sayings of Jesus and the Lost Gospel from Oxyrhynchus* (London: Egyptian Exploration Society, 1904), 9; Bernard P. Grenfell and Arthur S. Hunt, *The Oxyrhynchus Papyri IV* (London: Egyptian Exploration Society, 1904), 2; Hurtado, "The Greek Fragments of 'The Gospel of Thomas' as Artefacts," 25-26; Attridge, "Appendix," 97.

and presentation of the text is inferior to P.Oxy. 1. There are some corrections that are made above the line (lines 19 and 25), and the first line presents what appears to be a copying error that resulted from parablepsis. From the surviving evidence, P.Oxy. 654 was also a private copy with less concern about presentation and quality. Along the spectrum of private use texts, where a community may produce a higher quality text for private study to an individual who might own a personal copy, this text would exist on the lower end of the spectrum alongside poorly and cheaply produced texts.

Finally, P.Oxy. 655 is similarly a private copy written on a papyrus roll. The text was originally written in rather narrow columns of about 5 cm with only 12-16 letters per line.[24] The handwriting is also quite small, although the published editions do not note the exact size of the handwriting. Some of the surviving text does not appear to derive from the Gospel of Thomas, and those lines cannot be restored using the Nag Hammadi Coptic Gospel of Thomas.[25] This tantalizing discovery may indicate that P.Oxy. 655 contained another text that was copied alongside Thomas or that Thomas was still in the stage of literary development. The text is expertly prepared with small serifs adorning letters and a more consistently employed literary hand. The overall roll was quite small, perhaps suggesting that portability or mobility were issues. The scribe did attempt to achieve bilinearity, and again the ">" shape is used to fill up line endings to create what was likely a rigid right-hand margin. Unlike the previous two papyri containing sections of the Gospel of Thomas, this example is more professionally prepared and executed. Costs associated with production and copying were reduced by the small writing that is not elegant but certainly expert, and the overall size of the roll was limited by keeping the overall length in check.

Situating the Gospel of Thomas Fragments at Oxyrhynchus

Ron Cameron's charge that the study of Christian writings should be contextualized within "the intersection of complex textual and social histories" is particularly important for the purposes of this essay.[26] Returning to Riley's contributions to the scholarship

---

[24] Robert A Kraft, "Oxyrhynchus Papyrus 655 Reconsidered," *HTR* 54 (1961): 262.
[25] See Kraft, "Oxyrhynchus Papyrus 655," 254; Attridge, "Appendix," 121-22.
[26] Ron Cameron, "The Gospel of Thomas and Christian Origins," in *The Future of*

on the Gospel of Thomas, this final section will describe a later complex social history in which the Gospel of Thomas was copied, used, and transmitted. At this point it is important to note what this study is not: it is not a consideration of the forces of influence or conversations that led to the composition of the Gospel of Thomas; it is not an attempt to add nuance to the discussion about the original community that encouraged its production and used it, nor is it an argument for the earliness of lateness of the sayings of Jesus that it records. Instead, it is an intentional look at a later usage-community that received the Gospel of Thomas as part of its collection of religious texts, possibly even scriptural texts, that it subsequently copied, revered, and used. As the foregoing conversation has demonstrated, the Gospel of Thomas at Oxyrhynchus fell into the category of private texts with some markers indicating a public text, categories that now need further consideration.

The Gospel of Thomas, like other Christian texts, existed in multiple phases, beginning with an obscure pre-composition phase where traditions, possibly oral and/or written, were circulating among interested and sympathetic listeners. The text then moved into a written, pre-publication stage where the traditions were migrated into units, or sense units, that were connected to other units of the tradition.[27] At some point in this prepublication state, the title of author and the concept of authorship emerge as accurate descriptors. Even though the sayings may have been or were then currently described as sayings of Jesus, an author was required to create a physical written text with narrative framework—as simple as an introduction or conclusion. In some cases, the remnants of authorship can be traced to the author's or authors' hand, although it is clear that no Oxyrhynchite scribes engaged in authorship with respect to the Gospel of Thomas. It would appear that someone chose to connect the written production with a famous disciple of Jesus at an early state rather than place his or her own name on the production even though that individual acted as author.[28] The text

---

Early Christianity: Essays in Honor of Helmut Koester, ed. Birger A. Pearson (Minneapolis: Fortress Press, 1991), 388–89.

[27] Stephen J. Patterson, *The Gospel of Thomas and Jesus* (Sonoma, CA: Polebridge, 1993), 121–57.

[28] These early stages of growth are the focus of April D. DeConick, *Recovering the Original Gospel of Thomas: A History of the Gospel and Its Growth*, LNTS 286 (London: T&T Clark, 2005); idem, *The Original Gospel of Thomas in Translation*, LNTS 287 (London: T&T Clark, 2006).

then moved into the standard phases of production, which are defined in the Roman period as ἔκδοσις (publication), διάδοσις (distribution), and παράδοσις (dissemination or transmission).²⁹ Emmel adds a further stage to the history of this particular text, which is defined by translation into a new language to expand its dissemination.³⁰

By the time the Gospel of Thomas reached Oxyrhynchus, it was already in the stage of dissemination, where the text already had been irretrievably associated with a historical person, "οὗτοι οἱ λόγοι οἵ Ἰη(σοῦ)ς ὁ ζῶν καὶ ἔγραψεν Ἰούδας ὅ καὶ Θωμᾶς καὶ εἶπεν" *These are the words that the living Jesus spoke and Judas Thomas wrote* (P.Oxy. 654 ll. 1-2). In this way the transmission history of the text was unequivocally demarcated, and the obstacles to removing that footprint of authorship were very real for this later usage-community. The text had also been laid out into sense units, *nomina sacra* were used, and scribes took the time to fill in blank spaces at the end of lines to create a more visually appealing text, which seems to signal the practice of the *Vorlage* because the practice spans the three examples from Oxyrhynchus.³¹ Placing this information within the trajectories model indicates that if the Gospel of Thomas was composed in conversation with the canonical gospels, was then later redacted with ascetic, Encratite, or other interests, and was then later used by a Christian community in Oxyrhynchus, and also later reappearing in codex form at Nag Hammadi, then the threads of continuity between these stages should at least form a significant component of our understanding of that text. In other words, the Gospel of Thomas experienced a moment in its existence when it was incorporated into the body of religious texts used by a rather isolated Christian community that had a strong appreciation of canonical texts.

Michael Williams has approached this question of usage

---

²⁹ Larsen's discussion of these categories is much more detailed, and my brief synopsis builds on his conclusions, see, *Gospels Before the Book*, 6. Cf. Bernhard A. van Gronigen, "ΕΚΔΟΣΙΣ," *Mnemosyne* 16 (1963): 1-17.

³⁰ Stephen Emmel, "The Coptic Gnostic Texts as Witnesses to the Production and Transmission of Gnostic (and Other) Traditions," in *Das Thomasevangelium: Entstehung – Rezeption – Theologie*, eds. J. Frey, E. E. Popkes, and J. Schröter (Berlin and New York: Walter de Gruyter, 2008), 35.

³¹ A. M. Luijendijk, "Reading the *Gospel of Thomas* in the Third Century: Three Oxyrhynchus Papyri and Origen's *Homilies*," in *Reading New Testament Papyri in Context*, eds. C. Clivaz and J. Zumstein (Leuven: Peeters, 2011), 253-54.

from a different vantage point, arguing that the codex form of the Nag Hammadi collection becomes an important signifier of usage, "In other words, the very repackaging and ordering of the material resolved, as it were, theological diversity among the writings. Each writing had its own function and could be interpreted in terms of that function in relation to the other works within the codex. Once this is seen, it is fair to ask whether there is really all that much more theological diversity within the Nag Hammadi library (or at least within its sub collections) than within, say, Codex Sinaiticus, or the Septuagint, or even the New Testament itself."[32] Williams devotes a significant amount of discussion to establishing that each of the Nag Hammadi texts had a prehistory that was eventually absorbed into the covers of the codex. Differences, prehistory, nuance, and diversity were suddenly compressed in the interests of similarity, newly envisioned usages, and fourth century purposes.

To return to the question at hand, looking at the history of the Gospel of Thomas through the lens of its Oxyrhynchus life cycle hints at its embeddedness within a Christian community that placed it alongside other texts without canonical distinction, but possibly distinctions in usage. What started out to be a complex text with its own unique composition history appeared in written form in the late second century and early third century in an Egyptian city alongside a variety of other Christian texts.[33] Like Williams' and Emmel's concept of "library," community in this situation shapes the way the Oxyrhynchus Gospel of Thomas fragments are interpreted.[34]

The papyrological record is admittedly random, sporadic, and unpredictable, and therefore conclusions based on what is preserved, how it is preserved, and when it is preserved are made with an acknowledged sense of caution. Limiting the evidence to Oxyrhynchus but expanding it to include all Christian fragments up to the third century, a picture emerges that offers the new perspectives that can be recovered and will in turn bring this discussion to conclusion. By limiting the scope of this enquiry to

---

[32] Michael A. Williams, *Rethinking Gnosticism: An Argument for Dismantling a Dubious Category* (Princeton, NJ: Princeton University Press, 1996), 261.

[33] The date given by most editors of the text is the second or third century, see B. P. Grenfell and A. S. Hunt eds., *The Oxyrhynchus Papyri I* (London: Egypt Exploration Society, 1898), 1-3; Attridge, "Appendix," 95-128; Luijendijk, "Reading the *Gospel of Thomas* in the Third Century," 241-67.

[34] Emmel, "The Coptic Gnostic Texts as Witnesses," 37

first through third centuries, when the last of the three Greek Gospel of Thomas fragments appear at Oxyrhynchus, there are during that period approximately twenty-five papyrus witnesses to canonical texts: ten of the Gospel of John, five of the Gospel of Matthew, two each of the Gospel of Luke, the letter to Romans, and the epistle of James, and one each of Acts and Hebrews.[35] Additionally, there are papyrus witnesses to Genesis, Exodus, Psalms, and Esther, although they cannot be definitively connected to the Christian community. A full discussion of all of these texts is beyond the scope of this essay.

With one exception (P.Oxy. 1228, Gospel of John), all of the texts from the second and third centuries at Oxyrhynchus that would eventually be designated as canonical were written in codex form. The vast majority of those were written in a rather narrow codex size that was taller than it was wide, or within Turner's Groups 8–9.[36] Several of the early codices cannot be reconstructed with respect to their size, so the data set is incomplete, but only three codices diverge from the rather narrow and tall codex. In a few instances, page numbers, emerging punctuation, and a concern for aesthetic elegance are evident.

With respect to the non-canonical literature, an admittedly anachronistic designation that is helpful for the sake of comparison, it becomes readily apparent that the preference for the codex shifts dramatically toward the roll form, where there are 13 examples of papyrus rolls containing extra-canonical Christian texts written prior to the third century and 11 examples of papyrus codices. Among the codices, the preference for the tall narrow codex is manifest in a single example, there is also a miniature codex, while there is a mild preference for the square format codex. This observation is based on incomplete and limited data in some instances, and in a few cases the sizes of these early codices cannot be reconstructed. Despite the limitations, the comparisons provide a new way to restructure the existing evidence.

In the second century, Christians at Oxyrhynchus had in their library or held by members of the community gospel texts that were available in codex form. Surprisingly, the four gospel co-

---

[35] These texts can be found in Blumell and Wayment, *Christian Oxyrhynchus*, v–viii.

[36] Eric Turner, *The Typology of the Early Codex* (Philadelphia: University of Pennsylvania Press, 1977).

dices from the second century show a remarkable level of professional production, a level of quality that will diminish in the third century.[37] The scribal hands often have added serifs, bilinearity is a concern but it is not strictly observed, and moderate attention is paid to punctuation and division of pericopes (P.Oxy. 4405 + 2683). Christians also had what was likely a patristic text, written on a roll (PSI 1200 bis) and also a copy of Irenaeus *Adversus Haereses* in roll form (P.Oxy. 405), two copies of the *Shepherd of Hermas*, one on a roll and one in codex form (P.Oxy. 4706 and 3528), and a text with affinities to the Gospel of Peter (P.Oxy. 2949) all of which show a high level of professionalism and elegance. Additionally, they had access to several lower quality productions, with a miniature codex of the Gospel of Peter (P.Oxy. 4009), a codex of unknown size containing gospel like teachings (P.Oxy. 5072), and the Gospel of Thomas (P.Oxy. 1).[38]

The pattern immediately presents itself that gospel texts, ones that contain sayings or narratives about Jesus were disseminated during this period exclusively in codex form. They show varying degrees of professionality in production, with two of the Matthew codices (P.Oxy. 4403-4404 and P.Oxy. 3523) showing a high level of professional production, and one of the Matthew codices (P.Oxy. 4405 + 2683) slightly less professional. Overall, almost all Christian texts from this early period show a remarkably high degree of production, with only three examples of lower quality productions. This same community of Christians preferred to disseminate other Christian documents on roll form, including patristic texts and the *Shepherd of Hermas*. The high level of quality in production during this period has several possible explanations. During the first centuries of the community's existence, Christians at Oxyrhynchus may have obtained high quality copies from more wealthy Christian centers like Alexandria, or that traveling missionaries used higher quality texts in assisting the establishment of early Christ communities. Textual distinctions that would denote quality designations are completely absent, and the only significant

---

[37] Blumell and Wayment, *Christian Oxyrhynchus*, 20-89, 201-51, 285-324.

[38] Paul Foster, "Are There Any Early Fragments of the So-Called *Gospel of Peter*?" *NTS* 52 (2006): 1-28; Paul Foster, *The Gospel of Peter: Introduction, Critical Edition, and Commentary* (Leiden: Brill, 2010); Thomas Kraus and Tobias Nicklas, *Das Petrusevangelium und die Petrusapokalypse* (Berlin: de Gruyter, 2011); Thomas Kraus and Tobias Nicklas, *Das Evangelium nach Petrus: Text, Kontexte, Intertexte* (Berlin: de Gruyter, 2012).

distinction to be drawn is that some texts were produced more cheaply with less interest in elegance of production.

The third century continues these trends with some equally interesting observations: the quality of production among canonical texts drops in a number of important ways. Although there is a frequent concern for bilinearity, scribal hands are competent but often not professional, documentary influence can often be seen in scribal hands, and a strong pattern for producing a rather tall and narrow codex continues. One way that the drop-off in the quality of production may be explained is that it was a result of local artisans and scribes in the third century. This must remain a conjecture, but it does explain a period of decreased quality following a period of notable quality. With respect to other high-quality productions in the third century, P.Mich. 764 (a Christian homily), P.Lond.Christ. 2 [P.Egerton 3] + PSI inv. 2101 (Origen?), P.Oxy. 2072 (a locally produced Christian apology), P.Oxy. 4705 and 4707 (two copies of the *Shepherd of Hermas*), P.Oxy. 210 (a Christian gospel-like text), and P.Oxy. 1786 (a locally produced Christian hymn) are all produced in moderately high-quality copies. These texts are not all of the same quality, but for purposes of comparison they exist as more professionally produced texts. On the lower end of the quality spectrum are the two copies of the Gospel of Thomas (P.Oxy. 654–55), the *Gospel of Mary* (P.Ryl. 463), and a copy of the *Shepherd of Hermas* (P.Oxy. 3525).

If the anachronistic categories of canonical and extra-canonical texts can now be disregarded, then applying the categories developed in this essay provide the following portrait of the Oxyrhynchite textual tradition of the Christian community. With few exceptions, the sayings of Jesus whether in narrative or sayings form were disseminated and transmitted in codex form. The quality of the early productions was quite high in the second century, lower in the third, and then increasingly professional in the fourth. This may reflect a community that took over production of its own texts in the third century and no longer relied on the largess of Alexandria or other more established communities. Within that category of professionally produced texts, ones that have indications of being written for public usage in the liturgy or as community owned–produced objects, there are varying degrees of quality. At the high end of the development spectrum are the Gospels of Matthew and John and at the lower end is the Gospel of Thomas codex. In the third century the picture is quite similar. The gospels continued to be produced in codex form, with the majority of other Christian texts being reproduced in roll form. There are clear indications that the community put its resources behind the production of texts like P.Oxy. 406 (Patristic?), or P.Lond.Christ. 2 [P.Egerton 3] + PSI inv. 2101 (Origen?) as well as some of its gospel texts like the Gospel of John (P.Oxy. 1780).

The data strongly suggests that there were at least two types of usage, function, or approach to the production of religious texts within the Christian community. Earlier, the discussion adopted the terms of private and public usages of texts while trying to avoid making these categories qualitative. In this concluding discussion, the discussion will now shift to what these distinctions might mean and how it might shed light on the Gospel of Thomas in the second and third centuries in a Greek speaking Christian community in Egypt.

Private and Public Texts

The distinctions that have been drawn in the foregoing discussion regarding the quality of production of Christian texts are indications of real artifacts that have survived. What to make of those distinctions is less obvious, but some reasonable suggestions can be made. First, drawing upon the work of Roger Bagnall, it can be confidently asserted that purchasing a low-quality gospel text, one that contained a single gospel in a low-quality literary hand would have cost a hopeful buyer about $1,000 for someone with an income of $35,000 per year. That kind of price was beyond prohibitive, but not impossible to imagine. Clergy, on the other hand could afford such costs, and could further reduce the costs of ownership through local productions.[39] Second, there is a surprising lack of Pauline epistles, extra-canonical acts of the apostles, and of the Book of Acts.[40] Texts from the first three centuries overwhelmingly fall into the category of gospels, narratives, and sayings by or about Jesus. This was a Christ-centered community, and the textual productions from the earliest period show that gospel texts were produced with the highest quality standards. Third, in the earliest period some exceptionally high-quality texts were produced that contained patristic texts, one of Irenaeus and two of unknown au-

---

[39] Bagnall, *Early Christian Books in Egypt*, 50–69.

[40] There are no exceptions to this in the second century. For third century exceptions, see P.Oxy. 1597 (Acts); P.Oxy. 4497 (Romans); P.Oxy. 1355 (Romans); P.Oxy. 4498 (Hebrews); P.Oxy. 1171 (James); P.Oxy. 1229 (James).

thors. These texts, which connected the Oxyrhynchus community to the larger Christian world, are some of the finest textual examples that have survived from this community. They may have been carried to Oxyrhynchus, or they may represent a developing sense of identity where the local community put forward its best efforts to produce high quality texts from well-known ecclesiastical figures. Fourth, the *Shepherd of Hermas* and the Gospel of Thomas have examples that cross the spectrum of high quality and low-quality productions, as do the canonical gospels themselves. This suggests that those texts had multiple usages in the Oxyrhynchite churches.[41] With respect to the Gospel of Thomas, it was produced in one of the lowest quality productions from the first three centuries and P.Oxy. 654 was written on the reverse of a document recording a land lease or other transaction.

Scholars often speak of early Christ communities and how those communities passed on the sayings of Jesus. Oxyrhynchus almost certainly qualifies as one of these early Christ communities, and while almost nothing is known of its first century existence, it is clear that it became a Christ-centered community in the second century. It accepted into its canon, if that term can describe a circle of texts that were in use, gospel texts that are both known from the wider Christian world and those whose authors' names have been lost. Few qualitative distinctions can be made among those texts, but there are indications that some were reproduced in lower quality copies that probably signal private usage by Christians. The financial means to reproduce the highest quality texts from this period are almost certainly the products of the churches themselves who would have had the resources to make elegant copies that represented a growing community seeking to create religious boundaries and connections to the broad Christian world.

Conclusion

Without relying too heavily on the costs of production to make usage distinctions, it remains the most reasonable explanation of the evidence that the finest and most elegant copies of Christian texts from Oxyrhynchus were produced for Christian clergy or by the local churches themselves. The lowest quality copies may have been purchased by individuals who wished to own copies of texts that contained the sayings and teachings of Jesus.

---

[41] Blumell and Wayment, *Christian Oxyrhynchus*, 408–10.

The *Shepherd of Hermas* was also popular during the first centuries as was the Gospel of Thomas. Across this spectrum of quality is the added factor of adding lectional aids, the use of punctuation, and using slightly enlarged spaces to signal sense breaks and word divisions. All of those features added to the ability of a lector to use a text in the liturgy. In this regard, the Gospel of Thomas texts fall into the category of texts that were not likely to be read in public settings but were instead planned for private usage due to the small size of their scripts.

Scribes also show a nearly absolute interest in copying texts accurately while not taking liberties in composition, adaptation, or alteration of the texts they were copying. Certainly, there are textual variations from the first centuries, but they are often limited to single words and otherwise attested readings from other New Testament manuscripts. In other words, scribes were not creating text, they were copying texts. The community, however, did produce texts like an apology (P.Oxy. 2072), a dialogue against Jews (P.Oxy. 2070), a hymn (P.Oxy. 1786), and a prayer (P.Oxy. 407). Two of these texts are of a rather high quality, not the finest examples, but of a good quality nonetheless, while one was written on a reused papyrus and likely represents a private, personal prayer of an individual (P.Oxy. 407). Another was a draft copy (P.Oxy. 2070) that was intentionally not prepared for publication and circulation. Such distinctions continue to demonstrate the trends discussed in this essay, namely that some texts were produced in the least expensive ways while other texts and compositions were done with rather greater resources, interest, and creating more aesthetically pleasing products.

A Separate Son of Man[1]

L. Arik Greenberg

The issue of the Son of Man in the Christian canon has perennially been a difficult and thorny problem, one that has garnered wide reaching solutions upon which there is no scholarly consensus. Most scholarship has focused on two diametrically opposed solutions to the debate over the meaning and usage of the Son of Man within Christian literature. One major camp insists that Jesus used the term as a self-referential circumlocution (non-titular), devoid of apocalyptic meaning, and supposedly drawn from an extant contemporary idiom in Aramaic, in turn derived from idiomatic Biblical references to a human being as a "son of man"; this solution considers the subsequent association with apocalyptic imagery to be secondary, a product of the apocalyptic preaching of Jesus being associated with his historical usage of this idiom. The other major camp asserts that various Second Temple Period traditions gave rise to a messianic, apocalyptic figure known as the Son

---

[1] The roots of this article are quite old, planted several decades ago, and birthed under the guidance and nurturing of Greg Riley, whose life and work are being honored with this volume. Initially, in the Spring of 1993, Ron Cameron had suggested a novel reading of Q12:8–10, as presented herein, to his seminar on the Gospel of Mark and Christian Origins, of which I was a member; I am indeed indebted to him for the spark which lit the original fire that led to this article. Over the next decade of graduate work, I further explored the nature of the Son of Man, beginning with a seminar on the Son of Man, taught by the late James M. Robinson, and subsequently delivered early versions of this article in the New Testament Seminar, led by Greg Riley, who was one of my longtime professors and a member of my dissertation committee. I owe him a great debt of gratitude for the guidance he freely gave over the years, as well as embodying a spirit of inquiry and exploration, wreathed by a deep respect for the primary sources and the original adherents and tradents of those textual traditions. While some of the conclusions and reconstructions I offer in this article are speculative, it is with great reverence and gratitude that I offer this article as my contribution to the volume in honor of Dr. Riley, who encouraged and supported the kind of adventurous and passionate investigation that fueled this study.

of Man, and that Jesus either proclaimed himself, or was proclaimed, as this Son of Man; any signification of the term as a self-reference was secondary and consequential. The latter interpretation is often referred to as a titular interpretation. Modern scholars have debated for perhaps over a century which of these processes takes primacy.

In his monograph, *Pre-Existence, Wisdom, and the Son of Man: A Study of The Idea Of Pre-Existence In The New Testament*, R. G. Hamerton-Kelly provides a helpful and brief review of several competing scholarly treatments of the issue of the pre-existence of the Son of Man. He begins with an explication of the idea of pre-existence in Jewish thought, engaging the opinions of Harnack, Bultmann, and others. He then proceeds to address the idea of a pre-existent Son of Man and treats the rivalling work of Erik Sjöberg and H. E. Tödt. His assessment of their scholarly disagreements is summarized thusly:

> Nevertheless, Sjöberg's insistence that Jesus' use of the term 'Son of Man' does entail the idea of pre-existence, and, accordingly, reveals something important about his self-consciousness, is a valid emphasis. Our investigations will suggest that Sjöberg is essentially right at this point, and that H. E. Todt, who stridently and repeatedly denies that the idea of preexistence is present in any layer of the Synoptic Son of Man tradition, cannot be followed on this point. However, the idea of the hidden, pre-existent Son of Man seems to us a much smaller element in the total picture of Jesus in the gospels than Sjöberg claims, and many of the features he attributes to apocalyptic influence most probably derive from a 'Wisdom' tradition. The matter is more complicated than he suggests.[2]

While it is less certain that he is referring to an historically-chronologically pre-existent conception of a Son of Man—and rather referring to a theological entity that is pre-existent in heaven or in ideal—the latter is not far off from the former. For Jesus' references to a pre-existent Son of Man to find recognition in the minds of his audience, and not to fall on deaf ears, the community must

---

[2] See R. G. Hamerton-Kelly, *Pre-Existence, Wisdom, and the Son Of Man: A Study Of The Idea Of Pre-Existence In The New Testament* (Cambridge: Cambridge University Press, 1973), 11.

be aware of the idea to some degree within their theological landscape. For his ideas to take hold, they likely must resonate with something already within their lexicon, much like John's usage of Logos theology, an already extant category with rich connections to the contemporary cosmological and philosophical landscape.[3] That is to say, if some contemporary Jews held a belief in a pre-existent Son of Man as a messianic figure, then hearing Jesus' references in light of their own presuppositions—Jesus speaking directly to these preconceptions—then he likely viewed and presented himself as fulfilling pre-existent expectations that would satisfy his listeners.

According to Douglas R. A. Hare, Bultmann's thesis was one in which Jesus proclaimed the coming, apocalyptic Son of Man and in the post-Easter communities, Jesus was in turn identified with the figure he proclaimed, thus illustrating the idea of the proclaimer becoming the proclaimed. Hare's own monograph attempts to prove that the Son of Man was exclusively a non-titular usage which simply became misunderstood because of its propinquity to apocalyptic language and traditions. He performs very admirably in his discussion of each canonical gospel and the major usages of the term therein, proving that it is very possible to interpret each usage as either titular or non-titular, thus rendering the term ambiguous by nature.[4] Delbert Burkett more recently has produced an excellent and systematic review and reassessment of the scholarly literature in his *The Son of Man Debate*. In his monograph, he charts the major trends in Son of Man scholarship over

---

[3] For an extraordinary treatment of the development of a Logos theology among many of the philosophers within the early Roman imperial courts, as a mitigation of the unbridled expectation of divine honors within the ruler cult, and as a way for court philosophers to simultaneously save their necks and still embody the boldness of their forebears in the face of tyranny, see Glenn F. Chesnut, "The Ruler and the Logos in Neophythagorean, Middle Platonic, and Late Stoic Political Philosophy," *ANRW* 16.2:1310–1332. In this, he gives voice to the oft-ignored pagan roots of Logos Christology, amid the commonplace recognition of Philonic influence.

[4] For a clear picture of the debate as a whole, see Douglas R.A. Hare, *The Son of Man Tradition* (Minneapolis: Fortress Press, 1990). See also D. E. Aune, "Christian Prophecy and the Messianic Status of Jesus," in *The Messiah*, ed. James Charlesworth (Minneapolis: Fortress Press, 1992), 404–422. For an older, more traditional, yet engaging, work on the subject, see H. E. Tödt, *The Son of Man in the Synoptic Tradition* (Philadelphia: Westminster Press, 1965). As a side note, in this essay, I opt, with numerous other authors, to capitalize both nouns in the phrase "Son of Man," in part because of the conceptualization of the latter noun being a reference to the Heavenly Man, to be discussed later in this work.

the past century or so, similarly boiling the current state of scholarship down to the two major camps described above. He identifies himself as being largely in the camp of scholarship that prefers a more titular interpretation of the term, as used by Jesus and the Church, that it "originated as a messianic title applied to Jesus either by himself or by the early church."[5] His overview of the scholarship is quite helpful in coming to an understanding of the various lines and camps within the field. However, his own contribution to the discussion has little bearing on the passages that are key to my theories. While he mentions in passing the oft-neglected *Eugnostos*, and specifically that text's peculiar formulation of "Son of Son of Man", he either neglects to, or prefers not to, explore that text's employment of this formula, and how having an offspring or a successor to the Son of Man would impact upon the validity of the Christian tradition.[6]

On one hand, I am in disagreement with Hare's conclusions because he does not take heed of extracanonical traditions addressing the Son of Man, nor does he address critically enough one passage in particular upon which my own discourse will focus: Q 12:8-10. On the other hand, my assessment straddles a fine line between a titular and non-titular interpretation, recognizing the possibility that Jesus used an extant Aramaic idiom, but also that the term was of a titular and apocalyptic nature that was coined either by Jesus or the Church. While I do not insist upon the existence of a fully formed, "unified concept" of a pre-Christian Son of Man, it is imperative to recognize the possibility of numerous simultaneous and contributing factors and competing scenarios, participating in the creation of what one might term an incipient theology. Nevertheless, Burkett would likely categorize my somewhat liberal stance on the existence of a conceptualized pre-Christian, expected messianic Son of Man, being employed by Jesus (much in line with Bultmann), to qualify me as clinging to a now "marginal interpretation."[7]

---

[5] Delbert Burkett, *The Son of Man Debate: A History and Evaluation* (Cambridge: Cambridge University Press, 2000), 122.

[6] Burkett, 6.

[7] Burkett, 121. In this, he states, "Probably the majority of scholars have come to agree that no unified 'Son of Man' title or concept existed in pre-Christian Judaism (Chapter 7). Our examination of the relevant apocalyptic and rabbinic material confirmed this view (Chapter 9). Consequently, the view that Jesus referred to some other expected messianic figure as the Son of Man must now be considered a

It is my contention that the debate over the nature of the title/idiom "Son of Man" is perhaps far more complex than can be solved with a binary set of competing solutions; that the discussion has become derailed in the process, ignoring certain as-yet untried interpretations of central texts as well as altogether ignoring various non-canonical texts; and pursuant to this, I will offer an alternative, if somewhat speculative, model. Simply put, I stipulate that while there may very well have been a contemporary idiom of Son of Man, used as a circumlocution for "I", and that Jesus may have in fact identified himself this way, there was also likely an unrelated expectation of a coming Son of Man that preceded his ministry. As such, I contend that these two distinct usages of the Son of Man became enmeshed during the early days of the Jesus movement and the separate adherents to a Son of Man theology became subsumed and swallowed up by the Jesus movements, their Son of Man losing his separate identity as well. It may be that Jesus used the term in his discourse as an idiomatic expression, illustrative of things he was trying to say, but it soon became very easy to reimagine him as speaking of the apocalyptic Son of Man. Of course, it may be possible that Jesus, himself, knew of the pre-Christian Gnostic tradition which subscribed to a Son of Man messianology or theology. However, it is also possible that Jesus never intended to identify himself as this very figure, the apocalyptic Son of Man. The idea that a messianic expectation which, shaped by Danielic imagery, employed the ancient terminology of Son of Man as its nomenclature, preceded Jesus, and that he identified himself with this figure, is nothing new. However, many scholars have doubted that the terminology existed on its own, as a clear messianic reference, prior to Jesus' usage of it as such. As such, for one to prove that this terminology existed independently of Jesus, and prior to Jesus, one would have to look very differently at the primary sources, and even cleverly engage sources that have not been previously involved in the discussion. Taking a fresh look at one set of passages from Q, as well as passages from the Gnostic literature of Nag Hammadi, previously thought irrelevant to this discussion, this study seeks to do just that.

Within the pericope incorporating Q 12:10, there may be evidence to support a Son of Man separate from Jesus, perhaps even

---

marginal interpretation. The title 'Son of Man' in all of its occurrences in the Gospels can best be understood as referring solely to Jesus.

predating Jesus traditions. Inasmuch as there were Jewish sects which postulated a lesser Yahweh, such as in Enoch literature, or an emanation theology, such as in certain forms of Gnosticism and Adam speculation, it is plausible that there was a group which posited a theology in which an emanated entity known as the Son of Man was an offspring of Yahweh—the heavenly Adam. Ruled by literary ambiguity, this figure was linked with other, non-titular usages of the term, Son of Man. This community did not have wide acceptance in the mainstream forms of Judaism and thus sought refuge in the Q community, becoming inexorably linked to the figure of Jesus, who may or may not have used the term Son of Man to mean a variety of things, including the commonly mentioned, simple self-referential circumlocution.

My thesis relies heavily upon the proposition that Gnosticism has an early origin in the Mediterranean world, and is not merely a late and dependent perversion of the Christian message.[8]

---

[8] My supporting ideas come from the innovative views on the development of Christianity offered by Greg Riley, who has perennially argued that Zoroastrianism impacted the Judaean world after the Babylonian Exile and influenced the lines of development of Judaism and Christianity in ways that allow a wider definition of Gnosticism than has previously been utilized, one which assumes the general gnosticizing of Palestinian Christianities. His model allows Plato to be seen as gnostic since he has the basic elements which underlie all Gnostic sects: cosmological and anthropological dualism, a journey of the soul, and a gnosis which allows the human to know the underlying reality which he or she has forgotten. In turn, certain Gnostic texts which mention the Son of Man in his various roles, without connecting him to Jesus, will be shown to have pre-Christian origins.

Riley speaks most extensively about the early origins of Gnosticism, and its constructive influence upon nascent and incipient Christianity in his *The River of God* (San Francisco: HarperSanFrancisco, 2001), which I believe to be his magnum opus. As such, we are instructed that "The origins of these Gnostic texts go back before the Christian era into times of controversy between traditional Jews and other Jews highly educated in Greek philosophy." Riley, *River of God*, 47. He later states, regarding the early influence of Gnosticism, "The Gospel of John was written near the end of the first century, probably in Syria. It is the one Gospel of the canon that shares the most with Gnosticism, and is quite similar in worldview to its rough contemporary the *Gospel of Thomas*, also from Syria," and concludes, "Later Christianity owed much to these Gnostic teachers. Perhaps their most important contribution lay in further developing the concept and necessary language of emanation and applying these to Christian theology." Riley, 77–78.

Later in the same text, Riley shares a more extensive reconstruction of Gnostic history and its influence by Zoroastrianism, reiterating his earlier claim, "In the second century BCE, there began to develop among certain highly educated Jews the roots of what today is known as Gnosticism," Riley, 202 and concludes with

Two texts which may be seen as simultaneously pre-Christian and Gnostic are Eugnostos the Blessed and The Hymn of the Pearl. Just as Hymn Pearl must be removed from its context in the *Acts of Thomas*, a third text may be used as evidence if one allows a pre-Christian *Vorlage* to the Apocryphon of John. I will not, however, treat this text in my present discourse as it would be superfluous. The Son of Man in the Gnostic traditions is not primarily a judge, as he is in the Q passages to be discussed, and in *The Book of Enoch* as well, but a figure of celestial lineage, serving as one of the progenitors of the Gnostic redeemer. In order to avoid the slippery slope of philological debate over the meaning and origin of the term Son of Man as it manifests in the Greek texts, we must look at whose son the Son of Man is. In a simple glance at the term itself, one can already see that his father is "Man." A look at the texts of Ap. John and Eugnostos alike show us that the Son of Man is the son of the Primordial Man, the heavenly Adam. In order to engage fully in this discussion, we must address the importance of the tradition of Adam speculation within Gnostic and Gnosticizing circles. Without Adam speculation, the Son of Man simply becomes the υἱὸς ἀνθρώπου of the LXX and the בן אדם of the MT, simply a descendent of the earthly Adam, a human being. Kurt Rudolph in his book, *Gnosis*, treats the issue of Adam speculation very concisely in his initial explanation:

> The central position occupied by man in gnostic theology led to a particularly important complex of ideas, customarily described as the "doctrine of the God 'Man'". It is also known under the name of "Urmensch myth" or (from the Greek word for "man") "Anthropos myth". The basic idea lies in the close relation or kinship of nature between the highest God and the inner core of man. This relationship, evidently with an eye on the biblical statements, is understood as a relationship of copy to original, i.e. the (earthly) man is a copy of the divine pattern, which likewise often bears the name "man". One text refers to him as "the Father of truth, the Man of the greatness".... the highest being himself if the first or primal man (*anthropos*), who through his

---

what may be one of the most important statements about the origins of Christianity, "Gnostic writers were the first to combine the dualism of Zoroastrianism with the Monad of Greek philosophy, producing a comprehensive religious worldview that was a direct precursor of Christianity." Riley, 214–215.

appearance to the creator powers gives them a pattern or model for the creation of the earthly (and therefore second) man, in the other the highest God produces first of all a heavenly man of like nature (frequently called "son of man"), who is then the direct prototype of the earthly (and therefore third) man.[9]

This Adam speculation is central to the understanding of the Son of Man as a second Yahweh, the Son of God. This will be treated later as it fits into the discussion of the Son of Man as offspring of the heavenly Adam. As such, I will argue that there was dissent within the Q community over the nature and identity of the Son of Man and thus its usage and backing might have been fraught with controversy, making the discernment of its origins no simple endeavor.

Major amounts of energy have been focused on the passage of Daniel 7:13ff. due to its apocalyptic nature and the very ambiguous usage of *bar enash* in such a way that would make one wonder if the term is actually a title, or an idiom used to refer to oneself.[10] However, it is still uncertain as to whether the passage uses the term in a titular fashion, referring to this apocalyptic figure as "Son of Man" by name, or merely refers to the entity as being likened to a "son of man", an idiom which crops up over one hundred times in the Hebrew Bible.[11] Most of these instances are contained in the Book of Ezekiel, probably a post exilic composition, in which the term is used by the Lord to refer to Ezekiel. It is a somewhat diminutive term, conveying Ezekiel's humanity, his humility, and his ability to represent the entirety of humanity in his preachment.[12] However, most of the other circumstances in the Old Testament in which the term occurs bear a striking notion of counterpoint, poetically comparing man to the son of man in each passage. Most of

---

[9] Kurt Rudolph, *Gnosis* (San Francisco: HarperSanFrancisco, 1987), 92.

[10] See Reginald H. Fuller, *Foundations of New Testament Christology* (New York: Charles Scribners Sons, 1965), 34–43.

[11] For a lengthy discussion of the subject in which the author concludes that none of the pre-Christian literature can substantiate a titular usage of the term, see Geza Vermes, *Jesus and the World of Judaism* (Philadelphia: Fortress Press, 1983), 96.

[12] For a discussion of the Book of Ezekiel, see Bernard W. Anderson, *Understanding the Old Testament* (Englewood Cliffs, NJ: Prentice Hall, 1986), 429–430, 627. Anderson briefly mentions the usage of son of man in the Book of Ezekiel, where he mentions its intention "to highlight the mortal weakness and finite humanity of the prophet in contrast to the holy, majestic deity of Yahweh." Anderson, 627.

these are post-Exilic as well. Very few use the term when not in contrast to the term Man. However, one possibly pre-Exilic passage[13] is in Numbers 23:19, the first instance of its occurrence in the Bible:

> God is not a man, that he should lie, or a son of man, that he should repent. Has he said, and will he not do it? Or has he spoken, and will he not fulfill it?[14]

In Job 25:5-6, a post-Exilic rendering of a very old story, the usage is similar:

> Behold, even the moon is not bright and the stars are not clean in his sight; how much less man, who is a maggot, and the son of man, who is a worm!

The second occurrence in Job is similar. Most of the occurrences in the Psalms display the same rhetoric. However, one interesting thing is evident in Psalm 80:17. In this passage, which uses the counterpoint comparison of the majority of these Old Testament passages, the imagery is that of the Son of man seated at the right hand of God, an image which has become very powerful in the Son of Man mythology of the New Testament and associated literature. It runs thusly:

> Let your hand rest on the man at your right hand, the son of man you have raised up for yourself (Psalm 80:17).[15]

In this passage, we have the standard imagery of the Son of Man seated at the right hand of God. Whether or not this passage was intended to look upon the term as apocalyptic, this is pre-Christian imagery which has utilized the term in a way which recurs and, according to the suggestion of Riley, derives from the Baal cycle of Canaanite mythology.[16] I will most certainly return to this issue in more depth.

---

[13] John Van Seters suggests a different dating for this text. See *Prologue to History* (Louisville, KY: Westminster/John Knox Press, 1992), 227ff.

[14] Unless otherwise noted, all Biblical references are quoted from the Revised Standard Version, in order to provide a more literal translation of the text.

[15] Scriptural reference is in New International Version in this case, in order to highlight the usage of the idiomatic son of man terminology *in situ*.

[16] Fuller correctly recognizes apocalyptic as "having arisen out of the earlier prophetic eschatology, but [having] been extensively influenced by the dualistic eschatology of Iranian religion." Fuller, *Foundations of New Testament Christology*, 34.

There has been, also, much focus of energy upon the philological nature of the term. Scholars have attempted to prove one way or another, that the term either suggests the state of humanity or, on the other hand, suggests an apocalyptic nature which is inherently invested in the terminology. I contend that the case may simultaneously be both, but—along with numerous other scholars—that the term became unintentionally associated with the eschatological figure it was describing. Eduard Schweizer felt that Jesus used the term in his preachment for precisely these reasons.

> My hypothesis... supposes that Jesus took up the term 'son of man' just because it was not yet a definite title. It was a term stimulating the hearer to reflect and to answer the question, put by its usage, who Jesus really was. It described, first of all, the earthly 'man' in his humiliation and coming suffering. It depicted the messenger of God suffering for his people and calling it to repentance. It declared that this very 'man' would confront his hearers in the last judgement, so that their yes or no to the earthly Jesus would then decide their vindication or condemnation. It so contained the mystery of the one who like the poorest slave serves at table, and yet invites those on whom he is waiting to his heavenly meal..., in the fulfillment of time, when the insignificant grain of mustard seed will unexpectedly prove the greatest of all shrubs.[17]

This characterization of Jesus is intriguing. We may never be sure as to the exact intentions of the historical Jesus, but the current field of scholarship should stand ready to trace responsibly the heritage and development of various strands of Christian tradition. In such a way, Schweizer's Jesus serves as an example of what the earliest Christian apocalyptic tradents of the term may have meant or intended in their usage. The term was ambiguous enough to be placed upon an eschatologically loaded figure and it thus took on the accoutrements which surrounded the context. However, must we necessarily assume that the Son of Man was always identified with Jesus? Might there have been a corpus of literature prior to the Christian tradition which utilized the term in an apocalyptic light? I submit that although most scholars rely upon canonical works for their envisionment of the Son of Man, stepping outside of the canon only to give a biased look at Enoch or Ezra, there are certain

---

[17] E.D. Schweizer, "The Son of Man Again," *NTS*, 9, no. 3 (April, 1963), 259.

traditions which have been conveniently pigeon-holed as "late" so that scholars would not have to revise their understanding of Christian origins.

There are traditions, both pre- and post-Christian, which display a certain usage of the term which must necessarily be non-idiomatic/circumlocutionary and are exclusively titular in an apocalyptic fashion. By surveying these varied materials, we may see that the term was able to stand alone, prior to Jesus, as well as after him, and still bear apocalyptic significance without appealing to Christ Cult traditions. By looking at the non-Christian usages of this term, the evidence of a pre-Christian Son of Man contained even in the New Testament, and by showing the similarities between the imagery of the heavenly placement of the Son of Man and the heavenly placement of Baal at the right hand of El within the heavenly banquet, I argue for an alternative model to the standard assumptions that either Jesus used the term historically as a simple circumlocution which later became invested with the apocalyptic of his preaching, or that Jesus used the term from the very beginning as a way of identifying himself as the coming eschatological judge.

The exact moment in which the idiomatic usage of the terminology became melded with the apocalyptic title is an uncertain one, but I contend that the two usages were temporarily kept separate, and that the Jesus traditions may have been the originators of the dual usage of the idiom, allowing it both to signify the already extant apocalyptic and celestial figure, and also to stand as a circumlocution. It is important to isolate the particular stage in the tradition history of the Jesus traditions in which the term was still clearly a reference to such an apocalyptic demigod. This stage may be evident in a particular pericope of Q that I have briefly invoked, above, and it embodies the absolute necessity for us to recognize that the term could indeed stand apart from Jesus and yet simultaneously be placed on the tongue of Jesus. As such, Jesus may have indeed spoken of himself as a son of man (indicating an Ezekielic notion of humility); he may have even spoken of an apocalyptic Son of Man, as an external, otherworldly figure, separate from himself. Whether the historical Jesus identified himself as *the* apocalyptic Son of Man is likely immaterial to the current discussion, being as irretrievable as his very words have proven to be, among Q scholars.

Some conflict was likely centered around a misunderstanding of the proximity of these two usages of the previously idiomatic term, which in certain circles had already been used as a signifier for an apocalyptic figure. In an irretrievable moment, post-Easter groups of followers who were not initially part of Jesus' most intimate fold might have begun to interpret the two usages of the term as identical, feeling that Jesus' reference to the non-apocalyptic circumlocution was an identification of the apocalyptic Son of Man with himself. Such a misinterpretation may have caused some dissent and conflict between the rapidly expanding post-Easter Jesus communities.

The pericope in Q 12:10 may indicate a stage in which the development of the term as denoting an apocalyptic figure identical to Jesus may not yet have been an issue of importance to the Jesus communities, but the mere existence of such an apocalyptic figure was indeed a source of contention between factions within the community. This reconstruction of the social moment suggests that the Q community had not yet unanimously assumed the identity of Jesus and the apocalyptic Son of Man, or, for that matter, the existence of an apocalyptic Son of Man at all, and required no confession of belief in such an eschatological figure for inclusion in the community. The unquenchability of the debate required that persons must proclaim allegiance to Jesus as teacher or founder, but specific belief in the Son of Man, as a concept of an apocalyptic figure, was not required. This phraseology in Q 12:8-9 and then Q 12:10 was the source of later confusion, as it tended to allow the terminologies in their developing state to overlap. Thus, because of the ambiguity of these two adjacent passages, the next stage of the communities oversaw the identification of Jesus (who may very well indeed have used the term as a clever circumlocution) with the coming Son of Man.

Regardless of whether or not it is true that Matthew, in his rendering of Q 12:8-9, correctly understands the usage of the term as idiomatic in this particular pericope, this is a secondary and inauthentic saying which was cast by the post-Easter church in order to rationalize allegiance to the Jesus tradition as minimal criteria for membership in the group.[18] This logion could only have been created in the face of some moment of slander or persecution. Where is such persecution unequivocally evident and provable except in the post-Easter church? The following logion, Q 12:10, was

---

[18] The Son of Man in Luke 12:8 is shown by its Matthean parallel (10:32) to be helpful but negligible. I will address this below.

perhaps cast in the ensuing moment, in the face of the aforementioned conflict over the apocalyptic figure of the Son of Man. Equally as secondary, this saying was intended to express the necessity for reverence to the Holy Spirit as supreme, reducing belief in the concept of the Son of Man to the realm of an option. Thus, if one is caught slandering the Son of Man (as an apocalyptic figure) one will be forgiven; but if one cannot be reconciled to belief in the Holy Spirit, one will not be forgiven.

Q 12:8-10

It would be useful to the current discussion of the terminology to analyze this particularly confusing and ambiguous passage which employs the "Son of Man" (ὁ υἱὸς τοῦ ἀνθρώπου) in its various applications. Within the confines and context of this pericope, it is inconsistent with the narration to claim that Jesus and the apocalyptic Son of Man are identical. In Q 12:8-9, Jesus clearly states that what one does with respect to him, how one acknowledges or denies him, will be addressed accordingly in some unspecified, future eschatological moment. He claims that each will be repaid in kind and dealt with according to the mode of action of each. One who acknowledges Jesus will be likewise acknowledged. One who denies Jesus will be likewise denied. Forgiveness is not mentioned in this verse at all; it is not a consideration.

The subsequent conjunction "and" (καί) indicates that the statement follows a previous thought and is dependent upon the antecedent. If the common hypothesis presupposing the correctness of the Lukan order is true,[19] then we must trust that Q 12:10 continues the discursive domain of Q 12:8-9. In Q 12:10, the subject is that of what kind of transgression, or better yet a transgression against which party, is to be forgiven and who is thus to be shriven. If one interprets the "Son of Man" in this passage to be Jesus, then the previous proclamation is nullified, and the legislation is overturned. Can this logic possibly follow?

---

[19] In John S. Kloppenborg, et al., *Q Thomas Reader* (Sonoma: Polebridge Press, 1990), 23, we are told, "Though it seems inevitable that no two reconstructions of Q will be exactly alike, a broad scholarly consensus has been reached on some points. It is generally accepted, for example, that Luke has better preserved the sequence of the Q-material, whereas Matthew has tended to distribute it throughout his gospel."

> *If you slander me, you will be held accountable. And if you slander me, you will not be held accountable.*

It seems clear to me that at the point at which this pericope was composed, the "Son of Man" as an apocalyptic figure and Jesus were entirely separate. As the Lukan order demonstrates, this set of passages constitutes a three-part legislation about one's allegiances. The first is to Jesus, for which one will be held accountable. The second is to the Son of Man, for which one will not be held accountable. The third is to the Holy Spirit, for which (as with one's allegiance to Jesus) one also will be held accountable. In the mind of the skeptical exegete, the question may remain: does the mutual denial necessarily constitute the withholding of forgiveness? Or can it be a ceremonial denial and then a subsequent forgiveness? In his article on the reinterpretation of the roles of Jesus through the deeper understanding of Q, an article which appeared in the volume honoring James M. Robinson and his work, Burton L. Mack speaks of this very subject. He reminds us that

> One denies (or confesses) loyalty to Jesus; one speaks against (or for) the Son of Man. To 'speak a word against' is a rhetorical term that implies the making of a speech, not just the entering of a plea.[20]

To make a speech against Jesus would be tantamount to denying him publicly. This imagery is consistent with the forensic/legalistic imagery which is evoked by the resolution of this logion, that Jesus will deny the transgressor before God in heaven. We are not dealing with different crimes between these two verses (8/9 and 10).

Matthew understands the dialectical simultaneity of the two usages of the terminology, Son of Man, both idiomatic and apocalyptic, in the Q community, owing to the hypothesized ideological propinquity and dependence of the evangelist's community upon that of Q.[21] As such, compared with Luke's rendering of Q

---

[20] Burton L. Mack, "Lord of the Logia: Savior or Sage?" in *Gospel Origins & Christian Beginnings: in Honor of James M. Robinson*, ed. by James E. Goehring, et al. (Sonoma, California: Polebridge Press, 1990), 16.

[21] Burton Mack, in *The Lost Gospel: The Book of Q and Christian Origins* (San Francisco: HarperSanFrancisco, 1993), 173, states, "If one were to ask which of the narrative gospels most nearly represents an ethos toward which the community of Q may have tended, it would be the Gospel of Matthew."

12:8, which in Luke and its Markan parallel (Mark 8:38) is rendered as "Son of Man" (ὁ υἱὸς τοῦ ἀνθρώπου), Matthew's preservation (in Mt 10:32) of the clause as "I will acknowledge" (ὁμολογήσω) — rather than the Son of Man will acknowledge — may further suggest that Matthew understood that the Son of Man could be used as a circumlocutionary reference indicating Jesus, separate and distinct from the apocalyptic Son of Man, and thus wished to omit the reference to the Son of Man in verse 8, for fear that there might be some confusion over who was being invoked. Any reference to the "Son of Man" preserved by Luke in 12:8 would have shown that Luke most probably copied the verse accurately from Q but thought that its meaning was sufficiently apparent. Luke either knew that Jesus was claiming that the Son of Man would be his advocate in such a situation, or simply was unwilling to change the wording of the passage to fit what he may have thought it should say. If Luke's inclusion of the Son of Man in this passage was an editorial addition, then perhaps he was already eliding the differentiation between Jesus and the celestial Son of Man, something which Matthew was not yet willing to do. If it were intended that Jesus and the apocalyptic "Son of Man" were to be understood as identical, the aforementioned misgivings of the skeptic would have been addressed moreover by at least a conjunction or a qualification expressing a change of policy or the like. As it stands, the two — Jesus and the "Son of Man" — are discrete.

Allow me to address further the formal structure of the pericope. The tenor of the discourse changes at the point between Q verses 9 and 10. The addressing of the transgression against Jesus takes on a certain importance through the formal qualities of the verse. Two full verses are given to the subject. It is important enough to specify the two alternating situations (one of mutual denial, the other of mutual acknowledgement) and not merely to supply one and imply the other. The conjunction flow of the verses 8 through 10 runs as follows: (Proclamation) ... but (δὲ) ... and (καὶ) ... but (δὲ). If one takes a step back to observe these formal nuances, it follows that the καὶ beginning verse 10 shows that a different topic is being addressed. The hypothetical case of the skeptic (that the denial in vv.8 and 9 is merely ceremonial and is overshadowed by v.10 as if the "Son of Man" and Jesus were identical) would require that a conjunction of "but" or "nevertheless" or "still" or the like be entered in its place to allow one to think that way.

Of the differences between the Matthean and Lukan versions of this logion, John Dominic Crossan seems to follow along with Harnack's assumption of Matthean preservation of Q in his own offered text of the "Reconstructed *Sayings Gospel Q* 12:8–9."[22] Crossan, very reasonably, believes it was Luke who introduced the Son of Man reference to the core saying of Q 12:8–9, relying upon the assumption that it was probably done in light of the knowledge that the rough Markan parallel to this passage (8:38) carries a slightly differing form of this Q logion. We may assume that Mark either knew an oral form of what was included in Q or simply chose to alter it to fit his theology if one accepts the argument that Mark knew Q as a written text.[23] It is difficult to accept that Mark truly had any interest in using Q as a written document, judging by his lack of direct reference to it in his gospel. Nevertheless, it is probable that some form of this saying was common to the lexicon of both Mark and Q at the time of their respective compositions, permitting us to view them as independent attestation that the earliest core saying included a reference to the Son of Man. Crossan disagrees, saying that the attempt to prove that Mark retained the earliest form, including the invocation of the Son of Man, fails on account of its unlikeliness:

> [B]ecause of the formal matrix, it is God or the passive voice that one expects to find as the original protagonist, just as in the reconstructed *Sayings Gospel Q* version. Second, consider the two other versions from later strata of the tradition.

Crossan quotes Revelation 3:5 and 2 Timothy 2:12 as his examples of "later strata of the tradition", continuing:

> The last parts of those two verses contain, respectively, a positive and a negative edition of the aphorism, and in both cases the protagonist is Jesus as "I" or as "he."

He concludes that there are three distinct stages through which the core saying passed. The first is that which he offers as the double passive "will be acknowledged/will be denied" couplet of his rendering of the original Q. The second is Matthew, which identifies the giver of testimony as "I" or Jesus. The third is the Lukan version which he claims is affected by the Markan parallel, having

---

[22] John Dominic Crossan, *The Historical Jesus* (New York: Harper Collins, 1991), 248. Italics his.
[23] Mack, *The Lost Gospel*, 172.

been induced to identify the giver of testimony as the Son of Man.²⁴ However the redaction history of the text may actually stand, we must look at how Luke uses this device and how it affects his narration of Jesus' speech. Regardless of how the text stands in Q 12:8b, verse 8a proclaims that Jesus must not be denied, or else the transgressors will, themselves, be denied. If Jesus is the Son of Man mentioned in verse 10, the earlier saying cannot be true, because of the inherent contradiction in verse 10.

Leif Vaage, in his article for *Semeia* 55, rightly recognizes the disparity of the two sections of this pericope, but comes to a different conclusion than mine, though it is gratifying that he also perceives seemingly contradictory language in the passage as it stands.

> The stratigraphical location of 12:10 is very uncertain, though it is difficult to see how it could be argued that this saying once formed part of the document's formative stratum. It is most improbable that 12:10 originally followed 12:8-9, which it flatly contradicts. Q12:8-9 insists on the importance of confessing the son of man; 12:10 asserts precisely the opposite: that you can say whatever you want about him, the only relationship that counts is with the holy spirit. At the same time, it is evident that 12:10 is where it is in Q in order to comment on and qualify the preceding saying.²⁵

Vaage is unwilling to see that the Son of Man, here, is not Jesus. If it were, then the saying would in fact be contradictory. Why break up the continuity of the saying, being coherent in Luke and probably so in Q, if one has not yet considered all of the possibilities? The parallel to this saying in Gospel of Thomas 44 also preserves an independent witness to the fact that the core saying probably circulated together.²⁶ Vaage continues, saying that Q

---

²⁴ Crossan, *The Historical Jesus*, 247-248.

²⁵ Leif Vaage, "The Son of Man Sayings in Q," *Semeia* LXV (1991): 118.

²⁶ I agree with those scholars who believe it likely that the Gospel of Thomas is literarily independent of the Synoptics and circulated independently of them and may preserve an earlier stage of the tradition in many of its logia. Among them is Greg Riley, who intimates in *Resurrection Reconsidered*, that the Thomasine community, "produced widely used literature in the name of Thomas: the *Gospel of Thomas* survives in at least two recensions, in two languages, and contains very early forms of the sayings of Jesus, stamped with its own independent theological outlook...." *Resurrection Reconsidered* (Minneapolis: Fortress Press, 1995), 79. See

12:10 may have been added as a later gloss to verses 8 and 9, as "it is difficult to imagine that the person who contributed 12:8-9 to Q was also responsible for negating the import of that contribution in the next breath."[27] Vaage may be subject to circular reasoning in this particular case, because of his assumption that the Son of Man is necessarily by nature a self-reference to Jesus, which would necessitate a theory that the two halves of the saying cannot have been originally composed together.

If one is convinced by my study, above, of the formal flow of the passage, yielding a Son of Man who is personally and ontologically different from the Jesus who is speaking, then what would Luke's interpolation of the Son of Man do to the passage? This would make it even more obvious that the Son of Man is separate and distinct from Jesus. Jesus is remembered to have invoked him as heavenly advocate and judge. If I have convincingly argued above that the Son of Man in Q 12:10 is not Jesus, then how are we to understand the reference to this same Son of Man in this passage? The Son of Man is invoked as an unavenged victim in Q 12:10. When he is invoked by Luke in the preceding passage, he stands as a strong witness, one with the power to give testimony before God that will condemn or vindicate any defendant. In the following passage, he himself becomes the defender against slander. In Luke, this creates a motif of prosecutor becoming persecuted, which allows the Son of Man to have a definite apocalyptic and ontological function, but still to be negligible enough to stand as a minor doctrine, able to be denied by those who wish not to believe in his existence. The offense is minor, and the punishment is nil. Following the permissiveness of doctrine which 12:10 intends, Luke nevertheless strengthens the case for the Son of Man being present in the original form of verse 8.

Matthew, however, leaves out this first Son of Man reference, attributing the function of witness to Jesus himself. Jesus is given power in heaven as well as on earth which he does not have in the Lukan version. What is done to him on earth, he will tell to his Father in heaven. The Son of Man is then left powerless and contextless in the next verse. Perhaps at the time of Matthew's

---

also Helmut Koester, "Introduction: The Gospel According to Thomas," in *Nag Hammadi Codex II, 2-7*, 1. 38-39, ed. Bentley Layton, Nag Hammadi Studies 20 (Leiden: E.J. Brill, 1989), 38-39.

[27] Vaage, "The Son of Man Sayings in Q," 118.

composition in the post-Markan era, the Son of Man theology, as a separate movement, began to lose its popularity or its efficacy. This, we cannot yet know. Regardless, the Son of Man no longer has his role as witness in the Matthean version of this logion. Matthew has possibly erased the Son of Man from the logion in favor of Jesus being portrayed as the crux of the judgment scene, while Luke, in this one instance, remains faithful to the obtained Q text.

If Q 12:8–10 does indeed provide a glimpse of the dissent within the community over the belief in the Son of Man apocalyptic, then we are led to believe that the saying itself was probably composed by the faction in favor of the Son of Man apocalyptic. The phraseology bears the tendency of the author to consider the Son of Man a real figure, as real as our own existence, and offers amnesty to those who don't. As such, the writer seems also to have had a certain degree of authority in order to be able to grant this amnesty in such a categorical manner. One who had no such authority would have usurped it and simply denounced the Son of Man. Had the positions of authorship been reversed, that is, if the authors had been those who did not already believe in the existence of the Son of Man, then this saying might have betrayed more clearly the conceptual nature of the Son of Man. For example, in such a hypothetical scenario, the saying might have read "And whoever does not accept the Son of Man will be forgiven," or even "And whoever does not have faith in the Son of Man will be forgiven," but should include something which would further betray the fact that the author did not believe that Jesus spoke of such a Son of Man or that this Son of Man exists at all other than in the minds of those who support that theology.

A number of observations by H.E. Tödt are worth noting. Tödt has not reviewed or even considered the possibility of the Q movement proclaiming a Son of Man separate and distinct from Jesus, from the very beginning utilizing Jesus as the point of departure and treating him as the Son of Man. He insists upon interpreting Q passages in a post-passion and resurrection context despite the absence of both of these elements in the synoptic sayings source!

> Here, too, the name Son of Man does not designate the figure of a transcendent Perfecter, but in accordance with Matt. 11.19 and par. it designates Jesus acting on earth and being attacked by his opponents. The point of interest here is not how Jesus may be defended but how his opponents may be forgiven. They can be forgiven in so far as they

turned against the Son of Man merely in his activity on earth. But there is no forgiveness—in the post-Easter situation—for the one who sets himself in opposition to the manifest activity of the Holy Spirit. Two periods in the history of salvation are distinguished here in the saying from Q, the period of Jesus' activity on earth as the Son of Man and the subsequent period of the activity of the Holy Spirit.[28]

Tödt has interpreted the sayings of Q in a Markan context, seeing the Son of Man as representative of Jesus' "acting on earth with full authority". How, then, can the public Jesus of Q 12:8 be denied, the one committing this sin being punished; and the earthly Jesus as the Son of Man be denied, the one committing this sin being forgiven? As I demonstrated earlier, this does not work. The Son of Man spoken of in verse 10 cannot be the Jesus spoken about in verses 8 and 9, regardless of the authenticity of the Lukan interpolation of the Son of Man into verse 8, in which he acts merely as witness. The worthwhile qualities of Tödt's article are in his assessment of Mark's usage of the term. He rightly recognizes Mark's intentions of harmonization of the several Christological terms and concepts, with the resultant attitude that the Son of Man is not only the earthly Jesus but also the Jesus of the return. Mark has melded the apocalyptic Son of Man with the present Son of Man and even added the dimension of the suffering servant of God. Thus, forgiveness of those who blaspheme the Son of Man in this context would be unthinkable. Subsequently, Mark's rendering of the core from which Q 12:10 comes (Mark 3:28ff) reinterprets Son of Man as sons of men (meaning humans), leaving acts by and against humans in the realm of forgivable actions.[29]

In Q 6:23, we see once again a case in which Matthew understood the Q text—which may have included a Son of Man saying—to indicate the earthly Jesus of the public ministry, but which Luke tended to interpret apocalyptically. That Matthew preserves a different version of a saying, which included the Son of Man in its Lukan version, indicates that Matthew may have purposefully hidden evidence of a situation in which the Q people were being persecuted for their belief in the Son of Man theology. Matthew has interpreted this in light of his knowledge of Mark and

---

[28] Tödt, *Son of Man in the Synoptic Tradition*, 119.
[29] Tödt, 199–120.

states that the followers of Jesus will be persecuted on account of Jesus. Matthew, in light of Mark's portrayal of Jesus as the "suffering Son of Man," and through his attempt to elide the Son of Man theology formerly present or implicit within these logia, loosens his former Q ties to the Son of Man theology which was of so much trouble to the community that fashioned Q 12:8-10. This saying which Leif Vaage speaks about in his article on the stratigraphical location and significance of the Son of Man sayings in Q,[30] may be seen as an early, yet inauthentic saying of Jesus, called upon to grace the pages of the first edition of Q, but highly suspect when viewed in the context of Q1's theological positions, social tones and assumptions.[31] This saying shows us that even at a very early stage after the death of Jesus, the Son of Man theology may have begun to be teased out of (or interpolated into) its accidental invoking by Jesus. Here, some of the Q people are being hated, excluded, reproached, and cast out as evil, all on account of their propagation of the Son of Man theology.

Allow me to also take a brief look at the parallel of Q 12:8-10, preserved in Gospel of Thomas 44:

> Jesus said, "Whoever blasphemes against the Father will be forgiven, and whoever blasphemes against the son will be forgiven, but whoever blasphemes against the holy spirit will not be forgiven, either on earth or in heaven."[32]

This logion must be viewed fairly if one is to remain unbiased with regard to canonicity as a method of determining authenticity. With the finds of the Nag Hammadi Codices, we can no longer be content simply to study the Synoptics and argue ourselves into circular boxes, denying the existence of any valid extracanonical texts.[33] It is probable that Thomas preserves a more Gnosticizing recension of this logion, appealing to the topic of the Trinity, but what is evident still is that this was a highly copied and revised core saying which

---

[30] Vaage, "The Son of Man Sayings in Q," 107-109.

[31] See Mack, *The Lost Gospel*, for a treatment of the stratigraphy of Q, as well as John S. Kloppenborg, *The Formation of Q* (Minneapolis: Fortress Press, 1987), from which many of Mack's ideas derive.

[32] All translations of GTh by Marvin W. Meyer from Kloppenborg, et al., *Q Thomas Reader*, as above.

[33] I respectfully defer to James M. Robinson for the fire of my polemic against canonical incest. See Robinson, "The Study of the Historical Jesus After Nag Hammadi," *Semeia* XLIV (1988): 45-55.

found its way into the pages of Matthew, Luke, Thomas, and in a rough parallel, in Mark. Clearly, no Son of Man is present in this recension of the logion, and following my aforementioned observation about the frequency of this core's appearance, it can be said that it may actually harken back to an earlier and possibly authentic saying of Jesus which is now lost to us in its original form. It is questionable as to whether or not Thomas displays a tendency toward a Son of Man theology, and thus the ability to pick and choose, mix and match, which is so common of the early compilations of Jesus' sayings, shows up here in Thomas as his appeal to a Gnosticizing element (in light of his realized eschatology) devoid of any reference to an apocalyptic Son of Man. Therefore, the presence of the apocalyptic Son of Man in Q 12:10 and possibly in Q 12:8–9 suggests that whatever the authentic form of the core saying may be, the Q community utilized it to display their perception of persecution due to their Son of Man theology. Nevertheless, Thomas' reference to the Son will become evident as a possible recognition of the Son of Man theology present in other Gnostic texts, as I will discuss later.

The Baal Cycle

The issue of the imagery used in Mark 14:62 is of prime importance when we recognize that it may in fact derive from the Baal Cycle.[34] In the story of Baal and Yam, Baal has incurred the anger of his adversary, Yam, who is the sea, and messengers from Yam appear at the banquet of the gods to deliver an ultimatum to Baal. The council of the gods is admonished to hand Baal over to be taken captive by Yam. In this scene, upon which the messengers from Yam enter, we encounter Baal standing at the right hand of El, the king of the gods. While it must be admitted that there is no usage of the term Son of Man here in the story, the imagery used is clearly that which shows up in Daniel 7:13, Mark 14:62 and later in

---

[34] I express my gratitude to Greg Riley for bringing this rich mythology to my attention. Through his direction, his students and readers have come to realize the many parallels between Canaanite literature and the Israelite literature that shared imagery with it, and also with the later Christian corpus which did not relinquish its hold on the earlier imagery but accepted it as well as other imagery from the stream of Near Eastern culture. His constant reminder has been that nothing ever leaves the stream, things merely become resignified, reimagined and recycled. See Riley, *The River of God*, 57, 67.

Acts 7:55-56. The imagery is that of Baal, as is the language. Also, Baal is the son of El, prefiguring the Gnostic emanation theologies we will discuss later. Interestingly, the two aforementioned passages from the Christian canon display a moment in which such a confession enrages the listeners. I interpret this as an allegorization of a Son of Man theology, if you will, which evidently was quite unpopular. It may have originally been entirely independent of Christian tradition but became easily identified as part and parcel with it; the persecuted Son of Man theologians may have found refuge under the wings of the growing Christian communities. This is important as a possible early narrativization of the persecution which early Jesus communities must have undergone, Mark having allegorized the standard Q2 person in the role of the defendant Jesus. Due to the high priest's violent reaction to the confession of Jesus at his trial, it is tempting to believe that this is evidence of a very unpopular sect of adherents to a cult of the Son of Man. However, there is not enough evidence to argue convincingly for this and there may have been simply a strain or faction of Q followers who claimed a Son of Man theology, much in the same vein that is evident in *Enoch* and *Eugnostos,* as I will soon argue.

In this prelude to the passion, the high priest asks Jesus, during the trial before the Sanhedrin, if he is the "Christ, the Son of the Blessed?" And Jesus said, "I am; and you will see the Son of Man seated at the right hand of Power, and coming with the clouds of heaven." To this the high priest rent his clothes and cried blasphemy, demanding rhetorically of those witnessing Jesus' words to say if they had heard enough upon which to base a verdict? If we look closely at this passage, it is not merely Jesus' proclamation of his identity as the "Son of the Blessed" that has enraged the high priest so greatly, but rather, we may come to see, that it may also have been Jesus' invoking of the Son of Man and his description of the apocalyptic Parousia associated with this figure which has enraged the high priest. Of course, these may not be authentic words of Jesus, but they may give us a further clue to the kind of theological and political environment the early Christians may have lived in and helped to produce. This passage has fooled some, since the most common Christological term has been that one which gives the study of Christology its title—"Christ"—its connoted equivalent being "Son of God." However, to interpret this passage through the lenses of two thousand years of supracanonical myth-making is hardly a critical endeavor. As Burton Mack points out in *A Myth of Innocence,* the Markan usage of "Christ" is particularly contextually specific. It seems to me that the usage of this term in Mark is strangely offset by its counterbalance with the Son of Man title. While the term Christ is accepted and even promoted, its connotations gained from the Hellenistic Christ cult are downplayed, as Mack rightly points out. He also states that the appropriation of the term "Son of God" would not have been as offensive to the ears of a Palestinian audience as it might to us today.[35] Later, he states:

> As for the Jewish tradition, the term son of God could be used for the king, Israel as a people, special messengers such as angels, or even the ideal, true Israelite. In Alexandrian imagination, special figures of mediation between God and the people, such as the divine *logos* (Word), or Moses (mythologized), could be called a son of God or even a "second God." In the Wisdom of Solomon, the Righteous One knows himself to be a son of God (Wis 2:18), and the people waiting for their deliverance from Egypt are called the son of God (Wis 18:13).[36]

While Mack shows that the term Son of God did not necessarily presume so high a Christology as we automatically interpret today, it is possible that the high priest nevertheless was offended by Jesus' co-opting of the term. More complex relationships between terminologies may be at work here. As we shall see, the existence of Adam speculation and the existence of a lesser Yahweh theology, as well as Gnostic emanation theologies, all indicate levels of unorthodox factionalism which existed, positing alternate theological views. Some gained more acceptance than others; some even had influence on more hegemonic views. In Mark's recounting, the High Priest and the Sanhedrin seem to have been of a more mainstream theology which could not accept a lesser Yahweh called the Son of Man. Regardless, the possibility of a Son of Man theology undergirding the discussion portrayed in this scene, partly dependent upon Adam speculation and partly related to Enochian lesser Yahweh theology, is borne out by the existence of these other sects. The Son of Man may have been a controversial theology involving a lesser Yahweh, installed as judge of the eschaton. This theology may have gained partial acceptance by the Q people.

---

[35] Burton Mack, *A Myth of Innocence* (Philadelphia: Fortress Press, 1988), 281.
[36] Mack, 284.

The story of Baal and Yam very clearly holds the roots of the "eschatological banquet," or the "messianic banquet" as it is often called.[37] This story portrays Baal at the right hand of El. There is a direct correlation to the imagery of the Son of Man standing or sitting at the right hand of God. The situation has often been associated with apocalyptic motifs and has been identified as part of the messianic banquet. As such, Riley's viewpoint that ancient imagery and mythology continue to flow within the stream of culture seem to be validated by this parallel between the images of the New Testament and ancient ideas. J. C. L. Gibson points out what Riley has often stated:

> [Baal] is the prototype of a surprisingly large range of biblical images,... though only in the apocalyptic passage Isa. xxv 8, where in a magnificent figure the poet looks forward to a day when the swallower is himself swallowed, is there a veiled suggestion that the Hebrews knew of a mythical conflict between him and Yahweh.[38]

I recognize this to be self-evident; as such, we may take the New Testament imagery of the Son of Man located at the right hand of God as having its roots in the pre-Israelite tradition of Canaan. The right hand of God is a position of favor, of power, and of being chosen. Only the "anointed," the "chosen," the Messiah, would be granted such a position. God's son, as Baal was El's son, would be permitted a position of power such as this one.

---

[37] See J. Priest, "A Note on the Messianic Banquet," in *The Messiah: Developments in Earliest Judaism and Christianity*, ed. James Charlesworth (Minneapolis: Fortress Press, 1992), 222–238. In a brilliant article which attempts to look at some of the sources for this concept, an article which was included in the resulting tome of the Symposium on Judaism and Christian Origins, Priest alludes to earlier literature which concerns the origins of the banquet in apocalyptic literature, but does not mention the Canaanite roots of this concept. His disclaimer might be that source material was restricted by the Symposium to the time period 250 BCE to 200 CE, but out of such a remarkable article, one would think that brief mention would be made of the earliest Hebraic and Canaanite origins of the messianic banquet.

[38] J.C.L. Gibson, *Canaanite Myths and Legends* (Edinburgh: T.&T. Clark Ltd., 1978), 18–19. In this section, Gibson summarizes the stories which he transcribes and translates, giving brief commentary on the importance of these mythologies. He also mentions in a footnote to p. 19 that there is indeed a connection between these myths and the imagery in I Corinthians 15:26, 54. In this passage, it is Christ who has completed the cycle and has defeated and swallowed up Death.

Gnostic Traditions

Where, then, does this lead us? Who, then, is Man, if the Son of Man is his offspring? This is a question which is rarely posed. We must turn to some of the literature of contemporary groups in order to answer this question. First and foremost is the text called *Eugnostos*, two nearly identical versions of which are found in Nag Hammadi Codices III and V. For reasons obscure to me, this text is rarely referred to in the discussion of the nature of the Son of Man Christology and terminology. Perhaps this is due to a prejudice that anything identified with Gnosticism is necessarily late, derivative, and therefore irrelevant in discovering the meaning and origins of the Son of Man. *Eugnostos* nevertheless delivers an extensive cosmogonic genealogy, like many of the other texts in Nag Hammadi, and seems to focus intensely on the relationship of certain aeons to their consorts and their offspring, as well as to the nature of the universe. Particularly, we find mention of Immortal Man and the Son of Man, and even the Son of Son of Man! The name Sophia is mentioned several times, each time referring to one of the different consorts of these aeons. The relationship between the one known as Immortal Man and his offspring, Son of Man, is striking, especially when compared to parallel passages in the *Sophia of Jesus Christ*. It has been generally believed that the *Sophia of Jesus Christ* is a wholesale Christianization of the earlier work, *Eugnostos*, and is dependent upon it.[39] In the introduction to Douglas M. Parrott's translation of the text in the *Nag Hammadi Library in English*, he states:

> [*Eugnostos*] is without apparent Christian influence. With some minor omissions and one major one, it was used by a Christian gnostic editor as he composed *The Sophia of Jesus Christ*.... Thus the placing of the two tractates together in the present edition allows one to see the process by which a non-Christian tractate was modified and transformed into a Christian gnostic one.[40]

---

[39] One of the primary works of scholarship refuting this idea is an early article in German by Hans-Martin Schenke, which he later recanted. See Hans-Martin Schenke, "Nag Hammadi Studien II: Das System der Sophia Jesu Christi," *Zeitschrift fur Religions und Geistesgeschichte*, XIV (March, 1962): 263–278. Schenke believes that it is premature to conclude that the *Sophia of Jesus Christ* is dependent upon *Eugnostos* and that with the reverse it is not the case.

[40] Douglas M. Parrott, trans., "Eugnostos the Blessed and Sophia of Jesus Christ," in *The Nag Hammadi Library in English*, ed. James Robinson (San Francisco:

Parrott continues, speaking of both the provenance and the approximate date of *Eugnostos*, as the earlier of the two texts.

> The probable place of origin for *Eugnostos*, then, is Egypt. A very early date is suggested by the fact that Stoics, Epicureans and astrologers are called "all the philosophers." That characterization would have been appropriate in the first century B.C.E., but not later.[41]

While attempting to analyze, categorize and reconcile the various aeons and divine entities spoken of in *Sophia of Jesus Christ* (*SJC*),[42] on the other hand, it becomes mind-numbingly and stultifyingly confusing, and in some ways counter-productive is any attempt to coordinate the myriad epithets and often contradictory descriptions of these entities and generations of the divine and super-celestial realm. One is certainly reminded of the "endless genealogies" invoked by both Irenaeus and Tertullian, in reference to both the Valentinians and other Gnostics of their day, as well as the putative adversaries of Paul, as spoken of in 1 Tim 1:3-4. It is doubtful that in *SJC*, one will accurately and indisputably be able to differentially identify each aeon and its place within the genealogy—at least not without some deliberate and dubious equivocation for the sake of coherence—as there are numerous places where one aeon is given a name or epithet that is unworkably similar, if not identical, to that of another; and in some cases there are gross repetitions of a name in an entirely different context of the genealogy, blatantly contradicting the other. See, for instance, NHC III, 105:10 viz. 108:1, whereby in the former, "First Begetter Father is called 'Adam, Eye of light'", but in the latter passage, we see "The first aeon is that of Son of Man, who is called 'First Begetter,' who is called 'Savior', who has appeared. The second aeon (is) that of Man, who is called 'Adam, Eye of Light.'" As a prime example of the confusing schema of this text, it is vague and unclear as to whether the one known as "Man, who is called 'Adam, Eye of Light" is the same as the earlier "First Begetter Father," also called "Adam, Eye of Light," given also that they are in different orders and magnitudes in the genealogy.[43] The fact remains

---

HarperCollins, 1978), 220.
[41] Parrott, 221.
[42] Often referred to elsewhere as The Wisdom of Jesus Christ.
[43] See the following passages in Irenaeus and Tertullian. First, Irenaeus, *Against*

that Eugnostos includes no references to Jesus or the Christian tradition, and *Sophia of Jesus Christ* seems to pad all of the discourse with introductory questions by the disciples, placing the discursive material on the tongue of Jesus, thus making the text a sort of post-resurrection dialogue.

Some may argue that this is not enough evidence to assume that *Eugnostos* simply did not remove all Christian elements and references for its own benefit. Often, people are overly Christocentric in their assumption that Christianity was such a widespread phenomenon and such a household name that any concept or terminology shared by Christianity and the wider context of Near Eastern society would have naturally originated in Christian circles.[44] Certainly, one cannot base a reconstruction of tradition history upon a mere presumption of dependence. Nevertheless, there are places in the text which provide internal evidence for this theory of *SJC*'s dependence upon Eugnostos. The most significant of which, for our discussion, is that of NHC III 85:9-14 in Eugnostos and its parallel in *SJC*, as given in the manuscript of the Berlin Gnostic Codex (BG 108:1-10), where the version in the NHC III, 4, has lacunae. In both versions of Eugnostos in Nag Hammadi, the order of the progeneration is as follows: Father of the Universe (the ineffable, immortal and eternal), followed by Immortal Man, then

---

Heresies 1.1 (*ANF* 1:315): Inasmuch as certain men have set the truth aside, and bring in lying words and vain genealogies [Lat: *genealogias infinitas*] which, as the apostle says, "minister questions rather than godly edifying which is in faith," and by means of their craftily-constructed plausibilities draw away the minds of the inexperienced and take them captive, [I have felt constrained, my dear friend, to compose the following treatise in order to expose and counteract their machinations.]

It is also worthwhile to provide, at length, the text of Tertullian, *Against the Valentinians* 3 (*ANF* 3:505):Let, however, any man approach the subject from a knowledge of the faith which he has otherwise learned, as soon as he finds so many names of aeons, so many marriages, so many offsprings, so many exits, so many issues, felicities and infelicities of a dispersed and mutilated Deity, will that man hesitate at once to pronounce that these are "the fables and endless genealogies" which the inspired apostle by anticipation condemned, while these seeds of heresy were even then shooting forth?

See also 1 Timothy 1:3-4, in reference to "myths and endless genealogies which promote speculations rather than the divine training that is in faith".

[44] See above note about no longer being content simply to study the Synoptics and argue ourselves into circular boxes, denying the existence of any valid extracanonical texts. Q.v. Robinson, "The Study of the Historical Jesus After Nag Hammadi," 45-55.

Son of Man. Each of the offspring of the original Father—who is also spoken of as the monad—is termed an aeon. Intriguingly, Codex V adds a third aeon who is known as Son of Son of Man.

> The first aeon, then, is that of Immortal Man. The second aeon is that of Son of Man, who is called "First Begetter," (V13, 12-13 adds here: The third is that of son of Son of Man), who is called "Savior." (NHC III 85:9-14)[45]

The aforementioned first aeon is an emanation, a second principle that proceeds from the uncreated being, which was earlier spoken of variously as "He Who Is," "Father of the Universe," "Lord of the Universe," among other things. The being known as "Immortal Man" is an emanation of that "Unknowable" first principle.

The Berlin Gnostic Codex preserves a section of *SJC* which is parallel to the Eugnostos passage and is omitted in the NHC version of *SJC*.

> The fist aeon is that of Son of Man, who is called 'First Begetter,' who is called 'Savior,' who has appeared. The second aeon (is) that of Man, who is called 'Adam, Eye of Light.' (BG 108:1-11)[46]

The passage of aeons is reversed so as to have the Son of Man being the progenitor, and Man being the offspring. The Son of Son of Man is not mentioned in *SJC*. What can this mean? I contend that the most probable explanation for this reversal in genealogy is that a Christian (or Christianized) text would not likely allow there to be anyone or anything to come after Jesus Christ, or be more powerful than he. Therefore, the Son of Man—identified as Jesus Christ in all Christian texts—must be the end all and be all. None should come after him as a subsequent messiah. This further supports and presupposes the notion that Eugnostos is indeed the earlier work after all and is not merely non-Christian, but entirely pre-Christian. As such, *SJC* is dependent upon Eugnostos, altering the order of aeons to support the theology organic to, and ubiquitous within, the gospels. If it can be presumed, then, that *SJC* is a later Gnostic Christian text, then it is probably reliant upon gospel traditions for information such as the names of disciples and basic background knowledge which helped to shape the portrayal

---

[45] Parrott, "Eugnostos the Blessed and Sophia of Jesus Christ," *NHLE*, 236.
[46] Parrott, 236.

of this dialogue.

Douglas M. Parrott also believes that *SJC* may be a late first century CE text, and although his arguments are quite sound, this dating may still be a bit early for such a work.[47] Regardless of precise dating, the text's author must have some knowledge of the Christian tradition and thus will have the intention of making whatever Gnostic cosmology is contained in Eugnostos acceptable to a Christian theology. By the time of the gospels, the term Son of Man had already been identified with Jesus as something which he utilized to characterize himself. The intentions behind the usage of this term vary from gospel to gospel and even are varied within each. Nevertheless, if Parrott is correct in his suggestion that *SJC* may have been "produced to persuade non-Christian Gnostics to accept Christian Gnosticism," then Eugnostos' usage of the Son of Man would necessarily have been resignified to mean Jesus, in accordance with the gospels, which utilized the term exclusively on the tongue of Jesus and presumably almost entirely self-referentially.[48]

According to the Gospel of John, arguably a very gnosticizing gospel, the Logos was with God and was identified with God, as a second principle of the universe. Jesus, the Son of Man, was that Logos and was there with God in the beginning, participating in the act of creation. Thus mortal Man was created by the Logos, which in the case of *SJC* was the one also identified as the Son of Man, the first aeon. *SJC* identifies Man as the second aeon and proceeds to call him "Adam, Eye of Light." This is presumably the heavenly form of Adam, rather than the earthly Adam, yet the order of creation remains intact and congruent with the theology and Christology of John. In the Berlin Gnostic Codex, *SJC* 108:12, God the Father is identified as the "aeon over which there is no kingdom, (the aeon) of the Eternal Infinite God, the Self-begotten aeon of the aeons that are in it, (the aeon) of the immortals,..."[49] He encompasses within himself all things created by the aeons he emanated. This is completely complementary to the Johannine tradition. However, what remains is that *Eugnostos* includes an intermediary whom *SJC* has omitted. That entity is the Son of Son

---

[47] Douglas M. Parrott, ed. *Nag Hammadi Codices III, 3–4 and V, 1* (Leiden: E.J. Brill, 1991), 5–7, 18.
[48] Parrott, 5.
[49] Parrott, 143.

of Man. Additionally, in Eugnostos, Immortal Man, as the first aeon, takes the place of the Son of Man (*SJC*), pushing him down one rung of the ladder, and it may even be said that Immortal Man is characterized differently from "Man" of *SJC*. Eugnostos nevertheless originates the statement about the Eternal God who encompasses all of these aeons, preserved quite faithfully in *SJC*. How could a Christian possibly reconcile with the idea of God the Father, his son (Immortal Man), a grandson (Son of Man) and a great-grandson (son of Son of Man), when Jesus (as Son of Man) would have to be that grandson? Who, then, is Jesus' forebear, Immortal Man, if Jesus was the Johannine Logos who was with God in the beginning? Who then is to follow Jesus as the Son of Son of Man? In another place, Parrott states

> The notion of three divine men in the heavenly hierarchy appears to be based on Genesis 1–3 (Immortal Man = God; Son of Man = Adam [81,12]; Son of Son of Man, Savior = Seth). Because of the presence of Seth (although unnamed in the tractate), *Eugnostos* must be thought of as Sethian, in some sense. However, since it is not classically gnostic and lacks other elements of developed Sethian thought, it can only be characterized as proto-Sethian.[50]

While his assumption of the text's inherent connection to Sethianism may be unnecessary, the "heavenly hierarchy" has been correctly named and identified with God, Adam and Seth. His secondary modification of "proto-Sethian" may be more likely.

What we are left with, here, is an understanding of a cosmology delineating an unknowable God (whom Parrott oddly does not recount here in this three stage hierarchy, opting to speak of "Immortal Man" as identical to God in this treatise, an assessment that I observe to be erroneous on Parrott's part, as I have stated in a footnote above) who creates Immortal Man as the intermediary between himself and his creation. Immortal Man is in the form of

---

[50] Parrott, "Eugnostos the Blessed and Sophia of Jesus Christ," *NHLE*, 221. I disagree with Parrott's assessment of the specific identity of the Immortal Man as God. This is surely a confusing text, and after much analytical and genealogical re-reading, it is my understanding that Immortal Man cannot be seen as the same as the unknowable and ineffable God, "He Who is" (71:13–14), who is also termed "Father of the Universe" (73:2–3) and monad (78:17). For the Immortal Man that is spoken of in 76:19ff as the likeness of the Father. See my comment below about Parrott's omission of the monad from this litany of names.

humans, (and provides the form as well) as witnessed by the name, and as such can be identified as a demiurge of sorts.[51] He is the first aeon who is emanated and begins the act of creation and emanation of other aeons. As with the Hebrew Scriptures, Yahweh embodies the form in which humans are subsequently created.[52] So, Adam is then the Son of Man and his son, Seth, may be known as the Son of Son of Man. Not so for *Sophia of Jesus Christ*, in which God, I reiterate, is the originator of the heavenly Adam, Immortal Man, who is also identified as a "first aeon" called "Son of Man. The key difference here is which aeon in the supercelestial genealogy is identified as the Son of Man. In *SJC*, we are obviously to associate this Son of Man with Jesus Christ, who is the mediator of creation in the Christian tradition, following John and Colossians alike. But for Eugnostos, while Immortal Man is that first aeon, descended from the Unknowable, Father or Lord of the Universe, Immortal Man is *not* identified as the Son of Man but rather this being is the superior and predecessor of the Son of Man, the latter of whom is a second aeon, third in the lineage. In the clearly pre-Christian tradition of Eugnostos, the unknowable God, first identified as "Father of the Universe", is the progenitor of all the aeons, of whom Immortal Man is the first one, who is the heavenly Adam, and is even termed "Adam of the Light" (NHC III 81:12). Regardless of the specific order of the genealogy, Eugnostos provides a corroboration of the divine lineage which depicts God as the image or type of Man—Immortal Man, as such—and so the Son of Man as the Son of God, so to speak, holds a position quite similar to that of Baal in respect to El in the Canaanite myths of Ugarit, a client king subordinate to the suzerain.

As such, the text of Eugnostos provides evidence of a pre-Christian tradition which includes a theology—or mythology—of the Son of Man, not quite a cult, but indeed a strain which viewed this entity as a figure of prime importance in the process of emanation. In relation to the figure of the Heavenly Adam, the Son of Man is inextricably connected via this genealogy. Thus, Jesus

---

[51] This is not to be confused with the evil demiurge of classic Sethian texts like the *Apocryphon Ioannis*, but should be seen as a demiurge in the pure sense of one who assists in creation, a craftsman, as in Plato's *Timaeus*, 28a ff.

[52] See, for instance, Gen 1:26-28, Gen 5:1-3, and Gen 9:6 for Biblical sources referring to humans created in God's image. For several extracanonical texts referring to this, see 2 Enoch 44:1-32, 65:2; Wisdom of Solomon 1:13-14, 2:23; Sirach 17:1-42; and Esdras 8:44

would have to be simultaneously installed into the position of Son of Man if he were identified as the Son of God. Such a discussion may indeed raise questions of the Hellenistic origins or late Judaic provenance of the Son of God Christology as opposed to the Adamic-Gnostic origins of the Son of Man theology.[53] Nevertheless, in keeping with my assumptions about the evidence found in Q12:8-10, there is no reason to cling to the presumptions that any reference to Son of Man necessarily indicates or invokes Jesus, *a priori*, if there is no other evidence of the source of said reference being a Christian text. In Eugnostos we have a tradition which clearly utilizes the Son of Man as the second order emanation from the invisible deity, and is the offspring of the first aeon, the Immortal Adam. This order of genealogy would be incorrect for Christian usage without making some modifications. If the Son of Son of Man in Eugnostos was termed Savior, then Jesus—in order to maintain the Christian identity and provenance of this text—would have to be somehow identified with this figure or else the term Savior would have to be transposed upon the Son of Man and the third emanation or aeon (i.e. Son of Son of Man) would have to be nullified altogether. Since John witnesses that the early Christian traditions preferred that Jesus be the first emanation of the original divine spirit, it would be better that the Son of Man, when identified with Jesus, precede Adam in the emanations and be coexistent with God at the origin. Thus, we also see that the Son of Man cannot be a figure of earthly origin, but necessarily a heavenly figure. In the pre-Christian Gnostic traditions, the Son of Son of Man would have occupied the role of the redeemer who would enter the earthly realm as the Savior, which is the case according to Parrot's description of the genealogy in Eugnostos. The identification of the earthly Jesus with the heavenly Son of Man "throws a monkey wrench in the works" of this stable cosmology and would necessarily require the very reorganization and renaming that we see in *SJC*.

---

[53] In Birgir A. Pearson, *Gnosticism, Judaism and Egyptian Christianity* (Minneapolis: Fortress Press, 1990), 90, Pearson mentions an early usage of the concept of the Sons of God recorded in some pre-Christian circles. In Gen 6:2, the angels who have relations with the daughters of men are spoken of as "Sons of God".

Enoch Traditions

It would be worthwhile to look at the widely discussed Similitudes of Enoch, as well, because of the interesting characterizations of the Son of Man as a preexistent heavenly figure who will act as judge in the eschaton. This provides an interesting parallel to the apocalyptic usages of the Son of Man in Q and in this case bears witness to a possible pre-Christian apocalyptic Son of Man. The passages in this work bear a distinctive usage of the Son of Man.[54] Fuller makes one of the most brilliantly honest concessions to the critics of the early dating of this text, while pointing out the persistent value in it: "While therefore, we cannot be sure that the Similitudes themselves antedate the Christian era, we may treat them with some degree of confidence as evidence for a tradition in Jewish apocalyptic which is pre-Christian."[55] I agree with Fuller's assumption that the term is actually used as a title in the Similitudes. As noted earlier, most of the Old Testament passages which use the term to indicate humanity have it placed in contrast with the term Man as counterpoint.

To recap, the usages of the term which do not place it in contrast to the more mundane term for human are encountered in Ezekiel, which uses it as a diminutive epithet on the tongue of God, and in Daniel, which is the most formally similar to the usages in

---

[54] For a more extensive treatment of the presence of the Son of Man in the Enochic tradition, and for a discussion of the attribution of four specific epithets to the text's main protagonist, see J.C. VanderKam, "Righteous One, Messiah, Chosen One, and Son of Man in 1 Enoch 37–71," in *The Messiah: Developments in Earliest Judaism and Christianity* (ed. J.H. Charlesworth; Minneapolis: Fortress, 1992), 169-191. For a more recent treatment of the relationship between the Son of Man and the Qumran community, see Pierpaolo Bertalotto, "Qumran Messianism, Melchizedek, and the Son of Man," in *The Dead Sea Scrolls In Context (2 vols): Integrating the Dead Sea Scrolls in the Study of Ancient Texts, Languages, and Cultures* (eds. Armin Lange, Emanuel Tov, and Mathhias Weigold; Leiden: Brill, 2011), 325-339, pointing out significant parallels between Melchizedek in 11QMelch and the Son of Man in the Similitudes of Enoch. In this piece, Bertalotto explores the relationship between the Son of Man of 1 Enoch 46:1 and Melchizedek of the Qumranic tradition, concluding that both function similarly as an angelic embodiment of *Elohim*, arguably a second Yahweh as such. He intimates that these messianic figures are entirely independent of Christianity, as well as being pre-Christian in chronology and dating.

[55] Fuller, *Foundations of New Testament Christology*, 38.

Enoch.⁵⁶ In Daniel, the term may not actually be understood yet as a title. Many scholars debate endlessly over the usage in 7:13 and fail to address its occurrence in the very next chapter; in Daniel 8:17, the term crops up in the classic Ezekielic style. This shows that the author still has it in his consciousness that the term may be used as a diminutive epithet for humans. The term has not yet been afforded a wholesale changeover to an apocalyptic title.

In his discussion of the opinions of John Collins, John Dominic Crossan states that he feels the term in Enoch as well as Daniel is used as an idiomatic reference to a humanoid entity, rather than a title. He states, "Put crudely: Daniel 7:13 precedes the titular Son of Man, the titular Son of Man does not precede Daniel 7:13. And, once again, titular may be far too strong a word for that situation."⁵⁷ This is absolutely true. However, my sense is that Crossan jumps too quickly to his conclusion that the term is entirely non-apocalyptic. Daniel represents a tradition which may precede the widespread (if it was ever widespread) understanding of the term as titular. In Daniel, we see the term being used in an apocalyptic context to describe a being who is "like a human being". Daniel does not force anything further. Eugnostos does; Q12:8-10 does; the Similitudes of Enoch do as well. The resulting trend may be as a result of Daniel's ambiguity.

In the Similitudes, we see that the term is first invoked ambiguously, and then installed in several passages which use the term as a title. Even if the term bears a strong quality of namelessness, it becomes a title by its very application, in much the same way that an impromptu nickname for a third party, thought up among friends, will often persist and be used as a code name, having been invested with certain peculiar emotion and contextuality. The Similitudes offer a set of passages which allow this kind of rea-

---

⁵⁶ Fuller, 43, outlines the current debate over the possibility that Ezekiel may have served as the source or model for Jesus' usage of the term Son of Man. His views are that the usage in Ezekiel is not invested with any apocalyptic titular meaning as is its usage in New Testament texts. He concludes, "Therefore, when we come to examine the Son of man in the sayings of Jesus and in the development of his sayings in the early church, we shall assume that the term is throughout derived from the pre-Christian Jewish apocalyptic tradition." I do not believe that the standard literature labeled as Jewish apocalyptic bears as striking a titular usage of the term as does the Gnostic literature, but if Fuller were to expand his definition of apocalyptic to include Jewish Gnostic texts, I would agree with his argument entirely.

⁵⁷ Crossan, *The Historical Jesus*, 240-241.

soning to seem valid:

> And at that hour the Son of Man was named
> In the presence of the Lord of Spirits,
> And his name before the Head of Days.
>
> Yea, before the sun and the signs were created,
> Before the stars of the heaven were made,
> His name was named before the Lord of Spirits. (48:2ff)[58]

What was the Son of Man named? We are left with the strange notion that the Son of Man may have been named the "Son of Man". Is this absolutely impossible? Regardless, the Son of Man is nevertheless being referred to as "the Son of Man." There is no invocation of "One like a son of man" as we have seen in Daniel. The reference is very clearly referring to a specific Son of Man who has not been disclosed by personal name, but perhaps needs no introduction. The name itself is sufficient introduction, as is the barrage of passages describing his role and actions. In a later chapter, he is described in much the same way the Son of Man is described in Q12:8-10. He has been set up to rule and, more specifically, to judge.

> And (He will deliver them) to the angels for punishment,
> To execute vengeance on them because they have oppressed His children and His elect. (Enoch 62:11)

This role closely resembles that of the primary Q passage at hand. The Son of Man has the authority to judge and condemn before the angels. One of the most important quotes referring to the Son of Man is in Enoch 69:27 in which it is said that "the sum of judgment was given unto the Son of Man,..." This passage concisely summarizes the role of the Son of Man in the Enoch literature as well as in Q; he is a heavenly judge. Fuller, himself, summarizes the role of the Son of Man as the Similitudes portray him:

---

[58] All translations of Enoch are from Fuller, *Foundations of New Testament Christology*, unless otherwise specified. I would also note that in the first stanza of this quote, the parallelism of "Lord of Spirits" and "Head of Days" — surely both a reference to God, in order for this text to remain monotheistic — is rather strikingly reminiscent of the synonymous parallelism of earlier Old Testament references to the Son of Man, which is most perfectly exemplified by Numbers 23:19, and embodies that peculiar poetic style of Biblical Hebrew literature that states, and then re-states, the second element in a slightly modified form of the first.

Here emerges the most complete picture of the Son of man in the Jewish apocalyptic tradition. He is a preexistent divine being (48:2f.; 62:7). He is hidden in the presence of God from before all creation (48:2). He is revealed "on that day", i.e. at the End. He appears in order to deliver the elect from persecutions (62:7ff.). He judges the kings and rulers who have persecuted the elect (46:4; 62:11; 69:27). He presides as a ruler in glory over the elect as a redeemed community in eternity. Note especially the allusion to the Messianic banquet (69:29; 62:14).[59]

The issue of the Messianic banquet arises, as Fuller notes, in the statement about the Son of Man and the elect eating, and the statement about the Son of Man seated on the throne of glory. These images seem to derive from the Canaanite banquet imagery which was narrativized in the Baal cycle. Particularly, the council of the gods was seated, prepared to engage in a banquet, Baal standing at the right hand of El, when the messengers from Yam arrived.

The Similitudes of Enoch are present only in the Ethiopic Enoch and are absent in the Qumran manuscript of Enoch. Many have looked at this as evidence that the Son of Man is a later, Christian interpolation and cannot be looked upon as original to the text. Thus, the inclusion of the Son of Man seems that much more suspicious to some. Fuller argues for the continued relevance of the Similitudes regardless of their absence in the Qumran version:

> [T]he Son of Man in the Similitudes lacks the distinctively Christian differentia, viz., the identification with Jesus of Nazareth in his ministry (which, as we shall see, is a very early Christian use of the Son of man) and in his passion (which, as again we shall see, is, though not quite the earliest, at least a Palestinian feature). Second, the logia of Jesus, as again we shall see, seem to presuppose a reduced apocalyptic in which the future coming Son of man as eschatological judge was part of the traditional imagery.[60]

I am convinced by, and share in Fuller's reasoning for considering the Similitudes a valid, pre-Christian source of background for the Son of Man theology. I insist upon terming this a theology (and potentially a messianology) rather than a Christol-

---

[59] Fuller, *Foundations of New Testament Christology*, 39–40.
[60] Fuller, *Foundations of New Testament Christology*, 38.

ogy, since I have attempted to show that the entity of the Son of Man precedes Jesus of Nazareth and thus should not be looked upon as somehow describing his role as Christ; thus its continuous identification with Christology is somewhat tenuous. While the term may indeed be applicable to sectarian views of the Messiah, if you will, Christology has nevertheless become such a contextualized term that I would prefer to eschew it. As such, it describes a cosmological-theological complex which holds the Son of Man as a second in command to God, seated at his right hand, and thus modifies any theology which has deemed to accept this figure into its pantheon.

Thomas Traditions

The Thomas tradition also preserves a small amount of material about the Son of Man. While I have already discussed the parallel to Q 12:8–10, Logion 44, two other logia in the Gospel of Thomas are significant to this research. The first is saying 86, which preserves a nearly identical parallel to Q 9:58. There is no obvious or overt Son of Man Christology or theology in Thomas, but logia like this one force us to rethink our view on whether or not Thomas knew the Son of Man tradition as apocalyptic. It may be possible that Logion 86 follows 85 for good reason. As I have earlier discussed, the Son of Man tradition holds that this figure is the offspring of the heavenly Adam. ⲡϣⲏⲣⲉ ⲙ̅ⲡⲱⲙⲉ, the Coptic rendering of ὁ υἱὸς τοῦ ἀνθρώπου, still suggests this understanding of the offspring of the heavenly Man. In this pair of logia, it is attested that Adam (here, the mortal one) was not worthy of the disciple who has found the enlightenment esteemed by the Thomasine tradition. One would instantly wonder if "Son of Man", here, were not referring to Adam. However, this reference to Adam in 85 is negative.

The last reference to the Son of Man in Thomas, Logion 106, shows for sure that the Son of Man is indeed a positive figure. Retrospectively, we see that the Son of Man of 86 is what the disciples can become and as such, will not worry about the fact that they have no permanent home in this world. Logion 42, which says to "Be passersby", and Logion 21, which describes Jesus' disciples as willingly giving up their bodies and roles in the material world when the time comes, both speak of an ideal situation of enlightenment which allows the disciple to be unattached to this temporal realm. Upon inspection, Logion 106 shows a pluralization of the term at hand to produce "Sons of Man" (ⲚϢⲎⲢⲈ ⲘⲠⲢⲰⲘⲈ). This logion describes a situation in which the disciples are able to be heirs to the heavenly Man, having gained supernatural powers, indicated by their ability to move mountains by command. A similar statement of heritage is invoked in Logion 3.4. In this saying, placed very early in the gospel, it is said, "When you know yourselves, then you will be known, and you will understand that you are children of the living Father."[61] Both logia display similar terms, probably meaning the same thing. The disciple who has found enlightenment has become truly a Son of God, a Son of the Immortal Man.

The usage of Son of Man terminology in Thomas is quite different from that in the Synoptic Gospel traditions. The paucity of references to him may indicate that an emanation theology was not as important to Thomas as to other Gnostic traditions. It also may point to further evidence that the Q community directly dealt with the peculiar strain or faction which held ties to a Son of Man theology, while Thomas clearly did not; for the references to the Son of Man in Thomas are hardly judgment oriented, and only one could even be thought to refer directly to Jesus.

One other member of the Thomas tradition may shed some light on one aspect of the Son of Man tradition. It was once brought to my attention at the behest of Greg Riley that the character or element of the letter in the Hymn Pearl may in fact be an allegorization of the character of the Son of Son of Man who shows up in Eugnostos. This pre-Christian Gnostic hymn has circulated as a section of the *Acts of Thomas*. In the story, the protagonist has been sent to retrieve a pearl from a serpent which lives in Egypt. Upon arriving there, he is overcome by the heaviness of the food and lifestyle and gives up his search, forgetting his identity, his origin, and his destiny (three elements of the Gnostic quest, all of which appear in Gos. Thom. 50). His parents send him a letter which reminds him of his purpose in travelling to Egypt. This letter may be viewed as parallel to the "savior," which is mentioned as the Son of Son of Man in Eugnostos.

ⲠⲒⲘⲀⲈϢⲞⲘⲈⲦ ⲠⲀ ⲠϢⲎⲢⲈ ⲘⲠϢⲎⲢⲈ ⲘⲠⲢⲰⲘⲈ ⲠⲈ· ⲠⲎ ⲈⲦⲈ ϢⲀⲨⲘⲞⲨⲦⲈ ⲈⲢⲞϤ ϪⲈ ⲠⲤⲰⲦⲎⲢ

The third is that of the son of Son of Man, who is called Sav-

---

[61] Meyer's translation, as above, Kloppenborg, et al, 129-130.

ior.⁶²

As we have seen already in Eugnostos, it seems that the Son of Son of Man, in the earliest Gnostic traditions was the savior who was sent to redeem. Hymn Pearl upholds this theory in that the letter has been sent to awaken the prince, the son of the great king and queen. The letter acts as the savior for the redeemer who must be redeemed. This is, however, a topic much too deep for a discussion of my chosen parameters; it is sufficient here to simply illustrate the fleshing out of the Son of Man theology in its forms throughout the Gnostic traditions which recognized it.

Confessions in the Synoptics and Acts

At this juncture, it would be wise to reassess the employment of the Son of Man language within the Synoptics, as well as other canonical texts, such as Acts of the Apostles. Ragnar Leivestad, attempts to tackle the common assumption that the term is used as an apocalyptic title. His discussion of John's usage of the term is quite interesting, if somewhat inaccurate and myopic. Leivestad is only partially correct in saying that John recognizes "that the term is without meaning for those who are outside the Christian faith."⁶³ Seemingly, John wrestles with the Christian combining of metaphors and imagery in the term, as do Matthew and Mark. He addresses a post-Second Temple community which has probably lost contact with the earliest usages of the term. As we have seen, the Son of Man, as a concept, clearly has dual traditions comprising its roots. One set of traditions bears witness to the aforementioned idiomatic usage, meaning simply human being, while the other bears the notion of a title expressing the state of being the offspring of the heavenly Adam, i.e. the Son of God. I would go so far as to say that the intent behind these two terms (Son of Man and Son of God) are identical, for John, though deriving from different milieux.

Q seems to have a better grasp on the term as a pre-Christian theological category which was being drawn into the Christian circles of faith discussion, since it does preserve at least one memory (12:8–10) of the Son of Man being an entity separate

---

⁶² Translation by Parrott, *NHC III, 3–4 and V, 1,* 142.
⁶³ Ragnar Leivestad, "Exit the Apocalyptic Son of Man," *NTS* 13, no. 3 (April 1972): 253.

from Jesus. Thus, the gospels witnessed a moment of Palestinian tradition in which a strange theology concerning a lesser Yahweh, installed as Messiah, became quite notorious, perhaps even blasphemous. This theology became more palatable as a reference to the future form of Jesus, and the Son of Man adherents gained a home in Christian circles. We are thus left with a very strong imagery of the Son of Man who is "seated at the right hand of Power, and coming with the clouds of heaven."

In terms of the story Mark tells, it is my contention that this striking imagery certainly may have contributed to the high priest's dismay and terror, otherwise we would not have heard about it from Mark, the master storyteller. This horrified and incensed reaction of the high priest may give us further clues about the positions of the Jewish hierarchy—and perhaps other rival apocalyptic groups of the day—on the subject of the Son of Man. Between the time of the collation of the second stratum of Q, as reconstructed by numerous Q scholars,[64] and the time of the composition of Mark, the development of the Son of Man theology, as an apocalyptic tradition, met either with widespread distaste or widespread ambivalence. The later Q community seemed to have taken into their identity a persecution complex which may have resulted from their claiming the existence of and belief in a coming Son of Man. I am inclined to read the texts which speak of "persecution because of the Son of Man" as not primarily referring to persecution because of the community's allegiance to Jesus, but rather *perceived* persecution, or unpopularity, because of their association with an unpopular and questionable Son of Man theology.[65] The way the text is composed, it utilizes this dual meaning to its own benefit, playing up the persecution complex and drawing in those who did not believe in the Son of Man apocalyptic but certainly were willing to remember Jesus referring to himself as the idiomatic "Son of Man".

The text allows the gradual resignification of Jesus identified with the apocalyptic Son of Man. The stoning of Stephen is

---

[64] See my earlier footnote regarding the stratigraphy of Q in light of the scholarship of Mack and Kloppenborg.

[65] By persecution, I mean to say that the community perceived any derision or lack of acceptance as persecution, whether it consisted of purposeful and targeted malice or not.

certainly an example of this kind of persecution. What Stephen utters is that he sees "the heavens opened, and the Son of Man standing at the right hand of God" (Acts 7:56). While the crowd has been incensed by Stephen's diatribe against their unrighteousness, I contend that it is ultimately his invocation of the Son of Man that drives them to stone him. This remembrance of Stephen's confession is certainly an extraneous element—being one of the rare occurrences in the NT that the Son of Man is used as an external reference to Jesus, and not upon his own tongue—which possibly recalls an authentic confession of many adherents to the Son of Man theology, and perhaps even some early Christians. Leivestad has different ideas on the subject:

> What offends the audience and causes their outrageous reaction is, no doubt, the bold statement that he sees Jesus, the man whom the High Court had sentenced to death, standing at the right hand of God in heaven. Such a declaration is blasphemous. One may try to imagine how different the reaction might have been, if the people had been familiar with the Son of man ideas from I En.! Would it have been offensive to anybody if Stephen had described a vision of the heavenly Messiah? Surely not. But if the Son of man can only signify Jesus of Nazareth, then we understand why they stop their ears. That means that Stephen has put a son of man in the place of the Son of God. Such a blasphemy could not be tolerated.[66]

Leivestad's reasoning is quite sound and certainly no less valid than my own. It is entirely possible for the crowd to have stoned Stephen for his allegiance to an executed criminal, imagined as seated at the right hand of God. However, Leivestad does not acknowledge that the crowd may have been fairly well acquainted with the imagery of Enoch and simply might not have subscribed to such a theology.

It is axiomatic that not all Palestinians would have subscribed to the same vision of the Messiah. The author Leivestad makes the incorrect assumption that "son of man" was always idiomatic, that all Palestinians believed in a coming Messiah, and that all Palestinians agreed on the role and imagery of that Messiah.[67]

---

[66] Leivestad, "Exit the Apocalyptic Son of Man," 253.
[67] James Charlesworth, in his introduction to *The Messiah*, has adequately shown that through the huge amounts of research done by the members of the

According to the works investigated in this present study, such as Eugnostos and the Thomas traditions, the Son of Man was seen as the equivalent to the Son of God, the Son of the heavenly Adam. Not all adversaries of Stephen would necessarily be aware that "Son of Man" was a term of self-reference commonly employed by Jesus. They most probably would be familiar with Danielic or Ezekielic usage of the term first and foremost, and none of this would be terribly offensive, in my opinion. And if we entertain the idea that Acts is more a reflection of the late-first or early-second century audience for which it was written, as opposed to an accurate record of an early post-Easter occurrence, then it would be more likely that the target audience was more familiar with Enochian or Gnostic usages of the term, and such a dramatic reaction to Stephen's invocation of "Son of Man" may have been appropriate to the situation.

My belief is that the crowd is a narratological representation of the majorities who did not subscribe to the Son of Man theology and considered it unorthodox and distasteful for whatever their reasons were. If in certain Gnostic circles, God was identified as the Immortal Man, then the Son of Man can be equated with the Son of God. There is little evidence to support a widespread Jewish acceptance of Gnostic cosmologies and emanation theologies which posited the existence of a Son of Man / Son of God figure. The Christian tradition, having melded Jesus with this apocalyptic figure, narrativized such an historically possible scene as Stephen's confession to be congruous with the confessional traditions surrounding Jesus.

If read in a slightly different light than usual, it is possible to see that Q 6:22 preserves a memory of a time when adherents to the Son of Man theology, and soon after Christians as well, were being persecuted for their belief in the Son of Man. Mark 14:62 pre-

---

Symposium, no singular picture of an anticipated Messiah can be demonstrated throughout Palestinian Judaism of the Hellenistic and Roman era. However, in Justin Martyr's *Dialogue with Trypho*, it is evident that Trypho concedes that all Jews are awaiting the Christ. (89:1) This may be creative revision of history by Justin, attempting to suggest that all Jews were of the same monolithic mindset, conceiving of the Messiah in the same way. See James Charlesworth, ed., *The Messiah* (Minneapolis: Fortress Press, 1992), 3–35.

serves a memory of this also. The high priest has given Jesus sufficient opportunity to defend himself, but is met with a confession that the high priest "will see the Son of Man sitting at the right hand of Power, and coming with the clouds of heaven." I suggest that either this is an anachronistic placement of a confession of the Son of Man upon the tongue of Jesus, or it is witness to Jesus touting an early formulation of an unpopular and heretical Son of Man theology. It had become increasingly convenient to allow the *double entendre* inherent in the Son of Man terminology to remain and imply that Jesus knew himself to be the apocalyptic Son of Man. If this is Mark's intention, then it is highly effective. As for the adherents to the Son of Man theology, their movement continued, like the Q tradents, under the aegis of a larger cause.

Conclusion

It has been my contention that the Son of Man who was adopted into the Jesus traditions was first questioned and became the focus of some debate. Not everyone in the Jesus movements was required to believe in him. He thus retained his role of heavenly judge and served the Jesus communities in that capacity, theologically. Soon after, it became convenient to identify Jesus with that Son of Man and the traditions became entirely inseparable and indistinguishable. From simple philological evidence that there was a tradition recognizing a Son of Man tradition separate from Jesus (as per Q 12:8–10), and a controversy over allegiance to this figure, and from the recognition that Eugnostos preserves a pre-Christian Gnostic tradition mentioning and making use of the Son of Man theology, I have attempted to reconstruct a small piece of the development of the early Christian tradition.

Based upon the current study's analysis of the primacy and early dating of Eugnostos, among other texts like Enoch, then Q 12:8–10 can now be seen as acknowledging a moment in time when Jesus was preaching a Son of Man theology that was already pre-existent. Being that these demonstrate a real possibility that a Son of Man theology/messianology existed prior to Jesus, then the fact that the term—as per Schweizer[68]—was not yet a definite title can be seen as valid in that there was a certain ambiguity built into the term. It may have been simultaneously an idiom and a pre-

---

[68] See above, op. cit., Schweizer, "The Son of Man Again," 259.

Christian messianic title. As such, either Jesus or later generations remembered him as that Son of Man. Thus, Jesus in the Q passage in question is allowing the people of the early Jesus community to believe in that Christology or reject it, as they may desire. But they are first and foremost part of a larger movement, which is guided and governed by the Holy Spirit.

If the Son of Man existed as a theological, apocalyptic figure apart from the Christian traditions as evidenced in written literature, and can be proven to predate the Christian traditions, then it may be proposed that such ideas of the Son of Man influenced or were incorporated into the early Christian usages of the Son of Man. That is to say, if communities such as those that produced Enoch and Eugnostos knew of a Son of Man without knowing Jesus, then there may be evidence to support the idea that some early Jesus traditions also knew of this figure and reimagined Jesus as having spoken of him. As time wore on, the idiomatic, circumlocutive usages of this term, remembered as being employed by Jesus, became conflated with these apocalyptic references to the Son of Man theology, thus allowing the community to remember Jesus referring to himself as that very Son of Man.

My hope is that I have uncovered evidence, which is indeed within the literature, rather than having misinterpreted a few very opaque passages from the Christian and pre-Christian corpus. Incorrect assumptions and interpretations are among the Religious Studies scholar's worst enemies, and one hopes not to make too many (incorrect assumptions or enemies) in the process of trying to elucidate the most enigmatic of passages.

Eschatological Perspective in the Heikhalot Rabbati[1]

Marvin A. Sweeney

I

The Heikhalot Rabbati is one of the most intriguing and enigmatic documents of Rabbinic Judaism. Written at some point following the Tannaitic period in the 2nd/3rd through the 6th centuries CE, it portrays the ascent of R. Nehunyah ben HaQannah to heaven to appear before the throne of G-d in the seventh level of heaven. His purpose was to ask G-d why the Temple was destroyed, why the Rabbinic sages of Judaism were martyred, and why the population of Jerusalem and the land of Judah were exiled. R. Ishmael ben Elisha is the narrator of the Heikhalot Rabbati in which his teacher, R. Nehunyah, presents his account of his ascent to the heavenly throne.

Past scholarship has debated whether the Heikhalot Rabbati is fundamentally concerned with mystical experience or with the interpretation of Torah.[2] This paper argues that the Heikhalot Rabbati is fundamentally concerned with both, viz., Torah interpretation is a form of mystical experience insofar as it brings one closer to an understanding of divine presence and purpose in the world of creation. In terms of eschatological perspective, the Heikhalot Rabbati anticipates an ideal time when Torah—and thus divine presence and purpose—will be fully understood and applied to the

---

[1] Earlier versions of this paper were presented at the Annual Meeting of the Association for Jewish Studies, San Diego, December 18-20, 2016, and the Annual Meeting of the National Association of Professors of Hebrew and the Society of Biblical Literature, November 18-21, 2017.

[2] For critical discussion of the Heikhalot Rabbati, see Gerhom Scholem, *Major Trends in Jewish Mysticism* (New York: Schocken, 1961), 40–79; Ithamar Gruenwald, *Apocalyptic and Merkavah Mysticism* AGAJU 14 (Leiden: Brill, 1980), 150–73; Ra`anan S. Boustan, "The Study of Hekhalot Literature: Between Mystical Experience and Textual Artifact," *CBR* 6.1 (Oct. 2007): 130–60; Peter Schäfer, *The Origins of Jewish Mysticism* (Tübingen: Mohr Siebeck, 2009), 244–82; Boustan, *From Martyr to Mystic: Rabbinic Martyrology and the Making of Merkavah Mysticism*, TSAJ 112 (Tübingen: Mohr Siebeck, 2005).

sanctification of the world of creation. Indeed, it calls upon its readers to continue their endeavors in mysticism and Torah study to realize that ideal. The paper supports this thesis with examination of three key points. First, is the identity and role of the mystic *minyan* of martyrs who gave their lives for the sake of encounter with the divine and the application of Torah in the world. Second, is the recall of R. Nehunyah ben ha-Qanah to illustrate the principle that "those who descend in the chariot," i.e., those who return from the heavenly journey to apply knowledge of Torah to the sanctification of the world, in this case, the laws of *Niddah* as they are understood by the schools of Hillel and Shammai. And third, is G-d's response to the query of R. Nehunyah concerning the destruction of the Temple and the martyrdom of the Rabbis, which G-d admits was questionable, but nevertheless calls upon Jews to continue their studies of Torah ultimately to understand divine presence and purpose in the world.

II

The Heikhalot Rabbati begins with R. Ishmael's question, "What are those songs which he recites who would behold the vision of the *Merkavah*, who would descend in peace and would ascend in peace?" (sec. 81).[3] The following material then discusses the qualities of Torah knowledge and faithful observance of Judaism that would qualify them to sing such songs and to behold the divine *Merkavah*.

The identity of R. Ishmael is a key factor in understanding the Heikhalot Rabbati. R. Ishmael ben Elisha was one of the key Tannaitic sages of the 1st–2nd centuries CE.[4] He was likely the grandson of the High Priest of the Jerusalem Temple by the same name. He was forced into captivity in Rome, but he was ransomed by R. Nehunyah ben ha-Qanah whom Tractate Shabbat identifies as his teacher. R. Ishmael was a contemporary and colleague of R. Akiba ben Joseph. R. Ishmael is known for his 13 *Middot* for the

---

[3] For text editions, see Peter Schäfer, *Synopse zur Hekhalot-Literatur* TSAJ 2; (Tübingen: Mohr Siebeck, 1981); Avraham Yosef Wertheimer, *Batei Midrashot* (Jerusalem: Rav Kook, 1950) 1:63–135 (Hebrew). For an English translation, see Morton Smith, *Hekhalot Rabbati: The Greater Treatise concerning the Palaces of Heaven* (ed., D. Karr; 2009).

[4] Shmuel Safrai, "Ishmael ben Elisha," *EncJud* 9:83–86.

interpretation of the Torah, which emphasized the plain sense of scripture as the basis for interpretation rather than every minor feature of the text as argued by R. Akiba. He is also known as the author of the *Mekhilta de R. Ishmael*, an early Tannaitic halakhic Midrash on the legal material in the book of Exodus. He was one of the most prominent Rabbinic figures of his day, and he is included in the lists of the Ten Martyrs killed by the Romans in the aftermath of the Bar Kochba Revolt. Some maintain that he died with Bar Kochba at Betar, although some also maintain that the martyred figure was a namesake of the Sage.

R. Ishmael's question, "What are those songs which he would recite who would behold the vision of the Merkavah, who would descend in peace and who would ascend in peace?" indicates the primary agenda of the Heikhalot Rabbati, viz., to define those who would ascend to the seventh heaven to behold the vision of G-d. R. Ishmael's question presupposes Mishnah Hagigah 2:1, which reads as follows:

> The forbidden degrees (the discussion of forbidden sexual relations in Leviticus 18)
> May not be expounded before three persons nor the story of creation (Genesis 1)
> Before two, nor the chapter of the Chariot (Ezekiel 1) before one alone, unless he
> Is a sage who understands his own knowledge. Whoever gives his mind to four
> Things it would be better for him if he had not come into the world—what is above?
> What is below? What was before? And what will be? And whoever takes no
> Thought for the honor of his Maker (G-d), it would be better for him if had not
> Come into the world.[5]

The Mishnah passage is important here because it requires that those who would expound upon the forbidden degrees (forbidden sexual relations in Leviticus 18), the story of creation (Genesis 1:1–2:3), and the chapter on the Chariot (Ezekiel 1) before three people must be a sage who knows his own knowledge. Alt-

---

[5] Translation is my own adapted from Herbert Danby, *The Mishnah* (Oxford: Oxford University Press, 1977), 212-13.

hough the Mishnah warns that someone who does not consider the honor of his Maker (G-d) would be better left unborn, it seems clear that one who knows his own knowledge must be a rare sage.

The Gemara on this passage in bHagigah 14⁶b and Tosephta Hagigah 2:3-4 expound upon this principle with accounts of the story of those who attempted to enter *Pardes*, which holds that R. Akiba exemplifies one who knows his own knowledge.

> The Tosephtan passage, generally viewed as the earliest and most complete version of this text, reads as follows:
> Four entered *Pardes*: ben Azzai, be Zoma, Aḥer (i.e., Elisha ben Abuyah), and Rabbi Akiba. Ben Azzai looked and died. About him it is written, saying, "Precious in the eyes of the L-rd is the death of his saints (Ps 116:15). Ben Zoma
> Looked and was smitten (i.e., became demented). About him it is written, saying, "Have you found honey? Eat (only) what is sufficient for you, (lest you be filled With it and vomit it)" (Prov 25:16). Aḥer looked and cut the shoots (i.e., of Plants; became a heretic). About him it is written, saying, "Do not allow your Mouth to cause your flesh to sin" (Qoh 5:5). Rabbi Akiba entered in peace, And he went out in peace. About him it is written, saying "Draw me after you, Let us run, (the king has brought me into his chambers" (Song 1:4).[7]

Pardes, the Garden, is well known as an acronym for the four modes of Rabbinic exegesis, i.e., *Peshat* or the plain meaning of the text; *Remez* or the allegorical meaning of the text, *Derashah* or the expositional meaning of the text, and *Sod* or the mystical meaning of the text so that the attempt to enter *Pardes* portrays the proper study of Torah. Three sages attempt to enter, but they all fail. Shimon ben Azzai casts a look and dies. Shimon ben Zoma was smitten, i.e., became demented. And R Elisha ben Abuyah cut the shoots, i.e., abandoned Judaism and was forever after known as Aḥer, "another person." Only R. Akiba entered in peace and departed in peace. The meaning of the story becomes clear when one

---

[6] See my study, "*Pardes* Revisited Once Again: A Reassessment of the Rabbinic Legend concerning the Four Who Entered *Pardes*," *Form and Intertextuality in Prophetic and Apocalyptic Literature*, FAT 45; Tübingen: Mohr Siebeck, 2005), 269-82.

[7] Translation and notes mine from M. S. Zuckermandel, ed., *Tosephta, based on the Erfurt and Vienena Codices* (Jerusalem: Wahrmann, 1970), 234.

considers the background of each figure. Ben Azzai never married and therefore violated the first commandment, "Be fruitful and multiply." Ben Zoma never completed *Semikhah*, i.e., he was not ordained, and thus he never completed his halakhic education beyond homiletical midrashim. R. Elisha ben Abuyah witnessed the execution of his teacher as well as other halakhic anomalies and became an apostate to Judaism. By contrast, R. Akiba become a sage because of his marriage to Rachel, who demanded that he learn to read so that he might educate their children, set the basic parameters of both midrashic exegesis and halakhic study, and died as a martyr to Judaism with the words of the *Shema* on his lips. R. Akiba therefore exemplifies one who knows his own knowledge. In order to expound on such topics, one should be a sage like R. Akiba.

R. Akiba, because of his expertise in the study of Torah and his commitment to Judaism, exemplifies the sages who were qualified to behold the vision of the *Merkavah* in the Heikhalot Rabbati. He therefore joins R. Ishmael as a member of the mystic *minyan* that will hear the exposition of R. Nehunyah ben ha-Qanah concerning his journey through the seven levels of heaven to behold the *Merkavah* or the Presence of G-d. The eight other figures who comprise the mystic *minyan* are the traditional martyrs of Judaism who were murdered by the Romans in the aftermath of the Bar Kochba Revolt, viz., R. Judah ben Baba, R. Jeshbab the Scribe, R. Hananyah ben Teradyon, R. Hozpit the Interpreter, R. Elazar ben Shammua, R. Hanina ben Hakinai, R. Shimon ben Gamliel, and R. Eliezer ben Dama. Although versions of the Rabbinic lists of the Ten Martyrs vary in their identification of the victims, the point is clear, these figures were the major Sages of the time who gave their lives as martyrs for the sake of G-d and divine Torah.[8] They are the ones qualified to sing the songs that would allow them to behold the divine *Merkavah*. They are seated together before the crowd of eight thousand students in typical yeshivah fashion in which the sages are seated first and those of lesser accomplishment and status sit behind them.

We also must consider that the Ten Sages are the ten victims of their generation who are executed to atone for the ten sons

---

[8] For thorough discussion of the Rabbinic tradition concerning the Ten Martyrs of the Bar Kochba Revolt, see Gottfried Reeg, *Die Geschichte von den Zehn Martyren*, TSAJ 10 (Tübingen: Mohr Siebeck, 1985).

of Jacob who sold their brother, Joseph, into slavery in Egypt. As the faithful sages of their generation who knew their own knowledge, they were qualified to learn why they must be sacrificed.

III

R. Nehunyah ben ha-Qanah describes the mystic ascent to appear before the throne of G-d as a journey in which one must pass through the seven gates that permit entry into each of the seven palaces or levels of heaven. Each gate is guarded by eight angels, and the prospective mystic must know the name of each angel and correctly recite the incantation that addresses each angel, thereby allowing him to pass through the gate without harm. Insofar as there are eight angels for each of the seven gates, the mystic must learn one hundred and twelve angelic names and incantations to allow passage to ascend safely to the seventh level of heaven and to descend safely once the encounter has concluded. Such a procedure is analogous to the ascent of Adapa to heaven in the Babylonian tradition following his action to stop the south wind from blowing and thereby threatening his boat.[9] Such action enabled Adapa to appear before the throne of Anu to receive the bread and water of life. It is also analogous to the Greek magical papyrus in which the prospective mystic must correctly open the seals and recite the incantation that will allow him to pass through the seven levels of heaven in the Gnostic tradition.[10]

But a problem arises when R. Nehunyah arrives at the sixth level of heaven. The angels of the sixth heaven are especially fierce, thereby threatening to destroy any prospective mystic who is not absolutely pure enough to appear before the throne of G-d. When R. Ishmael states in section 224 that the gate keepers of the sixth level would destroy those who "do and do not" descend to the *Merkavah* because they act without permission, the *Havurah* of the Ten Martyrs and the Crowd of Colleagues that listen to R. Ishmael's narration of R. Nehunyah's journey want an explanation of what is meant by those who "do and do not" descend to the *Merkavah*. A halakhic means must be devised to recall R. Nehunyah from

---

[9] *ANET* 101–103.
[10] Hans Dieter Betz, *The Greek Magical Papyri in Translation including the Demotic Spells* (Chicago: University of Chicago Press, 1986).

his journey to the sixth level of heaven so that he can answer the question without coming to harm at the hands of the angels who guard the sixth gate. R. Nehunyah must be rendered minimally impure so that he will be dismissed from the heavenly setting but not killed by the guardian angels.

R. Ishmael devises a procedure to recall R. Nehunyah that requires detailed knowledge of the laws of Niddah, menstruation or feminine purity, to render R. Nehunyah minimally impure so that he might be recalled from the sixth palace without harm.[11] R. Ishmael takes a soft white woolen cloth, known as a *parhava*, which is used to check the purity of a woman following the conclusion of her menstrual cycle. He gives it to R. Akiba, who in turn gives it to a servant with instructions to lay the *parhava* by a woman who immersed herself in the *mikvah* at the conclusion of her cycle, but did not become pure, and then immersed herself again, so that she did become pure. The woman will then come before the Ḥavurah to declare the circumstances of her purity. One member of the Ḥavurah will declare her to be impure and forbid her to her husband, whereas the other members of the Ḥavurah will declare her pure and permitted to her husband. She is instructed to touch the *parhava* very lightly as one might remove a hair from one's eye. The *parhava* is then taken to R. Ishmael, who takes a branch of myrtle soaked in spikenard oil laid up in balsam to pick it up and place it upon the knees of R. Nehunyah. R. Nehunyah is then dismissed from the sixth level of heaven without harm, so that he might answer the question and then return to the sixth level of heaven to continue his journey.

What does this action mean? It is a means to distinguish between the strict halakhah of R. Shammai applied in heaven and the more lenient understanding of halakhah applied by R. Hillel on earth in cases of menstrual purity. The woman in question has a regular menstrual cycle and immersed herself in the *mikvah* without checking, which according to the strict halakhah of R. Shammai would leave her impure due to the uncertainty of her status. A second immersion would seemingly ensure her purity in the eyes of the school of R. Hillel, although R. Shammai would still not be

---

[11] For discussion of R. Ishmael's actions, see esp. Lawrence H. Schiffman, "The Recall of Rabbi Nehuniah ben ha-Qanah from Ecstasy in the Hekhalot Rabbati," *AJSRev* 1 (1976): 269–81.

persuaded. She then lightly touches the *parhava* rendering it minimally impure in the school of R. Shammai and pure in the school of R. Hillel. The *parhava* is then picked up with the myrtle branch soaked in spikenard oil laid up in balsam which is used in Jewish magical texts to protect one against impurity. It is laid upon the knees of R. Nehunyah to render him as minimally impure as possible in the eyes of the school of Shammai as applied to the halakhah of heaven, whereas R. Nehunyah remains halakhically pure in the eyes of the school of R. Hillel on earth. R. Nehunyah is able to return to earth to explain the difference between those who "do and do not" descend on the *Merkavah*. Those who do descend on the *Merkavah* are the *Ḥavurah* of the Ten Martyrs who witness R. Nehunyah's ascent and will make the journey themselves, whereas those who do not descend are the crowd of colleagues who listen and record the narration, but who will not yet make the journey themselves. R. Nehunyah then safely returns to the sixth level of heaven to continue his journey to appear before the throne of G-d.

R. Ishmael's use of the *parhava* functions as a test to see if a prospective mystic fully understands divine Torah on the laws of *Niddah*. The choice of subject takes up a halakhic issue that is far removed from the experience of the typical male *yeshivah baḥur*, but that entails a level of halakhic sophistication that distinguishes between the strict halakhic understanding of the school of R. Shammai that would see a woman with a known regular menstrual cycle as impure even after two immersions whereas the more lenient halakhic understanding of the school of R. Hillel would see a woman known for her regular cycle as pure after two immersions even without checking.

The basic premise underlying this example is that the prospective mystic must understand even the most arcane differences in halakhic understanding of an issue involving halakhic menstrual purity. In other words, the prospective mystic must be a sage who understands his own knowledge in keeping with the teachings of mHagigah 2:1.

IV

As R. Nehunyah prepares to pass through the gate of the sixth palace, he meets a special challenge in the form of the angels Katzpiel and Dumiel. Katzpiel represents the wrath of G-d insofar as the Hebrew root, *qṣp* means, "to be angry, wrathful. He therefore symbolizes the submission to the divine will by means of the observance of Torah. Dumiel represents the silence of G-d, insofar as the Hebrew root, *dwm*, means "to be silent." Dumiel is also named Abir Gahidariham, which represents a rendition of the four basic elements of creation in ancient Greek thought, i.e., *aer*, or air expressed with the Hebrew term, *avir*; *ge* or earth, *hudor*, or water, and *pur*, fire, expressed through the Hebrew term, *ham* or heat. As an angel that comprises the basic elements of the created world, Dumiel therefore symbolizes the submission to G-d through discernment of the natural world. The combination of Katzpiel, representing submission to G-d through Torah, and Dumiel, representing submission to G-d through understanding the world of creation, constitutes the two basic elements of Rabbinic thought in which the Torah is studied to understand the will of G-d, but in which Torah can also be learned by understanding the world of creation. The wisdom tradition of the Bible, particularly Proverbs 8, maintains that G-d created the world of creation by first creating wisdom, personified as a woman in Proverbs 8 and by then consulting her as the rest of creation proceeded. It is possible to learn Torah by studying creation in Rabbinic thought. Such study is known as gaining an understanding of *derek eretz*. Although an understanding of *derek eretz* constitutes a path to understanding the divine will, Rabbinic thought maintains that the Torah was revealed at Sinai to save time by offering a more expeditious way to learn the divine will.[12] In our case, the combination of Katzpiel and Dumiel therefore represents a full comprehension of the divine will as a basis for earning the right to appear before the throne of G-d.

When R. Nehunyah is able to show the divine seal that admits him through the seventh gate into the seventh palace, Katzpiel sheathes his sword, mounts R. Nehunyah on a wagon of radiance born by the storm wind, and ushers him into the seventh palace to appear before the throne of G-d. Dumiel constitutes an entirely different challenge. He poses two questions: the first is whether or not R. Nehunyah repeatedly studies the Torah, Prophets, and Writings, i.e., the Bible, and the second is whether or not R. Nehunyah repeatedly studies the Mishnah, Halakhah, Aggadah, and observes the entirety of the Torah tradition. If the answer is yes, Dumiel approves the admission, calls upon Gabriel to record the virtues of R. Nehunyah on parchment, and carries the document before the

---

[12] See Simon Schlesinger, "Derekh Erez," *EncJud* 5:1551.

wagon of radiance that bears R. Nehunyah. Based upon their satisfaction that R. Nehunyah has studied and observes the entire Torah tradition, Katzpiel, Dumiel, and Gabriel form a procession that escorts R. Nehunyah through the seventh gate and into the presence of G-d.

An objection emerges, however, when R. Shimon ben Gamliel is angered because R. Nehunyah did not reveal the names of the angels that guard the entrance to the seventh heaven. Not knowing the names could lead to death (sec. 238). R. Nehunyah responds that the name can be revealed, but the names of the guards of the seventh palace all have names that are derived from the names of G-d. He therefore agrees to reveal the names on the condition that the *Haverim* stand and bow at the mention of each name. Upon accomplishing such liturgical acknowledgment of the names of the angels guarding the seventh palace, the angel, Anaphiel, Branch of G-d, opens the gates of the seventh palace and allows the procession bearing R. Nehunyah to pass.

Upon entry into the seventh palace, R. Nehunyah will see the 512 eyes of the four *Hayot* of the divine throne chariot, the eyes of the Cherubim of the divine throne chariot, and the eyes of the *Ophanim*, or Wheels, of the divine throne chariot gazing upon him, each of which are like torches of fire and conflagrations of coals of juniper. These eyes clearly constitute a threat to R. Nehunyah, but Anaphiel supports R. Nehunyah, and invites him to fear not, enter and see the king of heaven, and not be destroyed. At that point, a horn sounds from above, and the cherubim and *ophanim* avert their gaze so that R. Nehunyah might enter. When he does so, the Throne of Glory hymn, which extols G-d's royal glory in creation, is sung by the angels.

The Sar haTorah section then follows. Here, R. Nehunyah appears before the throne of G-d, and he asks G-d the question which he came to ask concerning the great task that G-d had enjoined Israel to fulfill, i.e., to build a great house or Temple for G-d and to study the Torah. The question becomes, what tasks should Israel fulfill, insofar as he presupposes that the two tasks are a great burden on Israel. G-d's response is telling, and reads as follows:

> For there was great idleness to you from the exiles, and I desired the time when I would hear words of Torah from your mouths. For you did not do what was right
> (lit., you did not make beauty) when you diminished my glory [lit., when you took From my glory]. And I was an-

gry at you, and I stood, and I acted against my city [Jerusalem] and against my Temple and against the Rabbis [lit, the sons of Kallah, i.e., The Rabbinic assembly gathered to discuss points of halakhah], and I did not do what Was right [lit., I did not make beauty] when I stood against you and I sealed against You a decree of certain judgment which would stand forever and ever. And it was To me a matter of contention whether the measure of the punishment would be one Length of time or two or ten or thirty or with power until a hundred and more. But you reproached me; you were right [lit., beauty, you did]. I accept upon myself Your reproach. (Heikhalot Rabbati, Wertheimer, 29:2).[13]

G-d tells R. Nehunyah that Israel had a long period of idleness between the exiles from the land of Israel and that G-d longed to hear the words of Torah emanating from the mouths of Israel. G-d states, "you have not done well, and I have not done well," a shocking statement in that it charges that both Israel and G-d have done wrong and thereby brought punishment upon Israel. G-d explains that Israel differed from G-d and that in divine anger, G-d sealed a decree of judgment against Israel that might endure for all time due to what they had done. G-d concludes by stating that Israel has rebuked G-d and that they did well in doing so. G-d states, "already, I accept your rebuke."

But G-d continues by stating that the words of Israel are sweet in G-d's ears and that the words of Torah shall not depart from Israel's mouth. G-d claims to know what Israel wants, i.e., knowledge of Torah, a multitude of Talmud and halakhic teachings, and the increase of such knowledge among a multitude of students, academies, and *kallahs*, or Rabbinic conferences. The passage reads as follows:

> I know what you seek, and my heart recognizes what you desire. More Torah you seek And a multitude of students and much legal discussion to study halakhah [lit., to inquire Of Halakhah] you anticipate. And for the one who studies [lit., brings near, adduces] You wish for an increase of high Torah, high awesome wisdom, study in the streets, And *pilpul* [detailed exegesis] in the squares to increase Torah like the sand of the Sea and Halakhot like the dust of the earth (Heikhalot Rabbati, Wertheimer,

---

[13] Translation mine from Wertheimer, 1:113.

29:5).[14]

G-d's response therefore asserts that the role of Israel is to continue to learn Torah until such time as it understands the whole and thereby understands the full will of G-d. Such understanding constitutes the eschatological goals of Rabbinic Judaism.

V

Scholars have debated whether the Heikhalot Rabbati depicts mystical ecstatic experience or the study of divine Torah. As this paper attempts to demonstrate, it depicts both. The Heikhalot Rabbati focuses first and foremost on the study of Torah. The *minyan* that hears R. Nehunyah's account of his journey to the seventh level of heaven comprises the Ten Martyrs who gave their lives for the study of Torah during the Roman persecution of Jews following the failed Bar Kochba Revolt. The recall of R. Nehunyah from the sixth level of heaven requires detailed knowledge of the laws of *Niddah*, or menstrual purity, a topic farthest from the experience of the typical male *yeshivah baḥur*, so that a procedure that would render R. Nehunyah impure according to the heavenly halakhah of R. Shammai but pure according to the earthly halakhah of R. Hillel might be devised to recall him from the sixth palace without harm. G-d's response to R. Nehunyah's query about the task of Jews is that they must continue to be preoccupied with the study of Torah until they understand fully divine Torah and the will of G-d. Such an endeavor is consistent with the Rabbinic understanding of the eschatological task of Jews, i.e., the completion of the study of Torah to bring about the complete sanctification of creation. But such study of Torah must be seen as an ecstatic mystical journey as well, as one learns the *Peshat*, plain meaning of the text, the *Remez*, the allegorical meaning of the text, the *Darashah*, the homiletical meaning of the text, and *Sod*, the secret or mystical meaning of the text in order to enter into *Pardes*, the Garden, and thereby appear before the throne and presence of G-d.

---

[14] Translation mine from Wertheimer, 1:113–114.

Bibliography

Anderson, Bernard W. *Understanding the Old Testament*. Englewood Cliffs, NJ: Prentice Hall, 1986.
The Ante-Nicene Fathers. Edited by Alexander Roberts and James Donaldson. 1885-1887. 10 vols. Repr., Peabody, MA: Hendrickson, 1994.
Ariel, Donald T. "Identifying the Minds, Minters and Meanings of the First Jewish Revolt Coins." *The Jewish Revolt against Rome: Interdisciplinary Perspectives*, 373-97. Edited by Mladen Popović. Supplements to the Journal for the Study of Judaism. Leiden: Brill, 2011.
Ascough, Richar S., Philip A. Harland, and John S. Kloppenborg. *Associations in the Greco-Roman World: A Sourcebook*. Waco: Baylor University Press, 2012.
Attridge, Harold W. "The Gospel of Thomas. Appendix: The Greek Fragments." In *Nag Hammadi Codex II, 2-7 Together with XIII,2*, BRIT. LIB. ORG. 4926(1) and P.OXY. 1, 654, 655*, 96-97 Edited by B. Layton, 2 vols. Leiden: Brill, 1989.
Aune, D. E. "Christian Prophecy and the Messianic Status of Jesus." Pages 404-422 in *The Messiah*. Edited by James Charlesworth. Minneapolis: Fortress Press, 1992.
Bagnell, Roger. *Early Christian Books in Egypt*. Princeton: Princeton University Press, 2009.
Banister, Jamie A. " Ὁμίως and the Use of Parallelism in Romans 1:26-27." *Journal of Biblical Literature* 128.3 (2009): 569-590.
Barnstone, Willis, and Marvin Meyer, *The Gnostic Bible: Gnostic Texts of Mystical Wisdom from the Ancient and Medieval World*. New Seeds: Boston & London, 2006.
Bauer, Walter. *Rechtglaübigkeit und Kerzerei im ältesten Christentum*. Tübingen: Mohr Siebeck, 1934.
Bauer, Walter. *A Greek-English Lexicon of the New Testament and Other Early Christian Writings*. Translated and adapted by William E. Arndt and F. Wilbur Gingrich. Chicago: University of Chicago Press, 1957.

Beasley-Murray, George R. *John*. 2nd edition. WBC 36. Nashville, TN: Thomas Nelson, 2000.

Bertalotto, Pierpaolo. "Qumran Messianism, Melchizedek, and the Son of Man." *The Dead Sea Scrolls In Context (2 vols): Integrating the Dead Sea Scrolls in the Study of Ancient Texts, Languages, and Cultures*, 325-339. Edited by Armin Lange, Emanuel Tov, and Mathhias Weigold. Leiden: Brill, 2011.

Betz, Hans Dieter. *The Greek Magical Papyri in Translation including the Demotic Spells*. Chicago: University of Chicago, 1986.

Black, Matthew. *The Book of Enoch or I Enoch: A New English Edition*. Studia in Veteris Testamenti Pseudepigraphica. Leiden: Brill, 1985.

Black, Matthew, ed. *Apocalypsis Henochi Graece*. Pseudepigrapha Veteris Testamenti Graece 3. Leiden 1970.

Bilde, Per. *Flavius Josephus between Jerusalem and Rome: His Life, His Works and Their Importance*. Journal for the Study of the Pseudepigrapha Supplement Series 2. Sheffield, UK: JSOT, 1988.

Blosser, Benjamin P. *Become Like the Angels: Origen's Doctrine of the Soul*. Washington, D. C.: The Catholic University of America Press, 2012.

Blumell, Lincoln H. "Christians on the Move in Late Antique Oxyrhynchus." In *Travel and Religion in Antiquity*, 235-54. Edited by Philip Harland. Studies in Christianity and Judaism/Études sur le christianisme et le judaïsme 21. Waterloo, Wilfred Laurier University Press, 2011.

———. *Lettered Christians: Letters, and Late Antique Oxyrhychus* Leiden: Brill, 2012.

———. "PSI IV 311: Early Evidence for Arianism at Oxyrhynchus?" *Bulletin of the American Society of Papyrologists* 49 (2012): 279-99.

Blumell, Lincoln H. and Thomas A. Wayment. *Christian Oxyrhynchus: Texts, Documents, and Sources*. Waco: Baylor University Press, 2015.

Bock, Darrell L. *Luke Volume 2: 9:51-24:53*. Grand Rapids, MI: Baker Academic, 1996.

Bovon, Francois. *Luke 2: A Commentary on the Gospel of Luke 9:51 19:27*. H. Koester, Ed., D. S. Deer, Trans. Minneapolis, MN: Fortress Press, 2013.

Bovon, Francois and Gregoire Rouiller. *Exegesis: Problems of Method and Exercises in Reading (Genesis 22 and Luke 15)*. Trans. Donald G. Miller. Pittsburgh: The Pickwick Press, 1978.

Boustan, Ra`anan S. *From Martyr to Mystic: Rabbini Martyrology and the Making of Merkavah Mysticism.* TSAJ 112. Tübingen: Mohr Siebeck, 2005.

———. "The Study of Hekhalot Literature: Between Mystical Expe rience and Textual Artifact." *CBR* 6.1 (2007): 130–160.

Brakke, David. *The Gnostics: Myth, Ritual, and Diversity in Early Christianity.* Cambridge: Harvard University Press, 2010.

Brinkschröder, Michael. *Sodom als Symptom: Gleichgeschlechtliche Sexualität im christlichen Imaginären – eine religionsgeschichtliche Anamnese.* Religionsgeschichtliche Versuche und Vorarbeiten 55. Berlin: Walter de Gruyter, 2006.

Brooten, Bernadette. *Love between Women: Early Christian Responses to Female Homoeroticism.* Chicago Series on Sexuality, History, and Society. Chicago: University of Chicago Press, 1996.

Brown, Raymond E. *The Gospel according to John I-XII.* 2nd edition. AB 29. Garden City: Doubleday, 1985.

Bruce, F. F. *Romans: An Introduction and Commentary*, Tyndale New Testament Commentary. Downers Grove, Ill.: IVP, 1963.

Burge, G.M. "A Specific Problem in the New Testament Text and Canon: The Woman Caugt in Adultery (John 7,53–8,11)." *Journal of the Evangelical Theological Society* 27 (1984): 141–8.

Burkett, Delbert. *The Son of Man Debate: A History and Evaluation.* Cambridge: Cambridge University Press, 2000.

Cameron, Ronald D. "The Gospel of Thomas and Christian Origins." In *The Future of Early Christianity: Essays in Honor of Helmut Koester.* Edited by Birger A. Pearson. Minneapolis: Fortress Press, 1991.

Cameron, Ronald D. and Merrill P. Miller. *Redescribing Christian Origins.* Society of Biblical Literature Symposium Series 28. Atlanta: Society of Biblical Literature, 2004.

Casey, Maurice. *An Aramaic Approach to Q: Sources for the Gospels of Matthew and Luke.* New York: Cambridge University Press, 2002.

Catchpole, David R. "The Beginning of Q: A Proposal." *NTS* 38 (1992): 205–11.

Chapman, Dean W. "Locating the Gospel of Mark: A Model of Agrarian Biography." *Biblical Theology Bulletin* 25 (1995): 24–36.

Charles, R. H. *The Book of Enoch; or 1 Enoch. Translated from the Editor's Ethiopic Text and Edited with the Introduction, Notes and Indexes of the First Edition Wholly Recast, Enlarged and Rewritten, Together with a Reprint from the Editor's Text of the Greek Fragments*. Oxford: Clarendon Press, 1912.

———. *The Greek Versions of the Testaments of the Twelve Patriarchs. Edited from Nine MSS., together with the Variants of the Armenian and Slavonic Versions and Some Hebrew Fragments*. Oxford: Clarendon Press, 1908.

Charlesworth, James H., ed. *The Old Testament Pseudepigrapha. Volume 1: Apocalyptic Literature and Testaments*. New York: Doubleday, 1983.

———. *The Old Testament Pseudepigrapha and the New Testament: Prolegomena for the Study of Christian Origins*. Society for New Testament Studies Monograph Series. Graham Stanton, ed. Cambridge: Cambridge University Press, 1985.

———, ed. *The Messiah*. Minneapolis: Fortress Press, 1992.

Chesnut, Glenn F. "The Ruler and the Logos in Neopythagorean, Middle Platonic, and Late Stoic Political Philosophy." *ANRW* 16.2:1310–1333. Part 2, Principat, 16.2. Edited by Wolfgang Haase. New York: de Gruyter, 1978.

1 Clement. Apostolic Fathers I. Translated by Kirsopp Lake. 2 vols. Loeb Classical Library. Cambridge, Mass: Harvard University Press, 1985.

Coblentz Bautch, Kelley. "Decoration, Destruction and Debauchery: Reflection on 1 Enoch 8 in Light of 4QEn^b." *Dead Sea Discoveries* 15 (2008): 79-95.

Cohick, Lynn. *Women in the World of the Earliest Christians: Illuminating Ancient Ways of Life*. Grand Rapids, MI: Baker Academic, 2009.

Coxon,. Peter W. "Nephilim." Pages 618-620 in *The Dictionary of Deities and Demons in the Bible*. Second edition. Edited by Karel van der Toorn, Bob Becking, and Pieter W. van der Horst. Leiden: Brill, 1999.

Cranfield, C. E. B. *A Critical and Exegetical Commentary on The Epistle to the Romans*. 2 vols. International Critical Commentary. Edinburgh: T & T Clark, 1975.

Cross, Frank Moore. "Light on the Bible from the Dead Sea Caves." Pages 156-166 in *Understanding the Dead Sea Scrolls: A Reader from the Biblical Archaeology Review*. Edited by Hershel Shanks. New York: Vintage, 1993.

Crossan, John Dominic. *The Historical Jesus*. New York: HarperCollins, 1991.

Danby, Herbert. *The Mishnah*. Oxford: Oxford University Press, 1977.

Dicken, Frank E. "A King and Ruler Takes His Stand: 'Herod' as a Composite Character in Luke-Acts." Dissertation, University of Edinburgh, 2013.

DeConick, April D. "The Great Mystery of Marriage, Sex and Conception in Ancient Valentinian Traditions." *Vigilae Christiane* 57.3 (Aug. 2003): 307–42.

———. DeConick, April D. *Recovering the Original Gospel of Thomas: A History of the Gospel and Its Growth*. LNTS 286. London: T&T Clark, 2005.

———. *The Original Gospel of Thomas in Translation*. LNTS 287. London: T&T Clark, 2006.

Derrett, J. Duncan M. "The Woman Taken in Adultery." In *Law in the New Testament*, 156–88. London, Darton, Longman & Todd, 1970.

Desjardins, Michel. "Bauer and Beyond: On Recent Scholarly Discussions of *Hairesis* in the Early Christian Era." *SC* 8 (1991): 65–82.

Deutsch, Robert. "Coinage of the First Jewish Revolt against Rome: Iconography, Minting Authority, Metallurgy." In *The Jewish Revolt against Rome: Interdisciplinary Perspectives*, 361–71. Edited by Mladen Popović. Supplements to the Journal for the Study of Judaism. Leiden: Brill, 2011.

Dicken, Frank E. "A King and Ruler Takes His Stand: 'Herod' as a Composite Character in Luke-Acts." Dissertation, University of Edinburgh, 2013.

Dillon, Matthew and Lynda Garland. *Ancient Rome: Social and Historical Documents from the Early Republic to the Death of Augustus*. 2nd edition. New York: Routledge, 2015.

Dixon, Suzanne. "From Ceremonial to Sexualities: A Survey of Scholasrhip on Roman Marriage." In *A Companion to Families in the Greek and Roman Worlds*, 245–61. Edited by Beryl Rawson. Malden, MA: Wiley-Blackwell, 2011.

Dunderberg, Ismo. "Q and the Beginning of Mark." *NTS* (1995): 501-11.
Ehrman, Bart D. "Jesus and the Adulteress." *Biblical Theology* 40 (1989): 24-44.
Eisenman, Robert, and Michael Wise, *The Dead Sea Scrolls Uncovered: The First Complete Translation and Interpretation of 50 Key Documents withheld for over 35 Years*. Shaftesbury, UK: Element, 1992.
Emmel, Stephen. "The Coptic Gnostic Texts as Witnesses to the Production and Transmission of Gnostic (and Other) Traditions." In *Das Thomasevangelium: Entstehung-Rezeption-Theologie*, 33-49. Edited by E.E. Popkes and J. Schröter. Berlin: Walter de Gruyter, 2008.
Epp, Eldon J. "New Testament Papyrus Manuscripts and Letter Carrying in Greco-Roman Times." In *The Future of Early Christianity: Essays in Honor of Helmut Koester*, 35-56. Edited by Birger A. Pearson. Minneapolis: Fortress Press, 1991.
Evans, Craig A. *Mark 8:27-16:20*. Word Biblical Commentary 34B. Nashville, TN: Thomas Nelson, 2001.
Fitzmyer, Joseph A. *Romans: A New Translation with Introduction and Commentary*. The Anchor Bible 33. New York: Doubleday, 1993.
———. *The Gospel according to Luke X-XXIV: Introduction, Translation, and Notes*. New Haven: Yale University Press, 2008.
Fleddermann, Harry T. *Mark and Q: A Study of the Overlap Texts*. Bibliotheca Ephemeridum Theologicarum Lovaniensium CXXII. Leuven: Leuven University Press; Peeters, 1995.
———. *Q: A Reconstruction and Commentary*. Dudley, MA: Peeters, 2005.
Foster, Paul. "Are There Any Early Fragments of the So-Called *Gospel of Peter*?" *NTS* 52 (2006): 1-28.
France, R. T. *The Gospel of Mark: A Commentary on the Greek Text*. The New International Greek Testament Commentary. Grand Rapids, MI: W. B. Eerdmans Publishing Company, 2002.
Fuller, Reginald H. *Foundations of New Testament Christology*. New York: Charles Scribner's Sons, 1965.
Gamble, Harry Y. *Books and Readers in the Early Church: A History of Early Christian Texts*. New Haven: Yale University Press, 1995.
García Martínez, Florentino. *The Dead Sea Scrolls Translated: The Qumran Texts in English*. 2nd edition. Leiden: Brill; Grand Rapids: Eerdmans, 1996.

Gelardini, Gabriella. *Christus Militans: Studien zur politisch-militarischen Semantik im Markusevangelium vor dem Hintergrund des ersten judisch-romischen Krieges.* Supplements to Novum Testamentum 165. Boston: Brill, 2016.

———. "The Contest for a Royal Title: Herod versus Jesus in the Gospel According to Mark (6,14-29; 15,6-15)." *Annali di storia dell'esegesi* 28, no. 2 (2011): 93-106.

Gibson, J. C. L. *Canaanite Myths and Legends.* Edinburgh: T. & T. Clark Ltd., 1978.

Goodspeed, Edgar J. *Die altesten Apologeten. Texte mit kurzen Einleitungen.* Göttingen: Vandenhoeck & Ruprecht, 1914.

Gourgues, Michel. "'Moi non plus je ne condamne pas :' Le Mots et la théologie de Luc en Jean 8,1-11 (la femme adultère)." *Studies in Religion/Sciences Religieuses* 19 (1990): 305-18.

Green, Joel B. *The Gospel of Luke.* Grand Rapids: William B. Eerdmans Publishing Company, 1997.

Grenfell, Bernard P. and Arthus S. Hunt eds. *The Oxyrhynchus Paypri I.* London: Egypt Exploration Society, 1898.

———. *New Sayings of Jesus and the Lost Gospel from Oxyrhynchus.* London: Egyptian Exploration Society, 1904.

Gruenwald, Ithamar. *Apocalyptic and Merkavah Mysticism.* AGAJU 14. Leiden: Brill, 1980.

Guerra, Anthony J. *Romans and the Apologetic Tradition: The Purpose, Audience and Genre of Paul's Letter.* Society for New Testament Studies Monograph Series 81. Cambridge: Cambridge University Press, 1995.

Haenchen, Ernst. *John 2: A Commentary on the Gospel of John Chapters 7-21.* Translated by Robert W. Funk. Edited by Robert W. Funk and Ulrich Busse. Hermeneia. Philadelphia: Fortress, 1984.

Hamerton-Kelly, R. G. *Pre-Existence, Wisdom, and the Son of Man: A Study Of The Idea Of Pre-Existence In The New Testament.* Cambridge: Cambridge University Press, 1973.

Hare, Douglas R. A. *The Son of Man Tradition.* Minneapolis: Fortress Press, 1990.

Harnack, Adolf. *The Sayings of Jesus: The Second Source of St. Matthew and St. Luke.* Translated by J.R. Wilkinson. New York: G.P. Putnam's Sons, 1908.

Harrington, Daniel J. "The Reception of Walter Bauer's Orthodoxy and Heresy in Earliest Christianity During the Last Decade." *HTR* 73 (1980): 289-98.

Hays, Richard B. "Relations Natural and Unnatural: A Response to John Boswell's Exegesis of Romans 1." *Journal of Religious Ethics* 14.1 (1986): 184-215.

Heil, John Paul. "The Story of Jesus and the Adulteress (John 7,53–8,11) Reconsidered." *Biblica* 72.2 (1991): 182-91.

Hendel, Ronald S. "Of Demigods and the Deluge: Toward an Interpretation of Genesis 6:1-4." *Journal of Biblical Literature* 106 (1987): 13-26. doi.org/10.2307/3260551.

Hill, G.F. *Catalogue of the Greek Coins of Palestine (Galilee, Samaria, and Judaea)*. London: British Museum Dept. of Coins and Medals, 1914.

Hoehner, Harold W. *Herod Antipas*. Society for New Testament Studies Monograph Series 17. Cambridge: Cambridge University Press, 1972.

Horsley, Richard A. *Hearing the Whole Story: The Politics of Plot in Mark's Gospel*. Louisville, Ky: Westminster John Knox Press, 2001.

Hurtado, Larry W. "The Earliest Evidence of an Emerging Christian Material and Visual Culture: The Codex, the *Nomina Sacra* and the Staurogram." In *Text and Articfact in the Religions of Mediterranean Antiquity: Essays in Honour of Peter Richardson*, 271–88. Edited by Stephen G. Wilson and Michel Desjardins. Waterloo, Ont: Wilfrid Laurier University Press, 2000.

———. *The Earliest Christian Artifacts: Manuscripts and Christian Origins*. Grand Rapids: MI: Eerdmans, 2006.

———. "The Greek Fragments of 'The Gospel of Thomas' as Artefacts: Papyrological Observations on P.Oxy 1, P.Oxy 654, and P.Oxy 655." In *Das Thomasevangelium: Entstehung–Rezeption–Theologie*, 19–24. Edited by J. Frey, E.E. Popkes, and J. Schröter. Berlin: Walter de Gruyter, 2008.

Jensen, Morten Hørning. *Herod Antipas in Galilee: The Literary and Archaeological Sources on the Reign of Herod Antipas and Its Socio-Economic Impact on Galilee*. Wissenschaftliche Untersuchungen Zum Neuen Testament 2. Reihe 215. Tübingen: Mohr Siebeck, 2006.

Jewett, Robert. *Romans: A Commentary*. Assisted by Roy D. Kotansky. Hermeneia. Minneapolis: Fortress Press, 2007.

Jonas, Hans. *The Gnostic Religion*. Boston: Beacon Press, 2001.

Just, A.A. *Luke: Ancient Christian Commentary on Scripture*. Downer's Grove, IL: InterVarsity Press, 2005.

Käsemann, Ernst. *Commentary on Romans*. Translated by Geoffrey W. Bromily. Grand Rapids: Eerdmans, 1980.
Keck, Leander E. *Romans*. Abingdon New Testament Commentaries. Nashville: Abingdon, 2005.
Keener, Craig. *The Gospel of John: A Commentary*. Grand Rapids, MN: Baker Academic, 2003.
Keith, Chris. "Jesus Began to Write: Literacy, the *Pericope Adulterae*, and the Gospel of John." Ph.D. diss.; University of Edinburgh, 2008.
Kelly, J. N. D. *A Commentary on the Epistles of Peter and of Jude*. Harper's New Testament Commentaries. New York: Harper & Row, 1969.
King, Karen L. *What is Gnosticism?* Cambridge: Belknap Press, 2003.
Klijn, A. F. J. "The So-Called Hymn of the Pearl." *Vigilae Christianae* 14 (1960): 154-64.
Kloppenborg, John S. *Q Parallels*. Sonoma, CA: Sonoma Press, 1987.
Kloppenborg, John S. *The Formation of Q: Trajectories in Ancient Wisdom Tradition*. Studies in Antiquity and Christianity. Philadelphia: Fortress Press, 1987.
———. *Q Parallels*. Sonoma, CA: Sonoma Press, 1987.
———. "Evocatio Deorum and the Date of Mark." *Journal of Biblical Literature* 124, no. 3 (2005): 419-50.
Kloppenborg, John S., Marvin W. Meyer, Stephen J. Patterson, and Michael G. Steinhauser. *Q Thomas Reader*. Sonoma: Polebridge Press, 1990.
Koester, Helmut. "Introduction: The Gospel According to Thomas." In *Nag Hammadi Codex II*, 2-7, 1. 38-39. Nag Hammadi Studies, 20. Edited by Bentley Layton. Leiden: E.J. Brill, 1989.
———. *Introduction to the New Testament: History and Literature of Early Christianity*. 2 volumes. 2nd edition. Berlin: Walter de Gruyter, 2000.
Kraft, Robert A. "Oxyrhynchus Papyrus 655 Reconsidered." *HTR* 54 (1961): 253-62
Kraus, Thomas and Tobias Nicklas. *Das Petrusevangelium und die Petrusapokalypse*. Berlin: de Gruyter, 2011.
———. *Das Evangelium nach Petrus: Text, Kontexte, Intertexte*. Berlin: de Gruyter, 2012.
Lambrecht, Jan. "Q-Influence on Mark 8,34-9,1." In *Logia: Les Paroles Jésus – The Sayings of Jesus: Mémorial Joseph Coppens*, 277-304. Leuven: Uitgeverij Peeters, 1982.

———. "The Great Commandment Pericope and Q." In *The Gospel Behind the Gospels: Current Studies on Q*, 73–96. Edited by Ronald A. Piper. NovTSup 75. Leiden: Brill, 1995.
Larsen, Matthew. *Gospels before the Book*. Oxford: Oxford University Press, 2018.
Laufen, Rudolf. *Die Doppelüberlieferungen der Logienquelle und des Markusevangeliums*. Bonn: Peter Hanstein, 1980.
Layton, Bentley. *The Gnostic Scriptures: Ancient Wisdom for a New Age*. New York: Doubleday, 1987.
———. *The Gnostic Scriptures*. The Anchor Bible Reference Library. New York: Doubleday, 1995.
Leaney, A. R. C. *The Letters of Peter and Jude: A Commentary on the First Letter of Peter, a Letter of Jude and the Second Letter of Peter*. Cambridge Bible Commentary. Cambridge: Cambridge University Press, 1967.
Leivestad, Ragnar. "Exit the Apocalyptic Son of Man." *New Testament Studies* XVIII.3 (April 1972): 243–267.
Liew, Tat-siong Benny. *Politics of Parousia: Reading Mark Inter(Con)Textually*. Biblical Interpretation Series 42. Leiden, Boston: Brill, 1999.
Loader, William. *Enoch, Levi, and Jubilees on Sexuality: Attitudes toward Sexuality in the Early Enoch Literature, the Aramaic Levi Document, and the Book of Jubilees*. Grand Rapids: Eerdmans, 2007.
Lührmann, Dieter. *Die Redaktion der Logienquelle*. Wissenschaftliche Monographien zum Alten und Neuen Testament 33. Neukirchen-Vluyn: Neukirchener Verlag. 1969.
Lührmann, Dieter. "Markus 14.55-64: Christologie und Zerstörung des Tempels im Markusevangelium." *New Testament Studies* 27, no. 4 (1981): 457–74.
Luijendijk, AnneMarie. *Early Christians and the Oxyrhynchus Papyri*. Cambridge, MA: Harvard University Press, 2008.
———. "Sacred Sriptures as Trash: Biblical Papryi from Oxyrhynchus." *VC* 64 (2010): 217–54.
———. "Reading in the *Gospel of Thomas* in the Third Century: Three Oxyrhynchus Papyri and Origen's *Homilies*." In *Reading New Testament Papyri in Context*, 241–268. Edited by C. Clivaz and J. Zumstein. Leuven: Peeters, 2011.
MacDonald, Dennis R. *Does the New Testament Imitate Homer? Four Cases from the Acts of the Apostles*: New Haven: Yale University Press, 2003.

―――. *Two Shipwrecked Gospels: The* Logoi of Jesus *and Papias's* Exposition of Logia about the Lord. SBLECL 8. Atlanta: Society of Biblical Literature, 2012.
Mack, Burton L. "Lord of the Logia—Savior or Sage?" Pages 3–18 in *Gospel Origins & Christian Beginnings*. Festschrift J.M. Robinson. Edited by James E. Goehring, et al. Sonoma, California: Polebridge Press, 1990.
―――. *A Myth of Innocence: Mark and Christian Origins*. 1. paperback ed. Philadelphia: Fortress Press, 1991.
―――. *The Lost Gospel: The Book of Q & Christian Origins*. New York: HarperSanFrancisco, 1993.
Malherbe, Abraham. *Social Aspects of Early Christianity*. Baton Rouge, LA: Louisiana State University Press, 1977.
Maloney, Francis. *The Gospel of John*. Minneapolis: Liturgical Press, 1998.
Marchal, Joseph A. "The Usefulness of Onesimus: The Sexual Use of Slaves and Paul's Letter to Philemon." *Journal of Biblical Literature* 130.4 (2011): 749–70.
Marcus, Joel. "The Jewish War and The Sitz Im Leben of Mark." *Journal of Biblical Literature* 111 (1992): 441–62.
Marshall, I. Howard. *The Gospel of Luke*. Grand Rapids: William B. Eerdmans Publishing Company, 1978.
Mason, Steve. *A History of the Jewish War: A.D. 66–74*. Cambridge: Cambridge University Press, 2016.
Matiel-Gerstenfeld. *260 Years of Ancient Jewish: A Catalogue*. Tel Aviv: Kol, 1982.
McKeating, Henry. "Sanctions against Adultery." *JSOT* 11 (1979): 57–72.
McVann, Mark. "The 'Passion' of John the Baptist and Jesus before Pilate: Mark's Warnings about Kings and Governors." *Biblical Theology Bulletin* 38, no. 4 (2008): 152–57.
Meyer, Marvin E., ed. *The Nag Hammadi Scriptures*. New York: HarperCollins, 2007.
Meyer, Marvin E., trans. *Q Thomas Reader*. Sonoma, California: Polebridge, 1990.
Miller, James E. "The Practices of Romans 1:26: Homosexual or Heterosexual?" *Novum Testamentum* 37 (1995): 1–11.
Mosherer, Y. *A Treasury of Jewish Coins from the Persian Period to Bar Kokhba*. Jerusalem: Yad Ben-Zvi, 2001.

Nestle-Aland, *Novum Testamentum Graece*. 28th rev. ed. Edited by Barbara and Kurt Aland, Johannes Karavidopoulos, Carlo M. Martini, and Bruce M. Metzger in cooperation with the Institute for New Testament Textual Research, Münster/Westphalia. Stuttgart: Deutsche Bibelgesellschaft, 2012.

Nortje, S. "John the Baptist and the Resurrection Traditions in the Gospels." *Neotestamentica* 23 (1989): 349-58.

The New Oxford Annotated Bible: New Revised Standard Version with the Apocryphal/Deuterocanonical Books. Edited by Michael D. Coogan. Third Edition. Oxford: Oxford University Press, 2001.

Neirynck, Frans. "Recent Developments in the Study of Q." In *Logia: Les Paroles de Jésus – The Sayings of Jesus : Mémorial Joseph Coppens*, 29-75. Edited by Joël Delobel. Leuven: Uitgeverij, 1982.

Neyrey, Jerome H. *2 Peter, Jude: A New Translation with Introduction and Commentary. AB 37* New York: Doubleday, 1993.

Nickelsburg, George W. E. *1 Enoch 1: A Commentary on the Book of 1 Enoch, Chapters 1-36; 81-108*. Hermeneia. Minneapolis: Fortress Press, 2001.

Nolland, John. *Luke 9:21-18:34*. Dallas: Word Books, 1998.

O'Day, Gail R. "John 7:43-8:11: A Study in Misreading." *JBL* 111.4 (1992): 631-40

Öhler, Markus. "Greco-Roman Associations, Judean Synogogues and Early Christianity in Bithynia-Pontus." In *Authority and Identity in Emerging Christianities in Asia Minor and Greece*, 62-88. Edited by C. Breytenbach and J.M. Ogereau. Leiden: Brill, 2018.

Parrot, Douglas M., ed. *Nag Hammadi Codices III, 3-4 and V, 1*. Leiden: E. J. Brill, 1991.

Parrot, Douglas M., trans. "Eugnostos the Blessed and Sophia of Jesus Christ." *The Nag Hammadi Library in English*. Edited by James Robinson. San Francisco: HarperCollins, 1978.

Patterson, Stephen J. *The Gospel of Thomas and Jesus*. Sonoma CA: Polebridge, 1993.

Pearson, Birger A. *Gnosticism, Judaism, and Egyptian Christianity*. Minneapolis: Fortress Press, 1990.

———. *Ancient Gnosticism: Traditions and Literature*. Minneapolis, MN: Fortress Press, 2007.

Perkins, Pheme. *Gnosticism and the New Testament*. Minneapolis: Fortress Press, 1993.

Phillips, Anthony. *Ancient Israel's Criminal Law*. New York: Oxford University Press, 1970.

Porter, Calvin. "Romans 1.18-32: Its Role in the Developing Argument." *New Testament Studies* 40 (1994): 210–28.
Priest, J. "A Note on the Messianic Banquet." Pages 222–38 in *The Messiah*. Edited by James Charlesworth. Minneapolis: Fortress Press, 1992.
Rajak, Tessa. "Friends, Romans, Subjects: Agrippa II's Speech in Josephus's *Jewish War*." In *The Jewish Dialogue with Greece and Rome: Studies in Cultural and Social Interaction*, 147–59. Arbeiten Zur Geschichte Des Antiken Judentums Und Des Urchristentums 48. Leiden: Brill, 2001.
Reed, Annette Yoshiko. *Fallen Angels and the History of Judaism and Christianity: The Reception of Enochic Literature*. Cambridge: Cambridge University Press, 2005.
Reeg, Gottfriend, *Die Geschichte von den Zehn Martyren*. TSAJ 10. Tübingen, Mohr Siebeck, 1985.
Reiling, J. and J. L. Swellengrebel. *A Translator's Handbook for the Gospel of Luke*. London: E.J. Brill, 1971.
Richardson, Peter. *Herod: King of the Jews and Friend of the Romans*. Studies on Personalities of the New Testament. Columbia, SC: University of South Carolina Press, 1996.
Riley, Gregory J. "Doubting Thomas; Controversy Between the Communities of Thomas and John." Ph.D. diss., Harvard University, 1990.
———. *Resurrection Reconsidered: Thomas and John in Controversy*. Minneapolis: Fortress Press, 1995
———. *The River of God: A New History of Christian Origins*. New York: HarperCollins, 2001.
Robinson, James M. "The Study of the Historical Jesus After Nag Hammadi." *Semeia* XLIV (1988).
Robinson, James M. and Helmut Koester eds. *Trajectories Through Early Christianity*. Minneapolis, MN: Fortress Press, 1971.
Robinson, Thomas A. *The Bauer Thesis Examined: The Geography of heresy in the Early Christian Church*. Lewiston, NY: Edwin Mellen Press, 1917
Rollins, Sarah. "Does 'Q' Have Any Representative Potential?" *Method and Theory in the Study of Religion* 23 (2011): 73–77.
Rouillard, Hedwige. "Rephaim." Pages 692-700 in *The Dictionary of Deities and Demons in the Bible*. Second edition. Edited by Karel van der Toorn, Bob Becking, and Pieter W. van der Horst. Leiden: Brill, 1999.

Rudolph, Kurt. *Gnosis: The Nature and History of Gnosticism*. San Francisco: HarperCollins, 1987.

Safrai, Shmuel. "Ishmael ben Elisha." In *Encyclopedia Judaica*. Edited by Michael Berenbaum and Fred Skolnik New York: MacMillan, 2006.

Sanday, William, and Arthur Headlam, *A Critical and Exegetical Commentary on the Epistle to the Romans*, International Critical Commentary. New York, NY: Scribner's, 1902, 1968.

Schäfer, Peter, *Synopse zur Hekhalot-Literatur*. TSAJ 2. Tübingen: Mohr Siebeck, 1981.

———. *The History of the Jews in the Greco-Roman World: The Jews of Palestine from Alexander the Great to the Arab Conquest*. Rev. ed. London: Routledge, 2003.

———. *The Origins of Jewish Mysticism*. Tübingen: Mohr Siebeck, 2009.

Schenke, Hans-Martin. "Nag Hammadi Studien II Das System der Sophia Jesu Christi." *Zeitschrift fur Religions und Geistesgeschichte* XLIV (March, 1962).

Schiffman, Lawrence H. "The Recall of Rabbi Nehuniah ben ha-Qanah from Ecstasy in the Hekhalot Rabbati." *AJSRev* 1 (1976): 269-81.

Scholem, Gerhom, *Major Trends in Jewish Mysticism*. New York: Schocken, 1961.

Schüling, Joachim. *Studien zum Verhältnis von Logienquelle und Markusevangelium*. Würzburg: Echter Verlag, 1991.

Scopello, Madeleine. "Jewish and Greek Heroines." In *Images of the Feminine in Gnosticism*. Ed. Karen L. King. Harrisburg, PN: Trinity Press International, 1998.

Scott, J. Martin. "On the Trail of a Good Story: John 7.53–8.11 in the Gospel Tradition." In *Ciphers in the Sand*, 53–82. Edited by Larry J. Kreitzer and Deborah W. Rooke. Sheffield: Sheffield Academic Press, 2000.

Schmithals, Walter. *Das Evangelium nach Markus*. 2 vols. Ökumenischer Taschenbuchkommentar zum Neuen Testament. Gütersloh; Würzburg: Gütersloher Verlagshaus Mohn; Echter-Verlag, 1979.

Schweizer, Ed. "The Son of Man Again." *New Testament Studies* X (1964).

Sellew, Philip. "Thomas Christianity: Scholars in Quest of a Community." In *The Apocryphal Acts of Thomas*, 11–34. Edited by Jan Bremer. Leuven: Peeters, 2001.

Septuaginta. Edited by Alfred Rahlfs. 2nd rev. ed. Edited by Robert Hanhart. Stuttgart: Deutsche Bibelgesellschaft, 2006.

Smith, Abraham. "Tyranny Exposed: Mark's Typological Characterization of Herod Antipas (Mark 6:14-29)." *Biblical Interpretation* 14, no. 3 (July 1, 2006): 259–93.

Smith, Morton. *Hekhalot Rabbati: The Greater Treatise concerning the Palaces of Heaven*. Edited by D. Karr, 2009.

Stegemann, Ekkehard W. and Wolfgang Stegemann. *The Jesus Movement: A Social History of Its First Century*. Trans. O.C. Dean, Jr. Minneapolis: Augsburg Fortress Press, 1999.

Streeter, Burnett Hillman. *The Fourt Gospels: A Study of Origins*. London: MacMillan, 1942.

Stuhlmacher, Peter. *Paul's Letter to the Romans: A Commentary*. Translated by Scott J. Hafemann. Louisville, KY: Westminster/John Knox, 1994.

Sweeney, Marvin A. "*Pardes* Revisited Once Again: A Reassessment of the Rabbinic Legend Concerning the Four Who Entered Pardes." In *Form and Intertextuality in Prophetic and Apocalyptic Literature*, 269–282. FAT 45. Tübingen: Mohr Siebeck, 2005.

Tannehill, Robert C. *Luke: Abingdon New Testament Commentaries*. Nashville: Abingdon Press, 1996.

Telford, William. *The Theology of the Gospel of Mark*. New Testament Theology. Cambridge, UK ; New York: Cambridge University Press, 1999.

Tervahaufa, Ulla. *A Story of the Soul's Journey in the Nag Hammadi Library: A Study of Authenikos Logos. (NHC VI,3)*. Gottingen: Vandenhoeck & Ruprect, 2015.

Tödt, H. E. *The Son of Man in the Synoptic Tradition*. Philadelphia: The Westminster Press, 1965.

Tuckett, Christopher M. *The Revival of the Griesbach Hypothesis: An Analysis and Appraisal*. Cambridge: Cambridge University Press, 1983.

———. "Mark and Q." In *The Synoptic Gospels: Source Criticism and the New Literary Criticism*, 149–175. Edited by Camille Focant. Bibliotheca Ephemeridum Theologicarum Loveniensium CX. Leuven: Leuven University Press, 1993.

Turner, Eric G. *The Typology of the Early Codex*. Philadelphia: University of Pennsylvania Press, 1977.

Turner, John D. "Sethian Gnosticism: A Literary History." Pages 55-86 in Nag Hammadi, Gnosticism, and Early Christianity. Edited by Charles W. Hedrick and Robert Hodgson, Jr. Peabody, MA: Hendrickson, 1986.

Tuval, Michael. *From Jerusalem Priest to Roman Jew: On Josephus and the Paradigms of Ancient Judaism*. Wissenschaftliche Untersuchungen zum Neuen Testament. 2. Reihe 357. Tübingen: Mohr Siebeck, 2013.

Tyson, Joseph B. "Jesus and Herod Antipas." *Journal of Biblical Literature* 79 (1960): 239-46.

Vaage, Leif. "The Son of Man Sayings in Q: Stratigraphical Location and Significance." *Semeia* LV (1991): 103-29

van Gronigen, Bernhard A. "ΕΚΔΟΣΙΣ." *Mnemosyne* 16 (1963): 1-17.

van Seters, John. *Prologue to History*. Louisville, Kentucky: Westminster / John Knox Press, 1992.

VanderKam, J.C. "Righteous One, Messiah, Chosen One, and Son of Man in 1 Enoch 37-71." Pages in 169-191 in *The Messiah: Developments in Earliest Judaism and Christianity*. Edited by J.H. Charlesworth. Minneapolis: Fortress, 1992.

Vermes, Geza. *Jesus and the World of Judaism*. Philadelphia: Fortress Press, 1983.

Wallace, Daniel B. "The Story of Jesus and the Adulteress Reconsidered." *NTS* 39.2 (1993): 290-96.

———. *Greek Grammar Beyond the Basics: An Exegetical Syntax of the New Testament*. Grand Rapids: Zondervan, 1996.

Weiss, Bernhard. *A Manual of Introduction to the New Testament*. Translated by A.J.K. Davidson. New York: Funk & Wagnalls, 1889.

Wellhausen, J. *Einleitung in die drei ersten Evangelien*. 2nd edition. Berling: Reimer, 1911.

Wertheimer, Avraham Yosef. *Batei Midrashot*. Jerusalem: Rav Kook, 1950.

Williams, Michael A. *Rethinking "Gnosticism": An Argument for the Dismantling of a Dubious Category*. Princeton: Princeton University Press, 1996.

———. "Variety in Gnostic Perspectives on Gender." In *Images of the Feminine in Gnosticism*. Ed. Karen L. King. Harrisburg, PA: Trinity Press International, 1998.

Yarbro Collins, Adela. *Mark: A Commentary*. Edited by Harold W. Attridge. Hermeneia. Minneapolis, MN: Fortress Press, 2007.

Yonge, Charles Duke. *The Works of Philo Judaeus, the Contemporary of Josephus, Translated from the Greek*. London: H. G. Bohn, 1854-1890.

Zeichman, Christopher. "Military-Civilian Interactions in Early Roman Palestine and the Gospel of Mark." PhD diss., University of St. Michael's College, 2017.

Zeichmann, Christopher B. "Capernaum: A 'Hub' for the Historical Jesus or the Markan Evangelist?" *Journal for the Study of the Historical Jesus* 15, no. 1 (August 20, 2017): 147–65.

Zuckermandel, M.S. ed. *Tsephta, Based on the Erfurt and Vienena Codices*. Jerusalem: Wahrmann, 1970.

Modern Author Index

Anderson, B.W., 143
Ariel, D.T., 79
Attridge, H.W., 23, 124, 125, 126, 128
Aune, D.E., 138

Bagnell, R., 122, 132
Banister, J.A., 84, 85, 86, 87, 98, 99, 106, 109
Barnstone, W., 8
Bauer, W., 14, 119
Bautch, K.C., 113o

Beasley-Murray, G.R., 73
Bertalotto, P., 169
Betz, H.D., 186
Bilde, P., 35
Black, M., 93, 94, 109
Blosser, B.P., 7
Blumell, L.H., 119, 120, 121, 122, 129, 130, 133
Bock, D.L., 15, 16, 18, 20, 23, 25
Boustan, R.S., 181
Bovon, F., 8, 25
Brakke, D., 9
Brinkschröder, M., 83, 105, 107, 108, 109, 110
Brooten, B., 83, 84, 85, 86, 89, 90, 103, 107, 109, 110
Brown, R.E., 78
Bruce, F.F., 107
Bultmann, R., 137, 138
Burge, G.M., 72
Burkett, D., 137, 138

Cameron, R.D., 119, 125, 135
Casey, M., 62
Catchpole D.R., 54
Chapman, D.W., 31
Charlesworth, J., 108, 137, 159, 168, 176
Coggins, R.J., 19
Cohick, L., 75
Collins, J.J., 169
Coxon, P.W., 92

Cranfield, C.E.B., 84, 85, 87, 103, 105
Crossan, J.D., 152, 153, 171

DeConick, A.D., 22, 126
Derrett, D.M., 73, 75, 76, 78
Desjardins, M., 118, 119g lo
Deutsch, R., 79
Dillon, M., 78
Dixon, S., 76
Dunderberg, I., 54

Ehrman, B.D., 72
Emmel, S., 127, 128
Epp, E.J., 120
Evans, C. A., 27

Fitzmyer, J.A., 12, 19, 84, 87, 105, 107, 114
Fleddermann, H.T., 46, 47, 48, 49, 55, 56, 57, 62, 63, 64, 66, 67, 68
Foster, P., 130
France, R.T., 29
Fuller, R.H., 142, 143, 168, 170, 171
Funk, R.W., 73

Gamble, H.Y., 120
García Martínez, F., 93
Garland, L., 78
Gelardini, G., 27, 28
Gibson, J.C.L., 159
Gourgues, M., 73
Green, E.E., 74
Green, J.B., 12, 13, 18, 20, 21, 25
Grenfell, B.P., 124, 128
Gruenwald, I., 181

Haenchen, E., 73
Hamerton-Kelly, R.G., 136
Hare, D.R.A., 137, 138
Harland, P., 120

Harnack, A.,	45, 47,	Malherbe, A.,	120
136, 150		Marchal, J.A.,	84
Harrington, D.J.,	119	Marcus, J.,	27
Hays, R.B.,	84, 87, 88,	Marshall, I.H.,	16, 20, 24ie
97		Mason, S.,	35
Headlam, A.,	90, 91, 107	Matiel-Gerstenfeld, J.,	79
Heil, J.P.,	72	McKeating, H.,	75
Hendel, R.S.,	92, 93	McVann, M.,	28
Hill, G.F.,	79	Meyer, M.,	8, 10, 155,
Hoehner, H.W.,	28, 29, 31	173	
Horsley, R.A.,	27	Miller, D.G.,	25
Hunt, A.S.,	124, 128	Miller, J.E.,	84, 85, 86,
Hurtado, L.W.,	118, 123,	88, 105, 106. 107	
124		Miller, M.P.,	119
		Moloney, F.,	72
Jensen, M.H.,	30	Mosherer, Y.,	79
Jewett, R.,	84, 87, 88,		
103, 114		Neirynck, F.,	48, 49
Jonas, H.,	12	Neyrey, J.H.,	91, 98,
Just, A.A.,	20	100, 114, 115	
		Nickelsburg, G.W.E.,	93, 94,
Käsemann, E.,	84 87,, 89,	104, 105, 108, 109, 111, 112, 113, 120	
90, 105, 107		Nicklas, T.,	130
Keck, L.E.,	107	Nolland, J.,	11, 14, 16,
Keener, C.	73	18, 19, 21, 23, 24, 25	
Keith, C.,	72	Nortje, S.,	2u8
Kloppenborg, J.S.,	27, 61, 62,		
65, 101, 120, 147, 155, 173, 175		O'Day, G.R.,	73
King, K.,	9, 10, 13	O'Sullivan, M.,	74
Klijn, A.F.J.,	10	Öhler, M.,	120
Koester, H.,	117, 119,		
152		Parrott, D.M.,	160, 161,
Kraft, R.A.,	125	163, 164, 165, 167, 173	
Kraus, T.,	130	Patterson, S.J.,	126
Kreitzer, L.J.,	74	Pearson, B.,	6, 9, 120,
		126, 127, 167	
Lambrecht, J.,	37, 38, 39,	Perkins, P.,	13
66		Phillips, A.,	75er
Larsen, M.,	121, 127		
Layton, B.,	6, 7, 9, 10,	Piper, R.A.,	37
11, 123, 152		Porter, C.,	88
Laufen, R.,	47, 58, 62,	Priest, J.,	159
63, 66			
Leivestad, R.,	174, 176	Rajak, T.,	34, 35
Liew, T.B.,	27	Reed, A.Y.,	93, 95, 98,
Loader, W.,	113	99, 102, 112, 113	
Lührmann, D.,	27, 91, 100	Reiling, J.,	21
101, 102, 103, 104, 106, 109		Richardson, P.,	28
Luijendijk, A.,	118, 122,	Riley, G. J.,	1, 2, 3, 4,
127, 128		6, 45, 71, 117, 118, 119, 125, 135, 140,	
		141, 143, 151, 156, 159, 173	
MacDonald, D.R.,	37, 38, 39,	Robinson, J.M.,	117, 119,
91		137, 148, 155, 160, 162	
Mack, B.L.,	27, 148,	Robinson, T.A.,	119
150, 155, 158, 175		Rollens, S.,	117

Rooke, D.W.,	74	Zeichman, C.,	27, 31
Rouiller, G.,	25		
Rudolph, K.,	17, 142		

Safrai, S., 182
Sanday, W., 90, 91, 107
Schäfer, P. 32, 181, 182
Schiffman, L.H., 187
Schlesinger, S., 189
Schmithals, W., 27
Scholem, G., 181h
Schüling, J., 57, 61
Schweizer, E.D., 144, 178
Scopollo, M., 10, 13
Scott, J., 74
Scott, J.M.C., 75
Sellew, P., 117, 119
Shanks, H., 92, 93
Sjöberg, E., 136
Smith, A., 29
Smith, M., 182
Stegemann, E.W., 23
Stegemann, W., 23
Streeter, B.H., 45
Sweeney, M.A., 184

Tannehill, R., 7, 8
Telford, W., 27
Tervahaufa U., 8
Tödt, H.E., 136, 137, 153, 154
Tuckett, C.M., 37, 38, 48
Turner, E.G., 123, 129

Tuval, M., 35
Tyson, J.B., 28

Vaage, L., 151, 152, 155

van Gronigen, B.A., 127
van Seters, J., 143
VanderKam, J.C., 168
Vermes, G., 142

Wallace, D.B., 24, 72
Wayment, T.A., 122, 129, 130, 133
Weiss, B., 46
Wellhausen, J., 46, 47
Williams, M.A., 9, 10, 13, 127, 128
Winn, A., 27

Yarbro Collins, A., 28, 31

Ancient Author Index

Greek Authors

Artimedorus
  Onirocriticus
    1.80 — 109

Demosthenes
  Exordia
    4.28 — 30
    21.3 — 30

Homer
  Odyssey
    10 — 11
    11:554-558 — 5

Hyperides
  Against Demosthenes
    5.3 — 30

Plato
  Phaedo — 5
    79d — 5
    82a–c — 5

  Phaedrus
    246a-254e — 5

  Timaeus
    28a — 166

Strabo
  16.2.41 — 11

Xenophon
  Anabasis
    1.1.6 — 30

Latin Authors

Aulus Gellius
  Noctes Atticae
    10.23.1–5 — 78

Cicero
  Res Publica
    VI.18 — 7

Justinian
  Corpus Juris Civilis
    9.9.3 — 76

  Digest
    4.4.37 — 76
    48.5 — 76

  Institutiones Justiniani
    4.18 — 76

Quintilian — 80, 81
  Institutio Oratoria
    5.10.104–105 — 77
    7.2.52 — 77

Seneca — 79, 80
  Controveriae
    1.4 — 77

Valerius Maximus
  6.1.12–13 — 76

Vergil
  Aeneid 6 — 11

Hebrew Bible

Genesis
  1:1–2:3 — 183
  1:26–28 — 166
  1:27 — 103, 111, 113
  5:1–3 — 166
  5:2 — 113
  6:1–4 — 92, 98
  6:2 — 167
  7:2–3 — 113
  7:9 — 113
  9:6 — 166
  17:14 — 114
  17:23 — 114
  6:1–4 — 92, 93, 98
  34:24 — 114
  41:42 — 24

Exodus
  23:20 — 53, 54

215

Leviticus		**New Testament**	
1:3	113	**Matthew**	
1:10	113	2:2	29
3:1	113	3:7–12	49
3:6	113	3:11	50
4:23	113	9:32–34	58
11:7	15	10:32	146, 149
12:2	114	10:37	66
18:5	39	10:37–39	65
18:22	114	10:38	66
19:18	39	10:40	55
20:10	71, 75	11:2–15	53
		11:8	29
Numbers		11:10	53
23:19	143, 170	11:16–19	53
		11:25–27	38
Deuteronomy		12:22	58
6:5	39	13:16–17	38
11:8	39	14:1	29
22:22	71, 75	16:24–25	65
		19:3	79
Job		19:3–12	79
25:5–6	143	21:31–32	53
80:17	145	22:15–22	79
		22:18	79
Psalms		22:32	37
90:11–12 (LXX)	52	22:34–40	37, 39, 43
90:13 (LXX)	52	22:40	43
116:15	184	23:35–36	38
		24:37–39	101
Proverbs			
25:16	184	**Mark**	
		1:2	53
Ecclesiastes (Qohelet)		1:7–8	50
5:5	184	1:12–13	51
		3:22–26	59
Songs of Solomon		3:22–27	58
1:4	184	3:28	154
		6:14ff	28
Ezekiel		8:34–35	67
16:49–50	96	8:38	149, 150
		9:1	30
Daniel		9:1–6	55
7:13	142, 156, 169	10:2	79
8:17	171	10:2–12	79
		10:17	42
		10:18	42
Zephaniah		10:20–22	43
2:9	96	10:21	42
		10:29–30	65
		12:13–17	79
Malachi		12:15	79
3:1	54	12:27	37

12:28–31	40	20:20–26	79
12:28–34	37, 39, 43		
12:29	37, 42	John	
12:32–33	41	6:11	86
12:33	37, 43	8:3	71
12:34	41	8:3–4	74
13:9	28	8:4	77
13:26–27	30	8:5	71, 75
14:61	29	8:6	72, 74
14:62	30, 156, 177	8:7	72
15:2	29	8:8–9	72
15:18	28	8:10	72
15:32	29	8:11	72

Luke

		Acts	
3:7–9	49	7:55–56	157
3:16	50	7:56	175
3:16–17	49	12:19	30
7:18–19	53		
7:22–28	53	Romans	
7:27	53	1:16–17	109
9:7	29	1:17–18	88, 89
9:23–24	65	1:18	89, 90
10:1–16	54	1:18–23	84
10:24	29	1:18–25	87
10:25–28	37, 38, 39	1:18–32	88, 89, 91, 96, 107, 108, 109, 110, 116
10:30	11, 14		
10:30–35	8		
10:31–32	17		
10:33	19, 24	1:19–20	89, 107
10:34	19	1:21	89
10:35	23	1:21–23	88
11:14	58	1:23	104, 105
11:19–20	64	1:23–28	105
11:23–26	64	1:24–27	97
11:51	46	1:25	104
12:8	146	1:26	84, 85, 95, 104, 105, 107, 110, 111, 113, 114, 116
12:16–21	62		
15:11–32	8		
15:13	12, 15		
15:14	15		
15:15	16	1:26–27	83, 84, 85, 86, 87, 88, 89, 90, 91, 96, 100, 103, 104, 105, 106, 107, 109, 110, 113, 114, 115, 116
15:16	16		
15:17	20		
15:18–19	21		
15:20	19, 23		
15:22	24		
15:24	25		
16:16	53		
17:26–29	101, 106		
17:28	86, 101		
17:33	64, 66	1:27	86, 98, 105, 113, 114, 115
19:12	12		
19:38	29		
20:20	79	1:29–30	112

1 Corinthians
5:1–2   30
11:10   111
11:14   105
Galatians
3:28   103
1 Timothy
1:3–4   161, 162
2 Timothy
2:12   150

2 Peter
2:1   108
2:4–5   102
2:4–10   99
2:10   98, 100
2:18   97
2:20   98, 100
3:3–7   100

Jude
4   108
5–7   97–8, 114
7   99, 106
8   86, 98, 100
        106, 112
11   97

Revelation
3:5   150

Apocrypha
Baruch
3:26–28   93

Esdras
8:44   166

Judith
16:5–9   116

Sirach
16:7–8   102, 109
16:7–9   100, 102,
16:16–28   109, 118
17:1–42   166

Wisdom of Solomon
1:13–14   166
2:18   158
2:23   166

10:3–6   102
10:8–9   102
12:16   108
12:18   108
12:23–26   107
13:3   108
13:9   108
13–14   90–91
14:1–6   108
14:6   102, 108
14:12   89, 107
14:12–14   107
14:22–31   107
18:13   158

Pseudepigrapha
1 Enoch
1–19   93
1–36   93
2:1   110
2:1–2   105
2:2   108
2–5   109
5:1   109
5:4–7   109
6:1-8:4   94
7:1   94, 98, 100,
        110
8:1–2   94
9:5   117
9:8   110, 112
9:9   94, 112
10–11   93
15:3   95, 98, 100,
        102, 110
15:4   98, 100
15:5   111
15:7   111
15:9   93
16:3   112
46:4   170
48:2ff   170
62:7   170
62:11   170–71
62:14   171
69:27   170–71
69:29   172

2 Enoch
44:1–32   166
65:2   166

3 Maccabees
2.3–5   102–3
2:3:29   105

218

Apocalypse of Adam		109	*War*		
			1.66		30
Jubilees			2.167		31
	4.15	93, 113	2.247		33
	4.22-25	103	2.252		33
	5.1-11	103	2.345-401		33
	7.21	110	2.421		34
	7.21-23	103	2.500		34
	10.1-6	103	2.523-526		34
	16.5-6	103	4.14-15		34
	20.5	103	4.474		11

Testament of Naphtali
   3.2-4   87, 109
   3.2-5   105, 108
   3.3-5   109
   3.4   107, 110
   3.4-5   104, 105, 106
   3.5   110
   4.1   104-5

Philo
   Cherubim   7
   De Abrahamo   96, 109, 114
   De gigantibus   92

Testament of Reuben
   5.5-6   113

Dead Sea Scrolls
   4Q252   109

Josephus
   Antiquities

Rabbinic Literature
   Hekhalot Rabbati   183
      29:2   190-91
      29:5   191-92
      81   182

14.143	31	
14.272-75	31	
14.327	31	
14.384-85	32	
15.161-193	32	
15.195	32	
15.342f	33	
18.26-28	33	
18.36-38	33	
18.109-115	33	
18.111-125	33	
18.116-19	28	
18.236-273	33	
18.237	33	
18.240-55	29	
18.240-252	33	
18.240-556	33	
18.274-276	33	
18.297	33	
19.360	33	
20.104	33	
20.179	34	
20.211-212	34	
20.179-180	34	
20.252	33	

Mishnah Hagigah
   2:1   183, 188

Tosephta Hagigah
   2:3-4   184

Non-canonical Christian Literature
   1 Clement
      9.3-4   108
      11.1   108
      51.2   14

   Acts of Thomas   9, 117, 141, 173

   Clement of Alexandria
      Theo.,   7

   Damascus Document   109

   Eugnostos the Blessed (III.3)
      81:12   166
      85:9-14   164, 165
      105:10   161

Exegesis on the Soul (II.6)	8, 9, 10, 12, 13, 15, 16, 18, 21, 24	3527	123, 130
		3528	130
		4009	122, 130
		4365	122
Gospel of the Egyptians (III.2)	109	4403	122, 130
		4404	122, 130
Gospel of Thomas		4405	122, 130
3.4	172	4497	132
21	172	4498	132
42	172	4705	122, 131
44	155, 172	4706	123, 130
50	173	4707	123, 131
85	172	5072	123, 130
86	172		
106	174	*Paraphrase of Shem*	109
Hymn of the Pearl	8, 9, 10, 12, 13, 14, 15, 16, 17, 20, 22, 24, 141, 173	Q	
		4:1–13	51
		6:22	177
		6:23	154
		9:58	172
		10:16	55, 56
Irenaeus		10:21–24	38
Against Heresies		11:14–20	38
1.1	161	11:14–23	58
		11:15	59
Justin Martyr		11:17	59
2 Apology		11:17–20	59
5.2–6	115	12:8	147, 151, 153
Dialogue with Trypho			
89.1	176	12:8–9	146, 147, 150, 153, 156
P. Oxy			
1	123	12:8–10	138, 147ff, 172, 178
210	123, 131		
405	130	12:10	139, 146, 147, 151, 152, 154, 156
406	132		
407	134		
654	118, 123, 124, 125, 127, 131,		
		14:26	65, 67
		14:26–27	64
655	118, 123, 125, 131	14:27	67
1171	132	*Sophia of Jesus Christ*	
1229	132	71.13–14	165
1355	132	73.2–3	165
1597	132	78.17	165
1780	132	108.12	164
1786	131, 134		
2070	121, 134		
2072	131, 134		
2683	122, 130	Tertullian	
2949	123, 130	*Against the Valentinians*	
3523	124, 132	3	162
3525	122, 130		

www.ingramcontent.com/pod-product-compliance
Lightning Source LLC
Chambersburg PA
CBHW020329170426
43200CB00006B/320